T0296543

Geometric Tomography, second edition

Geometric tomography deals with the retrieval of information about a geometric object from data concerning its projections (shadows) on planes or cross-sections by planes. It is a geometric relative of computerized tomography, which reconstructs an image from X-rays of a human patient. The subject overlaps with convex geometry and employs many tools from that area, including some formulas from integral geometry. It also has connections to discrete tomography, geometric probing in robotics, and stereology.

This comprehensive study provides a rigorous treatment of the subject. Although primarily meant for researchers and graduate students in geometry and tomography, brief introductions, suitable for advanced undergraduates, are provided to the basic concepts. More than 70 illustrations are used to clarify the text. The book also presents 66 unsolved problems. Each chapter ends with extensive notes, historical remarks, and some biographies. This new edition includes numerous updates and improvements, with some 50 extra pages of material and 300 new references, bringing the total to more than 800.

Richard J. Gardner has been Professor of Mathematics at Western Washington University since 1991. He is the author of 70 papers and founded geometric tomography as a subject in its own right with the publication of the first edition of this book in 1995.

ENCYCLOPEDIA OF MATHEMATICS AND ITS APPLICATIONS

ENCYCLOPEDIA OF MATHEMATICS AND ITS APPLICATIONS

Geometric Tomography

Second Edition

RICHARD J. GARDNER

Western Washington University

CAMBRIDGE
UNIVERSITY PRESS

32 Avenue of the Americas, New York NY 10013-2473, USA

Cambridge University Press is part of the University of Cambridge.

It furthers the University's mission by disseminating knowledge in the pursuit of
education, learning and research at the highest international levels of excellence.

www.cambridge.org
Information on this title: www.cambridge.org/9780521684934

© Cambridge University Press 1995, 2006

This publication is in copyright. Subject to statutory exception
and to the provisions of relevant collective licensing agreements,
no reproduction of any part may take place without the written
permission of Cambridge University Press.

First published 1995, 2011
Second Edition 2006, 2012
Reprinted 2013

A catalogue record for this publication is available from the British Library

Library of Congress Cataloguing in Publication data

Gardner, Richard J.
Geometric tomography / Richard J. Gardner.– 2nd ed.
p.   cm. – (Encyclopedia of mathematics and its applications ; v. 58)
Includes bibliographical references and index.
ISBN-13: 978-0-521-86680-4 (hardback)
ISBN-10: 0-521-86680-4 (hardback)
ISBN-13: 978-0-521-68493-4 (pbk.)
ISBN-10: 0-521-68493-5 (pbk.)
1. Geometric tomography.   I. Title.   II. Series.
QA639.5.G37   2006
516.3´62 – dc22   2006002945

ISBN 978-0-521-86680-4 Hardback
ISBN 978-0-521-68493-4 Paperback

Cambridge University Press has no responsibility for the persistence or accuracy of
URLs for external or third-party internet websites referred to in this publication,
and does not guarantee that any content on such websites is, or will remain, accurate
or appropriate.

To Linda

But one man loved the pilgrim soul in you

CONTENTS

PREFACE TO THE SECOND EDITION

This second edition incorporates some 60 extra pages of material, including seven new figures, another 21 chapter notes, the new Sections 4.4 and A.4, and about 300 additional references. This expansion indicates the amazingly rapid development of geometric tomography over the decade since the first edition appeared. Despite this, the list of 66 open problems is of roughly the same length.

Many corrections have been made. The most significant amendment appears in Chapter 8, written for the first edition very shortly after the pioneering work was published. Alex Koldobsky's work described in Note 8.9 brought to light an error in the solution of the Busemann–Petty problem. The revised Chapter 8 contains the corrected solution, while the six new notes for Chapter 8 struggle to keep pace with the incredible activity around intersection bodies. The task of surveying recent developments would have been a great deal more difficult but for the publication of Koldobsky's fine book [465]. This takes an almost entirely analytic point of view, whereas in Chapter 8 the original geometrical approach is retained as far as possible.

The new Section 4.4 describes an algorithm constructed by the author and an electrical engineer, Peyman Milanfar. It employs another algorithm designed for the reconstruction of a convex body from its surface area measure, the topic of the new Section A.4. Two reasons lie behind the choice of this material. Firstly, the first edition was noticeably short of algorithms and devoid of any that apply to the sort of noisy measurements encountered in practice. There is still plenty to be done in this direction. Secondly, it is much easier to tailor one's own work to suit a book than those of others! But if time, energy, and publisher allowed a complete rewriting, there would be a very different book with the same title and author as this one, stating and proving in the text many results of others here only briefly mentioned in chapter notes.

The collaboration with Milanfar followed a web search for the phrase "geometric tomography" made in 1996. I discovered to my surprise that the term had
been used independently at least twice after I introduced it at the Oberwolfach
meeting on tomography in 1990. The usage in [683, Chapter 7] has essentially
no overlap with ours, but that of Thirion [802] is quite close in spirit; he defines
geometric tomography to be the process of reconstructing the external or inter-
nal boundaries of objects from their X-rays. The same web search led me to the
program of Alan Willsky, an electrical engineer at MIT, outlined in Note 1.5.

Much of the work represented by the additional references was presented at
various meetings on convex geometry or discrete tomography during the past
decade, but several international meetings have featured geometric tomography
specifically. Two Summer Schools on Local Stereology and Geometric Tomography were organized by Eva Vedel Jensen at the Sandbjerg Estate, Denmark,
the first in 2000 and the second in 2002. In 2004, Salvador Gomis organized the
Workshop on Geometric Tomography in Alicante, Spain. It is wonderful to have
such energetic and capable friends in beautiful locations.

Terminology and notation are constantly changing in mathematics. Currently
"origin symmetric" seems to be favored over "centered," and $\mathbb{R}^n$ is used more and
more instead of $\mathbb{E}^n$ as this part of geometry moves into the mainstream. One would
think that a notion as basic as volume would enjoy a standard notation, but V, Vol,
and vol, with or without subscripts, are all common. In the end I decided to retain
most of the terminology and notation of the first edition, so that at least the two
editions would be compatible. There are a couple of exceptions: the notation for
Radon and Fourier transform has been interchanged, and the definition of Fourier
transform is slightly different.

I am obliged to Hugh Murrell for permission to use his Mathematica program
that produced Figure 1.11, and to Ulrich Brehm for providing Figure 7.3. To the
list of people in the preface to the first edition to whom I owe thanks should be
added Robert Huotari, Alex Koldobsky, and Maria Moszyńska. I am especially
grateful to Paolo Gronchi, Markus Kiderlen, Wolfgang Weil, and Gaoyong Zhang
for their very helpful comments on this edition.

Department of Mathematics
Western Washington University
Bellingham, WA 98225-9063
E-mail: Richard.Gardner@wwu.edu
Web page: http://www.ac.wwu.edu/~gardner

PREFACE

The title of this book, *Geometric Tomography*, is designed to cover the area of mathematics dealing with the retrieval of information about a geometric object from data about its sections, or projections, or both. The term "geometric object" is deliberately vague; a convex polytope or body would certainly qualify, as would a star-shaped body, or even, when appropriate, a compact set or measurable set.

The word "tomography" originates from the Greek $\tau\acute{o}\mu o\varsigma$, meaning a slice. Mathematical computerized tomography is already a recognized subject with an enormously important application in the medical CAT scanner, with which an image of a section of a human patient can be reconstructed from X-rays. Mathematically, the object being reconstructed is a density function, and since it is known that the solution is not unique, no matter how (finitely) many X-rays are taken, the reconstructed picture is always an approximation. When density functions are replaced by geometric objects, there is some hope of a unique solution, and this gives geometric tomography a rather different flavor.

Despite this, there is a point where geometric tomography and computerized tomography merge, and both utilize integral transforms such as the Radon transform. Our definition of geometric tomography is also reminiscent of definitions of stereology or geometric probing. These subjects each have their own distinct viewpoint, while sharing common features with geometric tomography. For example, stereology focuses on random data and statistical methods, but also draws on integral geometry (as do two other related subjects, image analysis and mathematical morphology).

Via these connections and others discussed in the chapter notes, geometric tomography does not lack possible applications.

In computerized tomography, the term "projection" is routinely used for an X-ray. In this book, we adhere to the accepted mathematical definition of a projection as a shadow. In this sense, projections are of little or no interest in computerized tomography. In geometry, however, the well-known polar duality provides a link between sections and projections. There is, in fact, a remarkable correspondence between results concerning projections and those concerning sections through a fixed point, and it would be quite inappropriate to consider the one without considering the other.

Since projections are only shadows, the convex sets are a natural class of objects with which to work. Aristotle's argument, that the earth must be spherical since its shadows on the moon are circular during a lunar eclipse, is in the spirit of geometric tomography. Such matters can often be settled with standard geometrical arguments, but more sophisticated methods are required when only the areas, rather than the exact shapes, of the projections of a convex set are available. Then the Brunn–Minkowski theory, which includes Minkowski's theory of mixed volumes and which forms the core of classical convexity, becomes the ideal framework. In this way, geometric tomography absorbs concepts such as the support function of a convex body, sets of constant width and brightness, zonoids and projection bodies, and projection functions; results such as Aleksandrov's projection theorem and the solution to Shephard's problem; and tools such as Aleksandrov's area measures, the Aleksandrov–Fenchel inequality, and the cosine transform.

Geometric tomography overlaps with convexity, but is not subsumed under it. When the data concern sections through a fixed point, the sets that are star-shaped with respect to that point form a more appropriate class than the convex sets. Within the past three decades or so, a "dual Brunn–Minkowski theory" has arisen, including Erwin Lutwak's dual mixed volumes, and again providing a natural setting. As a consequence, geometric tomography assimilates concepts such as the radial function of a star body, sets of constant section, intersection bodies, and section functions; results such as Funk's section theorem and the solution to the Busemann–Petty problem; and tools such as the i-chord functions, the dual Aleksandrov–Fenchel inequality (a suitable form of Hölder's inequality), and the spherical Radon transform.

The items in the last sentences of the previous two paragraphs are in some sense dual to each other. There is a quite mysterious correspondence in concepts, results, and tools, which polar duality hardly begins to explain.

A parallel X-ray of a body (see Table 1 at the end of this preface) carries more information than a projection. Challenged by P. C. Hammer's 1963 problem, the author and Peter McMullen showed in 1980 that parallel X-rays in certain sets of four directions suffice to determine the shape of any convex body. Later, Aljoa Volčič proved that X-rays emanating from certain sets of four points suffice to determine the shape of any planar convex body. Again, there are two sets of results, for parallel and point X-rays, which are in some sense dual to each other. However, a point X-ray is just a special section function. This provides a bridge to the material described earlier, and dual mixed volumes and the spherical Radon transform are seen in action again.

When the areas of projections, or sections through a point, of a set do not determine it uniquely, one can still hope to estimate its volume. For projections of a convex body, such an estimate is provided by the isoperimetric inequality and Cauchy's surface area formula, but much better estimates are known. For example, Lutwak applied the affinely invariant Petty projection inequality to obtain an estimate in which equality holds for ellipsoids rather than balls. Other affine isoperimetric inequalities furnish corresponding estimates which similarly have the advantage of being invariant under affine transformations.

Geometric tomography houses a zoo of strange geometric bodies, powerful integral transforms, and exotic but highly effective inequalities. Teeming with open problems, it appears an extraordinarily fertile area for research. It resembles particle physics in that symmetry – for example, the duality alluded to before – sometimes allows missing theorems to be predicted, though proofs are not always easy to find. It is too much to expect a Grand Unified Theory, but a more satisfying synthesis is surely within reach.

Some of the open problems listed at the end of each chapter require advanced knowledge and may be very difficult, but others should be quite accessible to undergraduate students. With this in mind, considerable effort has been made to cater to those who have not attended courses in real analysis or convexity. We assume knowledge of calculus, linear algebra, and the basic geometry and topology of two and three dimensions, that is, terms such as scalar product, norm, subspace, interior, boundary, open set, compact set, and connected set. A student with these prerequisites should start with Chapter 1 and, with occasional reference to Chapter 0, be capable of understanding most of it, and nearly all if also familiar with complex numbers and the idea of a metric space. Later chapters involve more advanced topics, but gentle introductions are provided in Chapter 0 and the appendixes. The beginner can make inroads by consulting these and the illustrations, though even without them there should be much that can be absorbed. For example, Chapter 2 follows the same footpath as Chapter 1, with a few brambles created by the appearance of some measure theory. The rest of the book can be read independently of these first two chapters. Chapters 3 and 4 are largely classical convexity; the support function, mixed volumes, and area measures enter here, in such a way that these tools are motivated, rather than required in advance. With Chapters 5 and 6, a new route is followed, needing some measure theory, but mostly set in the plane and again mostly independent of the previous chapters. This route continues in Section 7.2, where dual mixed volumes can be seen at work for the first time. Though much of Chapters 7 and 8 mirrors Chapters 3 and 4, little hangs on the earlier material. Chapter 9, however, draws substantially from both Chapter 4 and Chapter 8.

Many different books could have been written with the title of this one. Here one theme unfolds, supported by the sort of detailed proofs appreciated by most students. Inevitably several important topics, such as Dvoretzky's theorem and others from the local theory of Banach spaces, and the Crofton intersection formulas of integral geometry, are relegated to the chapter notes. Furthermore, Euclidean space and the projective plane hold enough difficulties for us here, though many of the concepts introduced carry over to the more general homogeneous spaces, as in [388], for example.

To add some historical perspective, several biographies have been included in the chapter notes. The history of mathematics is a fascinating subject and a powerful but largely unexploited spur to the learning process. However, accurate historical writing requires special expertise together with careful examination of original documents. The author cheerfully admits his incompetence in this area; the biographies are merely thumbnail sketches pasted together from secondary material.

Though the main text of the book is almost entirely self-contained, the supporting material in Chapter 0 and the appendixes makes frequent reference to the literature. Such references have been limited, when possible, to suitable books (rather than journal articles). There is no escaping the fact that much of the heavy machinery from the Brunn–Minkowski theory eventually comes into play. In the early stages of writing no text contained all the necessary material, and it seemed that the pedestal might be too big for the statue. By great good fortune, Rolf Schneider's comprehensive – and pedagogically sound – treatise [737] on the Brunn–Minkowski theory appeared. The reader who wishes to consult Schneider's volume for more information will find that our notation and terminology are very similar to his.

There are many friends and colleagues to thank. The book evolved from notes of weekly lectures given in late 1989 and early 1990 at the Istituto Matematico "U. Dini" in Florence, during a visit to the Istituto di Analisi Globale e Applicazioni, directed by Professor C. Pucci. This lecture series was confined to Chapter 1 and small parts of Chapters 2 and 5, but was the spark that lit the fire. As work progressed, various assistance was lent by A. D. Aleksandrov, H. Antosiewicz, Keith Ball, John Beem, Yuri Burago, Stefano Campi, Branko Curgus, Hans Debrunner, Hans Goertz, Vladimir Golubyatnikov, Marco Longinetti, Luis Montejano, Frank Morgan, Alain Pajor, Washek Pfeffer, Hans Sagan, Steven Skiena, Alan Thompson, and Tohru Uzawa. Parts of drafts were read by Edoh Amiran, Don Chakerian, Lauren Cowles, Ken Falconer, Paul Goodey, Eric Grinberg, Peter Gritzmann, Helmut Groemer, Peter Gruber, Daniel Hug, Bob Jewett, Hans Kellerer, Joop Kemperman, Dan Klain, Vic Klee, Attila Kuba, Árpád Kurusa, Erwin Lutwak, Horst Martini, Peter McMullen, Frank Natterer, Rolf Schneider, Aljoa Volčič, Wolfgang Weil, and Gaoyong Zhang. Between them they made many valuable suggestions and caught copious mistakes and misprints. (There are infinitely many of these in every manuscript, since each time you look, you find another one.)

Most of the pictures were drawn with Aldus Freehand by Jill Skeels, in a project supported by a grant from the U.S. National Science Foundation. Others were made by the author, with Mathematica and Virtuoso, and four of these used programs written by Don Chalice (Figure 7.2, the original of which appears in the paper [96] of Ulrich Brehm), Alfred Gray (Figure 3.9, see [330, p. 427]), Branko Grünbaum (Figure 2.1), and Fred Pickel (Figure 8.3, intersection body of the cube). The book was written in LaTeX, part of the wonderful TeX package invented and donated to the world by Donald Knuth.

Several of those already mentioned deserve extra thanks. During a visiting year at the University of California at Davis, Don Chakerian lent his friendly help and encyclopedic knowledge of convexity. At this time the book was to be a joint work with Aljoa Volčič; though circumstances forced him to withdraw from the project, he wrote a first draft of Chapter 5, and our collaboration continued in research incorporated in the text. (This and some other research of the author, mostly also published elsewhere, were partly supported by a grant from the U.S. National Science Foundation.) Rolf Schneider kindly gave me a preprint of his book, which immediately became indispensable. Constant encouragement and expert advice from Erwin Lutwak became a pillar of support as the truth dawned of Gian-Carlo Rota's maxim: When you write a research paper, you are afraid that your results might already be known, but when you write an expository work, you discover that nothing is known.

function	symbol	data	symmetral	symbol
parallel X-ray of K in a direction u	$X_u K$	lengths of chords of K parallel to u	Steiner symmetral	$S_u K$
k-dimensional X-ray of K parallel to a k-dimensional subspace S	$X_S K$	volumes of sections of K parallel to S	k-symmetral	$S_S K$
ith projection function of K	$v_{i,K}$	volumes of projections of K on each i-dimensional subspace	exists only for $i = 1$ or $(n-1)$	
1st projection function, or width function, of K	$w_K = v_{1,K}$	lengths of projections of K on each line through the origin	central symmetral	$\triangle K$
$(n-1)$th projection function, or brightness function, of K	$v_K = v_{n-1,K}$	volumes of projections of K on each hyperplane through the origin	Blaschke body	∇K
ith section function of K, or i-dimensional X-ray of K at the origin	$\tilde{v}_{i,K}$	volumes of sections of K by each i-dimensional subspace	i-chordal symmetral	$\tilde{\nabla}_i K$
1st section function of K, or X-ray of K at the origin	$X_o K = \tilde{v}_{1,K}$	lengths of sections of K by lines through the origin	chordal symmetral	$\tilde{\triangle} K = \tilde{\nabla}_1 K$
section function, or $(n-1)$-dimensional X-ray of K at the origin	$\tilde{v}_K = \tilde{v}_{n-1,K}$	volumes of sections of K by each hyperplane through the origin	dual Blaschke body	$\tilde{\nabla} K = \tilde{\nabla}_{n-1} K$
X-ray of K at a point p	$X_p K = X_o(K-p)$	lengths of sections of K by lines through p	chordal symmetral at p	$\tilde{\triangle}_p K = \tilde{\triangle}(K-p)$

Table 1. Some functions and symmetrals of geometric tomography.

Note: In the table, K is a convex body; however, most of the functions and symmetrals are defined for more general sets. Each symmetral is a symmetric body yielding the same data from the corresponding function as K.

0

Background material

This chapter introduces notation and terminology and summarizes aspects of the theories of affine and projective transformations, convex and star sets, and measure and integration appearing frequently in the sequel.

Some passages are designed to ease the beginner into these areas, but not all the material is elementary. It is intended that the reader **start with Chapter 1**, and use the present chapter as a reference manual. For Chapter 1, *the requisite material is included in the first four sections of this chapter only*, and for Chapter 2, *the requisite material is included in the first five sections only*.

0.1. Basic concepts and terminology

This section is a brief review of some basic definitions and notation. Any unexplained notation can be found in the list at the end of the book.

Almost all the results in this book concern Euclidean n-dimensional space $\mathbb{E}^n$. The origin in $\mathbb{E}^n$ is denoted by o, and if $x \in \mathbb{E}^n$, we usually label its coordinates by $x = (x_1, \ldots, x_n)$. (In $\mathbb{E}^2$ and $\mathbb{E}^3$ we often use a different letter for a point and label its coordinates in the traditional way by x, y, and z.) The Euclidean norm of x is denoted by $\|x\|$, and the Euclidean scalar product of x and y by $x \cdot y$. The closed line segment joining x and y is $[x, y]$. Points are identified with vectors, and are always denoted by lowercase letters. For sets we usually employ capitals, although we also use lowercase for straight lines. Script capitals are used for classes of sets; an exception is the $\mathcal{S}$ we use for sets of directions in Chapters 1 and 2, but here we are really identifying a direction with the line through the origin parallel to it. The natural numbers, real numbers, and complex numbers have the usual symbols $\mathbb{N}$, $\mathbb{R}$, and $\mathbb{C}$. The letters i, j, k, m, and n denote integers unless it is stated otherwise (in parts of the book i often represents a real number), or unless we are working with complex numbers, when $i^2 = -1$

1

as usual. In particular, the default meaning of an expression such as $1 \leq i \leq n$ is $i \in \{1, \ldots, n\}$.

The *unit ball* in $\mathbb{E}^n$ is $B = \{x : \|x\| \leq 1\}$, with surface the *unit n-sphere* $S^{n-1} = \{x : \|x\| = 1\}$. When necessary we may write B^n instead of B. We attempt to reserve u for the members of S^{n-1}, the unit vectors. If $u \in S^{n-1}$, then $u^\perp$ is the $(n-1)$-dimensional subspace orthogonal to u, and l_u the 1-dimensional subspace parallel to u. Generally, S is used for a subspace, and $S^\perp$ for its complementary orthogonal subspace. The Grassmann manifold of k-dimensional subspaces of $\mathbb{E}^n$ is denoted by $\mathcal{G}(n, k)$. More often than not the topology on $\mathcal{G}(n, k)$ is unnecessary, and the symbol then simply denotes the corresponding set of subspaces.

Translates of subspaces are called *planes* or *flats*, or *hyperplanes* if they are $(n-1)$-dimensional. A hyperplane divides the space into two *half-spaces* (*half-planes* in $\mathbb{E}^2$). A *ray* is a semi-infinite straight line. If E is a set, the *linear hull* lin E and *affine hull* aff E of E are, respectively, the smallest subspace and the smallest plane containing E. The *dimension* dim E of a set E is the dimension of its affine hull.

We say that two planes are *parallel* if one is contained in a translate of the other, and *orthogonal* if, when translated so that they contain the origin, one contains the complementary orthogonal subspace of the other. (These terms are often used by other authors in a more restrictive way.) A *slab* is the closed region between two parallel hyperplanes.

Suppose that F_1, F_2 are planes in $\mathbb{E}^n$, of dimensions d_1 and d_2, respectively. Then by [85, Theorem 32.1], either $F_1 \cap F_2 = \emptyset$ or $\dim(F_1 \cap F_2) \geq d_1 + d_2 - n$. The planes F_1 and F_2 are in *general position* with respect to each other if either $d_1 + d_2 < n$, $F_1 \cap F_2 = \emptyset$, and there is no direction parallel to both planes, or $d_1 + d_2 \geq n$ and $\dim(F_1 \cap F_2) = d_1 + d_2 - n$. See [85, pp. 88–90] for more information. A finite set of points in $\mathbb{E}^n$ is said to be in *general position* if no more than $k + 1$ of them belong to any k-dimensional plane.

A few of our results are set in 2-dimensional projective space $\mathbb{P}^2$. Generally, *n-dimensional projective space* $\mathbb{P}^n$ can be defined as the space of 1-dimensional subspaces of $\mathbb{E}^{n+1}$. The points of $\mathbb{P}^n$ are labeled by *homogeneous coordinates* $w = (w_1, \ldots, w_{n+1})$, not all zero, so for real $t \neq 0$ the points w and tw are identified; see, for example, [85, p. 217]. In this way, $\mathbb{P}^1$ can be regarded as the unit circle S^1 with antipodal points identified. We can also identify $\mathbb{E}^n$ with $\{w : w_{n+1} \neq 0\}$, where the usual coordinates are given by $x_i = w_i / w_{n+1}$. The remaining set $H_\infty = \{w : w_{n+1} = 0\}$ is the *hyperplane at infinity* (strictly speaking, a copy of $\mathbb{P}^{n-1}$). In particular, $\mathbb{P}^2$ can be regarded as $\mathbb{E}^2$ with a *line at infinity* (strictly speaking, a copy of $\mathbb{P}^1$) adjoined.

Our terminology for set theory and topology is standard. If E is a set, then $|E|$, co E, cl E, int E, and bd E denote the *cardinality, complement, closure, interior,* and *boundary* of E, respectively; also, relint E is the *relative interior* of E, that is, the interior of E relative to aff E. The *relative boundary* of E is the boundary of E relative to aff E. The *symmetric difference* of E and F is

$$E \bigtriangleup F = (E \setminus F) \cup (F \setminus E).$$

A G_δ *set* is a countable intersection of open sets, and an F_σ *set* is a countable union of closed sets. A set is of *first category* if it is the countable union of nowhere dense sets, and *residual* if it is the complement of a set of first category. A set in a locally compact Hausdorff space is residual if it contains a dense G_δ set; see, for example, [700, pp. 158–60 and 200–1]. A *component* of a set is a maximal connected subset. A closed set is *regular* if it is the closure of its interior, and a *body* is a compact, regular set.

The *diameter* diam E of a set E is

$$\mathrm{diam}\, E = \sup \{\|x - y\| : x, y \in E\}.$$

If x is a point and E is a closed set, the *distance* between x and E is

$$d(x, E) = \inf \{\|x - y\| : y \in E\}.$$

If E and F are sets, and r is a real number, then

$$E + F = \{x + y : x \in E, y \in F\},$$

and

$$rE = \{rx : x \in E\}.$$

A set E is called *centered* if $-x \in E$ whenever $x \in E$, and *centrally symmetric* if there is a vector c such that the translate $E - c$ of E by $-c$ is centered. In the latter case c is called a *center* of E. The center of a nonempty bounded centrally symmetric set is unique.

If X is a subset of $\mathbb{E}^n$, or indeed any topological space, the *support* of a real-valued function f on X is the set cl $\{x \in X : f(x) \neq 0\}$. We denote by $C(X)$ the class of continuous real-valued functions on X. When X is an appropriate subset of $\mathbb{E}^n$, $C_e(X)$ denotes the even functions in $C(X)$, and $C_e^+(X)$ the nonnegative functions in $C_e(X)$.

If f and g are real-valued functions, we say that $f = O(g)$ on $A \subset \mathbb{R}$ if there is a constant c such that $|f(x)| \leq c|g(x)|$ for all $x \in A$. When $A = \mathbb{N}$, we sometimes say that $f = O(g)$ as $n \to \infty$, while $f = O(g)$ as $x \to 0$ means that $f = O(g)$ on $A = (0, a)$ for sufficiently small a.

0.2. Transformations

No single book seems to provide a completely satisfactory introduction to the various types of transformations of $\mathbb{E}^n$ and $\mathbb{P}^n$; somehow the required material falls between the texts on Euclidean or projective geometry currently available. Borsuk's book [85] is possibly the most comprehensive text for this purpose, but its notation is quite outdated.

If A is an $n \times n$ matrix, the inverse and transpose of A are denoted by A^{-1} and A^t. We call A *singular* or *nonsingular* according to whether $\det A = 0$ or $\det A \neq 0$, respectively; A^{-1} exists precisely when A is nonsingular. We also adopt the abbreviation A^{-t} for $(A^{-1})^t$. Note that if A is nonsingular, then A^t is also, and $(A^t)^{-1} = (A^{-1})^t$.

For transformations ϕ of $\mathbb{E}^n$ and $\mathbb{P}^n$, we shall permit ourselves the shorthand $\phi x = \phi(x)$. The reader may find Figure 0.1 useful in interpreting the definitions given below.

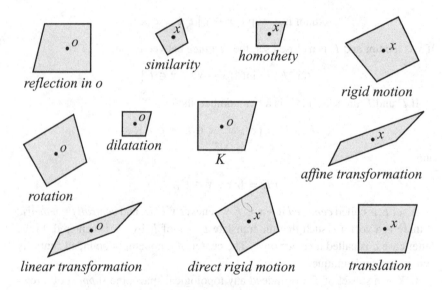

Figure 0.1. Transformations of a set K.

A *linear transformation* (or *affine transformation*) of $\mathbb{E}^n$ is a map ϕ from $\mathbb{E}^n$ to itself such that $\phi x = Ax$ (or $\phi x = Ax + t$, respectively), where A is an $n \times n$ matrix and $t \in \mathbb{E}^n$. (Here x is considered as a column vector, of course.) We call ϕ *singular* or *nonsingular* according to whether A is singular or nonsingular, respectively. The group of nonsingular linear (or affine) transformations is denoted by GL_n (or GA_n); its members are, in particular, bijections of $\mathbb{E}^n$ onto itself. The group of *special linear* (or *special affine*) transformations of $\mathbb{E}^n$ is denoted by SL_n (or SA_n, respectively). These are the members of GL_n (or GA_n) whose determinant is one. We shall write $\det \phi$ instead of $\det A$, and ϕ^{-1}, ϕ^t, and ϕ^{-t} for the affine transformations with corresponding matrices A^{-1}, A^t, and A^{-t}, respectively.

If A is the identity matrix, then $\phi x = x + t$, and the map ϕ is called a *translation*. Each affine transformation is composed of a linear transformation followed by a translation.

Any set of $n + 1$ points in general position in $\mathbb{E}^n$ can be mapped onto any second set of $n + 1$ points by a suitable affine transformation, and the latter is nonsingular if the second set is also in general position (see [595, Theorem 7, p. 16]).

If $\phi \in GA_n$, then ϕ takes parallel k-dimensional planes onto parallel k-dimensional planes (cf. [85, p. 156]).

An *isometry* of $\mathbb{E}^n$ is a map ϕ such that $\|\phi x - \phi y\| = \|x - y\|$; in other words, a distance-preserving bijection. Isometries are also called congruences, and the image and pre-image under an isometry are said to be *congruent*. Every isometry is affine (see, for example, [85, p. 150] or [839, p. 139]). Examples of isometries are the translations and the *reflections*, which map all points to their mirror images in some fixed point, line, or plane. (In particular, $\phi x = -x$ is the reflection in the origin.)

If $F = S + x_0$ (where $S \in \mathcal{G}(n, k)$, $x_0 \in \mathbb{E}^n$, and $1 \leq k \leq n - 1$) is a k-dimensional plane, and $x \in \mathbb{E}^n$, then there are unique points $y \in S$ and $z \in S^\perp$ such that $x = y + z$, and we can define a map taking x to $y + x_0 \in F$. This map is the (orthogonal) *projection* on the plane F. It is a singular affine transformation. If E is an arbitrary subset of $\mathbb{E}^n$, the image of E under a projection on a plane F is called the *projection of E on F* and denoted by $E|F$. Since $E|S$ is a translate of $E|F$ when $F = S + x_0$, we almost always work with the former.

If $\phi \in GL_n$, then

$$x \cdot \phi y = \phi^t x \cdot y, \tag{0.1}$$

for all x, $y \in \mathbb{E}^n$. The *orthogonal group O_n* of orthogonal transformations consists of those isometries of $\mathbb{E}^n$ that are also linear transformations; these are precisely the maps ϕ preserving the scalar product, that is, $\phi x \cdot \phi y = x \cdot y$. (An orthogonal matrix satisfies $A^t = A^{-1}$ and by (0.1) we have $\phi^t = \phi^{-1}$, hence the name.) It follows from this that orthogonal transformations have determinants with absolute value one. As is shown in [85, Theorem 50.6], every isometry is an orthogonal transformation followed by a translation, and for this reason isometries are sometimes also called *rigid motions*. The *special orthogonal group SO_n* of *rotations* about the origin consists of those orthogonal transformations with determinant one. A *direct rigid motion* is a rotation followed by a translation; these do not allow reflection.

A *dilatation* is a map $\phi x = rx$, for some $r > 0$. A *homothety* is a map $\phi x = rx + t$, for some $r > 0$ and $t \in \mathbb{E}^n$, that is, a composition of a dilatation with a translation (this is sometimes referred to as a direct homothety). A *similarity* is a composition of a dilatation with a rigid motion. We say two sets are *homothetic* (or *similar*) if one of them is an image of the other under a homothety (or similarity, respectively), or if one of the sets is a single point.

We find occasional use for projective transformations of $\mathbb{P}^n$. Such a transformation is given in terms of homogeneous coordinates by $\phi w = Aw + t$, where A is an $(n+1) \times (n+1)$ matrix and $t \in \mathbb{E}^{n+1}$, and where ϕ is called nonsingular if $\det A \neq 0$. Since we can regard $\mathbb{P}^n$ as $\mathbb{E}^n$ with a hyperplane H_∞ adjoined, we can also speak of a projective transformation of $\mathbb{E}^n$. In this regard, another formulation is useful. A *projective transformation* ϕ of $\mathbb{E}^n$ has the form

$$\phi x = \frac{\psi x}{x \cdot y + t}, \tag{0.2}$$

where $\psi \in GA_n$, $y \in \mathbb{E}^n$, and $t \in \mathbb{R}$, and ϕ is nonsingular if the associated linear map

$$\bar\psi(x, 1) = (\psi x, x \cdot y + t)$$

is nonsingular. If $y = o$, then ϕ is affine, but if $y \neq o$, ϕ maps the hyperplane $H = \{x : x \cdot y + t = 0\}$ onto H_∞. To avoid points in a set E being mapped into H_∞, we may insist that ϕ be *permissible* for E; this simply means that $E \cap H = \emptyset$.

Projective transformations map planes onto planes (neglecting the points mapping to or from infinity); see [595, pp. 19–20]. They also preserve cross ratio; a proof is given in [85, Corollary 96.11]. (The *cross ratio* of four points x_i, $1 \leq i \leq 4$ on a line is defined by

$$\langle x_1, \ldots, x_4 \rangle = \frac{(x_3 - x_1)(x_4 - x_2)}{(x_4 - x_1)(x_3 - x_2)},$$

where x_i also denotes the coordinate of the point x_i in a fixed Cartesian coordinate system on the line.) Affine transformations are also projective transformations, so the former also preserve cross ratio.

The sets E and F are called *linearly, affinely*, or *projectively equivalent* if there is a nonsingular transformation ϕ, linear, affine, or projective and permissible for E, respectively, such that $\phi E = F$. Suppose that E and F are bounded centered sets affinely equivalent via a nonsingular transformation ϕ. If $\phi o = p$, then p is the center of F; but since o is the unique center of F, we have $p = o$. Therefore ϕ is linear, proving that E and F are linearly equivalent.

0.3. Basic convexity

There are several possibilities for an introduction to the basic properties of convex sets. For the absolute beginner, the books of Lay [499] and Webster [827] are recommended. The first chapter of [595], by McMullen and Shephard, is terse, but very informative, as is the first chapter of [737], by Schneider. The text of [845], by Yaglom and Boltyanskiĭ, is set out in the form of exercises and solutions, with plenty of helpful diagrams. Chapters 11 and 12 of Berger's two-volume set [52], [53], contain some wonderful pictures, and Lyusternik's little book [554] is quirky but delightful. A list of books on convexity can be found in [737, p. 433].

A set C in $\mathbb{E}^n$ is called *convex* if it contains the closed line segment joining any two of its points, or, equivalently, if $(1 - t)x + ty \in C$ whenever $x, y \in C$ and $0 \le t \le 1$. A convex set, then, has no "holes" or "dents." A *convex body* is a compact convex set whose interior is nonempty; this definition conforms with general usage, but the reader is warned that in the important texts of Bonnesen and Fenchel [83] and Schneider [737] any compact convex set qualifies as a convex body. The *convex hull* conv E of a set E is the smallest convex set containing it.

If C is a compact convex set, a *diameter* of C is a chord $[x, y]$ of C such that $\|x - y\| = \operatorname{diam} C$.

A hyperplane H *supports* a set E at a point x if $x \in E \cap H$ and E is contained in one of the two closed half-spaces bounded by H. We say H is a *supporting hyperplane* of E if H supports E at some point.

A convex body is *strictly convex* if its boundary does not contain a line segment and *smooth* if there is a unique supporting hyperplane at each point of its boundary.

The intersection of a compact convex set with one of its supporting hyperplanes is called a *face*, and $(n - 1)$-dimensional faces are also called *facets*. An *extreme point* of K is one not contained in the relative interior of any line segment contained in K. The point x is called an *exposed point* of K if there is a supporting hyperplane H such that $H \cap K = \{x\}$. Every exposed point is extreme, but the converse is not true. Also, a compact convex set is the closure of the convex hull of its exposed points, implying that every compact convex set has at least one exposed point (see [737, Section 1.4], especially Theorem 1.4.7). A *corner point* of a compact convex set in $\mathbb{E}^2$ is one at which there is more than one supporting line.

If K_1 and K_2 are disjoint compact convex sets in $\mathbb{E}^n$, then there is a hyperplane H that (strictly) *separates* K_1 and K_2; that is, K_1 is contained in one open half-space bounded by H, and K_2 in the other. A proof can be found in [499, Theorem 4.12] or [737, Theorem 1.3.7]. (In infinite-dimensional spaces, this separation theorem is closely related to the Hahn–Banach theorem; see [52, Section 11.4].)

Every affine transformation preserves convexity. If ϕ is a projective transformation, permissible for a line segment, then it maps this line segment onto another line segment. Therefore ϕ preserves the convexity of convex bodies for which it is permissible.

A nonempty subset C of $\mathbb{E}^n$ is a *cone* with vertex o if $ty \in C$ whenever $y \in C$ and $t \ge 0$. A *convex cone* with vertex o is a cone with vertex o that is convex; such a set is closed under nonnegative linear combinations. A cone (or convex cone) with vertex x is of the form $C + x$, where C is a cone (or convex cone, respectively) with vertex o.

Let us define some special convex bodies. The unit ball B in $\mathbb{E}^n$ was defined already. A *ball* is any set homothetic to B, and an *ellipsoid* is an affine image of B. The centered n-dimensional ellipsoids whose axes are parallel to the co-ordinate axes are of the form

$$\left\{ x : \sum_{i=1}^{n} \frac{x_i^2}{a_i^2} \leq 1 \right\}.$$

If $0 \leq k \leq n$, a k-*dimensional simplex* in $\mathbb{E}^n$ is the convex hull of $k+1$ points in general position.

A *polyhedron* is a finite union of simplices; in $\mathbb{E}^2$, we shall use the term *poly-gon* instead. A convex polyhedron or *convex polytope* can also be defined as the convex hull of a finite set of points. We denote by $\mathcal{F}_k(P)$ the set of k-dimensional faces of a convex polytope P.

Important examples of convex polytopes are the *unit cube* $\{x : 0 \leq x_i \leq 1,\ 1 \leq i \leq n\}$ (and *centered unit cube* $\{x : |x_i| \leq 1/2, 1 \leq i \leq n\}$) in $\mathbb{E}^n$; the *parallelepipeds* or *parallelotopes*, affine images of the unit cube; the *boxes*, rectangular parallelepipeds with facets parallel to the coordinate hyperplanes; and the *cross-polytopes* (n-dimensional versions of the octahedron), each the convex hull of n mutually orthogonal line segments sharing the same midpoint. An n-dimensional *pyramid* P is the convex hull of an $(n-1)$-dimensional convex polytope Q (its *base*) and a point $x \notin \text{aff } Q$ called the *apex* of P.

A (right spherical) *cylinder* in $\mathbb{E}^n$ is the Cartesian product of an $(n-1)$-dimensional ball C and a line segment orthogonal to aff C. A (right spherical) *bounded cone* in $\mathbb{E}^n$ is the convex hull of an $(n-1)$-dimensional ball C and a point on the line orthogonal to aff C through the center of C.

If K is a convex body in $\mathbb{E}^n$, we denote by $r(K)$ and $R(K)$ the *inradius* and *circumradius* of K. These are the radii of the largest n-dimensional ball contained in K and the smallest ball containing K, respectively.

Topologically, a convex body is not very interesting. The surface of a convex body K in $\mathbb{E}^n$ is homeomorphic to S^{n-1} via a *radial map* f, defined by selecting a point $x_0 \in \text{int } K$ and letting

$$f(x) = (x - x_0)/\|x - x_0\|, \tag{0.3}$$

for each $x \in \text{bd } K$.

A real-valued function on $\mathbb{E}^n$ is *convex* if

$$f\big((1-t)x + ty\big) \leq (1-t)f(x) + tf(y),$$

for all $x, y \in \mathbb{E}^n$ and $0 \leq t \leq 1$, and *concave* if $-f$ is convex. (The terms concave up and concave down are sometimes used for convex and concave, respectively.)

0.4. The Hausdorff metric

Exactly what does it mean to say that a sequence of compact sets converges to another compact set? One must have a way of measuring the distance between two compact sets. This notion of distance must behave like the usual distance $d(x, y) = \|x - y\|$ between points, which has three fundamental properties: $d(x, y) \geq 0$, and equals zero if and only if $x = y$; $d(x, y) = d(y, x)$; and the triangle inequality

$$d(x, z) \leq d(x, y) + d(y, z).$$

Such a function is called a *metric*. We shall only define one metric for compact sets here, though there are several in common use (see Lemma 1.2.14 for another). The *Hausdorff metric* δ on the class of nonempty compact sets in $\mathbb{E}^n$ is defined by

$$\delta(E, F) = \max\{\max_{x \in E} d(x, F), \max_{x \in F} d(x, E)\}. \tag{0.4}$$

(A geometrically more appealing definition is given later.) It can be checked that δ satisfies the three conditions listed earlier. The proof, and basic properties of the metric space of compact sets in $\mathbb{E}^n$ defined in this way, may be found in [499, Section 14] or [737, Section 1.8].

Suppose that E is a nonempty set in $\mathbb{E}^n$ and $\varepsilon > 0$. Then

$$E_\varepsilon = E + \varepsilon B = \cup_{x \in E} (x + \varepsilon B) \tag{0.5}$$

is called an *outer parallel set* of E. When E is closed, E_ε is just the set of all points whose distance from E is no more than ε. (See [499, Section 14], [737, p. 134]; see also the illustration in the book [789, Fig. 1.1(b)] of Stoyan, Kendall, and Mecke, and the interesting accompanying discussion on the utility of this idea in the processing of images.) This convenient concept allows the following alternative definition of the Hausdorff metric:

$$\delta(E, F) = \min\{\varepsilon > 0 : E \subset F_\varepsilon \text{ and } F \subset E_\varepsilon\}. \tag{0.6}$$

This means that the Hausdorff distance between two convex bodies K_1 and K_2 is at most ε if K_1 is contained in the outer parallel body $K_2 + \varepsilon B$ of K_2, and K_2 is contained in the outer parallel body $K_1 + \varepsilon B$ of K_1.

The Hausdorff metric is the standard one in the study of convex sets. We denote by $\mathcal{K}^n$ (or $\mathcal{K}_0^n$) the space of nonempty compact convex sets (or convex bodies, respectively) in $\mathbb{E}^n$ with the Hausdorff metric. (The definition of a body in Section 0.1 implies the existence of interior points when the set is nonempty.) It is the default metric, always used unless stated otherwise, for example, when discussing continuity of a function defined on the class of compact convex sets. A specific, and important, example of this is the continuity of volume on $\mathcal{K}^n$; see [499, Theorem 22.6] or [737, Theorem 1.8.16]. (One should try not to be blasé about such statements. After all, length is not continuous in $\mathbb{E}^2$, since one can approximate

a closed line segment arbitrarily closely by polygonal arcs whose lengths are unbounded. According to Young [853, p. 303], this disturbed Lebesgue greatly when
he was at school! In fact, length is only semicontinuous in $\mathbb{E}^2$.)

A very frequently quoted theorem is the following one, whose proof may be
found in [499, Section 15] or [737, Theorem 1.8.6].

Theorem 0.4.1 (Blaschke's selection theorem). *Every bounded sequence of
compact convex sets has a subsequence converging to a compact convex set.*

(A sequence of sets is *bounded* if there is a ball containing each member of the
sequence.) In [737, Theorems 1.8.13 and 1.8.15], it is shown that each $K \in \mathcal{K}^n$
can be approximated arbitrarily closely from within or without by convex polytopes. This implies that the class of convex polytopes is dense in $\mathcal{K}^n$. It is also
known that both the class of smooth convex bodies and the class of strictly convex bodies are dense in $\mathcal{K}^n$; see [737, Theorem 2.6.1].

0.5. Measure and integration

Measure theory deals with the definition and generalizations of the intuitive
notions of length, area, and volume. The subject is amply supplied with well-
written books appropriate for the novice. Many a student has learned the basics
of Lebesgue measure and integration and the rudiments of general measure theory from [700], by Royden. At a slightly higher level, Munroe's book [639] is
to be recommended. Unfortunately, however, the *geometric* aspects of measure
theory are often ignored in the standard introductory texts. Exceptions are [839],
by Weir (see Chapter 6 of Volume 1), and [410], by Jones (see Chapter 3). Of
course, there are books on geometric measure theory proper, but here we can only
suggest a browse of the first three of chapters of the entertaining and exquisitely
illustrated introduction [637] by Morgan; we use no advanced geometric measure
theory in this book.

In practice one can get by without most of the complicated theory of abstract
measure. We summarize here the ingredients used in the sequel.

Consider, as a first example, area in the plane. Its essential properties are:

1. Familiar sets such as triangles, disks, and so on can be assigned a real number representing the area of the set.

2. The area of a countable union of disjoint sets is the sum of the areas of the
sets; that is, area is *countably additive*.

3. The area of a set does not change when it is moved by a translation; that is,
area is *translation invariant*. In fact, area is even invariant under isometries.

The same properties hold for a generalized notion of length in the real line,
or volume in space. Length and area are denoted by λ_1 and λ_2, respectively. For
Chapter 1, this is all one really needs.

Sooner or later, it becomes necessary to talk about the area of less familiar sets. It turns out that in order to retain the second and third properties, one has to give up the hope of assigning an area to *all* subsets of the plane (at least, if one wishes to use the commonly accepted axiom of choice). However, it can be shown that the concept of area can be defined so that all open sets can be assigned an area. Moreover, one can prove that the family of all sets that can be assigned an area forms a σ-*algebra*; that is, the family contains the empty set and is closed under the taking of complements and countable unions (and therefore also differences and countable intersections). Since the family of *Borel sets* is, by definition, the smallest σ-algebra containing the open sets, all Borel sets can be assigned an area.

Again, the same comments apply to generalized length in the real line and volume in space. Generalized length, area, and volume are examples of measures, and the sets that can be assigned a generalized length, area, or volume are called measurable sets. Among the measurable sets are those *of measure zero*, including all countable sets, but also many uncountable sets. For example, the Cantor ternary set in the real line has zero generalized length, and any line segment in the plane has zero area. Sets of measure zero (sometimes called *null sets*) are often neglected in measure theory, just as the number zero can be ignored in addition. For the types of measures encountered in this book, one is never too far from sanity when working with measurable sets, for it can be shown that each measurable set is the union of countably many closed sets and a (necessarily measurable) set of measure zero.

We are now ready for the formal definitions which abstract these ideas.

Let X be a set. A countably additive, extended real-valued function defined on a σ-algebra of subsets of X is called a *signed measure*; it is a *measure* if it is also nonnegative. The members of the σ-algebra are called *measurable sets*. We say a measure μ is σ-*finite* if X is a countable union of sets of finite μ-measure. A measure μ is said to be *concentrated* on a subset E of X if $\mu(X \setminus E) = 0$. If X is a topological space, and the σ-algebra consists of the Borel sets in X, the measure is called a *Borel measure*. An arbitrary measure in X is called *Borel regular* if Borel sets are measurable and every measurable set is contained in a Borel set of the same measure. A property is said to hold μ-*almost everywhere* or for μ-*almost all* $x \in X$ if there is a subset E of X with $\mu(E) = 0$ such that the property holds for all $x \in X \setminus E$.

We generally use lowercase Greek letters for measures. This is the convention adopted by most measure theorists, with the important exception of some who work in geometric measure theory, who use capital script letters, such as the $\mathcal{H}$ for Hausdorff measure (to be defined shortly). History has forced us to make, reluctantly, an exception for the area measures, defined in Section A.2.

After measures are defined, one can deal with the integral (some authors reverse this process). If μ is a measure in X, the μ-*measurable* extended real-valued functions are those for which the inverse image of an open set is a measurable set.

When X is a topological space, there is also the class of *Borel functions* on X, the extended real-valued functions for which the inverse image of an open set is a Borel set. Every continuous function is Borel, and if μ is a Borel measure, then every Borel function is μ-measurable. For certain functions f on X, a meaning can be given to

$$\int_E f(x)\, d\mu(x),$$

the *integral* of f over the measurable set $E \subset X$, in such a way that in the familiar case of a nonnegative f defined on $\mathbb{E}^n$, the integral gives the volume under the graph of f. Nonnegative functions are called *μ-integrable* on E if they are μ-measurable and the integral exists and is finite. An arbitrary function f is *μ-integrable* if both its *positive part* f^+ and its *negative part* f^-, defined by

$$f^+(x) = \max\{f(x), 0\} \text{ and } f^-(x) = \max\{-f(x), 0\},$$

are integrable. A bounded measurable function is integrable on any set of finite measure. All this can be found in Chapters 4 and 11 of [700], for example.

One theorem in the theory of integration is of outstanding importance: Fubini's theorem (see [700, Theorem 19, p. 307]) says that in all reasonable circumstances, the integral of a function on a product of two spaces equals both of the two iterated integrals. (This allows, for example, the volume of a measurable set in $\mathbb{E}^3$ to be calculated by integrating the areas of its sections by planes parallel to a given plane.)

The *n-dimensional Lebesgue measure* λ_n in $\mathbb{E}^n$ is often defined to be the unique Borel-regular, translation-invariant measure in $\mathbb{E}^n$ such that the unit cube has unit measure. This provides one definition of generalized length in the real line, area in the plane, and volume in space. Defined this way, however, λ_n is not the most important measure. This honor goes to *k-dimensional Hausdorff measure* $\mathcal{H}^k$ in $\mathbb{E}^n$, $0 \leq k \leq n$. This is the standard way of measuring k-dimensional volume in $\mathbb{E}^n$, so that, for example, one could use $\mathcal{H}^1$ to measure the perimeter of a disc, or $\mathcal{H}^2$ for the surface area of a ball. The definition of Hausdorff measure (see the texts of Morgan [637, p. 8] or Rogers [694, Chapter 2]) is somewhat technical, but not really more so than the very commonly adopted definition of Lebesgue measure in the real line via Lebesgue outer measure, as in Chapter 3 of [700], for example.

It is a convenient fact that the two measures λ_n and $\mathcal{H}^n$ agree in $\mathbb{E}^n$ (see [637, Corollary 2.8] or [694, Theorem 30]), provided the correct constant is included in the definition of $\mathcal{H}^n$. There is a similar agreement between $\mathcal{H}^{n-1}$ and $(n-1)$-*dimensional spherical Lebesgue measure* in S^{n-1}, the unique Borel-regular, rotation-invariant measure in $\mathbb{E}^n$ such that S^{n-1} has measure equal to the constant ω_n whose value is given by (0.10). Indeed, it is well known that $\mathcal{H}^{n-1}$ is Borel regular and rotation invariant (see [694, Theorem 27 and p. 58]), and the fact that $\mathcal{H}^{n-1}(S^{n-1}) = \omega_n$ follows from integration via the area formula in [637, 3.7, p. 25]. Therefore we allow ourselves to speak loosely of k-dimensional Lebesgue measure in $\mathbb{E}^n$ when we really mean k-dimensional Hausdorff measure, and use λ_k for integration in planes or spheres. Two abbreviations should be noted: We shall write dx (or du, etc.) for $d\lambda_k(x)$ (or $d\lambda_k(u)$, etc., as appropriate) when integrating over a k-dimensional plane or unit $(k+1)$-sphere S^k.

The measure $\mathcal{H}^0$ (we shall write λ_0) is just the counting measure, which counts the number of points in a set.

When no misunderstanding can arise – for example, when working with compact convex sets – we call the λ_k-measure of a k-dimensional body in $\mathbb{E}^n$ its *volume*. This is traditional in geometry.

Often we want to work with the equivalence classes of measurable sets modulo sets of measure zero, and here it is useful to write $E \simeq F$ when $\lambda_n(E \triangle F) = 0$.

Let $\phi \in GA_n$. Then $|\det\phi|$ is the factor by which ϕ changes volume, that is,

$$\lambda_n(\phi E) = |\det\phi|\lambda_n(E), \qquad (0.7)$$

for each λ_n-measurable set E in $\mathbb{E}^n$; see [839, pp. 142–4]. It follows that the members of SA_n, and more generally those maps in GA_n whose determinants are ± 1, are volume preserving. It also follows that if $r \geq 0$, then $\lambda_n(rE) = r^n\lambda_n(E)$. More generally, if $1 \leq k \leq n$, E is a λ_k-measurable set in $\mathbb{E}^n$, and $r \geq 0$, then

$$\lambda_k(rE) = r^k\lambda_k(E). \qquad (0.8)$$

One can also check that ϕ preserves the ratio of λ_k-measures of sets in parallel k-dimensional planes.

The volume of the unit ball in $\mathbb{E}^n$ is given by

$$\kappa_n = \lambda_n(B) = \frac{\pi^{n/2}}{\Gamma(1 + \frac{n}{2})}, \qquad (0.9)$$

with the convention $\kappa_0 = 1$, and its surface area is

$$\omega_n = \lambda_{n-1}(S^{n-1}) = n\kappa_n. \qquad (0.10)$$

The first computation is given in [570, pp. 324–5] and the second in [171, p. 125]; or see [746, p. 18]. To calculate special values of κ_n, one only needs $\Gamma(1+x) = x\Gamma(x)$, $\Gamma(1) = 1$, and $\Gamma(1/2) = \sqrt{\pi}$. It is interesting that κ_n increases with n to its maximum value $8\pi^2/15$ when $n = 5$, and then decreases, approaching zero.

Using (0.9) and (0.7), one shows that the n-dimensional centered ellipsoid $\{x : \sum_{i=1}^n x_i^2/a_i^2 \leq 1\}$ has volume

$$a_1 a_2 \cdots a_n \kappa_n. \qquad (0.11)$$

The volume of a parallelepiped is the λ_{n-1}-measure of its base times its height orthogonal to its base. The volume of the parallelepiped in $\mathbb{E}^n$ with vertices at o, $p_1, \ldots, p_n$ is also given by

$$|\det(p_{ij})|, \tag{0.12}$$

where $p_i = (p_{i1}, \ldots, p_{in})$, and the volume of the simplex in $\mathbb{E}^n$ with vertices at $o, p_1, \ldots, p_n$ is

$$\frac{1}{n!}|\det(p_{ij})|, \tag{0.13}$$

as in [85, p. 117]. We have the formula

$$\lambda_n(P) = \frac{1}{n} z \lambda_{n-1}(Q) \tag{0.14}$$

for the volume of a pyramid or bounded cone P with base Q and height (the distance from aff Q to the apex) z. This is easily obtained by integration and induction, as in [52, 9.12.4.4] for the simplex; Dehn's solution of Hilbert's third problem indicates that some form of limit argument is required (see the discussion in [53, 12.2.5.2], for example).

We occasionally need other Borel measures in $\mathbb{E}^n$ or S^{n-1}. A signed Borel measure μ in S^{n-1} is called *even* (or *odd*) if $\mu(-E) = \mu(E)$ (or $\mu(-E) = -\mu(E)$, respectively), for all Borel sets E.

Let μ be a measure in $\mathbb{E}^n$ and E a bounded set in $\mathbb{E}^n$ of finite positive μ-measure. The *centroid* of E with respect to μ is the point

$$c = \frac{1}{\mu(E)} \int_E x \, d\mu(x). \tag{0.15}$$

The centroid of E is contained in conv E; see [83, Section 6, p. 9].

There is another measure that is extremely important in geometry, and it occurs in this fashion. It is sometimes essential to be able to measure the size of a set of lines or planes, or to integrate a function defined on a set of lines or planes. We only need to do this for sets of subspaces, that is, lines and planes containing the origin, or generally for subsets of $\mathcal{G}(n, k)$. Moreover, our measure should be compatible with the appropriate geometric transformations, so that, for example, the measure of a subset E of $\mathcal{G}(n, k)$ should equal the measure of the set obtained by applying the same rotation about the origin to each member of E. For $k = 1$ (or $k = n - 1$), this is easy: Just identify each 1-dimensional subspace (or $(n - 1)$-dimensional subspace) S with the corresponding antipodal pair of points $\pm u$ in S^{n-1} such that the vector u is parallel to S (or orthogonal to S, respectively), and then use the measure λ_{n-1} in S^{n-1}. For $1 < k < n - 1$, however, one needs a new measure, which can be defined by the following general process.

Let X be a locally compact topological group. Then there is a nonzero Borel-regular measure μ in X that is also invariant under left translations by elements of X. This measure μ is called the *Haar measure* in X; it is unique up to multiplication by a constant, and is finite if X is compact. A detailed proof of its existence and uniqueness is given in the texts of Cohn [168, Chapter 9] and Munroe [639, Section 17], for example. However, for the special case of most interest here, this can be avoided. A clever direct construction due to Schneider and Weil [746, Satz 1.2.4, p. 21] shows that there is a Haar measure v_n in the compact group SO_n, normalized so that $v_n(SO_n) = 1$. Let $S \in \mathcal{G}(n, k)$, and let $f_k: SO_n \to \mathcal{G}(n, k)$ be defined by $f_k(\phi) = \phi S$ for each $\phi \in SO_n$. This map is surjective; the usual topology for $\mathcal{G}(n, k)$ is the finest topology for which f_k is continuous, and with this topology $\mathcal{G}(n, k)$ is a compact space. If $E \subset \mathcal{G}(n, k)$ is a Borel set, define

$$\mu_{n,k}(E) = v_n\big(f_k^{-1}(E)\big).$$

Then $\mu_{n,k}$, also referred to as the Haar measure in $\mathcal{G}(n, k)$, is the measure we need; note that it is normalized so that

$$\mu_{n,k}\big(\mathcal{G}(n, k)\big) = 1,$$

a fact we shall use several times without special comment. When integrating over $\mathcal{G}(n, k)$, we shall abbreviate $d\mu_{n,k}(S)$ by dS, and with this will have no further use for the symbol $\mu_{n,k}$.

Finally, we need a few more definitions. Let X be any set, and μ a measure in X. If $p \geq 1$ is a real number, $L^p(X)$ denotes the set of μ-measurable extended real-valued functions on X such that $\int_X |f(x)|^p \, d\mu(x) < \infty$, and $\|f\|_p = \big(\int_X |f(x)|^p \, d\mu(x)\big)^{1/p}$ is the L^p *norm*. Also, $L^\infty(X)$ is the space of essentially bounded μ-measurable functions on X, with the L^∞ *norm* given by $\|f\|_\infty = \text{ess sup} |f|$. (The function f is essentially bounded if it is equal μ-almost everywhere to a bounded function g. The *essential supremum* ess sup f of f is the infimum of the suprema of such g, and the *essential infimum* ess inf f of f is the supremum of the infima of such g. For continuous functions, the essential supremum and infimum reduce to the ordinary supremum and infimum, respectively.)

A sequence $\{\mu_n\}$ of finite Borel measures in a metric space X is said to *converge weakly* to the finite Borel measure μ in X if

$$\int_X f(x) \, d\mu_n(x) \to \int_X f(x) \, d\mu(x), \tag{0.16}$$

for each bounded $f \in C(X)$.

0.6. The support function

Perhaps the most widely applicable function connected with the study of convex sets is the support function, and the purpose of this section is to gather together some of its properties.

If K is a nonempty compact convex set in $\mathbb{E}^n$, the *support function* h_K of K is defined by

$$h_K(x) = \max\{x \cdot y : y \in K\}, \tag{0.17}$$

for $x \in \mathbb{E}^n$. From this definition it follows that if K_1 and K_2 are compact convex sets, then $K_1 \subset K_2$ if and only if $h_{K_1} \leq h_{K_2}$, and this implies that a compact convex set is determined by its support function.

If $u \in S^{n-1}$, then

$$H_u = \{x : x \cdot u = h_K(u)\} \tag{0.18}$$

is the supporting hyperplane to K with outer normal vector u. The support function $h_K(u)$ at a *unit vector* u gives the *signed* distance from o to H_u; see Figure 0.2.

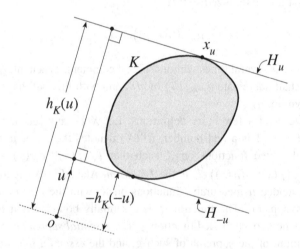

Figure 0.2. The support function.

As a function on $\mathbb{E}^n$, the support function is positively homogeneous, that is,

$$h_K(cx) = ch_K(x) \text{ for } c \geq 0, \tag{0.19}$$

and subadditive, that is,

$$h_K(x + y) \leq h_K(x) + h_K(y). \tag{0.20}$$

See [83, Section 15, p. 26]; a function having both these properties is called *sublinear*. An important result is that the converse is true: Every sublinear function on $\mathbb{E}^n$ is the support function of a unique compact convex set. See [737, Theorem 1.7.1] for three proofs. Every sublinear function is convex, and every convex function is continuous on the interior of its domain (cf. [499, Theorem 30.2] or [737, Theorem 1.5.1]); therefore the support function is convex and continuous on $\mathbb{E}^n$.

Equation (0.19) usually permits us to work with the restriction of h_K to the unit sphere, as we almost always do in this book, without special comment.

The support function has some other properties making it of fundamental importance in convexity. It is immediate from the definition that a compact convex set K is centered if and only if $h_K(u) = h_K(-u)$ for all $u \in S^{n-1}$. It is an easy exercise to show that if S is a subspace, then

$$h_{K|S}(u) = h_K(u), \tag{0.21}$$

for all $u \in S^{n-1} \cap S$. The support function also gives a convenient expression for the Hausdorff distance between two compact convex sets K_1 and K_2, namely, that

$$\delta(K_1, K_2) = \sup_{u \in S^{n-1}} |h_{K_1}(u) - h_{K_2}(u)| = \|h_{K_1} - h_{K_2}\|_\infty. \tag{0.22}$$

The simple proof is given in [737, Theorem 1.8.11].

If K_i is a compact convex set in $\mathbb{E}^n$, and $t_i \geq 0$, $1 \leq i \leq m$, then the vector sum

$$t_1 K_1 + \cdots + t_m K_m = \{t_1 x_1 + \cdots + t_m x_m : x_i \in K_i\}$$

is also called a *Minkowski linear combination*. The addition and scalar multiplication are also called *Minkowski addition* and *Minkowski scalar multiplication*. It is easily shown that this Minkowski linear combination is itself a compact convex set.

If $t_1, t_2 \geq 0$, then

$$h_{t_1 K_1 + t_2 K_2}(x) = t_1 h_{K_1}(x) + t_2 h_{K_2}(x), \tag{0.23}$$

for all $x \in \mathbb{E}^n$; for the simple proof, see [737, Theorem 1.7.5].

If K is the singleton set $\{x\}$, then $h_K(u) = u \cdot x$, for all $u \in S^{n-1}$. From this and (0.23) we see that if K is a compact convex set, then the support function of a translate of K is given by

$$h_{K+x}(u) = h_K(u) + x \cdot u, \tag{0.24}$$

for all $u \in S^{n-1}$. It also follows that if $K = [x, y]$ is a line segment, then $h_K(u) = \max\{u \cdot x, u \cdot y\}$; in particular, if $v \in S^{n-1}$, then

$$h_{[-v,v]}(u) = |u \cdot v|, \tag{0.25}$$

for all $u \in S^{n-1}$. Now (0.23) implies that when K is the centered cube in $\mathbb{E}^n$, with sides of length 2 and parallel to the coordinate hyperplanes, we have $h_K(u) = \sum_{i=1}^{n} |u_i|$, for $u = (u_1, \ldots, u_n) \in S^{n-1}$. Generally, one can show that K is a polytope if and only if h_K is piecewise linear.

Let $\phi \in GL_n$. Then, with (0.1),

$$
\begin{aligned}
h_{\phi K}(x) &= \max\{x \cdot y : y \in \phi K\} \\
&= \max\{x \cdot \phi z : z \in K\} \\
&= \max\{\phi^t x \cdot z : z \in K\},
\end{aligned}
$$

so

$$h_{\phi K}(x) = h_K(\phi^t x), \tag{0.26}$$

for all $x \in \mathbb{E}^n$, and

$$h_{\phi K}(u) = h_K(\phi^t u) = \|\phi^t u\| h_K\left(\frac{\phi^t u}{\|\phi^t u\|}\right), \tag{0.27}$$

for all $u \in S^{n-1}$.

Of course, $h_B(u) = 1$, for all $u \in S^{n-1}$; the support function of an ellipsoid can be obtained from this and (0.27).

0.7. Star sets and the radial function

The radial function is dual to the support function introduced in the previous section, but it appears much less frequently in the literature. Whereas it is natural to define the support function for convex sets, the radial function can be defined for the more general star sets. The purpose of this section is to explain the meaning of these terms.

A set L is *star-shaped* at o if every line through o that meets L does so in a (possibly degenerate) line segment. If L is nonempty, compact, and star-shaped at o, its *radial function* ρ_L is defined by

$$\rho_L(x) = \max\{c : cx \in L\}, \tag{0.28}$$

for $x \in \mathbb{E}^n \setminus \{o\}$ such that the line through x and o meets L. It is positively homogeneous of degree -1 that is,

$$\rho_L(cx) = c^{-1}\rho_L(x) \text{ for } c > 0. \tag{0.29}$$

As with the support function, this usually permits us to work with the restriction of ρ_L to the unit sphere, and we shall do this without further comment. We denote the domain of this restriction by D_L and its support by S_L.

Note that our definition of radial function differs in an important way from the usual one, in which the maximum is taken only over nonnegative c. One advantage of the new definition, introduced by Gardner and Volčič [283], is that it mirrors the definition (0.17) of the support function. The radial function $\rho_L(u)$ at a *unit vector u* gives the *signed* distance from o to the boundary of L along the line l_u through o parallel to u. See Figure 0.3.

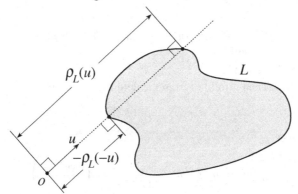

Figure 0.3. The radial function.

By a *star body* we mean a body such that ρ_L, restricted to S_L, is continuous. A *star set* is a set that is a star body in its linear hull.

We warn the reader that there are several definitions of the term "star body" currently in use. One insists that $o \in \operatorname{int} L$, which is clearly more restrictive than the definition here. Another, specifying that ρ_L is continuous, is a viable alternative for bodies containing the origin (see Chapter 8, especially the discussion at the beginning of Section 8.1). Our definition has the advantage that a convex body is always a star body. Sometimes, however, the extra assumption that S_L is centered is required, as in Theorem 7.2.3, for example.

Let L be a star set. If $o \in L$, then D_L coincides with S^{n-1}; otherwise, D_L is smaller. Let L be a star body. It is not difficult to see that since L is compact and regular, both D_L and S_L are also compact and regular in S^{n-1}. Since ρ_L is continuous on S_L, it is a bounded Borel function on D_L.

If L is a star set in $\mathbb{E}^n$, and $S \in \mathcal{G}(n, k)$, then $L \cap S$ need not be a star set, since it may not be regular. However, if $\dim(L \cap S) = k$, then $L \cap S$ is a star set.

If $x_i \in \mathbb{E}^n$, $1 \le i \le m$, then $x_1 \tilde{+} \cdots \tilde{+} x_m$ is defined to be the usual vector sum of the points x_i, if all of them are contained in a line through o, and o otherwise. Let L_i be a star body in $\mathbb{E}^n$ with $o \in L_i$, and $t_i \ge 0$, $1 \le i \le m$; then

$$t_1 L_1 \tilde{+} \cdots \tilde{+} t_m L_m = \{t_1 x_1 \tilde{+} \cdots \tilde{+} t_m x_m : x_i \in L_i\} \qquad (0.30)$$

is called a *radial linear combination*. The addition and scalar multiplication are called *radial addition* and *radial scalar multiplication*. (Lutwak [537] adds Minkowski's name to these terms.) Moreover,

$$\rho_{t_1 L_1 \tilde{+} t_2 L_2}(x) = t_1 \rho_{L_1}(x) + t_2 \rho_{L_2}(x), \qquad (0.31)$$

for all x.

One can measure distance between star bodies by means of the Hausdorff metric. However, in many respects the *radial metric* $\tilde{\delta}$ is more natural. This is defined by setting

$$\tilde{\delta}(L_1, L_2) = \sup_{u \in S^{n-1}} |\rho_{L_1}(u) - \rho_{L_2}(u)| = \|\rho_{L_1} - \rho_{L_2}\|_\infty, \qquad (0.32)$$

for star bodies L_1, L_2 in $\mathbb{E}^n$.

Let $\phi \in GL_n$. Then it follows from the definition of ρ_L that

$$\rho_{\phi L}(x) = \rho_L(\phi^{-1}x), \qquad (0.33)$$

for $x \in \mathbb{E}^n \setminus \{o\}$, so

$$\rho_{\phi L}(u) = \rho_L(\phi^{-1}u) = \frac{1}{\|\phi^{-1}u\|} \rho_L\left(\frac{\phi^{-1}u}{\|\phi^{-1}u\|}\right), \qquad (0.34)$$

for all $u \in S^{n-1}$.

Many examples of radial functions can be obtained from those of support functions via the important polar relation (0.36).

0.8. Polar duality

Polar duality is an important tool in geometry, and it will be used several times in this book. Though much is known about polar duality, it can be frustrating to search the literature for even the most basic facts, so these are collected together in this section.

If E is an arbitrary nonempty subset of $\mathbb{E}^n$, then the set

$$E^* = \{x : x \cdot y \le 1 \text{ for all } y \in E\} \qquad (0.35)$$

is called the *polar set* of E. The polar set is always closed and convex and contains the origin; see [499, p. 142]. Moreover, if K is a convex body and $o \in \text{int } K$, then the same is true of K^*, which we then call the *polar body* of K; see Figure 0.4. In this case $K^{**} = K$ (see [499, p. 142], [595, Section 2.2], or [737, Theorem 1.6.1]).

If K is a convex body in $\mathbb{E}^n$ such that $o \in \text{int } K$, the boundary of K^* can be calculated by the following important relation (see [499, Theorem 29.8] or [737, Remark 1.7.7]):

$$\rho_{K^*}(u) = 1/h_K(u), \qquad (0.36)$$

for all $u \in S^{n-1}$.

Suppose that K is a centered convex body in $\mathbb{E}^n$. Then (0.36) and the fact that $K^{**} = K$ show that the reciprocal of ρ_K (sometimes called the *gauge function* of K) is h_{K^*} and is therefore sublinear. In view of the fact that K is centered, the reciprocal of ρ_K actually defines a norm $\|\cdot\|_K$ on $\mathbb{E}^n$, for which K is the unit ball.

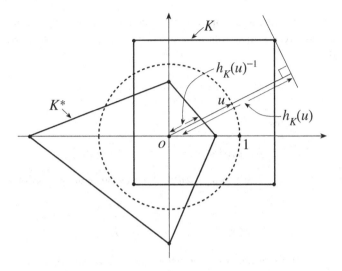

Figure 0.4. The polar body.

The dual of this Banach space is the one whose norm is given by h_K, which has the polar body K^* as its unit ball; see [737, Remark 1.7.8]. This is the source of an intimate and important connection between centered convex bodies and Banach spaces.

The polar set of the single point $\{x\}$, $x \neq o$, is the half-space $\{y : x \cdot y \leq 1\}$, and the polar body of the ball rB is the ball $r^{-1}B$. The following examples of polar bodies of convex polytopes are noted by Grünbaum [367, Section 3.4]. If P is an n-dimensional simplex containing the origin in its interior, then P^* is also. The polar body of the centered unit cube in $\mathbb{E}^n$ is a centered cross-polytope. A centered regular dodecahedron has a centered icosahedron for its polar body. Generally, the polar body of a convex polytope P is also a convex polytope, and polarity provides an inclusion-reversing bijection between the faces of P and the faces of P^* (see [595, Lemma 8, p. 65]).

Suppose that $\phi \in GL_n$, and that K is a convex body in $\mathbb{E}^n$ with $o \in \operatorname{int} K$. Then for each $u \in S^{n-1}$, by (0.27) and (0.34),

$$h_{(\phi K)^*}(u) = \frac{1}{\rho_{\phi K}(u)} = \frac{1}{\rho_K(\phi^{-1}u)} = h_{K^*}(\phi^{-1}u) = h_{\phi^{-t}K^*}(u).$$

Therefore

$$(\phi K)^* = \phi^{-t}K^*. \tag{0.37}$$

It follows immediately from (0.37) that centered convex bodies are similar, or linearly equivalent, if and only if their polar bodies are, respectively.

If E is an n-dimensional ellipsoid in $\mathbb{E}^n$, containing the origin in its interior, then E^* is also. To see this, note first that if $a = (a_1, \ldots, a_n) \in \operatorname{int} B$, then by (0.36) and (0.24),

$$\rho_{(B+a)^*}(u) = \frac{1}{h_{(B+a)}(u)} = \frac{1}{1 + a \cdot u},$$

for all $u \in S^{n-1}$. This gives

$$\rho_{(B+a)^*}(u) = 1 - \rho_{(B+a)^*}(u) u \cdot a.$$

If we let $x = (x_1, \ldots, x_n) = \rho_{(B+a)^*}(u) u$, the previous equation becomes

$$\sum_{i=1}^{n} x_i^2 = \left(1 - \sum_{i=1}^{n} a_i x_i\right)^2.$$

Since this equation is a quadratic, and since we know that $(B + a)^*$ is convex, it follows that $(B + a)^*$ is an ellipsoid (see [53, Proposition 15.4.7]). Now if $E = \phi(B + a)$, where $a \in \operatorname{int} B$ and $\phi \in GL_n$, then (0.37) implies that $E^* = \phi^{-t}(B + a)^*$ is an ellipsoid.

Polar duality provides a link between sections and projections. Indeed, suppose that K is a convex body in $\mathbb{E}^n$ with $o \in \operatorname{int} K$, and that S is a subspace. Then

$$K^* \cap S = (K|S)^*, \tag{0.38}$$

where the polar operation on the right is taken in S. One can see this by using (0.36) and (0.21) to conclude that, for any $u \in S \cap S^{n-1}$,

$$\rho_{K^* \cap S}(u) = \rho_{K^*}(u) = \frac{1}{h_K(u)} = \frac{1}{h_{K|S}(u)} = \rho_{(K|S)^*}(u);$$

or see [595, Theorem 15, p. 70] for a simple proof using only the definition of a polar body.

Despite (0.38), the use of polar duality in geometric tomography is severely limited by the fact that (as Figure 0.4 suggests) it is not an affine notion, but rather a projective one. To be more specific, consider the following result, proved in [595, Theorem 14, p. 67]. Let K be a convex body in $\mathbb{E}^n$ with $o \in \operatorname{int} K$, and let ϕ be a nonsingular projective transformation of $\mathbb{E}^n$, permissible for K, such that $o \in \operatorname{int} \phi K$. Then there is a nonsingular projective transformation ψ, permissible for K^*, such that $(\phi K)^* = \psi K^*$.

0.9. Differentiability properties

A real-valued function on an open subset U of $\mathbb{E}^n$ is said to be *of class* C^k if it is k-times continuously differentiable, that is, all partial derivatives of order k exist and are continuous. The class of such functions is signified by $C^k(U)$. The class $C^\infty(U)$ consists of those real-valued functions belonging to $C^k(U)$ for all $k \in \mathbb{N}$. A real-valued function f on U is *real analytic* if its Taylor series exists and converges to $f(x)$ at each $x \in U$. (See, for example, [570, Section 6.8].)

If $n \in \mathbb{N}$ and $1 \le i \le n$, let π_i be the real-valued function on $\mathbb{E}^n$ defined by $\pi_i(x) = x_i$, where $x = (x_1, \ldots, x_n)$. Suppose that f is a function from an open subset U of $\mathbb{E}^n$ into $\mathbb{E}^m$. Then we say f is of class C^k if each map $\pi_i \circ f$, $1 \le i \le m$, is in $C^k(U)$. Functions from an open subset of $\mathbb{E}^n$ into $\mathbb{E}^m$ that are of class C^∞ or real analytic are defined analogously.

Sometimes we want to speak, for example, about a function belonging to $C^\infty(S^{n-1})$, or the boundary of a convex body being of class C^2. Basically, the meaning of such terms is inherited from those defined in the previous paragraph, but precise definitions take a little work. These can be found in several books on differential geometry, but for the convenience of the reader we present them here.

A subset M of $\mathbb{E}^n$ is called an *m-dimensional submanifold of $\mathbb{E}^n$ of class C^k* if there is an *atlas* for M of class C^k. An atlas for M of class C^k is a family of pairs (U_r, f_r), called *charts*, such that

(i) each U_r is an open subset of M, and $\cup_r U_r = M$;

(ii) each f_r is a homeomorphism of U_r onto an open subset of $\mathbb{E}^m$;

(iii) if $U_r \cap U_s \ne \emptyset$, the map $f_s \circ f_r^{-1}$, from the open subset $f_r(U_r \cap U_s)$ of $\mathbb{E}^m$ into $\mathbb{E}^m$, is of class C^k.

Again, an *m-dimensional submanifold of $\mathbb{E}^n$ of class C^∞* and a *real-analytic m-dimensional submanifold of $\mathbb{E}^n$* are defined analogously.

The unit sphere S^2 is an example of a real-analytic 2-dimensional submanifold of $\mathbb{E}^3$. The reason for the somewhat technical definition given is that one cannot map the whole of S^2 onto an open subset of the plane in the appropriate way, so one has to use several charts; the sets U_r are patches on S^2 which cover it and which can be mapped onto open subsets of the plane. (In fact, the term "patch" is often used instead of "chart.") A picture of such patches covering a surface is one of many figures generated with Mathematica by Gray [330, p. 219].

We say that a convex body K is *of class C^k* or *of class C^∞* if bd K is of class C^k (or C^∞, respectively) as a submanifold of $\mathbb{E}^n$.

Now let M be an m-dimensional submanifold of $\mathbb{E}^n$ of class C^l, and suppose that $k \le l$. The real-valued function f on M is of class C^k if, for every chart (U_r, f_r) in an atlas for M, the real-valued function $f \circ f_r^{-1}$ on the open subset $f_r(U_r)$ of $\mathbb{E}^m$ is of class C^k. The class of such functions is denoted by $C^k(M)$. The real-valued functions on M of class C^∞, or real-analytic ones, are defined analogously, and the former class is signified by $C^\infty(M)$.

When E is an appropriate subset of $\mathbb{E}^n$, $C_e^k(E)$ and $C_e^\infty(E)$ denote the even functions in the classes defined earlier.

Occasionally we adopt the common practice of calling a function or body "sufficiently smooth." This simply means that it belongs to C^k for a sufficiently large k.

If U is an open subset of the reals, the notation $f \in C^{k+\varepsilon}(U)$, $0 < \varepsilon \leq 1$, means that $f \in C^k(U)$ and the kth derivative $f^{(k)}$ of f satisfies the following *Hölder condition*:

$$|f^{(k)}(x) - f^{(k)}(y)| \leq c\,|x - y|^\varepsilon, \qquad (0.39)$$

for some $c \geq 0$ and all $x, y \in U$. This allows some of the previous definitions to have meaning even when k is an arbitrary nonnegative real number.

In the early literature on convexity, it was standard procedure for authors to assume any convenient order of smoothness of the boundary of a convex body, often without explicit comment. As the years went by, it happened again and again that examples were discovered of convex bodies with surprisingly complicated boundary structure. Luckily, one can learn basic convexity without spending too much time on this topic, and in the advanced theory of convexity, Aleksandrov's magnificent theory of area measures (see Section A.2) often allows one to proceed without any special boundary assumptions.

Sooner or later, however, one has to make such assumptions. Unfortunately, there are many pitfalls and highly nonintuitive phenomena. For example, in [436], Kiselman has shown the existence of a convex body in $\mathbb{E}^3$ having a real-analytic boundary surface, though the boundary of its projection on some plane is only $C^{2+\varepsilon}$ for some $\varepsilon > 0$; and in [437], he proves that the boundary of a Minkowski sum of two planar convex bodies with real-analytic boundaries is $C^{20/3}$, and that this result is the best possible! Moreover, basic results and even definitions tend to be involved and require some knowledge of differential geometry. Until recently, there was no adequate treatment available, but Chapter 2 of Schneider's book [737] now provides a lucid and extremely valuable guide, plundered for the following summary.

Let K be a convex body in $\mathbb{E}^n$. The support function h_K is differentiable on $\mathbb{E}^n \setminus \{o\}$ if and only if h_K is C^1, and also if and only if K is strictly convex; see [737, p. 107]. In this case $H_u \cap K$ (where H_u is defined by (0.18)) is a single point x_u for each $u \in S^{n-1}$, and

$$x_u = \operatorname{grad} h_K(u),$$

where grad denotes gradient (see [737, Corollary 1.7.3]). From this, the boundary of K can be computed. For $n = 2$, we can do this directly, as follows. If h_K is differentiable at $u = (\cos\theta, \sin\theta) \in S^1$, then $h_K(u) = x_u \cdot u$. Regarding h_K as a function of θ, we get

$$h'_K(\theta) = x'_u \cdot u + u' \cdot x_u,$$

where the primes denote differentiation with respect to θ. Now x'_u is parallel to the tangent to K at x_u, so $x'_u \cdot u = 0$. Also, u' is orthogonal to u. It follows that $|h'_K(\theta)|$ is the distance from x_u to the foot of the perpendicular from o to H_u, and that

$$x_u = \left(h_K(\theta)\cos\theta - h'_K(\theta)\sin\theta, \, h_K(\theta)\sin\theta + h'_K(\theta)\cos\theta\right). \qquad (0.40)$$

The terms "convex body of class C^k (or C^∞)" and "smooth" have already been explained. In [737, p. 104], it is noted that K is smooth if and only if it is of class C^1. The proof given there requires several basic differentiability properties of convex functions.

Let K be smooth and $x \in \mathrm{bd}\, K$. Suppose that u is the outer unit normal vector to K at x. The *Gauss map* g from $\mathrm{bd}\, K$ to S^{n-1} is defined by $g(x) = u$; it is continuous, and a homeomorphism if K is smooth and strictly convex (see [737, p. 78]).

The *tangent space* of K at x is the translate $H_u - x = u^\perp$ of the supporting hyperplane to K with outer normal vector u. Suppose now that K is of class C^2. Then g is C^1. The differential $W_x = dg_x$ of the Gauss map is a linear map from this tangent space to itself, called the *Weingarten map*. The eigenvalues of W_x are called the *principal curvatures* of K at x. (In $\mathbb{E}^3$, the principal curvatures at a point in $\mathrm{bd}\, K$ give the maximal and minimal bending of $\mathrm{bd}\, K$ at the point.) The principal curvatures are nonnegative (see [737, pp. 104–6]). Their product is the *Gauss curvature* (or Gauss–Kronecker curvature) of K at x.

The Gauss curvature at a point on an $(n-1)$-dimensional submanifold of $\mathbb{E}^n$ or *hypersurface* of class C^2 can be defined similarly, as in [452, Chapter 7, Section 5]. Then a compact hypersurface forms the boundary of a strictly convex body if and only if the Gauss curvature at each of its points is positive; see [452, Theorem 5.6].

If $r = r(\theta)$ is a planar C^2 curve, we have the well-known formula

$$\frac{2(r')^2 - rr'' + r^2}{\left((r')^2 + r^2\right)^{3/2}} \qquad (0.41)$$

for its curvature in polar coordinates.

If $k \geq 2$, we say that K is *of class C^k_+* (or C^∞_+) if K is of class C^k (or C^∞, respectively) and the Gauss curvature of K at each x is positive.

Suppose that $h_K \in C^2$. Since this implies that K is strictly convex, the *reverse spherical image map* from S^{n-1} to $\mathrm{bd}\, K$, taking u to x_u, is defined. Furthermore, its differential $\overline{W}_u = dx_u$ is also defined; this linear map from the tangent space $u^\perp$ of S^{n-1} at u to itself is called the *reverse Weingarten map*. The eigenvalues of $\overline{W}_u$ are called the *principal radii of curvature* of K at $u \in S^{n-1}$. We denote them by $R_i(u)$, $1 \leq i \leq n-1$, where the labeling ranks them by magnitude. (See [737, pp. 107–8], where the notation $r_i(u)$ is used instead. The corresponding eigenvectors are called *principal curvature directions*.) They are also the nonzero eigenvalues of the second differential of h_K at u, by [737, Corollary 2.5.2]. If K is of class C^2_+, they are also the eigenvalues of the inverse Weingarten map

$W_{x_u}^{-1} = dg_{x_u}^{-1}$, and coincide with the reciprocals of the principal curvatures of K at $g^{-1}(u)$.

A *principal curve* in bd K is a curve whose tangent vectors point in a principal curvature direction. Figure 3.9 shows some principal curves in the boundary of an ellipsoid.

Notice that K is of class C_+^2 if and only if K is of class C^2 and all the principal curvatures of K are everywhere positive, and also (since K must be smooth, by the preceding remarks) if and only if all the principal radii of curvature are everywhere finite and positive.

It is proved in [737, pp. 106–11] that K is of class C_+^2 if and only if $h_K \in C^2$ and K has positive finite principal radii of curvature, or equivalently, if and only if $h_K \in C^2$ and the Gauss curvature of K exists and is positive everywhere. The existence of the Gauss curvature is necessary, since it is possible that $h_K \in C^2$ and the Gauss curvature of K is positive everywhere it exists, yet K is not even smooth. (In $\mathbb{E}^2$, for example, K may have a corner point x so $h_K(u) = h_{\{x\}}(u) = x \cdot u$ is linear for $u \in S^{n-1}$ in a neighborhood of $x/\|x\|$.) The positivity of the Gauss curvature is also essential; Hartman and Wintner [383, p. 480] have shown that h_K is not necessarily C^2 even if the boundary of K is real analytic.

Let $F_K^{(i)}$ be defined by

$$\binom{n-1}{i} F_K^{(i)} = \sum_{1 \le j_1 < \cdots < j_i \le n-1} R_{j_1} \cdots R_{j_i}, \qquad (0.42)$$

where the right-hand side is the ith elementary symmetric function of the principal radii of curvature of K. (In [737] the notation s_i is used instead of $F_K^{(i)}$.) Then (see [737, Corollary 2.5.3] or [83, Section 38]) $F_K^{(i)}(u)$ is the sum of the principal minors of order i of the Hessian matrix of h_K at u.

Using a result of Aleksandrov, it is possible to define, for almost all $u \in S^{n-1}$, the principal radii of curvature $R_i(u)$ of a general convex body K in $\mathbb{E}^n$ as the eigenvalues, corresponding to eigenvectors orthogonal to u, of the second differential of h_K at $u \in S^{n-1}$. Thus each $F_K^{(i)}$ can similarly be defined for almost all $u \in S^{n-1}$. Each R_i, and hence each $F_K^{(i)}$, is then integrable on S^{n-1} with respect to spherical Lebesgue measure; see [737, p. 118].

The *Laplacian* Δf of a function f on an open subset of $\mathbb{E}^n$ is defined, as usual, by

$$\Delta f = \sum_{i=1}^{n} \frac{\partial^2 f}{\partial x_i^2}.$$

The *Laplace–Beltrami operator* Δ_S on S^{n-1} can be defined as follows. If f is a function on S^{n-1}, it can be extended to a function g on $\mathbb{E}^n \setminus \{o\}$ by letting $g(x) = f(x/\|x\|)$ for each $x \ne o$. Then $\Delta_S f$ is defined to be the restriction of Δg to S^{n-1}.

If we set $i = 1$, we get (cf. [737, p. 110])

$$(n-1)F_K^{(1)} = R_1 + \cdots + R_{n-1} = (n-1)h_K + \triangle_S h_K.$$

In particular, if $n = 2$ and $u = (\cos\theta, \sin\theta)$, we obtain, as in [737, p. 110, eq. (2.5.22)],

$$R_1(u) = h_K(\theta) + h_K''(\theta). \tag{0.43}$$

In Section 0.4 it was noted that a convex body can be approximated arbitrarily closely in the Hausdorff metric by smooth convex bodies. Sometimes it is useful to have even stronger approximation theorems. For example, it follows from [737, pp. 158–60] that the class of C_+^∞ convex bodies is dense in $\mathcal{K}^n$, and further that each centrally symmetric member of $\mathcal{K}^n$ can be approximated by centrally symmetric C_+^∞ convex bodies.

1

Parallel X-rays of planar convex bodies

In this chapter our goal is to investigate the tomography of convex bodies in the plane. The requisite concepts of an X-ray and Steiner symmetral of a planar convex body are introduced in such a way that no knowledge of measure theory or Lebesgue integration is necessary. Furthermore, the reader can absorb the new ideas while avoiding the technicalities of higher-dimensional spaces. More general definitions are postponed until Chapter 2. (Occasional reference is made to these, but this is merely for cross-reference.) Granted some (but by no means all) of the background material in the first four sections of Chapter 0, and apart from references to a couple of auxiliary facts, the chapter is self-contained.

An X-ray of a convex body gives the lengths of all the chords of the body parallel to the direction of the X-ray. Corollary 1.2.12 states that there are four directions such that every convex body is determined, among all convex bodies, by its X-rays in these directions. Given a convex body, Theorem 1.2.21 says that there are three directions allowing the body to be distinguished from all others – "verified" – by the corresponding X-rays. A practical method by which every convex polygon can be "successfully determined" by three X-rays, the direction of each depending only on the previous X-rays, is provided by Theorem 1.2.23.

1.1. What is an X-ray?

We all know that dense material such as bone or teeth will show as light areas on a medical X-ray in a doctor's viewing box, while darker regions correspond to other less dense tissue. Each beam of the X-ray travels along a straight line, and its intensity after traversing the body depends on how much material it has passed through; high intensities result in a darker point on the X-ray picture, and low intensities show as a lighter point. If the beams in the X-ray are all parallel, then the X-ray picture contains information about the amount of material in the body lying on each straight line parallel to the direction of the X-ray.

There is a mathematical expression giving this quantity for each beam in the X-ray. If we measure the density of the body at each point, we get a *density function* f. Each beam of the X-ray travels along a line l, and the amount of material it passes through is then the integral of f along l. Some knowledge of the theory of integration is required to understand the meaning of this term, but if the body is sufficiently simple, this can be avoided.

The first simplification we make applies to every object in the main text of the book. We shall assume that the body being X-rayed is homogeneous, so that its density function has the value 1 at each of its points. (This eliminates the human body, of course, but leaves the possibility of approximating organic or other material of nearly uniform density.) This assumption alone is not enough to avoid integration theory, since the body may have a complicated shape. In fact we will consider quite general sets ("bounded, measurable sets") in parts of the book. Usually, however, we will insist that the body correspond more closely to our intuitive idea of a solid; in fact, the word body will be used in a precise sense given by the definition in Section 0.1. Specializing still further, we come to the class of planar convex bodies, whose structure is sufficiently simple that only basic notions of length and area are required here.

1.2. X-rays and Steiner symmetrals of planar convex bodies

A list of notation is provided at the end of the book, and basic definitions are given in Chapter 0, but let us briefly explain some of the terms now. The origin is denoted by o, and the unit circle by S^1. A direction is identified with the corresponding unit vector u, a point in S^1. The line through o parallel to u is denoted by l_u, and the line through o orthogonal to u by $u^\perp$.

All sets in this section will be subsets of the plane $\mathbb{E}^2$, unless stated otherwise.

The basic concept is that of a (parallel) *X-ray* $X_u K$ of a planar convex body K in the direction $u \in S^1$. This is the function, defined on $u^\perp$, giving the length of each chord of K parallel to u. Notice that this is in agreement with the idea that an X-ray of a body measures the amount of material met by each beam as it passes through the body, and that $X_u K$ is both continuous and concave on its support. See Figure 1.1.

We shall extend the definition of an X-ray to much more general sets in Chapter 2 (see Definition 2.1.1). It is convenient to mention one slight extension here, however. Suppose that $u \in S^1$ and that E is a connected planar body such that each line parallel to u meets E in a line segment. Then we call the function $X_u E$, defined on $u^\perp$, giving the lengths of these line segments the X-ray of E in the direction u.

The medical CAT scanner uses the data from a number of X-rays of a 3-dimensional object with possibly varying density, taken from different but

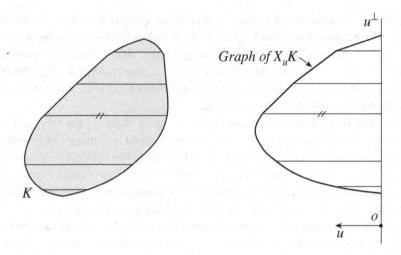

Figure 1.1. An X-ray of a convex body.

coplanar directions, to produce pictures of 2-dimensional slices of the object. This is quite remarkable, in view of the fact that no matter how many directions are used, there are always different objects that can produce the same set of X-rays; in short, no slice is uniquely determined. (This is a consequence of Theorem 2.3.3.) The pictures produced by the CAT scanner are approximations, and the mathematical techniques used to generate them are different in many respects from those of this book. (Note 1.1 provides a brief introduction to these techniques.)

Imagine a solid object of uniform density containing a single convex hole, whose shape we wish to determine by placing the object in a CAT scanner. Given the restricted nature of the hole, might we hope that each 2-dimensional slice of it is determined by a finite number of X-rays? The problem is to find which sets S of directions, if any, have the property that any planar convex body K is uniquely *determined* by its X-rays taken in the directions in the set. Here we assume a priori that the unknown set K is a convex body, so by "determined" we really mean that if K' is another convex body such that $X_u K' = X_u K$ for all $u \in S$, then $K = K'$. (See Definition 2.1.6 for a more general definition.) Any infinite set S has this property; this, and much more, follows from a basic uniqueness theorem for the X-ray transform (see Theorem C.1.1). We can therefore confine ourselves to finite sets of X-rays.

The X-ray of a convex body is related to a standard object in geometry known as the Steiner symmetral. If $u \in S^1$, the *Steiner symmetral* $S_u K$ of K in the direction u is the set obtained by translating, in the direction u, all the chords of K parallel to u so that they are bisected by $u^\perp$, and then taking the union of the resulting line segments. See Figure 1.2. Notice that K and $S_u K$ have the same X-rays in the direction u.

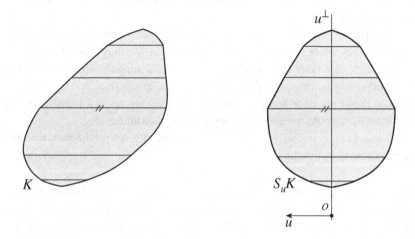

Figure 1.2. A Steiner symmetral of a convex body.

Theorem 1.2.1. *A Steiner symmetral of a convex body is also a convex body.*

Proof. Let K be a convex body, and u a direction. The fact that $S_u K$ is compact follows from the fact that since K is a convex body, $X_u K$ is continuous on its support.

To prove $S_u K$ is convex, let p_i, $i = 1, 2$, be points in $S_u K$, let $0 \le t \le 1$, and suppose that $p = (1 - t)p_1 + tp_2$. Let q_i, $i = 1, 2$, and q be the projections of p_i, $i = 1, 2$, and p, respectively, on $u^\perp$. Now $q = (1 - t)q_1 + tq_2$, and using the concavity of $X_u K$ on its support and the symmetry of $S_u K$ about $u^\perp$, we have

$$2\|p - q\| = 2\|(1 - t)(p_1 - q_1) + t(p_2 - q_2)\|$$
$$\le 2(1 - t)\|p_1 - q_1\| + 2t\|p_2 - q_2\|$$
$$\le (1 - t)X_u K(q_1) + tX_u K(q_2) \le X_u K(q),$$

from which $p \in S_u K$, as required. ∎

Observe that for a given K and u, both $X_u K$ and $S_u K$ contain the same information about K. In fact, for the purposes of this chapter, we can identify $X_u K$ and $S_u K$. Bear in mind, however, that $X_u K$ is a function, whereas $S_u K$ is a set.

A glance at Figures 1.1 and 1.2 will convince the reader that one X-ray will never determine a convex body. If K is not already symmetric about $u^\perp$, then $S_u K$ is different from K, but has the same X-ray as K in the direction u. On the other hand, if $K = S_u K$, the closed set K' of points between the graph of $X_u K$ and $u^\perp$ is a convex body with a different shape from K, but the same X-ray. Consequently, we may as well assume that there are at least two nonparallel directions in our set S.

At this point, a natural question arises. Suppose that we think of the X-rays as pictures, as they are on a doctor's desk or viewing box. We may know the direction in which they were taken, but not the position of their supports on the orthogonal line; in other words, we may only know the X-rays up to a translation. Of course, we could then only hope to determine the convex body up to a translation, but aside from this, would we encounter any additional difficulties?

The answer is, essentially, no. To see this, we need some preliminary remarks which will be useful in the sequel.

Lemma 1.2.2 (Cavalieri principle). *Suppose that E and E' are connected bodies such that each line parallel to a fixed direction u meets E and E' in line segments of equal length. Then $\lambda_2(E) = \lambda_2(E')$, that is, the areas of E and E' are equal.*

Proof. This follows on integration of the functions giving the lengths of the chords of E and E' parallel to u.   ■

In particular, K and $S_u K$ have the same area. Notice that Riemann integration is quite adequate for the previous lemma, as it is in the next. The Cavalieri principle is actually just a very special case of Fubini's theorem (see Section 0.5 for comments and a reference).

Lemma 1.2.3. *Suppose that E and E' are connected bodies such that each line parallel to a fixed direction u meets E and E' in line segments of equal length. Then the centroids of E and E' lie on the same line parallel to u.*

Proof. We can assume, without loss of generality, that u is parallel to the x-axis, so that $X_u E = X_u E(y)$ is the X-ray of E in the x-direction, and that $\lambda_2(E) = 1$. Let $E_y = \{x : (x, y) \in E\}$, and suppose that c is the centroid of E. Then by the definition of a centroid (cf. (0.15)),

$$c = \int_E (x, y)\, dx\, dy = \int_{-\infty}^{\infty} \int_{E_y} (x, y)\, dx\, dy.$$

The y-coordinate of c is therefore

$$\int_{-\infty}^{\infty} \int_{E_y} y\, dx\, dy = \int_{-\infty}^{\infty} y X_u E(y)\, dy.$$

(Here, we are just taking moments of the chords of E parallel to the x-axis.) The fact that $X_u E = X_u E'$ completes the proof.   ■

Theorem 1.2.4. *Let S be a set containing at least two nonparallel directions. The following statements are equivalent:*

(i) *Whenever K and K' are convex bodies, and $S_u K$ is a translate of $S_u K'$ for all $u \in S$, then K is a translate of K'.*

(ii) *Whenever K and K' are convex bodies, and $S_u K = S_u K'$ for all $u \in S$, then $K = K'$.*

Proof. Suppose that (i) holds. If $S_u K = S_u K'$ for all $u \in S$, then (i) implies that K is a translate of K'. For each $u \in S$, the centroids of K and K' lie on the same line parallel to u, by Lemma 1.2.3. Since S contains nonparallel directions, this implies that K and K' have the same centroid. Therefore $K = K'$, yielding (ii).

Assume (ii). Suppose that $S_u K$ is a translate of $S_u K'$ for all $u \in S$. Let K'' be the translate of K' with the same centroid as K, and let $u \in S$. Then $S_u K$ is a translate of $S_u K''$, so $X_u K$ is just a shift of $X_u K''$, in a direction parallel to $u^\perp$. Assuming that $u^\perp$ is the y-axis, and applying the expression obtained in the proof of Lemma 1.2.3 for the y-coordinate of the centroid of K, we see that a nonzero shift would imply that the centroids of K and K'' do not lie on the same line parallel to u. Therefore $X_u K = X_u K''$, and so $S_u K = S_u K''$, for all $u \in S$. By (ii), $K = K''$, and (i) follows. ∎

This theorem answers the earlier question in a satisfactory manner. Therefore we can return to the problem of finding finite sets S of directions such that the corresponding X-rays determine convex bodies in the original sense defined before.

Theorem 1.2.5. *For each $n \in \mathbb{N}$, there is a set of n mutually nonparallel directions such that there are different convex polygons with the same X-rays in these directions.*

Proof. Let T be a triangle whose base is on the x-axis. If T' is any other triangle with base on the x-axis of the same length as that of T, and of the same height as T, then T and T' have equal X-rays in the x-direction. Now consider a regular n-gon Q centered at o, and its rotation Q' by π/n about o. The convex hull of Q and Q' is a $2n$-gon; let u be a direction parallel to one of its edges. Using the previous remark about triangles, and symmetry, it is easy to see that Q and Q' have the same X-rays in the direction u. ∎

It is important to note right away that our problem is one of an affine nature. Let $\phi \in GA_n$ be a nonsingular affine transformation, and suppose that K and K' are two convex bodies with the same X-rays in the direction u. Since ϕ takes parallel lines to parallel lines, and preserves the ratios of lengths of parallel line segments (see Section 0.5), the images ϕK and $\phi K'$ have the same X-rays in a direction parallel to ϕu. This and Theorem 1.2.5 show that for uniqueness one should avoid subsets of the directions of the edges of an affinely regular polygon. (An *affinely regular polygon* is the image under some $\phi \in GA_n$ of a regular polygon.)

Suppose that K and K' are two different convex bodies with the same X-rays in a set S containing at least two nonparallel directions. Since they have the same centroid, their interiors must intersect. Suppose that C is a component (a maximal connected subset) of $\operatorname{int}(K \setminus K')$ and $u \in S$. Let uC be the set of all $x \in \operatorname{int}(K' \setminus K)$ such that the line $l_u + x$ meets C. (We shall always denote by l_u the line through o parallel to u, so $l_u + x$ is the line through x parallel to u.) Note that C and uC are disjoint. Moreover, since the X-rays of K and K' in the direction u are equal, C and uC also have chords of equal length on each line parallel to this direction. Therefore uC is a component of $\operatorname{int}(K' \setminus K)$. Let

$$ \mathcal{C} = \{u_{i_m} \cdots u_{i_1} C : m \in \mathbb{N}, u_{i_j} \in S\}. $$

Let us call $\mathcal{C}$ the *system of components associated to* C; it is the set of all (different) components obtained in this way from C by using a finite sequence of directions from S (see Figure 1.3).

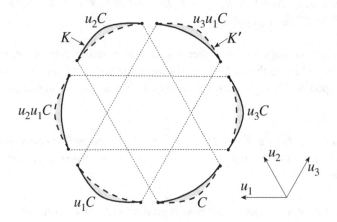

Figure 1.3. An associated system of components.

Lemma 1.2.6. *Let K and K' be two different convex bodies with the same X-rays in a set S of n mutually nonparallel directions. Let C be a component of $\operatorname{int}(K \triangle K')$ and $\mathcal{C}$ the system of components associated to C. Then $\mathcal{C}$ is finite, and $|\mathcal{C}| \geq 2n$.*

Proof. The components in $\mathcal{C}$, by definition different, are also disjoint. Moreover, by the Cavalieri principle (Lemma 1.2.2), they all have the same area. Since each of them is contained in the bounded set $K \cup K'$, the system $\mathcal{C}$ must be finite.

Let $m = |\mathcal{C}|$, and suppose that $\mathcal{C}$ consists of the components $C_1, \ldots, C_m$, ordered clockwise. If $u \in \mathcal{S}$, then $uC_1 = C_k$ for some $k \neq 1$, and k must be even, since otherwise there would be some C_j with $uC_j = C_j$. Moreover, if $u' \in \mathcal{S}$, $u' \neq u$, and $u'C_1 = C_{k'}$, then $k' \neq k$, since C_k and $C_{k'}$ must have different centroids, by Lemma 1.2.3. There are at most $[m/2]$ even k, so $[m/2] \geq n$, and $m \geq 2n$. ∎

Definition 1.2.7. Let $\mathcal{S}$ be a finite set of directions. A nondegenerate convex polygon Q is called $\mathcal{S}$-*regular* if it has the following property: If v is a vertex of Q, and $u \in \mathcal{S}$, then the line $l_u + v$ through v parallel to u meets a different vertex v' of Q. We say Q is *weakly* $\mathcal{S}$-*regular* if it has the preceding property, amended to allow $v' = v$ (which can only happen if $l_u + v$ supports Q at v).

Imagine looking at a distant forest of trees, each growing from a vertex of a convex polygon. This polygon is $\mathcal{S}$-regular if precisely half of the trees can be seen, when we look at the forest in any direction in $\mathcal{S}$. For example, a regular $2n$-gon has this property if $\mathcal{S}$ is the set of the directions of its edges. In the sequel we could confine ourselves to use of weakly $\mathcal{S}$-regular polygons. However, where possible, we shall use $\mathcal{S}$-regular polygons, since the appropriate generalizations of them in other contexts (e.g., the $\mathcal{S}$-switching components of Definition 2.3.1) are probably more natural and useful.

Lemma 1.2.8. *Let K and K' be two different convex bodies with the same X-rays in a set $\mathcal{S}$ of at least two mutually nonparallel directions, and suppose that C is a component of* int $(K \triangle K')$. *Then the centroids of the components in the system $\mathcal{C}$ associated to C form the vertices of an $\mathcal{S}$-regular polygon.*

Proof. Let V be the set of centroids of the components in $\mathcal{C}$. Suppose that $v \in V$, and that t is the line through the endpoints of the component in $\mathcal{C}$ of which v is the centroid. (That is, t passes through the endpoints of each of the two arcs, one in the boundary bd K of K and one in bd K', bounding this component.) Note that t separates the convex hull of C from the remaining points in V. Now the centroid of a set lies in its convex hull. (See Section 0.5 for a reference to a proof of this plausible statement.) Consequently, the line t also separates v from the other points in V. So V forms the vertex set of a convex polygon Q. Further, Q is nondegenerate, since by Lemma 1.2.6, $|V| = |\mathcal{C}| \geq 2n \geq 4$.

Now if $v \in V$ is the centroid of some $C \in \mathcal{C}$, and $u \in \mathcal{S}$, then by Lemma 1.2.3 the centroid v' of uC lies on the line $l_u + v$. Since $v' \in V$, this shows that Q is indeed $\mathcal{S}$-regular. ∎

Lemma 1.2.8 reduces the determination of convex bodies by X-rays to the classification of those sets $\mathcal{S}$ of directions admitting $\mathcal{S}$-regular polygons. The next lemma provides the key to this task. It requires the notion of the *midpoint polygon* $M(Q)$ of a polygon Q, the polygon whose vertices are the midpoints of the edges of Q. The Hausdorff metric is also mentioned; the reader not familiar with this can get by with the natural intuitive idea of a sequence of polygons converging to another polygon, or consult Section 0.4.

Lemma 1.2.9. *Let Q_0 be a convex n-gon with centroid at the origin. For each $k \in \mathbb{N}$, define $Q_k = \sec(\pi/n)M(Q_{k-1})$. Then the sequence $\{Q_{2k}\}$ converges (in the Hausdorff metric) to an affinely regular polygon.*

Proof. Suppose that the vertices of Q_0 are represented, in clockwise order, by $a_i \in \mathbb{C}, 0 \leq i \leq n-1$. Let $M^2(Q_0) = M(M(Q_0))$ denote the second midpoint polygon of Q_0, and in general let $M^k(Q_0) = M(M^{k-1}(Q_0))$ denote the kth midpoint polygon of Q_0. (Note that Q_k is not $M^k(Q_0)$, but rather a dilatate of the latter, with the dilatation factor chosen so that if Q_0 is a regular polygon, all the polygons Q_k have the same area.) The vertices of $M(Q_0)$ are represented by $(a_i+a_{i+1})/2$, and those of $M^2(Q_0)$ by $(a_{i-1}+2a_i+a_{i+1})/4, 0 \leq i \leq n-1$, where the indices i are taken modulo n. If the column vector $q_0 = (a_0, a_1, \ldots, a_{n-1})^t$ represents Q_0, then $q_2 = Aq_0$ represents $M^2(Q_0)$, where

$$
A = \begin{pmatrix}
\frac{1}{2} & \frac{1}{4} & 0 & \cdots & 0 & \frac{1}{4} \\
\frac{1}{4} & \frac{1}{2} & \frac{1}{4} & \cdots & 0 & 0 \\
\vdots & \vdots & \vdots & \ddots & \vdots & \vdots \\
\frac{1}{4} & 0 & 0 & \cdots & \frac{1}{4} & \frac{1}{2}
\end{pmatrix}.
$$

The matrix A is a *circulant* matrix, that is, a matrix of the form

$$
C = \begin{pmatrix}
c_0 & c_1 & \cdots & c_{n-1} \\
c_{n-1} & c_0 & \cdots & c_{n-2} \\
\vdots & \vdots & \ddots & \vdots \\
c_1 & c_2 & \cdots & c_0
\end{pmatrix}.
$$

Eigenvalues and eigenvectors of such a matrix are known. If $\omega = e^{2\pi i/n}$ is the principal nth root of unity, then the eigenvalues of C are

$$
b_j = \sum_{k=0}^{n-1} c_k \omega^{jk},
$$

with corresponding eigenvectors

$$
y_j = (1, \omega^j, \omega^{2j}, \ldots, \omega^{(n-1)j})^t,
$$

for $0 \leq j \leq n-1$. (See [492, Ex. 16, p. 166], where ample hints are given; it is easily shown that $CV_n = V_nD$, where V_n is the $n \times n$ Vandermonde matrix and D is the diagonal matrix with entries $b_j, 0 \leq j \leq n-1$, and the result soon follows.) For our matrix A, the eigenvalues are

$$b_j = \frac{1}{2} + \frac{1}{4}\omega^j + \frac{1}{4}\omega^{(n-1)j} = \cos^2 \frac{j\pi}{n},$$

for $0 \leq j \leq n - 1$, with corresponding eigenvectors y_j as before.

Now the eigenvectors are linearly independent, so we can write

$$q_0 = \sum_{j=0}^{n-1} z_j y_j,$$

where $z_j \in \mathbb{C}$. The centroid of Q_0 is o, so $\sum_{i=0}^{n-1} a_i = 0$; it follows that $z_0 = 0$. The column vector $q_{2k} = A^k q_0$ represents $M^k(Q_0)$, and

$$
\begin{aligned}
q_{2k} = A^k q_0 &= A^k \left(\sum_{j=1}^{n-1} z_j y_j \right) \\
&= \sum_{j=1}^{n-1} z_j (A^k y_j) \\
&= \sum_{j=1}^{n-1} z_j \left(\cos^{2k} \frac{j\pi}{n} \right) y_j.
\end{aligned}
$$

This means that the polygon Q_{2k} is represented by

$$\sum_{j=1}^{n-1} \left(\sec^{2k} \frac{\pi}{n} \cos^{2k} \frac{j\pi}{n} \right) z_j y_j.$$

Let $R = \lim_{k \to \infty} Q_{2k}$. For $1 < j < n - 1$, we have

$$\left(\frac{\cos(j\pi/n)}{\cos(\pi/n)} \right)^{2k} \to 0,$$

as $k \to \infty$. Therefore R is represented by

$$z_1 y_1 + z_{n-1} y_{n-1} = z_1 y_1 + z_{n-1} \bar{y}_1,$$

where $\bar{y}_1$ denotes the column vector whose entries are the complex conjugates of those of y_1. We now appeal to an expression for an affine transformation of $\mathbb{E}^2$, following from the definition, in terms of a complex variable $z = x + iy$; ϕ is affine if and only if there are complex numbers a, b, and c such that

$$\phi z = az + b\bar{z} + c,$$

for all z. From this, we see that R is an affine image of the regular polygon represented by $(1, \omega, \omega^2, \ldots, \omega^{n-1})^t$. ∎

Corollary 1.2.10. *If there is a weakly S-regular polygon, then S is a subset of the directions of the edges of an affinely regular polygon.*

Proof. Suppose that Q_0 is a weakly S-regular n-gon. Then for each k, the polygon Q_{2k} defined earlier is also weakly S-regular, and therefore $R = \lim_{k\to\infty} Q_{2k}$ is also. By Lemma 1.2.9, R is an affinely regular n-gon, and so S must be a subset of the directions of its diagonals or edges. But a set of directions of diagonals of an affinely regular n-gon is also a set of directions of edges of an affinely regular $2n$-gon. ∎

Theorem 1.2.11. *Convex bodies are determined by X-rays taken in any set of directions that is not a subset of the directions of the edges of an affinely regular polygon.*

Proof. Let S be such a set of directions, and suppose that K and K' are two different convex bodies with the same X-rays in the directions in S. There are two nonparallel directions in S, so by Lemma 1.2.8 there is an S-regular polygon Q_0. The theorem now follows from Corollary 1.2.10. ∎

Corollary 1.2.12. *Convex bodies are determined by certain sets of four X-rays, but by no set of three X-rays.*

Proof. Let s be the (finite) slope of an edge of the regular n-gon represented, as in Lemma 1.2.9, by $(1, \omega, \omega^2, \ldots, \omega^{n-1})^t$. Then $s = \tan(\pi j/n)$ for some integer j, so $f(s) = 0$, where f is the rational function such that $\tan n\theta = f(\tan\theta)$. This means that s is a root of a polynomial with integer coefficients – an algebraic number. It follows that the cross ratio

$$\langle s_1, \ldots, s_4 \rangle = \frac{(s_3 - s_1)(s_4 - s_2)}{(s_4 - s_1)(s_3 - s_2)}$$

of any four slopes s_i, $1 \le i \le 4$, of edges of a regular polygon is also an algebraic number. Choose $S = \{u_i : 1 \le i \le 4\}$ so that if $t_i = \tan\theta_i$, where $u_i = (\cos\theta_i, \sin\theta_i)$, the cross ratio $\langle t_1, \ldots, t_4 \rangle$ is a transcendental number, that is, not an algebraic number. Since cross ratio is preserved by affine transformations (see Section 0.2), we see that S satisfies the hypotheses of Theorem 1.2.11.

The case $n = 3$ of Theorem 1.2.5 provides two different equilateral triangles with the same X-rays in three equally spaced directions. These three directions may be transformed into any other three distinct directions by a suitable nonsingular affine transformation (see Section 0.2), so the images of the triangles have the same X-rays in the new directions. This completes the proof. ∎

There are sets of four directions, other than those provided by the proof of the previous corollary, such that convex bodies are determined by the corresponding

X-rays. In fact, all four directions can be chosen parallel to vectors with integer coordinates; see Note 2.2 for more details.

Theorem 1.2.11 classifies all the sets of directions such that the X-rays of a convex body, taken in those directions, determine it uniquely. The remaining sets of directions are affinely equivalent to those considered in Theorem 1.2.5, which constructs different polygons with the same X-rays in directions in these "bad" sets. It might be objected, however, that two congruent polygons are not really that different at all. The following example answers this objection, by proving that for each such bad set of directions, uniqueness fails completely.

Theorem 1.2.13. *Let S be a subset of the directions of the edges of an affinely regular polygon. The number of mutually noncongruent convex n-gons with the same X-rays in the directions in S increases at least exponentially with n.*

Proof. By invariance under affine transformations, it will suffice to consider the case $S = \{j\pi/m : 0 \le j \le m - 1\}$, and we may also assume that m is even.

Let F be the set of k points in S^1 represented by complex numbers z with $\arg z = \pi(2j + 1)/2mk, 0 \le j \le k - 1$. Suppose that $D \subset F$. We construct a set D' as follows. Let D_0 be the union of D and the reflection of $F \setminus D$ in the x-axis. For each j, let D_j be D_0 rotated about o by $2j\pi/m$. Finally, let $D' = \cup_{j=0}^{m-1} D_j$.

The example is based on the observation that if D and E are subsets of F, then D' and E' have the same projections in the directions in S. To see this, suppose that $x \in S^1$ and that $\arg x = \arg z + 2j\pi/m$ for some $z \in F$. Let t be the line through x parallel to the direction $p\pi/m \in S$ and suppose that $y = (t \cap S^1) \setminus \{x\}$. An easy calculation shows that

$$\arg y = -\arg x + \frac{(2p + m)\pi}{m} = -\arg z + \frac{(2p + m - 2j)\pi}{m}.$$

Since m is even, this means that $\arg y = \arg \bar{z} + 2j'\pi/m$, where $\bar{z}$ is the reflection of z in the x-axis. Now suppose that $D \subset F$. If $z \in F$, either $z \in D$ or $\bar{z}$ belongs to the reflection of $F \setminus D$ in the x-axis. It follows that one or the other (but not both) of the points x or y belongs to D'. Thus for any $D \subset F$, the projection of D' in the direction $p\pi/m$ is precisely the same as that of F'.

If $z \in S^1$, let $A(z)$ be the arc of S^1 centered at z and with length π/mk. Let $s(z)$ denote the closed line segment with the same endpoints as $A(z)$, and $t(z)$ the union of the two closed line segments with one endpoint at z and the other at an endpoint of $A(z)$.

Let $D \subset F$. Define $Q(D)$ to be the convex polygon

$$Q(D) = \text{conv}(\{s(z) : z \in D'\}, \{t(z) : z \in (F \setminus D)'\}).$$

(In Figure 1.4 we depict two such polygons, constructed using $m = 2, k = 2$, and $D = \{z \in F : \arg z = \pi/8\}$ and $\{z \in F : \arg z = \pi/8, 3\pi/8\}$, respectively.) By the observation above, it is clear that the polygons $Q(D)$ for $D \subset F$ all have the same X-rays in the directions in S. Furthermore, each such $Q(D)$ is an n-gon for $n = 3mk$.

Suppose that $D \subset F$. If E denotes the set obtained by reflecting D in the x-axis and then rotating by π/m about o, then $Q(E)$ is a rotation of $Q(D)$. Moreover, $Q(F \setminus D)$ (or $Q(F \setminus E)$) is the reflection of $Q(D)$ (or $Q(E)$, respectively) in the x-axis. However, any subset G of F apart from these four yields a polygon $Q(G)$ not congruent to $Q(D)$. We conclude that there are at least 2^{k-2} mutually noncongruent convex n-gons, each with the same X-rays in the directions in S. Since $k = n/3m$, this number grows exponentially with n. ∎

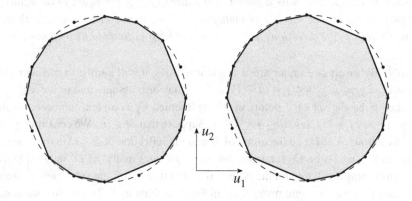

Figure 1.4. Noncongruent polygons with equal X-rays.

Despite the plethora of examples produced by the idea of Theorem 1.2.13, the feeling remains that these are very special cases, and that "most" convex bodies should be determined by their X-rays in just two directions. This is in fact true, in the following sense. We say a statement is true of most (in the sense of Baire category) convex bodies, if it is true of a residual set in $\mathcal{K}_0^n$, the space of convex bodies with the Hausdorff metric. (We refer the reader not familiar with the Hausdorff metric to Section 0.4.) It follows from Blaschke's selection theorem (Theorem 0.4.1) that $\mathcal{K}_0^n$ is locally compact. As was noted in Section 0.1, each dense G_δ set is therefore a residual set, so it suffices to prove that the statement is true for such a set.

To carry out this plan, we will need an alternative metric on the set of convex bodies. If K_1, K_2 are convex bodies, the *symmetric difference metric* δ^S is defined by

$$\delta^S(K_1, K_2) = \lambda_2(K_1 \triangle K_2).$$

Lemma 1.2.14. *On the set of convex bodies, the symmetric difference metric δ^S is equivalent to the Hausdorff metric δ.*

Proof. The term "equivalent" means that convergence in one metric is equivalent to convergence in the other. Suppose, firstly, that K_m, $m \in \mathbb{N}$, are convex bodies such that $\delta(K_m, K) \to 0$ for some convex body K. Using the continuity of λ_2 on $\mathcal{K}_0^2$ (cf. Section 0.4), we have

$$\delta^S(K_m, K) = \lambda_2(K_m \triangle K) = \lambda_2(K_m) + \lambda_2(K) - 2\lambda_2(K \cap K_m)$$
$$\to \lambda_2(K) + \lambda_2(K) - 2\lambda_2(K) = 0.$$

Conversely, suppose that $\delta^S(K_m, K) \to 0$ for some convex body K. We claim that the sequence of sets K_m is bounded. If this is not the case, we can assume, by taking a subsequence if necessary, that $\operatorname{diam} K_m \to \infty$. If $u_m \in S^1$ is the direction of a chord of K_m whose length is $\operatorname{diam} K_m$, then

$$(\operatorname{diam} K_m)\lambda_1(K_m|u_m^\perp)/2 \le \lambda_2(K_m) \le 2\lambda_2(K),$$

for sufficiently large m. This is because the quantity on the left is the area of a triangle whose base has length equal to that of the projection of K_m on $u_m^\perp$, and whose height equals $\operatorname{diam} K_m$; the intersection of K_m and any line parallel to u_m must have length at least that of the intersection of the triangle and this line, so the Cavalieri principle (Lemma 1.2.2) implies that the area of K_m must be at least that of the triangle. We deduce that $\lambda_1(K_m|u_m^\perp) \to 0$. Since

$$\lambda_2(K_m \cap K) \le (\operatorname{diam} K)\lambda_1(K_m|u_m^\perp),$$

we also have $\lambda_2(K_m \cap K) \to 0$. Therefore

$$\delta^S(K_m, K) = \lambda_2(K_m) + \lambda_2(K) - 2\lambda_2(K \cap K_m)$$
$$\ge \lambda_2(K) - 2\lambda_2(K \cap K_m) \ge \lambda_2(K)/2,$$

for sufficiently large m, a contradiction establishing our claim. We can now apply Blaschke's selection theorem (Theorem 0.4.1) to conclude that a subsequence of $\{K_m\}$ converges in the Hausdorff metric to some convex body K_0. By the first part of the proof, this means that the subsequence also converges in the symmetric difference metric, so $K_0 = K$. Now the whole argument can be applied to any subsequence of $\{K_m\}$ to show that it has a subsubsequence converging to K, and this suffices to show that $\delta(K_m, K) \to 0$, as required. ∎

Lemma 1.2.15. *Let* $n = 2$. *The Steiner map* $S_u \colon \mathcal{K}_0^n \to \mathcal{K}_0^n$ *taking a convex body* K *to its Steiner symmetral* $S_u K$ *is continuous.*

Proof. By the previous lemma, we can use the symmetric difference metric instead of the Hausdorff metric.

Let K_i, $i = 1, 2$, be convex bodies. It is enough to show that

$$\lambda_2(S_u K_1 \triangle S_u K_2) \le \lambda_2(K_1 \triangle K_2).$$

This will follow, by integration, if we prove that for closed intervals $I_i = [a_i, b_i] \subset l_u, i = 1, 2$, we have

$$\lambda_1(S_u I_1 \triangle S_u I_2) \le \lambda_1(I_1 \triangle I_2).$$

Assuming $b_1 - a_1 \le b_2 - a_2$, without loss of generality, the left-hand side is $d = (b_2 - b_1) + (a_1 - a_2)$. The right-hand side equals d if $I_1 \subset I_2$, and is larger otherwise.   ∎

It is at first surprising that Lemma 1.2.15 becomes false if $\mathcal{K}_0^2$ is replaced by the space $\mathcal{K}^2$ of all compact convex sets, with the Hausdorff metric. To see this, let H be the closed line segment joining o and the point $(0, 1)$, and H_n the closed line segment joining o and $(1/n, 1)$, $n \in \mathbb{N}$. Then $H_n \to H$, but the Steiner symmetrals $S_y H_n$ in the y-direction converge to $\{o\}$, whereas $S_y H$ is the closed line segment joining $(0, -1/2)$ and $(0, 1/2)$.

Lemma 1.2.16. *Let $\mathcal{D}$ be the set of all convex bodies determined by their X-rays in the coordinate directions. Then $\mathcal{D}$ is a G_δ set in $\mathcal{K}_0^2$.*

Proof. If K is a convex body not in $\mathcal{D}$, then there is a different convex body K' with the same X-rays as K in the coordinate directions. For each $n \in \mathbb{N}$, let $\mathcal{K}_n^2$ denote the set of $K \notin \mathcal{D}$ such that there exists a K' for which in addition the Hausdorff distance $\delta(K, K')$ satisfies $n \le \delta(K, K') \le n + 1$ or $1/(n + 1) \le \delta(K, K') \le 1/n$. Then $\mathcal{K}_0^2 \setminus \mathcal{D} = \cup_n \mathcal{K}_n^2$.

To show that $\mathcal{K}_n^2$ is closed, let $\{K_i\}$ be a sequence in $\mathcal{K}_n^2$ converging, in the Hausdorff metric, to a convex body K. There is an $r > 0$ such that $K_i' \subset rB$ for sufficiently large i, where B is the unit disc. By Blaschke's selection theorem (Theorem 0.4.1), there is a subsequence $\{K_{i_j}'\}$ converging to some compact convex K' as $j \to \infty$. Each K_{i_j}' has the same X-rays, and hence Steiner symmetrals, as K_{i_j} in the coordinate directions. Since $\lambda_2(K_{i_j}') = \lambda_2(K_{i_j})$ by the Cavalieri principle (Lemma 1.2.2), and $\lambda_2(K_{i_j}) \to \lambda_2(K) \ne 0$, it follows that $\lambda_2(K') \ne 0$, so $K' \in \mathcal{K}_0^2$. By Lemma 1.2.15, the map taking a convex body to its Steiner symmetral is continuous. Therefore K' has the same Steiner symmetrals, and hence X-rays, as K in the coordinate directions. Again by the continuity of the Steiner map, we have $n \le \delta(K, K') \le n + 1$ or $1/(n + 1) \le \delta(K, K') \le 1/n$. Consequently $K \in \mathcal{K}_n^2$, proving that $\mathcal{K}_n^2$ is closed. Thus the complement of $\mathcal{D}$ is an F_σ set in $\mathcal{K}_0^2$, and we are done.   ∎

Theorem 1.2.17. *Given any two nonparallel directions, the set of convex bodies determined by their X-rays in these directions is a residual set in $\mathcal{K}_0^2$.*

Proof. By affine invariance, we may take the two directions to be the coordinate directions. In view of Lemma 1.2.16, we need to show that the set $\mathcal{D}$ of that lemma is dense in $\mathcal{K}_0^2$. Let $K_0 \in \mathcal{K}_0^2$ be a convex body; by affine invariance again, we may suppose that it is inscribed in the unit square $[0, 1]^2$. We shall prove that there are convex bodies in $\mathcal{D}$ arbitrarily close to K_0 in $\mathcal{K}_0^2$.

Since the strictly convex bodies are dense in $\mathcal{K}_0^2$ (see Section 0.4), it is easy to see that there is a strictly convex body arbitrarily close to K_0 in $\mathcal{K}_0^2$, also inscribed in the unit square, meeting its left-hand edge at the single point $(0, a)$ and its right-hand edge at the single point $(1, b)$, where $a \neq b$; $a > b$ say. See Figure 1.5. By closely approximating part of the boundary of this convex body by a polygonal arc, we can replace the body by another convex body K with the same properties except that it is smooth above and polygonal below the line $y = a$. Let $x = c$ be the line through the point p other than $(0, a)$ where the smooth part of bd K meets the polygonal part.

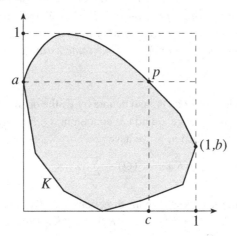

Figure 1.5. The body K of Theorem 1.2.17.

Suppose that $K' \neq K$ is a convex body with the same X-rays as K in the coordinate directions. Note that the Steiner symmetral $S_x K$ of K in the x-direction is smooth above and polygonal below $y = a$, and $S_y K$ is nonsmooth and nonpolygonal to the left of $x = c$ and polygonal to the right. Since the Steiner symmetrals of K' in the coordinate directions are the same as those of K, they share these properties of the Steiner symmetrals of K. Suppose that K' meets the left-hand edge of the unit square at a'. If $a' < a$, the polygonality of $S_x K'$ below $y = a$ would imply that $S_y K'$ is also polygonal close to the y-axis. If, on the other hand, $a' > a$, the smoothness of $S_x K'$ above $y = a$ would imply that $S_y K'$ is also smooth near the y-axis. Therefore $a = a'$, and this implies that $p \in K'$.

Let S denote the smooth part of bd K. If S is not contained in bd K', there are two disjoint open arcs A_1, A_1' of bd K, bd K' respectively, with common endpoints c_1, d_1 and with $A_1 \subset S$. Because the X-rays of K and K' are identical, there are corresponding arcs A_2, A_2' of bd K, bd K', also disjoint except at endpoints

c_2, d_2, such that c_i and d_i lie on the same vertical line, $i = 1, 2$. Clearly A_1 and A_1' are smooth arcs, while A_2 and A_2' are polygonal. But this is impossible if $S_y K = S_y K'$.

Consequently $S \subset \mathrm{bd}\, K'$. Now using, repeatedly if necessary, the fact that the Steiner symmetrals of K and K' are the same in the coordinate directions we deduce that $\mathrm{bd}\, K = \mathrm{bd}\, K'$. This proves that $K \in \mathcal{D}$, as required. ∎

If S is any finite set of at least two nonparallel directions, Theorem 1.2.17 shows that "most" convex bodies are determined by their X-rays in the directions in S. The next lemma provides a subclass with a definite structure.

Lemma 1.2.18. *Let S be a finite set of directions. Suppose that K is a convex body and that Q is a convex polygon inscribed in K (i.e., whose vertices are in $\mathrm{bd}\, K$) and with each of its edges parallel to a direction in S. Then Q is inscribed in any other convex body K' with the same X-rays as K in the directions in S.*

Proof. Label the edges of Q by e_i, and denote by E_i the open half-plane bounded by the line containing the edge e_i and not containing Q. For any set A, put $A_i = A \cap E_i$. Since Q is inscribed in K we have

$$\lambda_2(K) = \lambda_2(Q) + \sum_i \lambda_2(K_i).$$

Let K' have the same X-rays as K in the directions in S. By the Cavalieri principle (Lemma 1.2.2), $\lambda_2(K) = \lambda_2(K')$ and $\lambda_2(K_i) = \lambda_2(K_i')$ for each i.

Now $K' \setminus \cup_i K_i' \subset Q$, and

$$\lambda_2(K' \setminus \cup_i K_i') \geq \lambda_2(K') - \sum_i \lambda_2(K_i')$$

$$= \lambda_2(K) - \sum_i \lambda_2(K_i)$$

$$= \lambda_2(Q).$$

This shows that $(K' \setminus \cup_i K_i') = Q$ and hence $Q \subset K'$. Furthermore, Q must actually be inscribed in K', since otherwise

$$\lambda_2(K') < \lambda_2(Q) + \sum_i \lambda_2(K_i'),$$

contradicting the earlier equality. ∎

Definition 1.2.19. *Let S be a finite set of directions. We say that a convex body K is S-inscribable if $\mathrm{int}\, K$ is the union of interiors of convex polygons inscribed in K, each of whose edges is parallel to some direction in S.*

It follows from Lemma 1.2.18 that an S-inscribable convex body is determined among convex bodies by its X-rays in the directions in S. (Actually, more is true; see Theorem 2.3.7.) Applying this, we can see immediately that a disk is determined by X-rays in two orthogonal directions (see Figure 1.6).

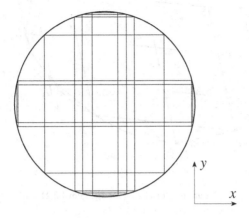

Figure 1.6. A disk as an inscribable convex body.

Lemma 1.2.18 will also be used in proving the next theorem. Suppose that we are shown a uniform object and asked to check if it contains a convex hole of a certain given shape and orientation. We might now choose directions for X-rays, depending on the convex body with this shape and orientation, in such a way that no other convex body will yield the same set of X-rays in this set of directions. We say that the convex body can be *verified* by X-rays in such a set of directions; see Definition 2.1.7 for a formal definition.

Definition 1.2.20. Let K be a convex body and u a direction. We call the map defined in bd K that interchanges the endpoints of the chords of K parallel to u the *u-map*, and denote it by f_u.

Note that the map f_u is defined on all of bd K, except for the relative interior of any line segment in bd K parallel to u. It is an involution of bd K if bd K contains no such line segments, in particular, if K is strictly convex.

Theorem 1.2.21. *A convex body can be verified by X-rays in a set of three directions.*

Proof. Let K be a convex body. The set of directions of line segments in bd K is at most countable. Choose the first direction u_1 different from all these directions. Then choose the second direction u_2 parallel to the chord k of K meeting both supporting lines to K parallel to u_1. Denote the endpoints of k by a and b. See Figure 1.7.

Let V be the closed set of all vertices of parallelograms inscribed in K whose edges are parallel to u_1 and u_2. (We include here the degenerate "parallelogram"

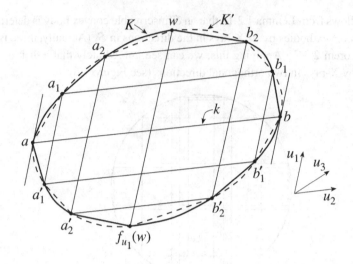

Figure 1.7. Illustration for Theorem 1.2.21.

k, as well as the chord k' meeting both supporting lines to K parallel to u_2, if k' is parallel to u_1.) Endpoints of the components of bd $K \setminus V$ form a countable set E. If $E = \emptyset$, then it follows from Lemma 1.2.18 that K can actually be verified by X-rays in the two directions u_1 and u_2. If $E \neq \emptyset$, choose the third direction u_3 not parallel to any line joining two points in E. Suppose that K' is any convex body with the same X-rays as K in the directions u_i, $1 \leq i \leq 3$. We shall prove that $K' = K$.

If $K' \neq K$, there is a component $C \subset (K \triangle K')$. Let $\mathcal{C}$ be the system of components associated to C. By Lemma 1.2.6, $\mathcal{C}$ is finite, so the endpoints of the components in $\mathcal{C}$ form the vertices of a convex polygon Q. If v is a vertex of Q, it means that v is an endpoint of a component C' in $\mathcal{C}$, so bd K and bd K' coincide at v. If $1 \leq i \leq 3$, the line $l_{u_i} + v$ must pass through an endpoint v' of the component $u_iC' \in \mathcal{C}$. This means that v' is a vertex of Q, but it can happen that $v' = v$. We conclude that the polygon Q is weakly $\{u_1, u_2, u_3\}$-regular.

Suppose that a_1 is the vertex of Q nearest to a in bd K on one side of the chord k. Let $a'_1 = f_{u_1}(a_1)$. Then a'_1 is the vertex of Q nearest to a in bd K on the other side of k. Similarly, there are two vertices b_1, b'_1 of Q, with $b'_1 = f_{u_1}(b_1)$, nearest to b in bd K, with b_1 on the same side of k as a_1. But now it is clear that $f_{u_2}(a_1) = b_1$ and $f_{u_2}(a'_1) = b'_1$, so the points a_1, a'_1, b_1, and b'_1 are all in E.

The same argument will now show that the four next nearest endpoints to a and b all belong to E, and so by induction they all do, except possibly for the last two, w and $f_{u_1}(w)$ say (see Figure 1.7). By Lemma 1.2.6, $|\mathcal{C}| \geq 6$, so Q has at least 6 vertices. From this and the weak $\{u_1, u_2, u_3\}$-regularity of Q, there must be two of these vertices, c, c' say, neither equal to w or $f_{u_1}(w)$, such that $c' = f_{u_3}(c)$.

Since c and c' belong to E, this contradicts the choice of u_3, and completes the proof. ∎

It transpires, then, that one less direction is needed for uniqueness when the directions are not chosen in advance. Somewhat surprisingly, two directions are not enough for verification, as we now show.

Theorem 1.2.22. *There is a convex polygon not verifiable by X-rays taken in any two directions.*

Proof. Let K be a centered convex body, and suppose that u_i, $i = 1, 2$, are non-parallel directions. Let ϕ denote the isometry taking a point p to the point ϕp on the line $l_{u_2} + p$ parallel to u_2 such that ϕp has the same distance as p from the line l_{u_1}. Let

$$K^r = \phi K = \{p' : p' = \phi p \text{ for some } p \in K\}.$$

We call the convex body $K^r = K^r(u_1, u_2)$ the *reflection* of K in l_{u_1} parallel to u_2.

Obviously K^r has the same X-ray as K in the direction u_2. Let $x \in u_1^\perp$. It is easy to see, from the definition of K^r, that $S_{u_1} K^r(x) = S_{u_1} K(-x)$. By the central symmetry of K, $S_{u_1} K(-x) = S_{u_1} K(x)$, so K^r also has the same X-ray as K in the direction u_1.

Now the centered unit square J almost provides the example we want. For any two directions except the coordinate directions, J^r is a different convex body (a parallelogram, in fact) with the same X-rays as J. For the coordinate directions, however, $J^r = J$. The required example must be slightly more complicated – a centrally symmetric hexagon K such that none of the diagonals are parallel to the edges (see Figure 1.8). In this case it is not difficult to check that $K^r \neq K$, for any two directions. ∎

Suppose that the shape of a hole is not known in advance, and it is to be found by means of X-rays. Rather than initially fixing all the directions, one might well decide to select them inductively. After taking the first X-ray, the X-ray picture could be consulted, and this information used to choose the direction for the second X-ray, and so on. We say that a convex body can be *successively determined* by X-rays in a set of directions if it can be distinguished from any other convex body by its X-rays taken in this way. This is, of course, a very natural and practical idea, lying somewhere between determination and verification; see Definition 2.1.8 for the formal definition and comments.

Theorem 1.2.23. *Convex polygons can be successively determined by three X-rays.*

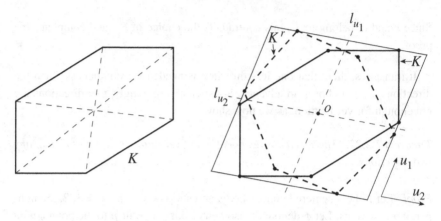

Figure 1.8. A convex polygon not verifiable by two X-rays.

Proof. The formal definition of successive determination (Definition 2.1.8) gives us the a priori knowledge that the object being X-rayed is a convex polygon. However, we only need assume that it is a convex body, since a convex body is a polygon if and only if one of its Steiner symmetrals is also.

Let Q be a convex polygon, and let u_i, $i = 1, 2$, be any two nonparallel directions. For $i = 1$ or 2, a line parallel to u_i meets a vertex of Q if and only if it meets a vertex of the corresponding Steiner symmetral $S_{u_i} Q$. Therefore the vertices of Q must belong to the finite grid F of points at the intersections of two families of parallel lines, namely, those parallel to u_i passing through the vertices of $S_{u_i} Q$, for $i = 1, 2$.

The third direction u_3 can now be chosen so that it is not parallel to any line joining two points in F. A point is a vertex of Q if and only if it is a point in F lying on a line parallel to u_3 through a vertex of $S_{u_3} Q$. ∎

Theorem 1.2.22 implies that the number three in Theorem 1.2.23 cannot be reduced to two.

We now turn to the question of reconstruction of convex bodies from their X-rays. Our purpose is to outline an algorithm that, for smooth and strictly convex bodies at least, has actually been implemented. Even in this case, however, the theoretical justification of the algorithm is as yet incomplete, and in fact leads to a fascinating open problem.

Lemma 1.2.24. *Suppose that K is a convex body and that u_i, $1 \leq i \leq 3$, are three mutually nonparallel directions. There are at least two (possibly degenerate) triangles inscribed in K whose sides are all parallel to the directions u_i.*

Proof. Let $p_3 \in \operatorname{bd} K$ and let k_1 be the chord of K through p_3 parallel to u_1. (Chords such as k_1 may be degenerate.) Denote by x and y the points where the supporting lines to K parallel to u_1 meet bd K. Then x and y divide bd K into two arcs, A_1 and A_2, and we suppose that $p_3 \in A_1$. See Figure 1.9.

Suppose that p_2 is the other endpoint of the chord k_1 (equal to p_3 if k_1 is degenerate). For $i = 2, 3$, let k_i be the chord of K with one endpoint at p_i and parallel to u_i. Finally, let p_1 be the intersection of the lines containing k_2 and k_3.

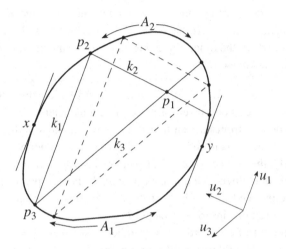

Figure 1.9. The body K of Lemma 1.2.24.

As p_3 varies over the arc A_1, the (possibly degenerate) triangle $T(p_3)$ with vertices p_i, $1 \le i \le 3$, always has the same orientation. When p_3 is close to x, $p_1 \in \text{int}\, K$, and when p_3 is close to y, $p_1 \in \text{co}\, K$ (or vice versa). By continuity, therefore, there is a position of p_3 such that $p_1 \in \text{bd}\, K$, and then $T(p_3)$ is one of the required inscribed triangles. The other, with the opposite orientation, is found by taking p_3 in the arc A_2. ∎

Note that if K is smooth, the triangles provided by Lemma 1.2.24 are not degenerate. If bd K contains no line segments parallel to the given directions, there will be exactly two such triangles; otherwise there may be infinitely many.

Lemma 1.2.25. *Let K be a convex body, and u_i, $1 \le i \le 3$, three mutually nonparallel directions. The inscribed triangles of Lemma 1.2.24 can be found from the X-rays of K in these three directions.*

Proof. For $1 \le i \le 3$, let x_i be a point in the support of $X_{u_i} K$. Suppose that $T = T(x_1, x_2, x_3)$ is the triangle whose boundary is formed by the lines $l_{u_i} + x_i$. Denote by E_i the closed half-plane bounded by $l_{u_i} + x_i$ and not containing T. For any set A, let $A_i = A \cap E_i$. Consider the function

$$f(x_1, x_2, x_3) = \lambda_2(T) + \sum_{i=1}^{3} \lambda_2\big((S_{u_i} K)_i\big),$$

computable from the X-rays in the directions u_i, $1 \leq i \leq 3$. Since, by the
Cavalieri principle (Lemma 1.2.2), $\lambda_2(K_i) = \lambda_2\big((S_{u_i}K)_i\big)$ for each i, we have

$$f(x_1, x_2, x_3) = \lambda_2(T) + \sum_{i=1}^{3} \lambda_2(K_i).$$

The right-hand side is not less than $\lambda_2(K)$, since the sets T and K_i, $1 \leq i \leq 3$,
cover K. Further, these sets overlap unless T is inscribed in K. Consequently the
values of x_i for which the triangle T is inscribed in K can, in principle, be found
by minimizing the known function f.   ∎

Now let $S = \{u_i : 1 \leq i \leq 4\}$ denote a set of four directions such that
X-rays of any convex body taken in these directions determine it, as in Corollary 1.2.12. Suppose from now on that K is a smooth and strictly convex body.
By Lemma 1.2.25, we can find several points on the boundary of K from the
X-rays, namely, the vertices of the four pairs of inscribed triangles whose sides
are parallel to some three of the four directions. In fact, to proceed with the algorithm, we only need to use the first three directions to compute the vertices
$V_1 = \{v_1, v_2, v_3\}$ of one inscribed triangle T.

Consider the set of all chords of K parallel to some u_i, $1 \leq i \leq 4$, having
one endpoint in V_1, and meeting bd T in a point other than this endpoint. There
will be four of these: the three sides of T and one chord k parallel to u_4, one of
whose endpoints is a vertex v of T. Let V_2 be the set of all endpoints of these
chords. Then V_2 consists of the vertices of T and the other endpoint of k. The X-
ray $X_{u_4}K$ gives the length of k, and the fact that $k \cap \text{int } T \neq \emptyset$ determines the side
of v on which k lies in the line $l_{u_4} + v$. Thus V_2 can be computed from our X-rays.
Continue this process inductively. Specifically, at the nth stage we consider the
set of all chords of K parallel to some u_i, $1 \leq i \leq 4$, with one endpoint in V_n, and
meeting the boundary of conv V_n in a point other than this endpoint. Let V_{n+1} be
the set of all endpoints of these chords. In this way we produce finite sets V_n of
points in bd K which can be found from the X-rays, with $V_n \subset V_{n+1}$ for each n.
(See Figure 1.10, in which the labeling denotes the smallest n such that the point
belongs to V_n.) Let $V = \cup_n V_n$.

Lemma 1.2.26. *The set V is infinite.*

Proof. If V is finite, then there is an n such that $V_{n+1} = V_n$. Let $Q = \text{conv } V_n$.
Note that since Q contains the vertices of the triangle T, it is nondegenerate. We
claim that Q is a weakly S-regular polygon, contradicting Corollary 1.2.10.

To prove this, let $v \in V_n$ and $1 \leq i \leq 4$. Let k be the chord of K parallel to
u_i and with one endpoint at v, and suppose that k meets bd Q in a point $w \neq v$.
Then $w \in V_{n+1}$, by the definition of V_{n+1}, and so $w \in V_n$ also.   ∎

Recall Definition 1.2.20 of the u-map f_u.

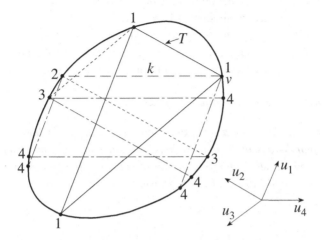

Figure 1.10. A possible reconstruction procedure.

Lemma 1.2.27. *For* $1 \leq i \leq 4$, *the set* cl V *is invariant under the map* f_{u_i}.

Proof. Fix i, $1 \leq i \leq 4$. We first show that $f_{u_i}(V) \subset$ cl V. To this end, let $v \in V$, so that $v \in V_n$ for all $n \geq N$, say. Let the supporting lines to conv V (or conv V_n) parallel to u_i be denoted by t_{ij} (or t_{ij}^n, respectively), $j = 1, 2$. The sets V_n increase with limit V, so the lines t_{ij}^n converge to t_{ij}.

If v does not belong to t_{ij}, then there is an $M \geq N$ such that for $n \geq M$, v does not belong to t_{ij}^n. Let $n \geq M$. Then the chord of K with one endpoint at v and parallel to u_i meets the boundary of conv V_n in a point other than v. Consequently the other endpoint of this chord is in V_{n+1}. But this point is $f_{u_i}(v)$, and $V_{n+1} \subset V$, so $f_{u_i}(v) \in V$, as required.

Suppose, then, that $v \in t_{ij}$ for some j. If v is not isolated in V, there is a sequence of points $v_m \neq v$ in V converging to v. For sufficiently large m, the chord of K with one endpoint at v_m and parallel to u_i meets the boundary of conv V_N, and hence that of conv V_n for $n \geq N$, in a point other than v_m. As before, this shows that $f_{u_i}(v_m) \in V$ for sufficiently large m. Now $f_{u_i}(v_m)$ converges to $f_{u_i}(v)$, so $f_{u_i}(v) \in$ cl V.

Suppose that v is isolated in V. There must be an i' such that the chord c of K containing v and parallel to $u_{i'}$ has its other endpoint v' in V. If v' is not isolated in V, there is a sequence of points $v_m' \neq v'$ in V converging to v'. The same argument as in the last paragraph, with i' replacing i, now shows that $f_{u_i'}(v_m')$ are points in V converging to v, a contradiction. So v' must also be isolated in V. Now if $w \in V$, then both v and w are in V_n for sufficiently large n, so there is a finite sequence of chords $c_k = [x_k, x_{k+1}]$, $1 \leq k \leq k_0$, of K, each parallel to a direction in S, such that $v = x_1$, $w = x_{k_0+1}$, and $x_k \in V_n$ for each k. By the previous argument and induction on k, w must also be isolated in V. But then all points in V are isolated, contradicting Lemma 1.2.26.

Therefore $f_{u_i}(V) \subset \operatorname{cl} V$. Now suppose that $x \in \operatorname{cl} V \setminus V$. Then there are $v_m \in V$ with v_m converging to x. So $f_{u_i}(v_m) \in \operatorname{cl} V$ and the points $f_{u_i}(v_m)$ converge to $f_{u_i}(x)$. We have $f_{u_i}(x) \in \operatorname{cl} V$, and the proof is complete. ∎

Summarizing, we have the following theorem.

Theorem 1.2.28. *Let S be a set of directions such that convex bodies are determined by their X-rays in these directions. For each smooth and strictly convex body K, an infinite set V of points in $\operatorname{bd} K$ can, in principle, be constructed from the X-rays of K in the directions in S. Furthermore, the closure of V is invariant under the u-map for each $u \in S$.*

The algorithm outlined here would be a valid solution of the reconstruction problem for smooth and strictly convex bodies if a positive answer could be obtained to Problem 1.4. See Note 1.3 for some comments.

Open problems

Problem 1.1. (See Note 1.2.) Characterize those convex bodies that can be determined by two X-rays.

Problem 1.2. (See Note 1.2.) When is a set of convex bodies a set of Steiner symmetrals of some convex body?

Problem 1.3. (See Theorems 1.2.21 and 1.2.23.) Can convex bodies be successively determined by three X-rays?

Problem 1.4. (See Note 1.3.) Let S be a finite set of directions such that X-rays of a convex body taken in those directions determine it. Suppose that K is a smooth and strictly convex body and that V is an infinite subset of $\operatorname{bd} K$ invariant under the u-map for each $u \in S$. Is $\operatorname{cl} V = \operatorname{bd} K$?

Problem 1.5. (See Note 1.4.) Obtain a stability result estimating the distance between two convex bodies whose X-rays in a set of n directions agree up to an error of ε.

Notes

1.1. *Computerized tomography.* Most major hospitals have a machine called a CAT (Computerized Axial Tomography) scanner (or CT scanner) that provides images of a 2-dimensional section of a patient, to be used by medical staff for diagnosis or surgery. Computerized tomography (also referred to as computer tomography or computed tomography) is the combination of mathematics, computer science, physics, and engineering involved in the image reconstruction, typically from several hundred X-rays taken in coplanar directions. Both parallel and point (or fan-beam) X-rays have been used, though modern scanners usually employ the latter.

Mathematically, the problem is to reconstruct an unknown density function $f(x, y)$ describing a 2-dimensional section of an object. (In practice, $f(x, y)$ is not the physical density, but a related quantity called the attenuation coefficient that measures the local loss of intensity of an X-ray beam at (x, y).) Each beam of the X-ray travels along a straight line l before arriving at a site where the attenuation of the beam is measured. The CAT scanner is calibrated so that from this attenuation, the line integral of f along l can be found. Suppose, for simplicity, that the X-ray is a parallel X-ray taken from a direction u and there is no error or noise in the data. Then the information obtained is given by the function

$$X_u f(s) = \int_{-\infty}^{\infty} f(s + tu)\, dt,$$

for $s \in u^{\perp}$, in other words, the X-ray transform of f in the direction u (see (C.4)).

The main mathematical tool behind computerized tomography is the central slice theorem (also called the projection slice theorem). This says that the restriction of the 2-dimensional Fourier transform $\hat{f}$ of f to $u^{\perp}$ is precisely the 1-dimensional Fourier transform $(X_u f)\hat{}$ of the X-ray $X_u f$. (See (C.8) for a more general statement.) Thus if $X_u f$ is known for all $u \in S^1$, then $\hat{f}$ is also known. By means of the inversion formula for the Fourier transform, the function f can be determined. In essence, all this was known to Radon, who in his 1917 paper [685] gave a direct method for inverting the X-ray transform in the plane. See Section C.1 for a discussion of these integral transforms and how they relate to each other.

While the previous paragraph provides a mathematical basis for computerized tomography, many issues remain. To begin with, only a finite number of X-rays can be taken, and this means that the reconstruction problem does not have a unique solution (see Theorems 2.3.3 and 7.3.1, [645, Theorem 3.7], and [777]). Moreover, each X-ray actually consists of only a finite number of beams, so in practice one has a finite set $X_{u_i} f(s_j)$, $1 \leq i \leq m$, $1 \leq j \leq n$ of numbers from which to obtain an approximation to f. The principal mathematical ingredients that deal with the problem of discretization are filtered back-projection (a modification of the central slice theorem), discrete convolution, and the Fourier sampling theorem. (The method must be adapted for point X-rays, but the basic techniques are the same.) The utility of the entire process rests on the fact that despite the lack of uniqueness, a sufficiently large number of X-rays permits a reconstruction to any desired degree of accuracy.

Two pioneers, A. M. Cormack, a physicist, and G. N. Hounsfield, an engineer, were awarded the 1979 Nobel prize in medicine for the "development of computer assisted tomography." Cormack's contribution was a theoretical analysis, published in 1963 and 1964, while Hounsfield described a complete system for computerized tomography in his 1968 application for a patent, granted in 1972. Neither Cormack nor Hounsfield employed filtered back-projection, however. The history of tomography is complicated and entangled with other applications of the Fourier inversion method, such as radio astronomy. Brief but helpful outlines with supplemental references are provided by Natterer [646], [648, p. 62]. Natterer and Ritman [647] give a much more thorough account that gives credit to many others for their part in the creation of computerized tomography and speculates on future developments.

The computerized tomography literature is vast. The reader who wishes to learn more could begin with the texts [211], by Epstein, and [414], by Kak and Slaney, both of which are accessible to advanced undergraduate students. The books of Natterer [645] and Natterer and Wübbeling [648] are set at a slightly higher level. Also ideal as an introduction is the article [640] of Murrell, which in six pages manages to set out in a very clear fashion not only the basic mathematics behind computerized tomography, but also a Mathematica program that reconstructs a test image called the Shepp–Logan image, after its inventors. This cleverly simulates a cross-section of a human head by a collage of ten ellipses of different densities (see Figure 1.11).

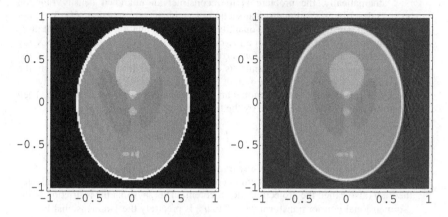

Figure 1.11. Shepp–Logan image : Original (left) and a reconstruction (right).

Murrell's program produces reconstructions such as that in Figure 1.11, where striations not present in the original can be seen, especially in the black area outside the main image. The suppression of such artifacts is of course a major consideration in computerized tomography, made more challenging if the data is very noisy or sparse. In this respect the very short summary above of the most widely used reconstruction algorithm is deceptive, since there are a huge number of variations on this theme, necessitated by the many various applications. In addition, there are quite different algorithms based, for example, on linear algebra.

The Shepp–Logan image is only one of many computer-generated models developed for the purpose of simulating structures, with a view to the evaluation and improvement of reconstruction algorithms in computerized tomography. Ellipsoids and other special geometrical shapes, often convex, are employed in very realistic 3-dimensional anatomical models, and even in 4-dimensional models of beating hearts. Work of Zhu et al. [865], [866] should lead the interested reader to the literature.

1.2. *Parallel X-rays and Steiner symmetrals of convex bodies.* Steiner [787] employed the symmetrals bearing his name in his contribution to the isoperimetric inequality (B.14), though according to Gruber's article [364] on the history of convexity, the idea of symmetrization first appeared in work of L'Huillier in the 1780s. Steiner symmetrization and other methods of symmetrization (see Definition 2.1.3, for example) occur frequently in geometry; see, for example, [83, Sections 40–2] or [718, Section 5]. For applications to mathematical physics, see Note 2.3.

The questions of when planar convex bodies can be determined, verified, or reconstructed from their X-rays arose from problems posed by Hammer [380], worth quoting in full:

Suppose there is a convex hole in an otherwise homogeneous solid and that X-ray pictures taken are so sharp that the "darkness" at each point determines the length of a chord along an X-ray line. (No diffusion, please.) How many pictures must be taken to permit exact reconstruction of the body if:

a. The X-rays issue from a finite point source?

b. The X-rays are assumed parallel?

For the planar counterpart we have shown that two perpendicular directions are insufficient for (a) and we conjecture that 3 directions are sufficient, although whether or not such directions must be strategically chosen is also open.

Hammer actually asked his questions in 1961, a year before the (independent) appearance of Theorem 1.2.21, proved by Giering [288]. The short proof given here is basically that of Gardner [256], to which P. McMullen also contributed; a slightly different proof, also allowing more possibilities for the choice of the three directions, was found by Volčič [820]. All proofs, however, use Giering's simple, yet crucial and ingenious, Lemma 1.2.18 (proved for two directions in [288] and generalized in [822]). The technique behind Theorem 1.2.22 is also due to Giering [288, p. 241], but the explicit example of the hexagon appeared in [256, Example 2.5].

The solution to the problem of determination, Theorem 1.2.11, was published by Gardner and McMullen [277]. The proof in that paper used the endpoints rather than the centroids of an associated system of components, and so employed weakly S-regular, rather than S-regular, polygons. A central ingredient is the beautiful Lemma 1.2.9. This has been rediscovered several times, but was apparently first proved by Darboux [177], in essentially the same way as we prove it here.

If S is a set of two nonparallel directions, a body E is called Q-convex with respect to S if $p \in E$ whenever E contains points in each of the four closed regions formed by the two lines through p parallel to the directions in S. If S is any set of mutually nonparallel directions, E is Q-convex with respect to S if it is Q-convex with respect to each pair of directions in S. Examples of Q-convex bodies with respect to a set S of directions include any connected body whose intersection with each line parallel to S is empty or connected. This is shown by Brunetti and Daurat [105], who also prove that Q-convex bodies with respect to a set S of directions are determined by X-rays taken in the directions in S if and only if convex bodies are also so determined. It follows that Theorem 1.2.11 extends to the class of bodies that are Q-convex with respect to the set of directions in which the X-rays are taken.

It is possible to generalize Theorem 1.2.11 in various other ways. For example, Theorem 5.4.6 is an extension to the projective plane. Kurusa [485] sketches a version holding in spaces of constant curvature and another applying to X-rays involving certain weight functions.

The simple examples of Theorem 1.2.5 are mentioned in [288]. The more complicated ones of Theorem 1.2.13 are constructed following the technique of Volčič in [819], where it is also shown that given a set of directions as in Theorem 1.2.13, there is a family, with cardinality equal to that of the real numbers, of mutually noncongruent convex bodies, all with the same X-rays in these directions.

Volčič and Zamfirescu [823] prove Theorem 1.2.17, thereby confirming a conjecture of Gruber [362]. The auxiliary Lemma 1.2.14 is taken from [763], and Lemma 1.2.15 is a weak form of [187, Satz 3]. Both lemmas remain true in higher dimensions, with essentially the same proofs. Theorem 1.2.17 also holds in $\mathbb{E}^n$, $n \geq 2$; see Theorem 2.2.4.

The interactive notion of successive determination was conceived by Edelsbrunner and Skiena, and Theorem 1.2.23 is from their paper [203].

The partial results of Giering [289], concerning triangles and quadrilaterals, suggest that Problem 1.1 may be difficult. (This contrasts with the known characterizations of planar sets of finite measure determined by two X-rays; see Theorem 2.3.17 and Note 2.3.)

Problem 1.2 was raised in [277]. There are two versions of the question, depending on whether or not the set of directions of the symmetrals is stipulated in advance. Obviously the bodies must all be symmetric about a line. Falconer [214] gives such consistency conditions for a finite set of X-rays, but his result corresponds to the case of X-rays of a smooth convex body whose density is allowed to vary.

Milanfar [612] has proved that a nondegenerate convex m-gon is determined by its X-rays in any $2m - 2$ mutually nonparallel directions; see Note 2.5 for more details.

1.3. *Exact reconstruction.* By means of the standard Fourier-transform algorithm of computerized tomography (see Note 1.1), an arbitrarily close approximation to a convex body can be constructed from its X-rays, if these are taken in a sufficiently large set of directions. However, Hammer's X-ray problem, stated in Note 1.2, clearly expresses the hope for a algorithm by which convex bodies can be reconstructed from their X-rays in a *fixed* finite set of directions.

This chapter discussed in some detail one of several attempts to find such an algorithm. The method of finding some points in the boundary from X-rays, Lemma 1.2.25, is due to Volčič [822]. Kramer and Németh [474] proved Lemma 1.2.24 for the special case of smooth and strictly convex bodies; the general case is presented in [822]. Michelacci [609] found a different proof of Lemma 1.2.24, based on the fact that an orientation-reversing homeomorphism of the circle has exactly two fixed points. The reconstruction algorithm was implemented by Manni [569]. Some good results were obtained using X-rays in four equally spaced directions for the convex bodies bounded by the curves given in polar coordinates by $r(\theta) = 1 + \cos^2\theta$ and $r(\theta) = 8 + \sin\theta + \cos 3\theta$. Here, the initial boundary points were found, as in Lemma 1.2.25, by minimizing the function $f(x_1, x_2, x_3)$. An alternative approach is suggested by Michelacci [608]. This finds the inscribed triangle by a method using fixed-point theory which is apparently faster than minimization.

Problem 1.4 was asked independently by the author and A. Volčič. It seems to implicate the theory of homeomorphisms of the circle, as follows. If bd K is a circle, the composition of any two u-maps is a rotation, so such compositions for general K constitute natural generalizations of rotations of the circle. Further, the u-map on bd K can be regarded, via the radial map (0.3), as a homeomorphism of S^1 of a special type. From the general theory of such homeomorphisms, it follows that cl V must either be bd K or a Cantor set. For a single homeomorphism of S^1, Denjoy showed that the latter possibility can occur, but not if the homeomorphism belongs to $C^2(S^1)$ (see, for example, [385, p. 81]). We can ensure that each u-map is in $C^2(\text{bd } K)$ by restricting attention to convex bodies of class C^2. However, we require a Denjoy-type theorem applying not just to one homeomorphism, but to a group of homeomorphisms of the appropriate type.

A completely different algorithm for the reconstruction of convex bodies from their X-rays was found by Kölzow, Kuba, and Volčič [471]. The idea, roughly speaking, is to discretize and then apply a method deriving from those used in reconstructing binary matrices from their column and row sums (see [476]). The algorithm provides a tree of convex bodies which are approximate solutions. When the solution is unique, the convex bodies in one infinite branch of this tree will converge to the solution. The problem is that there seems to be no effective way of knowing when one is close to the desired solution; see [471, Examples 1 and 2]. Brunetti and Daurat [104] also propose an algorithm based on discretization, specifically by means of approximation by convex lattice sets; they note, however, that this would require a stronger version of their algorithm for reconstructing such sets (see Note 2.2).

Yet another algorithm, constituting a rather complete answer to Hammer's problem, was formulated recently by Gardner and Kiderlen [273]. See Note 1.5, where the algorithm is discussed in the context of reconstruction of convex bodies from possibly noisy data.

Milanfar [612] has proved that a nondegenerate convex m-gon can be reconstructed from its X-rays in any $2m - 2$ mutually nonparallel directions; see Note 2.5 for more details.

1.4. *Well-posedness and stability.* In [818], Volčič shows that the problem of reconstructing a convex body from its X-rays in four directions so as to guarantee a unique solution is well posed. This involves showing the continuity of the inverse of the map taking a convex body in $\mathcal{K}_0^2$ to the quadruple, in $(\mathcal{K}_0^2)^4$, of its four Steiner symmetrals in these directions.

The following stability result is proved by Longinetti [517]: If the X-rays of two convex bodies K, K' are equal in a set of n mutually nonparallel directions, then

$$\lambda_2(K \triangle K') \le \frac{\tan(\pi/n)}{8n} \lambda_1 \Big(\text{bd}\,(K \cap K')\Big)^2,$$

with equality if and only if K is a regular n-gon and K' is K rotated by π/n about its center. Note that an upper bound for $\lambda_1\Big(\text{bd}\,(K \cap K')\Big)$ can easily be estimated from the X-rays. To prove his theorem, Longinetti uses the idea of an associated system of components, and in fact the observation in Lemma 1.2.6 that $|\mathcal{C}| \ge 2n$ is his. From this theorem it can be seen that the L^2 distance between the characteristic functions of convex bodies such as K and K' is of order n^{-1}, to be compared with the corresponding order $n^{-1/2}$ for the case of bounded integrable functions defined on a bounded set; see [557].

Splendid though this result of Longinetti is, it suffers from the disadvantage that it is not affine invariant. In the earlier paper [516], he suggests an affine-invariant estimate in which $\lambda_1\Big(\text{bd}\,(K \cap K')\Big)$ is replaced by $\lambda_2(K \cap K')$. This involves a function defined on the class of convex n-gons, whose precise value has recently been found by Gronchi and Longinetti [361]. However, more has to be done to establish the corresponding general stability result.

Further stability estimates are given by Longinetti in [515]. Here it is assumed that the X-rays are not known exactly, but only up to an error ε. The main results are as follows: Suppose that K, K' are convex bodies such that for all directions u, $\|X_u K - X_u K'\|_\infty < \varepsilon$, so K and K' are "ε-equichordal." Then $\lambda_2(K \triangle K') \le C\varepsilon^2$, where $C \le 14.2$ is a constant independent of K and K'. Also, if $K \cap K' \ne \emptyset$, then $\delta(K, K') < 3\varepsilon$.

It is possible that methods of Sonnevend [778] may be helpful in answering Problem 1.5.

1.5. *Reconstruction of convex bodies from possibly noisy data.* Since it may be undesirable or costly to take too many X-rays, workers in cardiac cineangiography have attempted to reconstruct objects from X-rays taken in two orthogonal directions. For example, Chang and Chow [159] and Chang and Wang [160] approximate sections of a heart by planar convex bodies symmetric with respect to two orthogonal axes, which are then to be reconstructed from the two X-rays. It follows easily from the reflection technique of Theorem 1.2.22 that the solution is not unique unless the directions of the X-rays coincide with those of the axes of symmetry. Nevertheless, some pictures can be produced by approximating further by polygons and trying various combinations of the grid of points corresponding to that in the proof of Theorem 1.2.23.

In the early 1980s, A. S. Willsky, an electrical engineer, initiated a research program, similarly motivated by applications. The PhD thesis [698] of Rossi was the first of several supervised by Willsky that develop and implement algorithms specially suited for reconstructing planar convex (and more general) sets. None of these algorithms solve the theoretical problem of obtaining arbitrarily accurate images of convex sets from X-rays in a fixed finite set of directions. However, they do produce images from a limited set of X-rays that are often sufficient for detecting some particular feature of the object, such as the approximate position or orientation of an ellipse or a rectangle. Moreover, they can work even when the data is inaccurate and when noise is present. The general idea is to take advantage of a priori information that the unknown object belongs to a restricted class of homogeneous objects, an idea completely compatible with the theoretical results of this chapter. As motivation, Rossi gives references for applications to geophysics, oceanography, meteorology, nondestructive testing (locating cracks in nuclear reactor cooling pipes, etc.), and medicine (nearly homogeneous regions such as kidneys and airspaces between organs), in addition to others already mentioned in the text, where a large number of high-resolution measurements may be infeasible and where a high-resolution image may not be needed.

He also points out that some of the standard algorithms of computerized tomography, such as the filtered back-projection algorithm employing the Fourier transform (see Note 1.1), tend to be ineffective when few X-rays are available, and very sensitive to measurement errors or noise. See also the article [699] of Rossi and Willsky.

Publications by Belcastro, Karl, and Willsky [50], Greschak [333], Milanfar, Karl, and Willsky [613], [614], and Prince and Willsky [680], [681] continued this program and provide a wealth of references to related work. Techniques include utilization of special features in pictures of Radon transforms ("sinograms"), modeling the sinogram by a Markov random field, interactive methods in the spirit of successive determination, and the use of moments (cf. Note 2.5).

In most of this work, measurement noise is modeled by adding to the exact data a Gaussian random variable. Thus noisy X-rays of a planar convex body K in directions $u_i, 1 \leq i \leq m$, would be given by

$$Y_{u_i}(x) = X_{u_i} K(x) + N_i(x),$$

for $x \in u_i^\perp$ and $1 \leq i \leq m$, where the $N_i(x)$'s are independent normal random variables with zero mean and equal variance. Gardner and Kiderlen [273] propose an algorithm for reconstructing K from $Y_{u_i}, 1 \leq i \leq 4$, where the u_i's are four directions for which a unique solution is guaranteed if the X-rays are given exactly. Assuming that K is contained in a known ball, it is shown in [273] that the algorithm produces a sequence of polygons that converge, almost surely, to K. The method is based on a least-squares approach, similar to that often employed by participants in Willsky's program.

1.6. *Geometric probing.* In his thesis [768] and surveys [770] and [771], Skiena writes: *Geometric probing considers problems of determining a geometric structure or some aspect of that structure from the results of a mathematical or physical measuring device, a probe.* Since sections and projections are not mentioned in this definition, geometric tomography appears to be just a special type of geometric probing. However, most of the probes considered by Skiena do involve sections or projections; furthermore, the objects of interest in [770] and [771] are usually polyhedra, since contributions to the field have often come from computer science, image processing, and computational geometry.

Geometric probing, as summarized in [770] and [771], appears to have been initiated by the paper [169] of Cole and Yap. Motivated by the idea of providing a robot with tactile sensors, the authors consider the *finger probe*, returning the first point of intersection between a directed line and an object. It is assumed that we know that the object contains o in its interior, for without this information any finite set of finger probes might miss the object completely. With this understanding, Cole and Yap show that $3n$ finger probes suffice, and $3n - 1$ are necessary, to successively determine a convex n-gon in $\mathbb{E}^2$ among all convex polygons; while for verification, $2n$ finger probes are necessary and sufficient. (Cole and Yap mention that an algorithm similar to theirs was implemented on the IBM RS-1 robot.) It is observed in [770] and [771] that the upper bounds can be lowered if it is known in advance that the polygon is an n-gon.

Skiena also examines *X-ray probes*. An X-ray probe returns the length of an intersection of the object with a single line; in other words, it is a single value of a parallel X-ray. As in the case of finger probes, it is assumed that the object contains o in its interior. Edelsbrunner and Skiena [203] prove that $5n+19$ X-ray probes suffice to successively determine a convex n-gon among all convex polygons. On the other hand, $3n - 3$ X-ray probes are necessary; see Lindenbaum and Bruckstein's article [508]. For verification of a convex n-gon, again among all convex polygons, it is noted in [770] that $[3n/2] + 7$ X-ray probes suffice, while n are necessary.

Other types of probes are also contemplated in [770] and [771] (see also work of Dobkin, Edelsbrunner, and Yap [190]; Gritzmann, Klee, and Westwater [342]; and Notes 2.1, 3.10, and 7.7), and long lists of references is provided. Some open problems can be found in [769] and [771].

Figure 1.12. Jakob Steiner.

1.7. *Jakob Steiner (1796–1863)*. (Sources: [113], [450].) Jakob Steiner was born in Utzensdorf, in the canton of Bern, Switzerland. The youngest of eight children of a small farmer and tradesman, Steiner did not learn to write until he was 14 years old. Mainly self-educated, at the age of 17 he became a pupil and then a teacher at the school of J. Pestalozzi in Yverdon. Here Steiner admired Pestalozzi's successful use of the Socratic method, and acquired a lifelong interest in education, as well as in research. He left Yverdon in 1818, moving to Heidelberg and then to Berlin in 1821, where he enrolled as a student for a couple of years, at the same time as Jacobi. During this time, Steiner earned a living by private tuition (one of his pupils was the son of Wilhelm von Humboldt). His very liberal views and eccentric methods (according to Kline [450, p. 847], he darkened the room when training doctoral candidates) may have been the cause of some initial difficulties as a teacher. Nevertheless, Steiner was appointed extraordinary professor at the University of Berlin in 1834. He never married.

Steiner was a prolific author who wrote several books and many articles, including 62 for the newly founded Crelle's journal (*Journal für die reine und angewandte Mathematik*). These included many theorems to be proved (Steiner often stated theorems without proof) and research problems. The collection of 85 theorems and problems in his *Vorlesungen über synthetische Geometrie* (*Lectures on Synthetic Geometry*), posthumously published in 1867, were an enormous stimulus to other mathematicians. Steiner was instrumental in unifying much of the geometry known at that time. He was a proponent of synthetic as opposed to analytical methods in a dispute involving several geometers, and once threatened to stop contributing to Crelle's journal if it continued to publish J. Plücker's analytical papers. Steiner was one of the first to use inversion in a circle, and he developed the projective treatment of conic sections. He proved that every construction with straightedge and compass can be carried out with straightedge and "rusty compass" of fixed radius. He wrote on a general theory of the centroid, introducing the idea of a curvature centroid, now also called the Steiner point. Several other fundamental concepts bear his name, such as Steiner curves, Steiner surfaces, and Steiner triple systems. In his long two-part paper, *Über Maximum und Minimum bei den Figuren in der Ebene, auf der Kugelfläche und im Raum überhaupt* (*On maxima and minima of figures in the plane, on the surface of the sphere, and in space generally*), he gave five proofs of the isoperimetric inequality in the plane, some using Steiner symmetrization. (However, as Dirichlet pointed out to Steiner, all of these assumed the existence of the extremum. This was established by F. Edler, H. A. Schwarz, and Weierstrass, and again later by Minkowski (cf. Note 4.13) and Blaschke (cf. Note 6.4).) His formula (A.30) for the volume of the outer parallel body of a convex body is a precursor of Minkowski's theory of mixed volumes.

2

Parallel X-rays in n dimensions

The notion of a parallel X-ray of a planar convex body admits several extensions, and the aim of this chapter is to study some of them. One can consider X-rays of convex bodies in higher dimensions. For this Lebesgue measure and integration can be avoided, as in Chapter 1, but when convex bodies are replaced by compact sets, for example, this is no longer possible. In addition, we would also like to discuss higher-dimensional X-rays, in which sections by parallel lines are replaced with sections by parallel planes. In view of this, we begin the chapter with the definitions of the X-ray and k-dimensional X-ray of a bounded Lebesgue measurable set in $\mathbb{E}^n$, and the corresponding generalizations of the Steiner symmetral.

Although a working knowledge of Lebesgue measure and integration is required for a full understanding of this chapter, much of it can be assimilated with only intuitive ideas of length, area, and volume. Some, but not all, of the background material in the first five sections of Chapter 0 is relevant; this includes a brief introduction to the theory of Lebesgue measure and integration.

Theorem 2.2.5 yields an efficient algorithm for successive determination of a convex polyhedron in $\mathbb{E}^3$ by only two X-rays. If S is a subspace, an X-ray of a measurable set parallel to S gives the volumes of the intersections of the set with all translates of S. The S-additive sets are introduced in Definition 2.3.9. Theorem 2.3.11 indicates the utility of this concept by showing that any such set can be determined, among all measurable sets, by its X-rays parallel to the subspaces in S.

Throughout this chapter, "measurable subset of $\mathbb{E}^n$" will always mean "λ_n-measurable subset of $\mathbb{E}^n$."

2.1. Parallel X-rays and k-symmetrals

We begin with the following definition, generalizing the notion of an X-ray of a planar convex body.

Definition 2.1.1. Let $u \in S^{n-1}$ be a unit vector, and E a bounded measurable subset of $\mathbb{E}^n$, $n \geq 2$. The *X-ray of E in the direction u* is defined for λ_{n-1}-almost all $x \in u^{\perp}$ by

$$X_u E(x) = X_u 1_E(x) = \lambda_1(E \cap (l_u + x)),$$

where l_u is the line through o parallel to u, 1_E is the characteristic function of E, and X_u is the X-ray transform (see (C.4)).

More generally, let $1 \leq k \leq n - 1$, and suppose that $S \in \mathcal{G}(n, k)$ is a k-dimensional subspace. The *k-dimensional X-ray of E parallel to S* is defined for λ_{n-k}-almost all $x \in S^{\perp}$ by

$$X_S E(x) = X_S 1_E(x) = \lambda_k(E \cap (S + x)),$$

where X_S is the k-plane transform (see (C.7)).

In words, the k-dimensional X-ray of E parallel to S is the function giving the λ_k-measure of the intersection of E with each k-dimensional plane parallel to S; if $k = 1$, this is just the ordinary X-ray.

If E is a Borel set, then $X_S E$ is a Borel function, defined for all $x \in S^{\perp}$. For an arbitrary measurable set E, Fubini's theorem assures us that $X_S E$ is defined for λ_{n-k}-almost all $x \in S^{\perp}$.

In Chapter 1, we noted that when K is a planar convex body, $X_u K$ is continuous and concave on its support. For a convex body in $\mathbb{E}^n$, $n \geq 3$, $X_u K$ is still concave on its support, but no longer need be continuous on its support. (To see this, let $n = 3$, and let K be the convex hull of a disk $D \subset u^{\perp}$ and a line segment of length 1 parallel to u and containing a point $p \in \text{bd } D$. Then $X_u K = 0$ at all points in bd D except p, while $X_u K(p) = 1$.) However, the following lemma is true.

Lemma 2.1.2. *Let E be a compact set in $\mathbb{E}^n$, and let $S \in \mathcal{G}(n, k)$, $1 \leq k \leq n - 1$. Then $X_S E$ is upper semicontinuous.*

Proof. Let x_m, $m \in \mathbb{N}$, be a sequence of points in $S^{\perp}$ converging to a point x. The compactness of E implies that

$$\limsup_{m \to \infty}(E \cap (S + x_m)) \subset E \cap (S + x).$$

Therefore

$$\limsup_{m \to \infty} X_S E(x_m) = \limsup_{m \to \infty} \lambda_k(E \cap (S + x_m))$$

$$\leq \lambda_k\left(\limsup_{m \to \infty} E \cap (S + x_m)\right)$$

$$\leq \lambda_k(E \cap (S + x)) = X_S E(x).$$

(The first inequality is a well-known property of Lebesgue measure; see [410, p. 60], for example.)   ∎

Note that for $k > 1$, $X_S K$ need not be concave on its support, even when K is a convex body. For example, let $K \subset \mathbb{E}^3$ be a bounded cone, the convex hull of the unit disk in the xy-plane and the point $(0, 0, 1)$. The 2-dimensional X-ray of K parallel to the xy-plane is the function $f(z) = \pi(1 - z)^2$, $0 \le z \le 1$, and $f(z) = 0$ otherwise. However, for $S \in \mathcal{G}(n, k)$, the function $(X_S K)^{1/k}$ is concave on its support when K is a convex body; this is just a restatement of the renowned Brunn–Minkowski inequality (B.10).

Next, we need a generalization of the concept of a Steiner symmetral of a planar convex body.

Definition 2.1.3. Suppose that E is an arbitrary set in $\mathbb{E}^n$, $n \ge 2$, u is a direction, and that l_u is the line through o parallel to u. For each $x \in u^\perp$, let $c(x)$ be defined as follows. If $(l_u + x) \cap E$ is empty or not λ_1-measurable, let $c(x) = \emptyset$. Otherwise, let $c(x)$ be the (possibly degenerate) closed line segment parallel to u, with center x and length $\lambda_1\big(E \cap (l_u + x)\big)$. The union of all the line segments $c(x)$ is called the *Steiner symmetral of E in the direction u* set, and is denoted by $S_u E$.

A more general concept is the following one. Suppose that E is an arbitrary set in $\mathbb{E}^n$, $1 \le k \le n - 1$, and that $S \in \mathcal{G}(n, k)$. For each $x \in S^\perp$, let $B(x)$ be defined as follows. If $(S + x) \cap E$ is empty or not λ_k-measurable, let $B(x) = \emptyset$. Otherwise, let $B(x)$ be the (possibly degenerate) closed k-dimensional ball in $S + x$ with center x and λ_k-measure equal to that of $E \cap (S + x)$. The union of all the balls $B(x)$ is called the *k-symmetral of E parallel to S*, and is denoted by $S_S E$. When $k = n - 1$, $S_S E$ is also called the *Schwarz symmetral* of E.

Of course, the k-symmetral is just the Steiner symmetral if $k = 1$.

Figure B.1, designed to illustrate the Brunn–Minkowski inequality, also serves to depict a Schwarz symmetral in $\mathbb{E}^3$. The intimate connection between the Brunn–Minkowski inequality and the k-symmetral is utilized in the following significant generalization of Theorem 1.2.1.

Theorem 2.1.4. *Let E be a set in $\mathbb{E}^n$, and $1 \le k \le n - 1$. A k-symmetral of E is compact, or measurable, or a convex body, according to whether E is compact, or measurable, or a convex body, respectively.*

Proof. Let $S \in \mathcal{G}(n, k)$. If $x \in S^\perp$, then $S_S E \cap (S + x)$ is a k-dimensional ball. The λ_k-measure of this ball, by definition $X_S E(x)$, is also $\kappa_k r^k$, by (0.8), where r is the radius of the ball, so $r = \kappa_k^{-1/k} X_S E(x)^{1/k}$.

Suppose that E is compact, and let p_m, $m \in \mathbb{N}$, be points in $S_S E$ converging to a point p. Let x_m, $m \in \mathbb{N}$, and x be the projections of p_m, $m \in \mathbb{N}$, and p, respectively, on $S^{\perp}$. Using Lemma 2.1.2, we obtain

$$
\begin{aligned}
\|p - x\| &= \lim_{m \to \infty} \|p_m - x_m\| \\
&\leq \limsup_{m \to \infty} \kappa_k^{-1/k} X_S E(x_m)^{1/k} \\
&\leq \kappa_k^{-1/k} X_S E(x)^{1/k},
\end{aligned}
$$

proving that $p \in S_S E$, so $S_S E$ is compact.

The fact that $S_S E$ is measurable if E is measurable follows from the compact case, since we can approximate any measurable set by compact sets; we omit the details.

Let K be a convex body. Only the convexity of $S_S K$ remains to be established. Let p_1 and p_2 be points in $S_S K$, $0 \leq t \leq 1$, and $p = (1 - t)p_1 + tp_2$. Suppose that x_i, $i = 1, 2$, and x are the projections of p_i, $i = 1, 2$, and p, respectively, on $S^{\perp}$. Then, using the symmetry of $S_S K$ and the Brunn–Minkowski inequality (B.10) with n replaced by k, we have

$$
\begin{aligned}
\|p - x\| &= \|(1 - t)(p_1 - x_1) + t(p_2 - x_2)\| \\
&\leq (1 - t)\|p_1 - x_1\| + t\|p_2 - x_2\| \\
&\leq (1 - t)\kappa_k^{-1/k} X_S K(x_1)^{1/k} + t\kappa_k^{-1/k} X_S K(x_2)^{1/k} \\
&\leq \kappa_k^{-1/k} X_S K(x)^{1/k},
\end{aligned}
$$

from which $p \in S_S K$, as required. ∎

The X-ray $X_u K$ of a convex body determines its Steiner symmetral $S_u K$, and vice versa. For a measurable set E, the relationship is slightly more subtle. If we know $S_u E$, then we also know $X_u E$, but knowledge of $X_u E$ only determines $S_u E$ up to a set of measure zero. Specifically, $S_u E(x)$ is determined for all x such that $X_u E(x) > 0$; when $X_u E(x) = 0$, however, we can have either $S_u E \cap (l_u + x) = \emptyset$ or $\{x\}$. The analogous statements remain true for the k-dimensional X-ray and the k-symmetral.

When working with measurable sets, it makes sense to neglect sets of measure zero, and say that two sets have equal k-dimensional X-rays parallel to S if these agree λ_{n-k}-almost everywhere on $S^{\perp}$. (For convex bodies, this is equivalent to exact equality.) Then $X_u E$ and $S_u E$ may be identified, for our purposes, by the previous paragraph.

We state the following generalization of Lemma 1.2.2, also just a special case of Fubini's theorem, as a lemma.

Lemma 2.1.5. *Let E and E' be bounded measurable sets in $\mathbb{E}^n$ with the same X-rays parallel to a subspace S. Then $\lambda_n(E) = \lambda_n(E')$.*

In particular, a measurable set has the same measure as any of its k-symmetrals.

We now define three different ways in which X-rays can be used to distinguish one set in a class from other sets in the same class. In the definitions, we are identifying X-rays with the corresponding symmetrals, according to our previous remark. The symbol $\simeq$, defined in Section 0.5, denotes equivalence up to a set of measure zero, and for $S \in \mathcal{G}(n, k)$, $X_S E \simeq X_S E'$ means that $X_S E(x) = X_S E'(x)$ for λ_{n-k}-almost all $x \in S^{\perp}$.

Definition 2.1.6. Let $\mathcal{E}$ be a class of bounded measurable sets and $\mathcal{S}$ a fixed set of subspaces in $\mathbb{E}^n$. We say that $E \in \mathcal{E}$ is *determined* by the X-rays parallel to the subspaces in $\mathcal{S}$ if whenever $E' \in \mathcal{E}$ and $X_S E \simeq X_S E'$ for all $S \in \mathcal{S}$, we have $E \simeq E'$.

Definition 2.1.7. Let $\mathcal{E}$ be a class of bounded measurable sets in $\mathbb{E}^n$. We say that $E \in \mathcal{E}$ can be *verified* by the X-rays parallel to the subspaces in a set $\mathcal{S}$ if $\mathcal{S}$ can be chosen (depending on E) such that if $E' \in \mathcal{E}$ and $X_S E \simeq X_S E'$ for all $S \in \mathcal{S}$, then $E \simeq E'$.

Definition 2.1.8. Let $\mathcal{E}$ be a class of bounded measurable sets in $\mathbb{E}^n$. We say that a set $E \in \mathcal{E}$ can be *successively determined* by X-rays parallel to the subspaces S_i, $1 \leq i \leq m$, if these can be chosen inductively, the choice of S_i depending on $X_{S_j} E$, $1 \leq j \leq i - 1$, such that if $E' \in \mathcal{E}$ and for $1 \leq i \leq m$ we have $X_{S_i} E' \simeq X_{S_i} E$, then $E' \simeq E$.

When E and E' are convex bodies, we can replace equality modulo sets of measure zero with exact equality in each of these definitions.

We also say that sets in $\mathcal{E}$ are *determined by m X-rays* if there is a set $\mathcal{S}$ of m 1-dimensional subspaces (or directions) such that each set in $\mathcal{E}$ is determined by the X-rays parallel to the subspaces in $\mathcal{S}$ (or directions). The phrases "$E \in \mathcal{E}$ can be *verified by m X-rays*" and "$E \in \mathcal{E}$ can be *successively determined by m X-rays*," and the corresponding phrases for k-dimensional X-rays, are defined analogously.

When each subspace in $\mathcal{S}$ is 1-dimensional, we are back to ordinary X-rays. Let us confine ourselves to this situation for now. As we saw in Chapter 1, determination means that the directions of the X-rays are fixed in advance, whereas verification is a "checking" procedure in which the directions depend on a given set. The interactive notion of successive determination lies between these two: If the sets in $\mathcal{E}$ can be determined by X-rays in a set of directions, then each such set can be successively determined by the same X-rays; and if the latter is true, then each set in $\mathcal{E}$ can be verified by X-rays in these directions. Unless one is working with a class of sets having some special orientation, there will be no favorable way to choose the first direction for successive determination, and it must therefore be regarded as given rather than chosen. The next theorem, together with Corollary 1.2.12 and Theorem 1.2.23, shows that successive determination is indeed a different concept from determination or verification.

Theorem 2.1.9. *Let $\mathcal{P}$ be the class of parallelograms in $\mathbb{E}^2$. Each $P \in \mathcal{P}$ can be verified by two X-rays, but no $P \in \mathcal{P}$ can be successively determined by two X-rays.*

Proof. Any parallelogram is verified by the two X-rays in the directions parallel to its edges.

Let P be a parallelogram and u_1 any direction not parallel to one of its edges. Suppose that u_2 is a direction not parallel to u_1. Then the reflection $P^r(u_1, u_2)$, defined in the proof of Theorem 1.2.22, is a parallelogram different from P with the same X-rays in the directions u_1 and u_2. ∎

2.2. X-rays of convex bodies in $\mathbb{E}^n$

In this section we return to X-rays of convex bodies and attempt to generalize some of the results of Chapter 1 to higher dimensions. We first note the following corollary to Theorem 1.2.11.

Corollary 2.2.1. *Convex bodies in $\mathbb{E}^n$, $n \geq 2$, are determined by certain sets of four X-rays, taken in directions contained in the same 2-dimensional subspace.*

Proof. Let J be a 2-dimensional subspace, and choose the directions in J according to Corollary 1.2.12. If K is a convex body, the X-rays of K in these directions determine each section of K by a 2-dimensional plane parallel to J, and therefore determine K itself. ∎

The trouble with Corollary 2.2.1, as with Corollary 1.2.12, is that it is in practice impossible to produce a suitable set of four directions. This is because the "bad" sets of four directions (those not providing uniqueness) are, by Theorem 1.2.5, dense in the class of all sets of four directions; a small perturbation may destroy uniqueness. Because of this, it is of interest to consider X-rays of convex bodies in $\mathbb{E}^n$ in noncoplanar sets of directions. We first look at some examples.

Theorem 2.2.2. *There is a set of six directions in $\mathbb{E}^3$ in general position (i.e., with no three coplanar) and two different convex polyhedra with the same X-rays in these directions.*

Proof. Let us begin by constructing two different convex polyhedra with the same X-rays in the coordinate directions. Color the vertices of the unit cube in $\mathbb{E}^3$ alternately black and white. Let T_1 and T_2 be the two tetrahedra formed by taking the convex hull of the black and of the white vertices, respectively. By symmetry, T_1 and T_2 have the required property.

Now consider any convex polyhedron P inscribable in the sphere S^2 whose vertices can be colored alternately black and white. Let P_1 and P_2 be the convex hulls of the black and of the white vertices, respectively. Suppose that ϕ denotes reflection in a plane H containing the origin. If ϕ takes the black vertices to the white ones, and vice versa, then clearly P_1 and P_2 have the same X-rays in the direction orthogonal to the plane H.

The semiregular polyhedron $(4, 6, 10)$ (see Figure 2.1(i)) has the required properties (see [229, pp. 112–13]). Its 180 edges split into 15 classes of mutually parallel edges. The 15 directions of these edges are the directions of the 15 pairs of parallel edges of an icosahedron. From this one can see that there are six directions (for example, those indicated in Figure 2.1(i)) in general position, orthogonal to appropriate planes of reflection. The convex hulls of the black and the white vertices are two polyhedra combinatorially equivalent to (but not equal to) the polyhedron $(3, 3, 3, 3, 5)$; one of these is depicted in Figure 2.1(ii). These polyhedra have the same X-rays in the aforementioned set of six directions. ∎

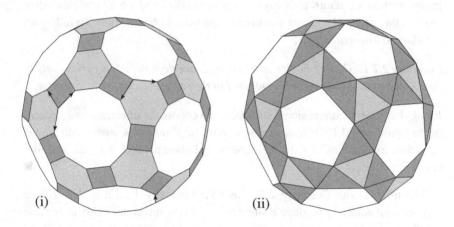

(i)   (ii)

Figure 2.1. The polyhedra of Theorem 2.2.2.

The last theorem provides a result in stark contrast to Corollary 2.2.1.

Corollary 2.2.3. *Convex bodies in $\mathbb{E}^3$ are not determined by X-rays in any set of four noncoplanar directions.*

Proof. Consider any set S of four directions in $\mathbb{E}^3$ in general position. Any set of four of the six directions of Theorem 2.2.2 may be mapped onto the directions in S by a suitable $\phi \in GA_n$ (see Section 0.2). The corresponding images ϕP_1 and ϕP_2 of the polyhedra of that example have the same X-rays in the directions in S. If S' is a set of four directions that are neither coplanar nor in general position, then three of them lie in a plane. Now let us again consider Theorem 2.2.2 (though simpler examples exist). Take three directions parallel to adjacent edges in one of the hexagonal facets of $(4, 6, 10)$, together with another of the six directions considered earlier not parallel to this facet. Then the two polyhedra, combinatorially $(3, 3, 3, 3, 5)$, also have the same X-rays in these directions. Again, a suitable affine transformation maps these four directions onto the directions in $\mathcal{S}'$. ∎

Recall Definition 1.2.7 of an $\mathcal{S}$-regular polygon. We can define an $\mathcal{S}$-regular convex polyhedron in $\mathbb{E}^3$ in exactly the same way, and conclude that $(4, 6, 10)$ is $\mathcal{S}$-regular, where $\mathcal{S}$ is the set of six directions shown in Figure 2.1. The question of which sets of directions in general position are such that X-rays in these directions determine convex bodies is evidently related to that of which sets of directions in general position admit $\mathcal{S}$-regular convex polytopes. See the list of open problems for this chapter; a positive answer to Problem 2.1 would provide a "stable" set of directions for determining convex bodies in $\mathbb{E}^3$ by X-rays.

The next theorem extends Theorem 1.2.17 to $\mathbb{E}^n$, $n \geq 2$.

Theorem 2.2.4. *Most (in the sense of Baire category) convex bodies in $\mathbb{E}^n$, $n \geq 2$, are determined by their X-rays in any pair of nonparallel directions.*

Proof. By applying an affine transformation, if necessary, we may take the two directions to be those of the first and second coordinate axes, and denote them by x_1 and x_2. Let J be a 2-dimensional plane parallel to these directions. Denote by $\mathcal{K}_J$ the set of convex bodies K either not meeting J, or meeting J and such that $K \cap J$ is determined among all 2-dimensional convex bodies in the plane J by X-rays in J in the x_1- and x_2-directions. Then $\mathcal{K}_0^n \setminus \mathcal{K}_J$ consists of those K meeting J for which there is a convex body K' such that $K' \cap J \neq K \cap J$ and $K' \cap J$ has the same X-rays as $K \cap J$ in the x_1- and x_2-directions. A proof very similar to that of Lemma 1.2.16 shows that $\mathcal{K}_0^n \setminus \mathcal{K}_J$ is an F_σ set in $\mathcal{K}_0^n$, and so $\mathcal{K}_J$ is a G_δ set.

We shall show that $\mathcal{K}_J$ is also dense in $\mathcal{K}_0^n$. Let $\mathcal{U}$ be an open set in $\mathcal{K}_0^n$; we must exhibit a $K \in \mathcal{K}_J \cap \mathcal{U}$. If there is a $K \in \mathcal{U}$ not meeting J, this is clearly true. We may suppose, then, that $K \cap J \neq \emptyset$ for all $K \in \mathcal{U}$. Since $\mathcal{U}$ is open in $\mathcal{K}_0^n$, we can select a $K \in \mathcal{U}$ such that $K \cap J$ is of the type considered in Theorem 1.2.17, that is, as in Figure 1.5. Then $K \cap J$ is determined by its X-rays in the x_1- and x_2-directions, so $K \in \mathcal{K}_J$.

Consequently, $\mathcal{K}_J$ is a residual set in $\mathcal{K}_0^n$. Let J_m, $m \in \mathbb{N}$, be 2-dimensional planes parallel to J, whose union is dense in $\mathbb{E}^n$. Then $\cap_m \mathcal{K}_{J_m}$ is also a residual set in $\mathcal{K}_0^n$. Now we are finished, since $K \in \cap_m \mathcal{K}_{J_m}$ if and only if K is determined by its X-rays in the x_1- and x_2-directions. ∎

We have seen in Corollary 2.2.1 that four X-rays suffice for determination of convex bodies in $\mathbb{E}^n$. It follows that they also suffice for successive determination or verification. We now show that for successive determination of convex polyhedra, one can do much better.

Theorem 2.2.5. *A convex polyhedron in* $\mathbb{E}^3$ *can be successively determined by two X-rays.*

Proof. Let P be a convex polyhedron in $\mathbb{E}^3$, and u a direction. If z is any point, let z_u be the projection of z on $u^\perp$. Denote the vertices of P by V, so $V|u^\perp = \{v_u : v \in V\}$. Let W_u be the set of vertices of the Steiner symmetral $S_u P$. Then $W_u|u^\perp$ can be found from the X-ray of P in the direction u. More generally, we can determine from this X-ray the set union E_u of the projections on $u^\perp$ of the edges of $S_u P$.

We would like to know $V|u^\perp$. Observe that although $V|u^\perp \subset W_u|u^\perp$, the inclusion may be strict. In fact, a point x belongs to $(W_u|u^\perp) \setminus (V|u^\perp)$ if and only if it is the intersection of the relative interiors of the projections on $u^\perp$ of two edges of P not contained in the same plane. In this case there is a neighborhood U of x in $u^\perp$ such that

$$U \cap E_u = U \cap (t_1 \cup t_2),$$

where t_1 and t_2 are lines in $u^\perp$ with $t_1 \cap t_2 = \{x\}$.

Let Y_u be the set of all $x \in W_u|u^\perp$ with the properties of the previous sentence. Note that it can happen that $x \in (V|u^\perp) \cap Y_u$. In this case the edges of P meeting at v must project into two straight lines t_1, t_2 in $u^\perp$ with $t_1 \cap t_2 = \{x\}$.

We shall denote by $\mathcal{L}_u$ the finite set of all lines $l_u + x$ for $x \in W_u|u^\perp$.

Let us begin by taking the first X-ray of P in an arbitrary direction u_1. We can then choose a direction u_2 so that for all $x \in u_2^\perp$ the line $l_{u_2} + x$ intersects at most one of the lines in $\mathcal{L}_{u_1}$. Let $W = (\cup \mathcal{L}_{u_1}) \cap (\cup \mathcal{L}_{u_2})$. Then W is a finite set of points containing V.

Let $w \in W$. If w is not a vertex of P, we have $w_{u_j} \in Y_{u_j}$ for $j = 1, 2$. We claim that if $w \in V$, then it is not possible that $w_{u_j} \in Y_{u_j}$ for $j = 1, 2$. To see this, suppose that $w_{u_1} \in Y_{u_1}$. Let t_i, $i = 1, 2$, be the corresponding lines, as before. For $i = 1, 2$, let H_i be the plane parallel to u_1 and containing t_i. Each edge of P meeting w must be contained in H_1 or H_2, so there are exactly four such edges. Let us label these edges in clockwise order around w as e_k, $1 \leq k \leq 4$. Then e_1 and e_3 belong to H_1, say, and e_2 and e_4 belong to H_2. Suppose that $w_{u_2} \in Y_{u_2}$. Let t_i' be the corresponding lines in $u_2^\perp$, $i = 1, 2$, and H_i' the plane parallel to u_2 and containing t_i'. Then H_1' contains two of the edges e_k. If H_1' contains an adjacent pair of these edges, then it contains a facet of P, and the other two edges cannot project into the same line in $u_2^\perp$. Consequently $H_1' = H_1$ or H_2. Applying the same argument to H_2', we see that u_2 must be parallel to u_1, which is impossible. This proves our claim.

We conclude that if $w \in V$, then $w_{u_j} \in Y_{u_j}$ for at most one value of j. Thus we can identify the vertices of P among the points in W, and so (because P is convex) determine P itself. ∎

Very little is known about the determination of convex bodies by k-dimensional X-rays for $k > 1$. New difficulties arise. Suppose, for example, that K is a convex body in $\mathbb{E}^3$, not symmetric about the xy-plane. Let $K' = S_z K$ be the Steiner symmetral of K in the z-direction. For each plane S containing the z-axis, $X_S K = X_S K'$, showing that even an uncountable family of 2-dimensional X-rays may not determine a convex body. (On the other hand, there are uniqueness results concerning the k-plane transform implying that there are infinite families of k-dimensional subspaces such that X-rays parallel to these subspaces determine a convex body; see Section C.1.)

Another difficulty arises in attempts to employ the method of Theorem 2.2.5 to successively determine a convex polytope in $\mathbb{E}^n$ by k-dimensional X-rays with $k > 1$. This method depends on locating the position of the vertices of the polytope from the X-rays, but for $k > 1$ a new phenomenon arises.

Theorem 2.2.6. *There is a convex polyhedron P in $\mathbb{E}^3$ with a vertex v "invisible" to the 2-dimensional X-ray of P parallel to the yz-plane.*

Proof. Let P be the pyramid in $\mathbb{E}^3$ formed by taking the convex hull of the unit square in the xy-plane and the point $v = (1/2, 1/2, 1)$. The X-ray of P parallel to the yz-plane is $f(x) = 2x(1 - x)$. But this is also the X-ray parallel to the yz-plane of a different polyhedron P', namely, the tetrahedron P' with vertices at o, $(0, \sqrt{2}, 0)$, $(1, 0, 0)$, and $(1, 0, \sqrt{2})$. Notice that P' has no vertices between the planes $x = 0$ and $x = 1$. Therefore there is no way to tell, from this X-ray of P alone, that P has a vertex in the plane through v orthogonal to the x-direction. ∎

For k-dimensional X-rays with $k > 1$, only verification of convex polytopes is more or less completely understood; we refer the reader to the open problems and Note 2.1.

2.3. X-rays of bounded measurable sets

The uniqueness theorems obtained in Chapter 1 and the previous section of this chapter for the class of convex bodies raise the question of whether similar theorems might be proved for more general classes of sets. In this section we shall mainly consider the class of all measurable sets; we always assume that these are bounded.

We begin with a useful property of some finite sets.

Definition 2.3.1. Let $\mathcal{S}$ be a finite set of subspaces in $\mathbb{E}^n$. A union $F \cup G$ of disjoint nonempty finite sets F and G is called an $\mathcal{S}$-*switching component* if for each x and each $S \in \mathcal{S}$, $|F \cap (S + x)| = |G \cap (S + x)|$.

As with X-rays, when $\mathcal{S}$ consists of 1-dimensional subspaces, we usually refer to directions parallel to these subspaces, rather than to the subspaces themselves.

To illustrate Definition 2.3.1, let F and G be the two sets of alternate vertices of a regular (or, more generally, affinely regular) $2m$-gon Q in the plane; then $F \cup G$ is an $\mathcal{S}$-switching component, where $\mathcal{S}$ is the set of directions of the edges of Q. In contrast to Corollary 1.2.10, however, $\mathcal{S}$-switching components exist for any finite set of directions.

Lemma 2.3.2. *Let $\mathcal{S}$ be an arbitrary finite set of directions in $\mathbb{E}^n$. There exists an $\mathcal{S}$-switching component.*

Proof. Let $\mathcal{S} = \{u_1, \ldots, u_m\}$ be a set of directions. If $m = 1$, let $F_1 = \{o\}$, v_1 be any vector parallel to u_1, and $G_1 = (F_1 + v_1)$. Then $F_1 \cup G_1$ is a $\{u_1\}$-switching component.

If the lemma is true for $m = k$, let $F_k \cup G_k$ be the corresponding switching component. Suppose that v_{k+1} is a vector parallel to u_{k+1} such that $(F_k \cup G_k) + v_{k+1}$ is disjoint from $F_k \cup G_k$. Let $F_{k+1} = F_k \cup (G_k + v_{k+1})$ and $G_{k+1} = G_k \cup (F_k + v_{k+1})$. Then $F_{k+1} \cup G_{k+1}$ is a switching component for $m = k + 1$. ∎

Note that the switching component constructed in Lemma 2.3.2 is simply a projection on $\mathbb{E}^n$ of the vertices of an m-dimensional parallelepiped.

Theorem 2.3.3. *Measurable (even compact) sets in $\mathbb{E}^n$ are not determined by X-rays in any finite set of directions.*

Proof. Let $\mathcal{S}$ be a set of directions and $F \cup G$ an $\mathcal{S}$-switching component. Let $r > 0$ be small enough that the sets $rB + x$ for $x \in F \cup G$ are all disjoint. Here, as always, B denotes the unit ball. Put $E = \cup\{rB + x : x \in F\}$ and $E' = \cup\{rB + x : x \in G\}$. Then E and E' are disjoint and nonempty compact sets (finite unions of closed balls) with the same X-rays in the directions in $\mathcal{S}$ (see Figure 2.2). ∎

We see immediately that Theorem 1.2.11 does not extend to the class of finite unions of convex bodies. In fact, the idea used in Theorem 2.3.3 yields rather more.

Theorem 2.3.4. *Polygons in $\mathbb{E}^2$ that are star-shaped at the origin are not determined by X-rays in any finite set of directions.*

Proof. Let $\mathcal{S}$ be an arbitrary finite set of directions in the plane. By Lemma 2.3.2, there is an $\mathcal{S}$-switching component $F \cup G$. We may assume that no line through two points of $F \cup G$ contains o (in particular, $o \notin F \cup G$). Let T be a triangle

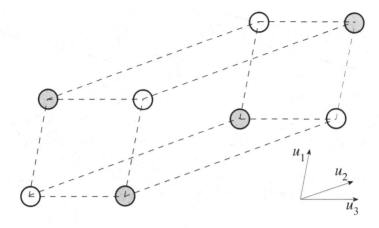

Figure 2.2. Measurable sets with equal X-rays.

containing o such that no side of T is parallel to a line through o and some point of $F \cup G$. Now choose $r > 0$ small enough that the set $rT + (F \cup G)$ has the following properties: (i) If $x \in F \cup G$ and t is a line meeting both the triangles rT and $rT + x$, then t does not meet $rT + x'$ for any $x' \in F \cup G$ with $x' \neq x$, and (ii) if e is an edge of a triangle $rT + x$ for some $x \in F \cup G$, visible from a vertex of the triangle rT, then e is also visible from o. (An edge of a convex polygon is *visible* from an exterior point p if for each point q in the edge, the line segment $[p, q]$ contains no other points of the polygon.) Let

$$C = \cup \{\operatorname{conv} \{rT, rT + x\} : x \in F \cup G\},$$

and

$$P = \operatorname{cl}(C \setminus \cup \{rT + x : x \in F\}),$$

$$Q = \operatorname{cl}(C \setminus \cup \{rT + x : x \in G\}).$$

Then P and Q are the required polygons (see Figure 2.3); they are star-shaped at o and have equal X-rays in the directions in S. ∎

Theorem 2.3.4 shows that Theorem 1.2.11 cannot be extended much beyond the class of convex bodies; see Problem 2.6, however.

When can a convex body be determined among measurable sets by its X-rays? Here we are no longer granted a priori the information that the set is convex. Another question then arises: Might we be able to deduce that a measurable set is convex if we know that its Steiner symmetrals are convex? This is so, if all the Steiner symmetrals are convex (see Note 2.3), but not if this is only true for a finite number of them, as the next theorem demonstrates. The question of determining

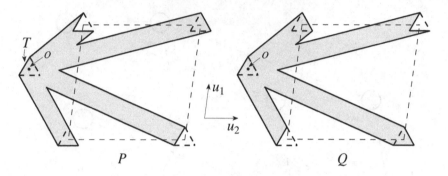

Figure 2.3. Star-shaped polygons with equal X-rays.

convex bodies among measurable sets is therefore essentially different from that considered in Chapter 1.

Theorem 2.3.5. *For any finite set of directions in the plane, there is a nonconvex measurable set whose Steiner symmetrals in those directions are convex polygons.*

Proof. Let $S = \{u_i : 1 \leq i \leq m\}$ be a set of at least two directions. Suppose that s is a closed line segment not parallel to any direction in S, and for each i, let P_i be the closed infinite strip whose bounding lines are parallel to u_i and each contain one endpoint of s. Construct a convex polygon Q with $4m$ sides, containing s in its interior, such that for each i there are two sides of Q with the same orthogonal projection on $u_i^{\perp}$ as s, and each of the remaining $2m$ sides is not contained in any P_i. Find a triangle T contained in P_i for each i (and therefore in int Q), one of whose sides is s, sufficiently thin to have the following property: For each i, there is a triangle $T_i \subset P_i$ such that T_i and T have equal X-rays in the direction u_i, int $T_i \cap$ int $Q = \emptyset$, and $Q \cup \cup_i T_i$ is a convex polygon. Then the set $E = (Q \cup \cup_i T_i \setminus \text{int } T)$ (illustrated for $m = 4$ in Figure 2.4) is a nonconvex set whose Steiner symmetrals in the directions in S are convex polygons. ∎

It would not be difficult to extend the previous theorem to $\mathbb{E}^n$, $n \geq 2$.

Lemma 2.3.6. *Let S be a finite set of directions in $\mathbb{E}^2$. Suppose that K is a planar convex body and that Q is a convex polygon inscribed in K and with each of its edges parallel to a direction in S. Then Q is contained, up to a set of measure zero, in any measurable set E with the same X-rays as K in the directions in S.*

Proof. The proof is the same as that of Lemma 1.2.18, with E replacing the convex set K' of that lemma. (The convexity is not needed for the slightly weaker conclusion.) ∎

Recall Definition 1.2.19 of S-inscribable convex bodies.

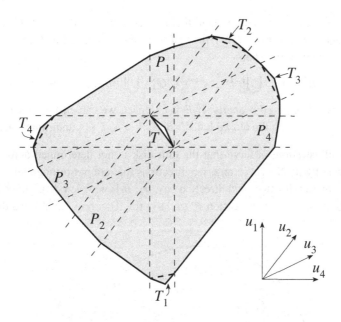

Figure 2.4. A nonconvex set with some convex Steiner symmetrals.

Theorem 2.3.7. *Let S be a finite set of directions in $\mathbb{E}^2$. An S-inscribable convex body is determined among measurable sets by its X-rays in the directions in S.*

Proof. Let K be S-inscribable. Apart from a set of measure zero, K is contained in the union of inscribed polygons Q_n, $n \in \mathbb{N}$, each having every edge parallel to a direction in S. If E is a set with the same X-rays as K in the directions in S, then by Lemma 2.3.6, $Q_n \subset E$, up to a set of measure zero, for each $n \in \mathbb{N}$. Therefore $K \subset E$, up to a set of measure zero, and the fact (Lemma 2.1.5) that K and E have the same measure completes the proof. ∎

The last theorem provides an appealing geometric condition for uniqueness among measurable sets. It is restricted to the plane, however, and we wish to obtain higher-dimensional results. The next trivial theorem provides the only known simple way of concluding that a convex body is *not* determined among measurable sets by its X-rays.

Theorem 2.3.8. *Let S be a finite set of subspaces of $\mathbb{E}^n$, of dimensions between 1 and $n - 1$. Suppose that E is a measurable set such that there is an S-switching component $F \cup G$ with $F \subset \operatorname{int} E$ and $G \subset \operatorname{int}(\operatorname{co} E)$. Then there is a measurable set $E' \not\simeq E$ with the same X-rays as E parallel to the subspaces in S. If E is compact, we can ensure E' is also compact.*

Proof. Choose $r > 0$ small enough that the sets $rB + x$ for $x \in F \cup G$ are all disjoint and if $x \in F$ (or $x \in G$), the ball $rB + x$ is contained in E (or contained in co E, respectively). Then

$$E' = E \cup \bigcup \{rB + x : x \in G\} \setminus \bigcup \{rB + x : x \in F\}$$

clearly has the required property. If E is compact, we can ensure that E' is also compact by replacing the closed balls $\{rB + x : x \in F\}$ by their interiors.   ∎

As an illustration, observe that the unit disk is not determined among measurable sets by its X-rays in any two directions that are not orthogonal (see Figure 2.5). (In fact, for two such directions u_1, u_2, B is not even determined among convex bodies, since its reflection $B^r(u_1, u_2)$ (cf. Theorem 1.2.22) is a different ellipse.)

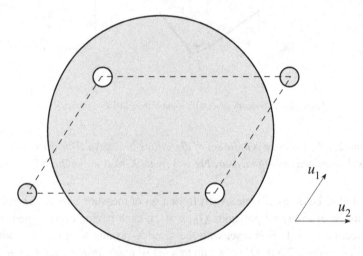

Figure 2.5. A disk is not determined by two nonorthogonal X-rays.

Definition 2.3.9. If S is a subspace of $\mathbb{E}^n$, $n \geq 2$, a *ridge function orthogonal to* S is a function constant on each translate of S.

Let $\mathcal{S} = \{S_i : 1 \leq i \leq m\}$ be a set of subspaces of $\mathbb{E}^n$, $n \geq 2$, of dimensions between 1 and $n - 1$. A measurable set E in $\mathbb{E}^n$ is called $\mathcal{S}$-*additive* if

$$E \simeq \left\{ x \in \mathbb{E}^n : \sum_i f_i(x) > 0 \right\},$$

where $>$ denotes either $\geq$ or $>$ and f_i is a measurable ridge function orthogonal to S_i and bounded on compact sets.

In short, additive sets are (up to a set of measure zero) the positive or nonnegative sets for finite sums of ridge functions of the required type. A simple example is the unit disk, additive with respect to the coordinate directions by virtue of its representation

$$B = \{(x, y) \in \mathbb{E}^2 : \left(\frac{1}{2} - x^2\right) + \left(\frac{1}{2} - y^2\right) \geq 0\}.$$

Lemma 2.3.10. *If E is an S-additive set in $\mathbb{E}^n$, and $\phi \in GA_n$, then ϕE is S'-additive, where $S' = \{\phi S - \phi o : S \in S\}$.*

Proof. Suppose that E is S-additive, that is, $E \simeq \{x \in \mathbb{E}^n : \sum_i f_i(x) > 0\}$, in the notation of Definition 2.3.9. Let $\phi \in GA_n$. Then

$$\phi E \simeq \left\{\phi x : \sum_i f_i(x) > 0\right\}$$

$$= \left\{x : \sum_i f_i(\phi^{-1}x) > 0\right\}$$

$$= \left\{x : \sum_i g_i(x) > 0\right\},$$

where $g_i(x) = f_i(\phi^{-1}x)$ for all x. Now g_i is constant on each translate of ϕS_i, proving the lemma. ∎

The importance of additive sets stems from the next theorem.

Theorem 2.3.11. *An S-additive set is determined among measurable sets by its X-rays parallel to the subspaces in S.*

Proof. Suppose that $E \simeq \{x \in \mathbb{E}^n : \sum_i f_i(x) > 0\}$ is S-additive, as in Definition 2.3.9. Let E' be measurable, with the same X-rays as E parallel to the subspaces in S. For each i,

$$\int_E f_i(x)\, dx = \int_{\mathbb{E}^n} f_i(x) 1_E(x)\, dx$$

$$= \int_{S_i^\perp} f_i(x_i) X_{S_i} E(x_i)\, dx_i$$

$$= \int_{S_i^\perp} f_i(x_i) X_{S_i} E'(x_i)\, dx_i = \int_{E'} f_i(x)\, dx,$$

where we have used Fubini's theorem and the fact that f_i is a ridge function orthogonal to S_i. Let us now subtract the common part $E \cap E'$ from the range of the first and last integrals, and sum over i. We obtain

$$\int_{E \backslash E'} \sum_i f_i(x)\, dx = \int_{E' \backslash E} \sum_i f_i(x)\, dx.$$

The S-additivity of E implies either that the left side is nonnegative and the right side is negative, or that the left side is positive and the right nonpositive, unless $\lambda_n(E' \setminus E) = 0$ or $\lambda_n(E \setminus E') = 0$, respectively. In either case we have $E' \simeq E$, since $\lambda_n(E') = \lambda_n(E)$, by Lemma 2.1.5. ∎

With Theorem 2.3.11 in hand we have two ways of showing that a planar convex body is determined among measurable sets by its X-rays, that is, by demonstrating that it is inscribable or additive. The next lemma and its corollary indicate the power of the latter approach.

Lemma 2.3.12. *Let E be a body in $\mathbb{E}^2$ symmetric with respect to the y-axis and such that each horizontal line that meets it does so in a (possibly degenerate) line segment. Then E is S-additive when S is the set consisting of the x-direction together with any other two nonparallel directions.*

Proof. Our assumptions imply that there is a bounded Borel function g such that $E = \{(x, y) \in \mathbb{E}^2 : x^2 \le g(y)^2\}$. Then E is clearly additive with respect to the two coordinate directions. Let $y + a_i x = 0$, $a_i \ne 0$, $i = 1, 2$, $a_1 \ne a_2$, be lines through the origin. Now

$$g(y)^2 - x^2 = \left(g(y)^2 - \frac{1}{a_1 a_2} y^2 \right) + \frac{1}{a_1(a_2 - a_1)}(y + a_1 x)^2$$
$$+ \frac{1}{a_2(a_1 - a_2)}(y + a_2 x)^2.$$

The right-hand side is a sum of ridge functions of the appropriate type, orthogonal to the directions parallel to the x-axis, $y + a_1 x = 0$, and $y + a_2 x = 0$, as required. ∎

The preceding assumption, that E meets each horizontal line in a line segment, cannot be removed, as the annulus E of Figure 2.6 demonstrates.

Corollary 2.3.13. *Let S be a set of three mutually nonparallel directions. An ellipse in $\mathbb{E}^2$ is S-additive and therefore determined among measurable sets by its X-rays in the directions in S.*

Proof. The unit disk is S-additive, by Lemma 2.3.12. The corollary now follows from Lemma 2.3.10 and Theorem 2.3.11. ∎

The next goal is to give a complete classification of those planar convex bodies determined among measurable sets by their X-rays, in the special case when S consists of the two coordinate directions.

Lemma 2.3.14. *Let S be the set of coordinate directions in $\mathbb{E}^2$. A convex body K is S-inscribable if and only if it has the following property: Each $z \in \operatorname{bd} K$ is the vertex of a box $R(z)$ inscribed in K. (Here $R(z)$ may be degenerate if it contains*

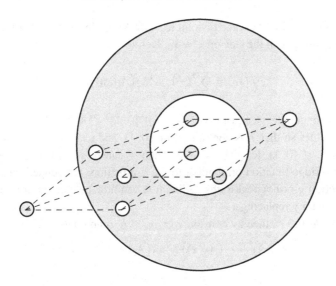

Figure 2.6. An annulus not determined by three X-rays.

the intersection with K of each line through its vertices parallel to a coordinate direction.)

Proof. Suppose that the property in the statement of the theorem does not hold. The set V of all points in bd K that are vertices of (possibly degenerate) boxes inscribed in K is closed. If $z \in$ bd $K \setminus V$, there is an $r > 0$ such that $(rB + z)$ does not meet any such box. Now if $z' \in$ int $K \cap (rB + z)$, then z' cannot be in the interior of a box inscribed in K, and so K is not S-inscribable.

Conversely, suppose that the property in the theorem holds. For any $z \in$ bd K, denote by $R(z)$ a corresponding (possibly degenerate) box inscribed in K with z as a vertex. Let $z' \in$ int K, and let k_0 be the chord of K containing z' and parallel to the x-axis. Suppose that z_0 is an endpoint of k_0. Let $z \in$ bd K be such that if k is the chord of K with endpoint at z and parallel to the x-axis, then $z' \in$ int (conv $\{k, R(z_0)\}$). Then if z is sufficiently close to z_0, we also have $z' \in$ int $R(z)$, as required. ∎

Lemma 2.3.15. *Let S be the set of coordinate directions in $\mathbb{E}^2$. An S-inscribable convex body is S-additive.*

Proof. By the easily established affine invariance of inscribability, and that of additivity (Lemma 2.3.10), we may assume that K is inscribed in the unit square.

For x_0 (or y_0) in $[0, 1]$, let $R(x_0)$ (or $R(y_0)$, respectively) denote a (possibly degenerate) box inscribed in K with one side on the line $x = x_0$ (or $y = y_0$, respectively); such boxes exist, by Lemma 2.3.14. Suppose that $R(x_0)$ is unique. Let $A_i(x_0)$, $1 \le i \le 4$, be the (possibly degenerate) convex subsets of K bounded by bd K and a side of $R(x_0)$ and disjoint from int $R(x_0)$, in clockwise order around $R(x_0)$, beginning with the side on $x = x_0$. Define

$$f(x_0) = \sum_{i=1}^{4}(-1)^{i+1}\lambda_2\big(A_i(x_0)\big).$$

If $R(x_0)$ is not unique, we must have $\{x = x_0\}\cap[0,1]^2 \subset K$. We define $f(x_0) > 0$ for these values so that f is continuous on $(0,1)$, and extend to $[0,1]$ by continuity. For $x \notin [0,1]$, let $f(x) = -2$. Then f can be regarded as a bounded measurable ridge function orthogonal to the y-direction. A bounded measurable ridge function g orthogonal to the x-direction can now be defined in exactly the same way, with y replacing x.

If $(x, y) \in [0,1]^2$, then by convexity $(x, y) \in K$ if and only if

$$\lambda_2\big(A_1(x)\big) + \lambda_2\big(A_3(x)\big) \geq \lambda_2\big(A_2(y)\big) + \lambda_2\big(A_4(y)\big)$$

and

$$\lambda_2\big(A_1(y)\big) + \lambda_2\big(A_3(y)\big) \geq \lambda_2\big(A_2(x)\big) + \lambda_2\big(A_4(x)\big).$$

From this it follows that $K = \{(x, y) \in \mathbb{E}^2 : f(x) + g(y) \geq 0\}$, so K is S-additive. ∎

The ridge functions produced by the method of Lemma 2.3.15 have a clear geometrical interpretation. The value of their sum at a point (x, y) can be seen from Figure 2.7. If $(x, y) \in K$, it is the area of a strip of paper, folded where necessary, covering the region between the inscribed boxes $R(x)$ and $R(y)$ (the doubly shaded parts are counted twice). If $(x, y) \in [0, 1]^2 \setminus K$, it is minus this value.

Lemma 2.3.16. *Let S be the set of coordinate directions in $\mathbb{E}^2$. If a convex body K does not admit an S-switching component $F \cup G$ with $F \subset$ int K and $G \subset$ co K, then K is S-inscribable.*

Proof. Suppose that K is not S-inscribable. By Lemma 2.3.14, there is a $z_1 \in$ bd K that is not the vertex of a (possibly degenerate) box inscribed in K. Without loss of generality we may assume that the chord k_1 of K with one endpoint at z_1 and parallel to the y-axis is nondegenerate; let z_2 be its other endpoint. Again, we may assume that the chord k_2 with one endpoint at z_2 and parallel to the x-axis is nondegenerate; let its other endpoint be z_3. Denote by z_4 the point such that $\{z_i : 1 \leq i \leq 4\}$ form the vertices of a box R. By our assumption, R is not inscribed in K, so $z_4 \notin$ bd K.

Suppose that $z_4 \in$ co K. Let z_2', z_4' be points on the line through z_2 and z_4, with $z_2' \notin K$ and $z_4' \in$ int $R \setminus K$, and let z_1', z_3' denote the other vertices of the box whose diagonal is $[z_2', z_4']$. If z_i' is sufficiently close to z_i, $i = 2, 4$, then z_1' and z_3'

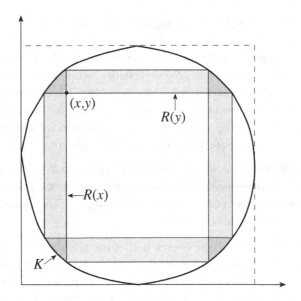

Figure 2.7. Illustration for Lemma 2.3.15.

belong to the interior of K. We can now take $F = \{z_1', z_3'\}$ and $G = \{z_2', z_4'\}$. The case when $z_4 \in \operatorname{int} K$ can be treated similarly. ∎

Theorems 2.3.8 and 2.3.11 and Lemmas 2.3.15 and 2.3.16 together yield the following satisfying theorem.

Theorem 2.3.17. *Let K be a planar convex body and $\mathcal{S}$ the set of coordinate directions in $\mathbb{E}^2$. The following statements are equivalent:*

(i) *K is $\mathcal{S}$-inscribable.*

(ii) *K is $\mathcal{S}$-additive.*

(iii) *K is determined among measurable sets by its X-rays in the directions in $\mathcal{S}$.*

(iv) *K does not admit an $\mathcal{S}$-switching component $F \cup G$ with $F \subset \operatorname{int} K$ and $G \subset \operatorname{co} K$.*

Remark 2.3.18. (i) Theorem 2.3.17 actually remains valid for measurable sets when the definitions of $\mathcal{S}$-inscribability and $\mathcal{S}$-switching component are suitably modified.

(ii) By Theorems 2.3.8 and 2.3.11, the implications (ii)⇒(iii)⇒(iv) remain true for general finite sets $\mathcal{S}$ of directions. Also, the proof of the implication (i)⇒(ii) in Lemma 2.3.15 can be refined to show that it holds for any finite set $\mathcal{S}$ of directions. But unfortunately it is not true that for general finite sets $\mathcal{S}$ of directions, an $\mathcal{S}$-additive convex body is $\mathcal{S}$-inscribable. This follows from Corollary 2.3.13, on observing that the unit disk is not $\mathcal{S}$-inscribable when $\mathcal{S}$

comprises the three directions with angles 0, $\pi/6$, and $5\pi/6$ from the positive x-axis.

We shall now investigate the extent to which Theorem 2.3.17 remains true in higher dimensions. The concept of S-inscribability seems to be of limited interest in $\mathbb{E}^n$, $n \geq 3$, and we shall ignore it here (see Note 2.3, however). By Theorems 2.3.8 and 2.3.11, the implications (ii)$\Rightarrow$(iii)$\Rightarrow$(iv) of Theorem 2.3.17 are still valid, but the following two examples show that even for 2-dimensional X-rays parallel to the coordinate planes in $\mathbb{E}^3$ neither implication can be reversed.

Theorem 2.3.19. *Let S be the set of coordinate planes in $\mathbb{E}^3$. There is a convex polyhedron P not determined among measurable sets by its 2-dimensional X-rays parallel to the planes in S and also not admitting an S-switching component $F \cup G$ with $F \subset \operatorname{int} P$ and $G \subset \operatorname{co} P$.*

Proof. Let P be a pyramid, the convex hull of the unit square in the xy-plane and the point $(1, 1, 1)$. We shall omit the proof that P is not determined among measurable sets by its X-rays parallel to the coordinate planes; see Note 2.3. (It is a simple exercise to show that P is determined among convex bodies.) If $F \cup G$ is an S-switching component as in the statement of the theorem, then G must be contained in the unit cube. Let $a = (a_1, a_2, a_3)$ be the point in F with the smallest z-coordinate. Then there is a $b = (b_1, b_2, b_3)$ in G with $b_3 = a_3$. Since $b \notin P$, either $b_3 > b_1$ or $b_3 > b_2$. Without loss of generality, we shall assume the former. Then there is a point $a' = (a'_1, a'_2, a'_3)$ in F with $a'_1 = b_1$. But $a' \in P$, so $a'_3 \leq a'_1$ and hence $a'_3 < a_3$, a contradiction to the choice of a. ∎

The second aforementioned example is more complicated to describe. We shall need some auxiliary results. The first of these is a stronger form of Theorem 2.3.11 depending on an extension of the class of sums of ridge functions to a larger class.

Definition 2.3.20. Let $S = \{S_i : 1 \leq i \leq m\}$ be a set of subspaces in $\mathbb{E}^n$, $n \geq 2$, of dimensions between 1 and $n - 1$. Let $\mathcal{F}(S)$ be the set of all functions f on $\mathbb{E}^n$, possibly taking values $+\infty$ and $-\infty$, with the following property: There is a sequence of functions $f^{(k)} = \sum_{i=1}^{m} f_i^{(k)}$ on $\mathbb{E}^n$, where each $f_i^{(k)}$ is a ridge function orthogonal to S_i and *integrable* on compact subsets of $S_i^\perp$, such that

$$\lim_{k \to \infty} \int_{\mathbb{E}^n} f^{(k)} g \, dx = \int_{\mathbb{E}^n} fg \, dx, \qquad (2.1)$$

for all bounded measurable g such that the right-hand side is well defined, that is, at least one of the functions $(fg)^-$ or $(fg)^+$ is integrable. (Note that we do not require f to be integrable.)

Theorem 2.3.21. *Let $S = \{S_i : 1 \leq i \leq m\}$ be a set of subspaces in $\mathbb{E}^n$ of dimensions between 1 and $n-1$. Suppose that $E \subset \mathbb{E}^n$ is such that $E \simeq \{x : f(x) > 0\}$ for some $f \in \mathcal{F}(S)$. (Here, as earlier, $>$ is either $\geq$ or $>$.) Then E is determined among measurable sets by its X-rays parallel to the subspaces in S.*

Proof. Let $\{f^{(k)}\}$ be a sequence of functions converging to f in the sense of equation (2.1), so that $f^{(k)} = \sum_{i=1}^{m} f_i^{(k)}$, where each $f_i^{(k)}$ is a ridge function orthogonal to S_i and integrable on compact subsets of $S_i^{\perp}$. Suppose that E' has the same X-rays as E parallel to the subspaces in S. Then, as in the proof of Theorem 2.3.11, we obtain

$$\int_E f^{(k)}(x) \, dx = \int_{E'} f^{(k)}(x) \, dx,$$

for each k, and so

$$\int_{\mathbb{E}^n} f^{(k)}(x) 1_{E \backslash E'}(x) \, dx = \int_{\mathbb{E}^n} f^{(k)}(x) 1_{E' \backslash E}(x) \, dx.$$

Taking limits, we see that

$$\int_{\mathbb{E}^n} f(x) 1_{E \backslash E'}(x) \, dx = \int_{\mathbb{E}^n} f(x) 1_{E' \backslash E}(x) \, dx.$$

The conclusion that $E' \simeq E$ now follows exactly as in the last paragraph of the proof of Theorem 2.3.11. ∎

Remark 2.3.22. The hypotheses of the previous theorem may seem complicated, but they merit attention. Remarkably, the converse of Theorem 2.3.21 holds, under our blanket assumption that sets such as E are bounded; see Note 2.6.

Lemma 2.3.23. *Let $S = \{S_i : 1 \le i \le m\}$ be a set of subspaces in $\mathbb{E}^n$ of dimensions between 1 and $n - 1$. Suppose that E is a set of the form*

$$E \simeq \{x : f(x) > 0, \text{ or } f(x) = 0 \text{ and } g(x) > 0\},$$

where $f = \sum_i f_i$, $g = \sum_i g_i$, and for each i, f_i and g_i are measurable ridge functions orthogonal to S_i and bounded on compact sets. Then E is determined among measurable sets by its X-rays parallel to the subspaces in S.

Proof. For $k \in \mathbb{N}$, let

$$f^{(k)}(x) = k^3 f(x) + k^2 g(x) - k,$$

for all $x \in \mathbb{E}^n$. Each member of this sequence of functions is a sum of measurable ridge functions orthogonal to the subspaces S_i and bounded on compact sets. The function $f_0(x) = \lim_k f^{(k)}(x)$ is $+\infty$ almost everywhere on E, since the k^3 term dominates if $f(x) > 0$ and the k^2 term dominates if $f(x) = 0$ and $g(x) > 0$. Similarly, $f_0(x) = -\infty$ almost everywhere on co E. Suppose that g_0 is a bounded measurable function such that $(f_0 g_0)^+$ is integrable. Since $f_0 g_0$ is $+\infty$ almost everywhere it is positive, $f_0 g_0 \le 0$ almost everywhere, so $g_0 \le 0$ almost everywhere on E and $g_0 \ge 0$ almost everywhere on co E. It follows that either (i) both g_0 and $f_0 g_0$ are zero almost everywhere or (ii) $\int_{\mathbb{E}^n} f_0 g_0 \, dx = -\infty$.

We claim that equation (2.1) is satisfied when f and g are replaced by f_0 and g_0, respectively. This is clear if (i) holds. If (ii) holds, then g_0 must either be negative on a nonnull subset of E or positive on a nonnull subset of co E. It is now easy to check that

$$\int_{\mathbb{E}^n} f^{(k)} g_0 \, dx = k^3 \int_{\mathbb{E}^n} f g_0 \, dx + k^2 \int_{\mathbb{E}^n} g g_0 \, dx - k \int_{\mathbb{E}^n} g_0 \, dx \to -\infty$$

as $k \to \infty$. (One has to consider the various cases. For example, suppose that g_0 is negative on a nonnull subset of the portion $\{x : f(x) > 0\}$ of E. Then the first integral on the right-hand side dominates.) This proves the claim.

Similar considerations apply to bounded measurable functions g_0 such that $(f_0 g_0)^-$ is integrable. We conclude that $f_0 \in \mathcal{F}(\mathcal{S})$. Theorem 2.3.21 now completes the proof of the lemma. ∎

Theorem 2.3.24. *Let $\mathcal{S}$ be the set of coordinate planes in $\mathbb{E}^3$. There is a nonconvex polyhedron P that is determined by its 2-dimensional X-rays parallel to the planes in $\mathcal{S}$, but is not $\mathcal{S}$-additive.*

Proof. Let $K = [0, 2) \times [0, 2) \times [0, 1)$. If $x \in \mathbb{E}^3$, let $x = (x_1, x_2, x_3)$. For $i = 1, 2$, let $f_i(x_i) = 0$ for $0 \le x_i < 1$ and $f_i(x_i) = 3/2 - x_i$ for $1 \le x_i < 2$. Set $f(x) = f_1(x_1) + f_2(x_2)$ and $g(x) = x_1 + x_2 + x_3 - 2$. Let $P_0 = P_1 \cup P_2$, where $P_1 = \{x \in K : f(x) > 0\}$ and $P_2 = \{x \in K : f(x) = 0 \text{ and } g(x) > 0\}$. The set P_1 is the union of $[1, \frac{3}{2}) \times [0, 1) \times [0, 1)$, $[0, 1) \times [1, \frac{3}{2}) \times [0, 1)$, and $T \times [0, 1)$, where T is the triangle with vertices $(1, 1)$, $(2, 1)$, and $(1, 2)$ in the xy-plane; cl P_2 is a tetrahedron contained in the unit cube (see Figure 2.8). By Lemma 2.3.23, P_0 is determined among measurable sets by its 2-dimensional coordinate X-rays, so $P = \text{cl } P_0$ is also.

We shall now prove that P is not $\mathcal{S}$-additive. It will suffice to show this for P_0, because $\lambda_3(\text{bd } P) = 0$. Suppose that P_0 is $\mathcal{S}$-additive, so there are bounded measurable functions h_i on $[0, 2)$, $i = 1, 2$, and h_3 on $[0, 1)$, and a set N with $\lambda_3(N) = 0$, such that for $x \notin N$, $x \in P_0$ if and only if $\sum_i h_i(x_i) \ge 0$. (The case where $\ge$ is replaced by $>$ is handled similarly.) Then for $x \in [0, 1]^3 \setminus N$, we have

$$\sum_i h_i(x_i) \ge 0 \Leftrightarrow \sum_i x_i > 2. \tag{2.2}$$

Let $v = (1, 1, 0)$. Then $x \in P_0 \cap ([0, 1]^3 + v)$ if and only if $(x_1 - 1) + (x_2 - 1) < 1$. This implies that if $N' = N - v$ and $x \in [0, 1]^3 \setminus N'$, then

$$h_1(x_1 + 1) + h_2(x_2 + 1) + h_3(x_3) \ge 0 \Leftrightarrow x_1 + x_2 < 1. \tag{2.3}$$

Let $a = \text{ess inf } h_3(x_3)$ and $b = \text{ess sup } h_3(x_3)$. Suppose that $a = b$, that is, h_3 is constant almost everywhere on $[0, 1)$. Since $\lambda_3(N) = 0$, there is, by Fubini's theorem, a subset Z of $[0, 1)$ with $\lambda_1(Z) = 0$ such that for each $x_3 \in [0, 1) \setminus Z$, $(x_1, x_2, x_3) \notin N$ for almost all $(x_1, x_2) \in [0, 1)^2$. Choose x_3' and x_3'' in $[0, 1) \setminus Z$

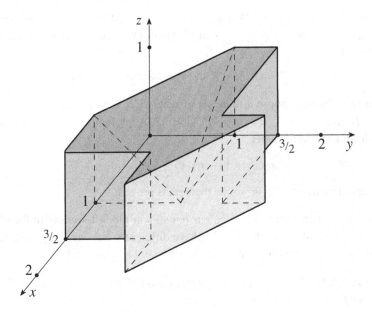

Figure 2.8. The polyhedron P of Theorem 2.3.24.

such that $h_3(x_3') = h_3(x_3'') = a$ and $0 < x_3' < x_3''$. Then we can also choose x_1 and x_2 with $(x_1, x_2) \in [0, 1]^2$ and $2 - x_3'' < x_1 + x_2 < 2 - x_3'$. Now, by (2.2), we have

$$x_1 + x_2 + x_3' < 2 \Rightarrow h_1(x_1) + h_2(x_2) + a < 0$$

and

$$x_1 + x_2 + x_3'' > 2 \Rightarrow h_1(x_1) + h_2(x_2) + a \geq 0,$$

which contradict each other. Therefore $a < b$.

Because $\lambda_3(N') = 0$, there is, by Fubini's theorem, a subset D of $[0, 1)^2$ with $\lambda_2(D) = 0$ such that if $(x_1, x_2) \notin D$ then $(x_1, x_2, x_3) \notin N'$ for almost all $x_3 \in [0, 1)$. For each such (x_1, x_2, x_3) for which, in addition, $x_1 + x_2 < 1$ (or ≥ 1), we have $h_1(x_1 + 1) + h_2(x_2 + 1) + h_3(x_3) \geq 0$ (or < 0, respectively), by (2.3). Taking the infimum (or supremum, respectively) over x_3, we conclude that for $(x_1, x_2) \notin D$,

$$h_1(x_1 + 1) + h_2(x_2 + 1) + a \geq 0 \text{ or } h_1(x_1 + 1) + h_2(x_2 + 1) + b \leq 0, \quad (2.4)$$

when $x_1 + x_2 < 1$ or $x_1 + x_2 \geq 1$, respectively.

Set $c = b - a > 0$. By Fubini's theorem again, there is a $Z' \subset [0, 1)$ such that $\lambda_1(Z') = 0$ and if $x_2 \in H = [0, 1) \setminus Z'$, then $(x_1, x_2) \in [0, 1)^2 \setminus D$ for almost all $x_1 \in [0, 1)$. Suppose that x_2' and x_2'' belong to H and that $x_2' < x_2''$. Then there is an $x_1 \in [0, 1)$ such that (x_1, x_2') and (x_1, x_2'') do not belong to D

and $1 - x_2'' < x_1 < 1 - x_2'$. Now $x_1 + x_2' < 1$ and $x_1 + x_2'' > 1$, so (2.4) implies that $h_1(x_1 + 1) + h_2(x_2' + 1) \geq -a$ and $h_1(x_1 + 1) + h_2(x_2'' + 1) \leq -b$. Adding, we see that

$$h_2(x_2'' + 1) - h_2(x_2' + 1) \leq -b + a = -c < 0.$$

Let $m \in \mathbb{N}$, and choose a finite set of reals $x_2^{(j)}$, $1 \leq j \leq m - 1$, such that $x_2^{(j)} \in H$ and $x_2' < x_2^{(1)} < \cdots < x_2^{(m-1)} < x_2''$. Applying the previous argument to each adjacent pair, we get

$$h_2(x_2'' + 1) - h_2(x_2' + 1) \leq -mc.$$

This cannot be true for all m, so the theorem is proved. ∎

The last two examples reinforce our remarks in the last section, to the effect that k-dimensional X-rays for $k > 1$ behave differently from ordinary X-rays.

We now turn to the question of verification (cf. Definition 2.1.7).

Theorem 2.3.25. *For $n \geq 2$, there is a measurable (even compact) set in $\mathbb{E}^n$ that is not verifiable by any finite set of X-rays.*

Proof. The theorem is based on the following observation. Let S be a finite set of directions and let $F \cup G$ be the S-switching component provided by Lemma 2.3.2. Suppose that E is a finite union of disjoint closed balls of equal radius centered at points in F, as in Theorem 2.3.3, so that $F \subset \operatorname{int} E$ and $G \subset \operatorname{co} E$. If we make a sufficiently small perturbation to the directions in S to obtain a new set of directions S', there is clearly an S'-switching component $F' \cup G'$ with $F' \subset \operatorname{int} E$ and $G' \subset \operatorname{co} E$. Then by Theorem 2.3.8 there is a different measurable set E' with the same X-rays as E in the directions in S'. Note that any image of E under a homothety will have the same property as E.

Let us identify sets of m (not necessarily distinct) directions with points in the compact metric space $(S^1)^m$ with the product topology. Given an $S \in (S^1)^m$, we can apply the previous observation to conclude that there is a measurable set E and a neighborhood U of S in $(S^1)^m$ such that for each $S' \in U$, there is a different measurable set E' with the same X-rays as E in the directions in S'. (Note that the inductive construction of the switching component in Lemma 2.3.2 can still be applied when the directions are not distinct.) Furthermore, there is an $r > 0$ such that E and each E' for $S' \in U$ are contained in the ball rB.

By compactness, there is a finite set $S_1, \ldots, S_k$ of points in $(S^1)^m$ such that the associated neighborhoods $U_1, \ldots, U_k$ cover $(S^1)^m$. For $1 \leq i \leq k$, let E_i be the measurable set, and $r_i B$ the ball, corresponding to S_i. For each i, let v_i be a vector such that the balls $r_i B + v_i$, $1 \leq i \leq k$, are disjoint. Set $C_m = \cup_{i=1}^k (E_i + v_i)$. Then C_m has the property that for any set S of m directions, there is a different measurable set $C_m'(S)$ with the same X-rays as C_m in these directions. Moreover, there is a ball $r_m' B$ containing C_m and each such $C_m'(S)$.

Choose positive numbers s_m decreasing to 0, and vectors w_m, such that the balls $s_m r'_m B + w_m$, $m \in \mathbb{N}$, are disjoint and converge to a single point. Let $C = \mathrm{cl}\left(\cup_m (s_m C_m + w_m)\right)$. For any finite set of directions, there is a measurable set different from C with the same X-rays in these directions, meaning that C cannot be verified by a finite set of X-rays. Finally, note that C is compact, and the construction guarantees that for each finite set of directions, there is a different compact set with the same X-rays. ∎

The following result utilizes the notion of inscribability to obtain a verification theorem for convex polygons.

Theorem 2.3.26. *Every convex polygon in* $\mathbb{E}^2$ *can be verified among measurable sets by three X-rays.*

Proof. Let Q be a convex polygon. Consider three adjacent edges of Q, labeled e_i, $1 \le i \le 3$, in order around bd Q. Let u_2 be a direction parallel to e_2. For $i = 1, 3$, take u_i to be any direction not parallel to u_2, but parallel to a line supporting Q at the vertex at the intersection of the edges e_2 and e_i. We can assume that Q lies in the upper half-plane with its base e_2 on the x-axis. See Figure 2.9.

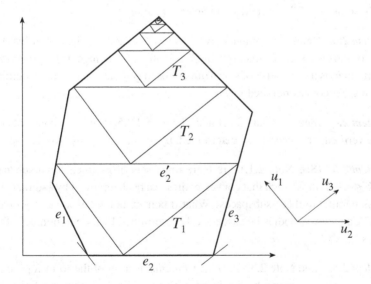

Figure 2.9. The polygon Q of Theorem 2.3.26.

Let E be a measurable set with the same X-rays as Q in the directions u_i, $1 \le i \le 3$. A continuity argument produces a triangle T_1 with one vertex in e_2, with the others in bd $Q \setminus e_2$, and with sides parallel to the directions u_i (so that the uppermost side, e'_2 say, is parallel to e_2). Denote by P_1 the closed part of Q between e_2 and e'_2. Clearly, if $z \in \text{int } P_1$, then z is contained in the interior of a convex quadrilateral inscribed in Q (and in P_1), each of whose edges is parallel to some u_i. Then Lemma 2.3.6 shows that $P_1 \subset E$, up to a set of measure zero.

By subtracting the X-rays of P from those of Q, we obtain the X-rays of the polygon $Q' = \text{cl}(Q \setminus P)$. Note that Q' has its base on the horizontal edge e'_2, and that we can repeat the previous argument with Q replaced by Q', obtaining another triangle T_2 inscribed in Q'. The closed part P_2 of Q' between e'_2 and the top edge of T_2 is contained in $E \setminus P$, and therefore in E, up to a set of measure zero. Continuing, we note that the process may not terminate after finitely many steps, but otherwise the similar triangles T_m, $m \in \mathbb{N}$, will converge to a unique vertex of Q with largest y-coordinate, as in Figure 2.9. After a countable sequence of steps we will have shown that $Q \subset E$, up to a set of measure zero. The fact (Lemma 2.1.5) that $\lambda_2(E) = \lambda_2(Q)$ completes the proof. ∎

Open problems

Problem 2.1. (See Theorem 2.2.2.) Are convex bodies in $\mathbb{E}^3$ determined by X-rays in any set of seven directions in general position?

Problem 2.2. (See Corollary 2.2.3.) Are convex bodies in $\mathbb{E}^3$ determined by X-rays in some set of five directions in general position?

Problem 2.3. Define S-regular polytopes in $\mathbb{E}^n$ exactly as in Definition 1.2.7. Classify those sets S of directions in $\mathbb{E}^n$ admitting an S-regular polytope in $\mathbb{E}^n$. In particular, is there a set S of seven directions in general position in $\mathbb{E}^3$ admitting an S-regular convex polyhedron?

Problem 2.4. (See Corollary 2.2.1 and Theorem 2.2.5.) Can each convex body in $\mathbb{E}^3$ be verified, or successively determined, by three, or even by two, X-rays?

Problem 2.5. (See Note 2.1.) Are there finite sets of k-dimensional subspaces, $1 < k \le n-1$, in $\mathbb{E}^n$ such that convex bodies can be determined by k-dimensional X-rays parallel to these subspaces? What if convex polytopes are considered instead? Can convex bodies be successively determined by a finite number of such X-rays?

Problem 2.6. (See Note 2.3.) A *convex annulus* is a set of the form $K_1 \setminus \text{int } K_2$, where K_1, K_2 are convex bodies in $\mathbb{E}^2$ with $K_2 \subset \text{int } K_1$. Is there a finite set of directions such that convex annuli are determined by X-rays in these directions?

Problem 2.7. (See Theorem 2.3.17.) Is it true that a convex body in $\mathbb{E}^2$ that is determined among measurable sets by its X-rays in the directions in a set S is S-additive?

Problem 2.8. Is there a finite set of directions such that each convex body in $\mathbb{E}^2$ is determined among measurable sets by its X-rays in these directions?

Problem 2.9. Is there a convex body with the properties of the polyhedron P of Theorem 2.3.24?

Problem 2.10. (See Theorem 2.3.4 and Note 2.3.) Can star-shaped polygons be successively determined by a finite number of X-rays? If not, is it true that for each $m \in \mathbb{N}$ there is a star-shaped polygon in $\mathbb{E}^2$ that is not verifiable (i.e., among star-shaped polygons) by X-rays in any set of m directions?

Problem 2.11. Given a convex body K in $\mathbb{E}^2$, can three directions be found such that the X-rays of K in these directions distinguish K from any other measurable set?

Problem 2.12. (See Corollary 2.3.13 and Note 2.3.) If $1 < k \leq n - 1$, what is the least number of k-dimensional X-rays needed to determine each ellipsoid in $\mathbb{E}^n$ among measurable sets?

Problem 2.13. (See Note 2.3.) Characterize those sets of functions that are the X-rays, taken in a finite set of directions, of some measurable set.

Notes

2.1. *Parallel X-rays and k-symmetrals of convex bodies.* The following sentence is taken from the abstract of [179]: In radar identification of perfectly conducting targets, a need arises to determine a convex body from its cross-sectional areas normal to the lines of sight for a number of view angles.

The Schwarz symmetral of a convex body, sometimes called the "Schwarz rounding," was introduced by Schwarz [756]. Like the Steiner symmetral, the Schwarz symmetral has often been used in geometry; see [83, Sections 40–2] or [718, Section 5]. Wills [841] establishes some basic properties of the k-symmetral of a convex body. See also Note 2.3 for further references.

It was noted just before Lemma 2.1.2 that the X-ray of a convex body in $\mathbb{E}^n$, $n \geq 3$, need not be continuous on its support. The X-ray of a convex body K is continuous on its support if and only if K has "Property (P)" as defined by Brown [102]. The latter makes sense in normed linear spaces and is of interest in approximation theory; see, for example, the work of Huotari [405]. It is worth noting that the class of convex bodies in $\mathbb{E}^n$ with this property includes all strictly convex bodies and all convex polytopes.

Theorem 2.1.9 is taken from [770, Section 4.3], by Skiena. A. Volčič, J. Wills, and the author discovered Theorem 2.2.2 during discussions. Theorem 2.2.4, like its planar version in Chapter 1, was proved by Volčič and Zamfirescu [823].

Giering [288] considered the problem of verifying convex bodies in $\mathbb{E}^3$ by X-rays. He showed that four X-rays suffice, but, as we noted, this is already implied by Corollary 2.2.1.

Gardner and Gritzmann [266] prove Theorem 2.2.5; it resembles a result of Golubyatnikov [295], but is different in a fundamental way (see [266] for a comparison). It is also proved in [266] that convex polytopes in $\mathbb{E}^n$ can be verified by almost any set of $[n/(n-k)] + 1$ k-dimensional X-rays, but that there are zonotopes in $\mathbb{E}^n$ not verifiable by any set of $[n/(n-k)]$ k-dimensional X-rays. (The phrase "almost any" is with respect to the appropriate Haar measure.) Even this result on verification of polytopes is not easy to prove. The main idea, essentially already seen in Theorem 1.2.23, is to use the differentiability properties (or the lack of them) of an X-ray parallel to a subspace S to locate the projections of the vertices on $S^\perp$. Unless S is suitably chosen, however, this is not always possible when $k > 1$ (as we saw in Theorem 2.2.6, a more general form of which is also noted in [266]). One of the main ingredients of the proof is a formula first discovered by Bieri and Nef [64] (see also work of Lawrence [498] and Filliman [232]) for the volume of the intersection of a simple polytope P with a half-space, the boundary of which is in general position with respect to the vertices of P. The second part of the preceding result on verification of polytopes, showing that the number of X-rays is the best possible, elaborates on the idea (originally due to Giering) of Theorem 1.2.22.

In [266], heavy use is made of the fact that a k-dimensional X-ray $X_S P$ of a convex polytope P, parallel to a subspace S in general position with respect to the vertices of P, is a piecewise polynomial of degree at most k and is $(k-1)$-times continuously differentiable. In short, it is a *spline*. The case when P is a simplex is of particular importance, and when P is a cube, the term *box spline* has been used; box splines are of considerable interest in spline theory and computer-aided design. See, for example, [182].

Martini [575] (see also [578] and [582]) characterizes simplices by their Steiner symmetrals in certain sets of directions.

Problem 2.5 was posed by D. G. Larman at the 1982 Oberwolfach Conference on Convex Bodies.

In the terminology of geometric probing (cf. Note 1.6), Skiena [768] finds that

$$f_0(P) + 4f_2(P) + (n+2)f_{n-1}(P) + 46$$

X-ray probes determine a convex polytope P in $\mathbb{E}^n$, where $f_k(P)$ is the number of k-dimensional faces of P.

2.2. *Switching components and discrete tomography.* The term *switching component* in Definition 2.3.1 was used previously in the reconstruction of binary matrices from their row and column sums. (See, for example, Chang's paper [158]. The idea appeared in earlier work under different names; for example, Ryser [714, Section 6.3] uses the word "interchange.")

Suppose that E is a finite set and that u is a direction in $\mathbb{E}^n$. The *discrete (parallel) X-ray of E in the direction u* is the function giving the number of points in E on each line parallel to u; this corresponds to replacing λ_1 by λ_0 in Definition 2.1.1. If S is a finite set of m directions and P is an S-regular polygon, as in Definition 1.2.7, then the two sets of m alternate vertices of P have the same discrete X-rays in the directions in S. (Corollary 1.2.10 characterizes the sets S of directions for which this construction is possible.) Such examples were known to Rényi [689], who also proved that a set of m points in $\mathbb{E}^2$ or $\mathbb{E}^3$ is determined by any set of $(m+1)$ discrete X-rays in mutually nonparallel directions. Heppes [393] extended this result to $\mathbb{E}^n$.

Lemma 2.3.2, an observation of Lorentz [523], shows that for each set of m directions, there are two different sets of 2^{m-1} points with equal discrete X-rays in these directions; in other words, $[\log_2 m] + 1$ X-rays are always insufficient to determine m points. On the other hand, Bianchi and Longinetti [61] prove, roughly speaking, that apart from examples of the S-polygon type, sets of m points in $\mathbb{E}^2$ are determined by k mutually nonparallel X-rays, providing

$$k > m + (3 - \sqrt{m+9})/2.$$

They also show that any set of four directions in $\mathbb{E}^2$ admits different sets of five points with equal discrete X-rays in those directions.

The results just described could be regarded as early work in *discrete tomography*, a term introduced by L. A. Shepp at a mini-symposium he organized at DIMACS in September, 1994. The focus of discrete tomography is on the reconstruction of *lattice sets*, that is, finite subsets of the integer lattice $\mathbb{Z}^n$, from their discrete X-rays parallel to *lattice directions* (i.e., directions parallel to vectors with integer coordinates). Lattice sets are natural models for atoms in a crystal, and the original motivation for discrete tomography came from a genuine application in high-resolution transmission electron microscopy (HRTEM). Indeed, the DIMACS meeting was inspired by work of a physicist, P. Schwander, whose team developed a technique that effectively allows the discrete X-rays of a crystal to be measured in certain lattice directions parallel to vectors with small integer coordinates. The process is described in detail by Schwander et al. [755]. The algorithms of computerized tomography (cf. Note 1.1) are useless, since the high energies required to produce the discrete X-rays mean that no more than about three to five discrete X-rays of a crystal can be taken before it is damaged. Discrete tomography has been a very active field since its inception, due to attempts to find suitable algorithms and to the stimulus provided by applications to crystallography and other areas, such as data compression and data security (see, for example, [268]). The books [394] and [395] edited by Herman and Kuba provide a good overview of discrete tomography. The short survey below is not comprehensive, but should indicate the main developments.

As with computerized tomography, the main interest is in the planar case. A simple modification of Lemma 2.3.2 shows that given any finite set of planar lattice directions, there are different lattice sets in $\mathbb{Z}^2$ with the same discrete X-rays in these directions. The problems caused by this fundamental lack of uniqueness are compounded by the need to reconstruct crystals containing very large numbers of atoms.

From an algorithmic point of view, there are three separate problems, each involving a given class $\mathcal{E}$ of finite subsets of $\mathbb{Z}^2$ and a given set of m mutually nonparallel lattice directions. The consistency problem is to decide whether, for a given set of discrete X-ray data (that may have been corrupted by errors of measurement or noise), there is a set in $\mathcal{E}$ whose discrete X-rays match the data. The uniqueness problem asks whether, given an $E \in \mathcal{E}$, there is a different set in $\mathcal{E}$ with the same discrete X-rays as E. The reconstruction problem is the task of reconstructing a set in $\mathcal{E}$ consistent with the data, if one exists.

The work of Chang [158] and Ryser [714, Section 6.3] on reconstructing binary matrices from column and row sums yields polynomial-time algorithms solving all three of these problems when $\mathcal{E}$ is the class $\mathcal{F}^2$ of all finite lattice sets, $m = 2$, and the directions are the horizontal and vertical directions. Gardner, Gritzmann, and Prangenberg [269] prove that when $\mathcal{E} = \mathcal{F}^2$ and $m \geq 3$, the consistency and uniqueness problems are $\mathbb{NP}$-complete and the reconstruction problem is $\mathbb{NP}$-hard. So there is a dilemma: It is extremely unlikely that there exists a very efficient algorithm to reconstruct lattice sets from m discrete X-rays if $m \geq 3$, while discrete X-rays in two lattice directions will almost never provide a unique solution.

However, efforts to obtain satisfactory algorithms have continued. Salzberg [717] offers an algorithm based on the maximum likelihood principle, while Fishburn et al. [244] pursue a linear programming approach. Hajdu and Tidjeman [377] base an algorithm on ideas in their earlier paper [376], in which they use algebraic techniques to show that switching components can always be generated by projections of vertices of parallelotopes. Gritzmann, de Vries, and Wiegelmann [339] (see also the paper of Gritzmann et al. [343]) examine greedy algorithms and others, sometimes with a view to approximate reconstruction. The greedy algorithms are the most successful presently known for the general case, though Batenburg [46] proposes a network-flow algorithm with smoothness priors that seems to perform better in certain situations.

Special classes of lattice sets have been considered. Gardner and Gritzmann [267] examine *convex lattice sets*, that is, lattice sets whose convex hulls contain no additional lattice points. Using Corollary 1.2.10 and p-adic number theory, they establish a lattice version of Corollary 1.2.12 by showing that convex lattice sets in $\mathbb{Z}^2$ are determined by discrete X-rays taken in certain sets of four lattice directions. In fact, they prove that any set of four lattice directions whose slopes, when arranged in increasing order, have a cross ratio not equal to 4/3, 3/2, 2, 3, or 4, will have this property. A specific such set consists of directions parallel to $(1, 0)$, $(2, 1)$, $(0, 1)$, and $(-1, 2)$. It is also shown in [267] that convex lattice sets in $\mathbb{Z}^2$ are determined by *any* set of seven discrete X-rays taken in mutually nonparallel lattice directions. Three lattice directions are never enough, and in general, six lattice directions are not enough; Figure 2.10 depicts two convex lattice sets (black circles belonging to one, grey circles to the other, and half-black and half-grey circles to both) with the same discrete X-rays parallel to the vectors shown on the left. This is a consequence of the existence of the corresponding lattice S-regular polygon illustrated in the center of Figure 2.10.

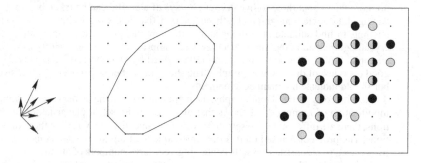

Figure 2.10. Convex lattice sets with equal discrete X-rays in six lattice directions.

Daurat [180] extends the uniqueness result of Gardner and Gritzmann to a class of subsets of $\mathbb{Z}^2$ he calls Q-convex sets, containing the class of convex lattice sets. (These constitute a discrete version of the Q-convex bodies defined in Note 1.2.) In a major contribution to the area, Brunetti and Daurat [103] provide a polynomial-time algorithm for reconstructing Q-convex sets from discrete X-rays. (See also [106], in which these authors and Del Lungo find a polynomial-time algorithm for reconstructing approximations to sets ("strongly Q-convex sets") in a smaller class from their approximate discrete X-rays.) As a corollary, they find a polynomial-time algorithm for reconstructing a convex lattice set in $\mathbb{Z}^2$ from its discrete X-rays in the sets of directions found by Gardner and Gritzmann that provide uniqueness. Barcucci et al. [32] and Dulio and Peri [198] obtain results complementary to those of Gardner and Gritzmann in [267]. A different extension of the results in [267], to so-called cyclotomic model sets, is motivated by the need to reconstruct quasicrystals; see the work of Baake and Huck [18].

Various authors focus on polyominoes, which for our purposes can be defined as lattice sets in $\mathbb{Z}^2$ such that each point has at least two horizontal or vertical neighbors in $\mathbb{Z}^2$. They show that the problems above have polynomial-time solutions for some classes of polyominoes, but are $\mathbb{NP}$-complete or $\mathbb{NP}$-hard for others. See the work of Barcucci et al. [30], [31], [33], Brunetti et al. [108], and Woeginger [844].

It is easy to determine subsets of a fixed rectangle, say $A = \{(i, j) \in \mathbb{Z}^2 : 0 \le i \le k, 0 \le j \le l\}$; a single discrete X-ray parallel to a lattice direction with positive slope less than $1/k$ or greater than l will suffice. If k and l are not very small, however, such directions are not permissible for the application to HRTEM. With this in mind, Hajdu [375] finds that given A, there are sets $\{(a_i, b_i)\}$ of four vectors in $\mathbb{Z}^2$ with $\sum_i |a_i| < k$ and $\sum_i |b_i| < l$, such that the discrete X-rays in these directions determine subsets of A.

For some results on stability, see the articles of Alpers and Gritzmann [7], Alpers, Gritzmann, and Thorens [8], and Brunetti and Daurat [104].

It is natural to define k-dimensional discrete X-rays of lattice sets in $\mathbb{Z}^n$, by analogy with Definition 2.1.1, and these are treated by Brunetti, Del Lungo, and Gerard [107] and by Gritzmann and de Vries [338]. Fishburn et al. [243] investigate the relation between lattice sets determined by their discrete X-rays in coordinate directions (or parallel to coordinate hyperplanes in higher dimensions) and a discrete analogue of the additive sets of Definition 2.3.9. (They employ the term *bad configuration* for a switching component.) These discrete additive sets are studied by Onn and Vallejo [652] and Vallejo [812], [813], [814]. For other related work, see the PhD thesis of Beauvais [49].

Another extension is the "polyatomic case" considered by Gardner, Gritzmann, and Prangenberg [270], where modified discrete X-rays measure the number of atoms of each of $c \geq 2$ different types in a lattice. They show that the corresponding consistency, uniqueness, and reconstruction problems are $\mathbb{NP}$-complete when $m = 2$ and $c \geq 6$. Subsequently Chrobak and Dürr [166] proved that this result holds for $c \geq 3$, but the case $c = 2$ is perhaps the principal open theoretical problem in the area. (Costa et al. [170] obtain a partial result.)

Zopf and Kuba [867] cite several papers concerning modified discrete X-rays, defined by line sums containing weights, that model absorption.

See also Note 5.6 for work on discrete point X-rays.

2.3. *Parallel X-rays and k-symmetrals of measurable sets.* Measurements approximating 2-dimensional X-rays in $\mathbb{E}^3$ are taken in NMR (nuclear magnetic resonance) computerized tomography; see, for example, the article of Shepp [764].

The Steiner symmetral, the Schwarz symmetral, and many other types of symmetral have frequently been applied for sets more general than convex sets. For example, applications to electrostatic capacity, torsional rigidity, and other topics of mathematical physics can be found in the survey of Talenti [800], as well as in the books of Bandle [28, Chapter 2] and Pólya and Szegö [676]. The k-symmetral of a measurable set was considered by Dinghas [187].

In discussing only measurable sets we have in mind the additional pathology that might otherwise occur. For example, there is the famous nonmeasurable subset of the plane constructed by Sierpiński [766], meeting every line in at most two points. This set is invisible to X-rays, yet it is not of measure zero.

Theorem 2.3.3 predates most other work on X-rays and is due to Lorentz [523], while Theorem 2.3.4, on star-shaped polygons, was proved by Gardner [260]. Theorem 2.3.5, based on ideas of A. Volčič, also appeared in [260]. It complements the result of Falconer [213] stating that a compact set in $\mathbb{E}^n$ of positive measure is convex if its Steiner symmetrals in every direction are also convex.

As we remarked in Chapter 1, the notion of inscribability, for a set of two directions, goes back to Giering's paper [288]; Theorem 2.3.7 was noted by Volčič [822].

The simple but basic Theorem 2.3.8 was employed independently by Kuba and Volčič [478] and by Fishburn et al. [242]. The latter paper introduces the additive sets. Our Definition 2.3.9 is slightly more general, since the definition of additive set in [242] employs only nonnegative ridge functions. The powerful Theorem 2.3.11 is also taken from [242].

The question of when two X-rays determine a disk among measurable sets was posed by Horn [398, p. 57] in his book on robot vision. The answer actually follows from Lorentz's work in [523], with alternative proofs in both [478] and [242]. Corollary 2.3.13, stating that three X-rays determine ellipses, was proved by Gardner [259]. The latter remains true for ellipsoids in $\mathbb{E}^n$, with similar proof; compare the open Problem 2.12. (It is interesting that the explicit formula obtained by Lee and Shih [501] for the volume of the intersection of an ellipsoid in $\mathbb{E}^n$ with a hyperplane was motivated by radar target-shape estimation; cf. the quote at the beginning of Note 2.1.)

Lorentz, in the 1949 paper [523], raised and answered the question of characterizing those measurable sets E in the plane determined by their X-rays in the two coordinate directions. (He used the term *cross function* instead of X-ray.) Lorentz employed Riesz's result that a nonnegative integrable function f has a nonincreasing rearrangement g, that is, a nonincreasing function such that for each real c, the sets $\{x : f(x) \geq c\}$ and $\{x : g(x) \geq c\}$ have the same measure. Suppose that g_x and g_y are the nonincreasing rearrangements of the X-rays $X_x E$ and $X_y E$ of E in the coordinate directions. Lorentz proved that E is determined if and only if g_x and g_y are inverses (when these are suitably defined) of each other. He also obtains a characterization of the pairs of functions that are the X-rays of some measurable set in two fixed directions. These characterization problems are open for finite sets of more than two directions; see Problem 2.13 and the work of Kellerer [420].

Lorentz's results can be generalized to modified X-rays defined by line integrals containing an extra term to model absorption; see the article [867] of Zopf and Kuba.

Lorentz's work was utilized in both [478] and [242], which together give the generalization of Theorem 2.3.17 mentioned in Remark 2.3.18(i). The equivalence of S-inscribability (S-additivity) and determination by X-rays for coordinate directions in the plane was demonstrated in [478] ([242], respectively). In [478, Theorem 5.2] there is also an explicit formula for a planar measurable set determined by its X-rays in the coordinate directions, in terms of these X-rays. It seems probable that a reconstruction algorithm could be based on this formula, at least for sets of sufficiently simple structure, such as convex polygons or convex bodies.

Even when a measurable set E is not determined by its X-rays in the coordinate directions, there may be a subset F of it which is, in the sense that F is contained, up to a set of measure zero, in any measurable set E' with the same X-rays as E. Kuba and Volčič [479] investigate the structure of E in terms of such an invariant part F.

It was M. Longinetti who suggested to the author the problem of investigating how far Theorem 2.3.17 carries over to arbitrary finite sets of directions. The results of Remark 2.3.18(ii) are proved in [259], and the proof given here of Lemma 2.3.15 is also taken from that paper.

The only known proof that the pyramid P of Theorem 2.3.19 is not determined by its X-rays parallel to the coordinate planes is quite long. The example was discovered by the authors of [242], and the proof in that paper uses probability to show the existence of a different measurable set with the same X-rays. It would be of interest to devise an alternative proof providing an explicit example of such a measurable set.

Gardner [260] proved Theorems 2.3.25 and 2.3.26 on verification. The first part of Problem 2.10 is a problem of Skiena [769, Problem 12]. The second part of Problem 2.10 has a negative answer when star-shaped polygons are replaced by simple polygons; this is proved in [260], using some of the ideas of Theorem 2.3.25.

Problem 2.6 was posed by P. M. Gruber and A. Volčič at the 1984 Oberwolfach Conference on Convex Bodies. Bianchi and Longinetti [61] obtain some results on the determination by X-rays of planar convex bodies containing a finite number of disjoint circular holes. Their method is to combine Theorem 1.2.11 with their theorem on determination of finite sets of points, mentioned in Note 2.2. Bocconi [76] shows that convex annuli with polygonal holes can be successively determined by seven X-rays.

In this and the previous chapter we have always assumed that the directions from which the X-rays are taken are known. If this is not the case, additional difficulties arise; see the work of Basu and Bresler [44], [45], Goncharov [303], and Kuba [477].

2.4. *Blaschke shaking.* Blaschke defined a sort of antisymmetrization process sometimes called *shaking* and used it to solve Sylvester's problem in the plane (see Note 9.4). Let $u \in S^{n-1}$ and let E be a bounded measurable subset of $\mathbb{E}^n$. The shaking process in the direction u, applied to E, results in a set $B_u E$, defined as in Definition 2.1.3, except that $c(x)$ instead has one endpoint at x and the other at x plus a positive multiple of u. Of course, $B_u E$ and the Steiner symmetral $S_u E$ have the same X-ray in the direction u. If K is a convex body, $B_u K$ is the convex body contained in one of the halfspaces bounded by $u^\perp$ such that for each line l parallel to u, $B_u K \cap l$ is a line segment of length equal to that of $K \cap l$ and with one endpoint on $u^\perp$; this is just the closed region under the graph of $X_u K$, as in Figure 1.1. Campi, Colesanti, and Gronchi [145] prove that there are directions $u_1, \ldots, u_{n+1}$ such that given any compact set C in $\mathbb{E}^n$, the compact sets obtained by successive shaking in these directions converge, in the Hausdorff metric, to the simplex of the same volume as C and with outer unit normal vectors $u_1, \ldots, u_{n+1}$. They use this result to obtain another proof of the Brunn–Minkowski inequality (B.10). (A discrete analog of this process, called *compression*, was defined by D. L. Kleitman and used by Gardner and Gronchi [271] to obtain versions of the Brunn–Minkowski inequality for the integer lattice $\mathbb{Z}^n$.)

2.5. *Reconstruction of polygons and polyhedra from possibly noisy X-rays.* Let $u = (\cos\theta, \sin\theta)$, $0 \le \theta < 2\pi$. A version the central slice theorem (cf. (C.8) and Note 1.1), following easily from the definition (C.4) of the X-ray transform, states that

$$\int (X_u f)(t) F(t)\, dt = \int\int f(x, y) F(x\cos\theta + y\sin\theta)\, dx\, dy,$$

for suitable functions $f(x, y)$ and $F(t)$. The central slice theorem in $\mathbb{E}^2$ corresponds to the special case $F(t) = e^{-it}$. If we take $F(t) = t^k$, we get a known formula (see, for example, [774, (3.1)]) that, combined with the binomial theorem, gives

$$\int (X_u f)(t) t^k\, dt = \int\int f(x, y) \sum_{j=0}^{k} \binom{k}{j} \cos^{k-j}\theta \sin^j\theta \mu_{k-j,j}\, dx\, dy,$$

for $k = 0, 1, \ldots$, where

$$\mu_{p,q} = \int\int f(x, y) x^p y^q\, dx\, dy$$

is a *real moment* of f. In his PhD thesis [612], Milanfar employs this connection between X-rays and moments, a simple relationship between real moments and complex moments

$$c_k = \int\int f(x, y) z^k\, dx\, dy,$$

where $z = x + iy$, and an extension of an earlier result of P. J. Davis, who proved that the vertices of a triangle (and hence the triangle itself) can be reconstructed from its first four moments c_j, $j = 0, \ldots, 3$. Specifically, Milanfar proves that the vertices of a nondegenerate, simple m-gon P are determined by its moments c_j, $0 \le j \le 2m - 3$, and he concludes from this that they are also determined by the X-rays of P taken in any $2m - 2$ mutually nonparallel directions. Moreover, the method provides an algorithm for finding the vertices of P.

When P is convex, it is determined by its vertices, so convex m-gons are determined by, and can be reconstructed from, their X-rays in any $2m - 2$ mutually nonparallel directions. In general, however, there may be different simple polygons with the same vertices; see [612] or the article [615] of Milanfar et al. for discussions, and the paper [614] by Milanfar, Karl, and Willsky for more about the moment method, which is also viable for noisy data (cf. Note 1.5). Further reconstruction methods and references can be found in the article of Soussen and Mohammad-Djafari [781].

Using purely geometric methods, Meijer and Skiena [597] show that $2m + 2$ X-rays are sufficient and, if $m \geq 16$, $[\log m] - 2$ X-rays are necessary to successively determine simple m-gons.

2.6. *Ridge functions and the additivity conjecture.* When a subspace S is $(n - 1)$-dimensional, a ridge function orthogonal to S is often called a *plane wave*, as in the books of John [408] and Helgason [389], for example.

The authors of [242] posed the question (the "additivity conjecture") of whether a measurable set in $\mathbb{E}^3$ that is determined by its 2-dimensional X-rays parallel to the coordinate planes S is also S-additive. The polyhedron of Theorem 2.3.24, showing this is false, was constructed by Kemperman [425]. Kemperman actually does considerably more, by studying the following more general problem. Suppose that μ is a σ-finite measure on a space X. A subset E of X can be identified with the measure μ_E defined by $\mu_E(A) = \mu(E \cap A)$. One can now examine which E's (or rather their associated measures μ_E) are essentially determined by their images (marginals) under a finite or infinite collection of measurable maps π_i from X to spaces Y_i. A special case is where $X = \mathbb{E}^2$, $\mu = \lambda_2$, and $\pi_i, i = 1, 2$, are the orthogonal projections from $\mathbb{E}^2$ onto the ith coordinate axis Y_i, when such E are determined by their X-rays in the coordinate directions. Kemperman makes the corresponding generalization of the additive sets, and shows that Theorem 2.3.11 remains true in this framework. He then defines a hierarchy of sets "additive of degree m," where degree 1 corresponds to the usual additive sets, and proves that any set in this hierarchy is determined. Lemma 2.3.23 is a special case of this result, corresponding to $m = 2$. The polyhedron of Theorem 2.3.24 is exhibited in [425] as an example of a set that is additive of degree 2, but not additive.

Independently of Kemperman, a counterexample to the additivity conjecture was found by Kellerer [422]. Kellerer takes the same more general viewpoint as Kemperman, but we caution the reader that in [422] ridge functions that are integrable on compact sets, rather than measurable and bounded on compact sets, are used throughout. Kellerer's example is actually stronger in the sense that it is a set that is determined, but that lies outside the entire hierarchy of additive sets of degree m; on the other hand, it is not a polyhedron. Definition 2.3.20, Theorem 2.3.21, and the proof of Lemma 2.3.23 given here are also due to Kellerer. In fact, Kellerer [422, Theorem 2.5] uses the Hahn–Banach theorem to prove the result of Remark 2.3.22 that the converse of Theorem 2.3.21 is true. Evidently the additivity conjecture was not far off the mark.

2.7. *X-rays of bounded density functions.* Many of the results in this chapter extend from measurable sets to density distributions defined on a bounded set and bounded in value by 0 and 1. For example, A. Volčič has observed that the proof of Theorem 2.3.11 also shows that an S-additive set is determined among all such densities by its X-rays (i.e., X-ray transforms; cf. (C.4)) in the directions in S. Surprisingly, the restriction to densities bounded by 0 and 1 is necessary; a result of Kellerer [421, Satz 2.3] implies that for each $\varepsilon > 0$ and compact planar set E, there is an $F \subset E$ such that 1_E and $(1 + \varepsilon)1_F$ have the same X-rays in the coordinate directions. The following concrete example is due to Volčič. Consider the function f bounded by $4/\pi > 1$, defined by

$$f(x, y) = \frac{4}{\pi}\sqrt{(1 - x^2)(1 - y^2)},$$

for $(x, y) \in [-1, 1]^2$, and $f(x, y) = 0$, otherwise. Since

$$\int_{-\infty}^{\infty} f(x, y)\, dx = \frac{4}{\pi}\sqrt{1 - y^2} \int_{-1}^{1} \sqrt{1 - x^2}\, dx = 2\sqrt{1 - y^2},$$

for $-1 \leq y \leq 1$, this function has the same X-ray in the direction of the x-axis as the unit disk. Similarly, it also has the same X-ray in the direction of the y-axis as the unit disk. One can obtain examples whose upper bounds are arbitrarily close to 1 by taking convex combinations of f and the characteristic function of the unit disk.

Gutman et al. [368] establish the following important connection between measurable sets and densities bounded by 0 and 1. Let S be a finite set of subspaces of $\mathbb{E}^n$, each with dimension between 1 and $n - 1$, and suppose that f is a measurable function defined on a bounded subset E of $\mathbb{E}^n$, with $0 \leq f \leq 1$; then there is a measurable subset F of E such that $X_S F(x) = X_S f(x)$ for almost all $x \in S^\perp$, for each $S \in \mathcal{S}$. (The special case corresponding to ordinary X-rays in the coordinate directions was proved earlier by Kellerer [421]. A different but related result is due to Falconer [216, Corollary 2.4].) The bounds on f are clearly necessary; consider, for example, a function equal to $1 + \varepsilon$ on the unit square, for $\varepsilon > 0$, and zero elsewhere. To prove the theorem, one considers the set $M(f)$ of all measurable functions g on E with $0 \leq g \leq 1$ and with the same X-rays as f parallel to the subspaces in S. One observes that $M(f)$ is a nonempty compact convex subset of L^∞ with the weak* topology, and concludes by the Krein–Milman theorem that $M(f)$ contains at least one extreme point g. If g is not essentially the characteristic function of a measurable set F, then for some $0 < \varepsilon < 1/2$ the set $D = \{x : \varepsilon \leq g(x) \leq 1 - \varepsilon\}$ is of positive λ_n-measure. Now the idea of Theorem 2.3.3 is used to produce a bounded measurable function h on D, not almost everywhere zero, with zero X-rays parallel to each $S \in \mathcal{S}$. In fact h takes the value 1 on one part of a sort of "measurable-set switching component" contained in D, and -1 on the other part. The functions $g \pm \varepsilon h$ belong to $M(f)$; since g is their average, g cannot be an extreme point, and this contradiction proves the result.

Figure 2.11. Johann Radon.

2.8. *Johann Radon (1887–1956)*. (Sources: [254], [362], [396], [725].) The Austrian mathematician Johann Radon was born in Tetschen, Bohemia (now Decin, Czech Republic). Apart from half-sisters, he was the only child of the head bookkeeper at a local savings bank. Despite attacks of asthma, Radon had a distinguished school career, including a strong emphasis on the classic languages. At one time he wanted to be a philosopher, or even an opera singer; as an adult, he possessed a beautiful baritone voice and became an accomplished violinist and lute player. Radon attended a wide variety of courses at the University of Vienna, obtaining his doctorate in 1910. After a semester in Hilbert's seminar in Göttingen, and a short stay in Brünn (now Brno, Czech Republic), he taught in Vienna, obtaining an assistantship at the Technical University in 1912. He was exempt from military service in the first world war due to his poor vision. In 1916 he married, and had three sons and a daughter; tragically, one

son died in infancy, one became incurably ill and died in 1939, and the last was killed
on the Russian front in 1943.

In 1919 Radon joined the newly founded University of Hamburg as associate professor, at the suggestion of his friend Wilhelm Blaschke (cf. Note 6.4). He moved on
in 1922, taking professorships in Greifswald and Erlangen before arriving in Breslau
(now Wrocław, Poland) in 1928. Here Radon stayed until he and his family were
forced to leave, losing nearly everything, to escape the Russian army in 1945. He accepted a position at the University of Vienna in 1947, and served as rector in 1954.
He also became a member, and later secretary, of the Austrian Academy of Sciences,
and from 1948 to 1950 was chairman of the Austrian Mathematical Society. Once
described by Paul Funk (cf. Note 7.16) as "very retiring but very nice," Radon was
apparently of a cheerful disposition and enjoyed company, despite his shyness.

Radon only published 45 papers, but several of these were incredibly influential.
His work began in the calculus of variations. His paper *Theorie und Anwendungen der
absolut additiven Mengenfunktionen* (*Theory and applications of absolutely additive
set functions*), published in 1913 and 144 pages long, was of paramount importance
for measure theory, where the most widely applicable class of measures are now called
Radon measures. One of its results, generalized by O. Nikodym, became the famous
Radon–Nikodym theorem of real analysis. The paper [685] has established Radon as
the father of tomography, in which the Radon transform (see Appendix C) is an analytical tool of astounding power, having early applications in astronomy and electron
microscopy, and underpinning the work of A. M. Cormack and G. N. Hounsfield (see
Note 1.1). (The Radon transform continues to have extraordinary significance in integral geometry; the interesting article [292] of Gindikin indicates developments far
beyond the scope of this book.) Radon is also renowned for his work on orthogonal
matrices and Foucault's pendulum, as well as for contributions to Riemann geometry
and conformal geometry. In convexity, his name is associated with Radon curves and
with the indispensable Radon partition theorem.

3

Projections and projection functions

What can one say about an object, given some information about its projections (shadows)? A substantial part of the classical theory of convexity has grown out of attempts to answer this question. In the first section of this chapter, the given information includes the shape of the shadows. For example, Theorem 3.1.3 implies that two compact convex sets in $\mathbb{E}^n$, $n \geq 3$, must be homothetic if their projections on any hyperplane are also homothetic, whereas Theorem 3.1.8 shows that "homothetic" cannot be replaced by "similar." The width function is then introduced, followed by the more general projection functions. Such functions do not give the shape, but only the length, area, or volume, of the shadows. The latter yield surprisingly strong information, thanks to Aleksandrov's projection theorem, Theorem 3.3.6, implying that different centered convex bodies have different projection functions. However, the existence of certain special sets – the central symmetral, the Blaschke body, and sets of constant width and brightness – place limits on how much can be divined from projection functions.

The beautiful and powerful results in this chapter provide a convincing demonstration of the tools of the classical theory of convexity. The first of these is the support function. We suggest that the reader not familiar with this refer to Section 0.6 and ignore the proof of Theorem 3.1.8; Section 3.2 furnishes a gentle introduction to the support function at work. In Section 3.3, mixed volumes and area measures are encountered for the first time. Initially, it may help to skip the proofs and interpret the statements of the theorems in three dimensions. Bear in mind that V_1 gives the length of a line segment; for a 2-dimensional convex body, V_1 is a constant multiple of the average or mean width and V_2 is the area, while for a 3-dimensional body, V_1 has the same meaning, V_2 is just half the surface area, and V_3 is the volume. Appendix A is designed to be both a summary and an exposition of the theory of mixed volumes. Also in Section 3.3, we see in action special cases of the celebrated Aleksandrov–Fenchel inequality, an extensive and profound generalization of the isoperimetric inequality. Such inequalities are collected together in Appendix B.

3.1. Homothetic and similar projections

In Chapters 1 and 2 we saw how even a finite number of X-rays, when taken in appropriate directions, can distinguish a convex body from any other. The support of the X-ray of body E in a direction u is just the orthogonal projection of E on the subspace $u^\perp$ orthogonal to u. Consequently, the projection gives strictly weaker information about E. Nevertheless, one can identify a set, given enough information about its projections. It is the purpose of this chapter to collect some results of this type.

All projections will be orthogonal projections. We denote the projection of a set E on a subspace S by $E|S$. See Figure 3.1.

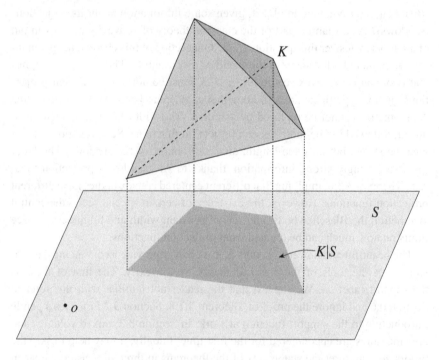

Figure 3.1. A projection of a tetrahedron.

Let us begin with some elementary observations. A convex annulus in $\mathbb{E}^2$ (cf. Problem 2.6) is obviously not determined by the entire set of its projections, since the latter can recognize only its outer boundary. We shall therefore restrict ourselves to convex sets, though projections of more general sets (such as the "k-convex" bodies defined in Note 3.1) can still be considered. Many of the results

extend to compact convex sets, so we shall generally work with those, rather than
the full-dimensional convex bodies. We say that a compact convex set is *determined* by a family of projections if it can be distinguished from other compact
convex sets by these projections.

Suppose that $1 \leq k \leq n - 1$, $S \in \mathcal{G}(n, k)$, and K is a compact convex set in
$\mathbb{E}^n$. If we know the projection $K|S$, and if $1 \leq j < k$ and T is a j-dimensional
subspace contained in S, then we also know $K|T$, since the latter is equal to
$(K|S)|T$. The lower the dimension of the subspace, the weaker the information
content of the projection. Generally, then, one should attempt to prove positive results for projections on low-dimensional subspaces, and to find counterexamples
for projections on high-dimensional ones. Assume, for example, that (P) is some
property of compact convex sets, preserved by taking projections, and suppose
that $n \geq 3$ is fixed. If one can prove for $k = 2$ that a compact convex set K in $\mathbb{E}^n$
has (P) whenever all its projections on k-dimensional subspaces have (P), then it
follows that this is true for this same n and all k with $2 \leq k \leq n - 1$.

In most cases, however, an alternative approach is possible. Let (P) be any
property of compact convex sets, not necessarily preserved by projections. Suppose that we can prove for *all* $n \geq 3$ and $k = n - 1$ that a compact convex set
K in $\mathbb{E}^n$ has (P) whenever all its projections on k-dimensional subspaces have
(P). Then this statement is true for all n and k with $2 \leq k \leq n - 1$. To see this,
assume that all projections of K on 2-dimensional subspaces have (P), and let S
be any 3-dimensional subspace. Identifying S with $\mathbb{E}^3$, and applying our result for
$k = n - 1$ with $n = 3$, we see that the projection of K on S has (P). It then follows,
by induction on the dimension of subspaces, that K has (P). So in attempting to
prove theorems of this type, we may confine ourselves to $k = 2$ or to $k = n - 1$,
whichever is convenient.

We give two proofs of the first simple theorem. The second proof, clearly
shorter, applies the support function; the reader not familiar with this extremely
useful concept is encouraged to consult Section 0.6 and Figure 0.2.

Theorem 3.1.1. *Let $1 \leq k \leq n-1$, and let K be a compact convex set in $\mathbb{E}^n$. Then
K is determined by all its projections $K|S$, $S \in \mathcal{G}(n, k)$. In fact, K is determined
by its projections on all 2-dimensional subspaces containing a given line through
the origin.*

Geometric proof. For the first statement it suffices to take $k = 1$. Suppose that
we know the projection $K|l_u$ of K on the line l_u through o parallel to u, for each
$u \in S^{n-1}$. Let

$$C = \cap\{(K|l_u) \times u^\perp : u \in S^{n-1}\}.$$

Then $K \subset C$. Let $x \in \text{co} K$, and suppose that H is a hyperplane separating x
from K. If $v \in S^{n-1}$ is orthogonal to H, then $H \cap l_v$ separates the projection of

x on l_v from $K|l_v$, so x does not belong to the cylinder $(K|l_v) \times v^\perp$. This means that $x \in \text{co}\, C$, and so $C = K$.

Let us consider the second part of the theorem. Let l_v be a fixed line through o, and suppose that we know $K|S$ for each $S \in \mathcal{G}(n, 2)$ such that S contains l_v. If $u \in S^{n-1}$, let $S \in \mathcal{G}(n, 2)$ contain u and v. Then $K|l_u = (K|S)|l_u$, so K is determined by the first statement of the theorem. ∎

Analytic proof. For the first statement it suffices to take $k = 1$. Let $u \in S^{n-1}$. If we know $K|l_u$, then we know $h_K(u)$ and $h_K(-u)$. Since convex bodies are determined by their support functions (see Section 0.6), K is determined. Again, the second part of the theorem follows from the first. ∎

Theorem 3.1.2. *Let* $1 \le k \le n - 1$. *Given any finite set* $\mathcal{S}$ *of k-dimensional subspaces, there is a convex polytope P in $\mathbb{E}^n$ not determined by its projections $P|S$ for $S \in \mathcal{S}$.*

Proof. It suffices to prove this for $k = n - 1$. Let $\mathcal{S} \subset \mathcal{G}(n, n - 1)$ be finite, and construct any n-dimensional convex polytope P with no facet orthogonal to any subspace in $\mathcal{S}$. Suppose that

$$Q = \cap\{(P|S) \times S^\perp : S \in \mathcal{S}\}.$$

Then Q is a convex polytope, $P \subset Q$, and Q has the same projections as P on the subspaces in $\mathcal{S}$. If x is in the relative interior of a facet of P, then by the construction of P, $x \in \text{int}\left((P|S) \times S^\perp\right)$, whenever $S \in \mathcal{S}$. Therefore $Q \ne P$. ∎

The proof of Theorem 3.1.2 actually shows rather more. If P is a convex polytope in $\mathbb{E}^n$, then it is generally not possible to choose a finite set of subspaces in such a way that the corresponding projections distinguish P from every other convex polytope. In the language of Chapter 1, convex polytopes cannot be *verified* by any finite set of projections. Further a priori information about a convex polytope is necessary in order to obtain such a result; see the discussion on geometric probing in Note 3.8. Without this, however, no close analogue of Theorem 1.2.21 can be proved for projections, and we shall concentrate on examining what can be done with knowledge of all the projections.

Theorem 3.1.3. *Suppose that K_1 and K_2 are compact convex sets in $\mathbb{E}^n$, and that $2 \le k \le n - 1$. If $K_1|S$ is homothetic to (or a translate of) $K_2|S$ for each $S \in \mathcal{G}(n, k)$, then K_1 is homothetic to K_2 (or a translate of K_2, respectively).*

Proof. Since equivalence up to homothety or translation is preserved by projection, it suffices (cf. the remarks at the beginning of this chapter) to find a proof for $k = 2$, that is, for projections on 2-dimensional subspaces.

Choose points p_1, q_1 in bd K_1 such that $[p_1, q_1]$ is a diameter of K_1. We may assume that $p_1 = o$ and $q_1 = (1, 0, \ldots, 0)$ so that the hyperplanes $x_1 = 0$ and $x_1 = 1$ support K_1 and intersect K_1 at the unique points p_1 and q_1, respectively. After translating K_2, we may assume that it is supported by the hyperplane $x_1 = 0$ at o and also by $x_1 = r, r > 0$.

Suppose that S is a 2-dimensional plane containing the x_1-axis. The projections $K_i|S$ are supported at o by the line $t = S \cap \{x_1 = 0\}$, $i = 1, 2$. For $i = 1$, o is the unique point with this property; since $K_1|S$ is homothetic to $K_2|S$, this is also true for $i = 2$, and the homothety must be a dilatation. Next, observe that if $q_2 = (r, 0, \ldots, 0)$, then the lines parallel to t supporting $K_i|S$ do so at q_i, $i = 1, 2$. Therefore the dilatation factor is r, and $K_2|S = r(K_1|S)$. It follows that

$$(K_2|S) \times S^\perp = r\big((K_1|S) \times S^\perp\big).$$

As in the proof of Theorem 3.1.1, for $i = 1, 2$, K_i is the intersection, over all $S \in \mathcal{G}(n, 2)$ containing the x_1-axis, of the cylinders $(K_i|S) \times S^\perp$. Therefore $K_2 = rK_1$, and K_1 is homothetic to K_2. For the case where the projections are translates, we have $r = 1$, and K_1 is a translate of K_2. ∎

Lemma 3.1.4. *A compact convex set K is centrally symmetric with center c if and only if $-K = K - 2c$.*

Proof. We have $x \in -K$ if and only if $-x \in K$. If K has a center c, $-x \in K$ is equivalent to $x + 2c \in K$, or $x \in K - 2c$. Therefore $-K = K - 2c$.

Conversely, suppose that $-K = K - 2c$, and let $x \in K$. Then $-x \in -K$, so $-x + 2c \in K$. Therefore K has a center at c. ∎

Corollary 3.1.5. *Suppose that $2 \le k \le n - 1$, and that all projections $K|S$, $S \in \mathcal{G}(n, k)$, of a compact convex set in $\mathbb{E}^n$ are centrally symmetric. Then K is centrally symmetric.*

Proof. Let $K_1 = K$ and $K_2 = -K$. For each $S \in \mathcal{G}(n, k)$, $K_2|S = -(K_1|S)$. As $K_1|S$ is centrally symmetric, $K_2|S$ is a translate of $K_1|S$, by Lemma 3.1.4. By Theorem 3.1.3, $K_2 = -K$ is a translate of $K_1 = K$. By Lemma 3.1.4 again, K is centrally symmetric. ∎

Corollary 3.1.6. *Suppose that $2 \le k \le n - 1$, and that every projection $K|S$, $S \in \mathcal{G}(n, k)$, of a compact convex set K in $\mathbb{E}^n$ is a k-dimensional ball. Then K is an n-dimensional ball.*

Proof. In Theorem 3.1.3, let $K_1 = K$ and $K_2 = B$. We conclude that K is homothetic to B, and so it must be an n-dimensional ball. ∎

It is natural to try to extend Corollary 3.1.6 to ellipsoidal projections. This seems to require a separate proof, though one very similar in approach to that of Theorem 3.1.3.

Theorem 3.1.7. *Suppose that* $2 \leq k \leq n - 1$, *and that every projection* $K|S$, $S \in \mathcal{G}(n, k)$, *of a compact convex set* K *in* $\mathbb{E}^n$ *is a* k-*dimensional ellipsoid. Then* K *is an ellipsoid.*

Proof. In view of the remarks at the beginning of this chapter, it is enough to consider the case $k = n - 1$.

If the dimension of K is less than n, we can find an $S \in \mathcal{G}(n, n-1)$ such that a translate of S contains K. Then $K|S$ is a translate of K, and so K is a (degenerate) ellipsoid. We may therefore assume that K is a convex body.

Choose points p, p' in bd K such that $[p, p']$ is a diameter of K. We may assume that $p = (-1, 0, \ldots, 0)$ and $p' = (1, 0, \ldots, 0)$ so that the hyperplanes $x_1 = -1$ and $x_1 = 1$ support K and are orthogonal to $[p, p']$.

Let H be a supporting hyperplane to K parallel to $[p, p']$. Suppose that $S \in \mathcal{G}(n, n-1)$ contains $[p, p']$ and is orthogonal to H. Then $K|S$ is an ellipsoid, and $[p, p']$ is one of its axes. The plane $H \cap S$ supports $K|S$ at a single point contained in $x_1 = 0$. Therefore $H \cap K$ lies in $x_1 = 0$. This implies that $K \cap \{x_1 = 0\}$ is the same as its projection $E = K|\{x_1 = 0\}$, an ellipsoid. Moreover, E is easily seen to be centered.

Let ϕ be an affine transformation of $\mathbb{E}^n$ keeping $[p, p']$ fixed and mapping E onto a ball D in $x_1 = 0$ centered at o. Since affine transformations take ellipsoids onto ellipsoids, we see that all the projections of ϕK on 2-dimensional subspaces are ellipses. Let $[q, q']$ be a diameter of D. The previous argument, with ϕK and $[q, q']$ replacing K and $[p, p']$, shows that the hyperplane through o orthogonal to $[q, q']$ intersects ϕK in an ellipsoid E'. (Note that we did not use the fact that $[p, p']$ is a chord of K of maximal length, only that the supporting hyperplanes to K at p and p' are orthogonal to $[p, p']$.) Now E' has $[p, p']$ as an axis. So every 2-dimensional subspace containing $[p, p']$ intersects ϕK in an ellipse, with $[p, p']$ as one axis and a diameter of D as the other. We conclude that ϕK is an ellipsoid of revolution. This means that K is an ellipsoid, as required. ∎

As we remarked earlier, if two compact convex sets are homothetic, then the same is true for their projections on a subspace. This is not generally true if the two are only affinely equivalent, or even similar. For example, the projection of a tetrahedron on a fixed plane in $\mathbb{E}^3$ may look quite different from that of a rotation of the tetrahedron. Despite this, it seems possible at first that there might be a version of Theorem 3.1.3 applicable to affine transformations or similarities instead of homotheties. Though extra conditions can be imposed to achieve this (see Note 3.1), the next startling example shows that this is not generally the case.

Theorem 3.1.8. *There are centered, coaxial convex bodies of revolution K_1 and K_2 in $\mathbb{E}^n$, $n \geq 3$, such that for each $S \in \mathcal{G}(n, 2)$, $K_1|S$ is similar to $K_2|S$, yet K_1 is not affinely equivalent to K_2.*

Proof. The support function h_{K_1} of K_1 will be defined for nonzero $x \in \mathbb{E}^n$ by

$$h_{K_1}(x) = \|x\| \exp\left(\frac{x \cdot Ax}{\|x\|^2}\right),$$

where A is initially any real symmetric $n \times n$ matrix, with eigenvalues a_i, $1 \leq i \leq n$ satisfying the condition $\max |a_i - a_j| \leq 1/2$. The support function h_{K_2} of K_2 is defined similarly, where the matrix A is replaced by $-A$ (whose eigenvalues satisfy the same condition). We must check that h_{K_1} and h_{K_2} are indeed support functions, but it is more convenient for us to assume this temporarily, and proceed as follows. First, we check that the projections of K_1 and K_2 have the required properties. We shall then restrict A slightly, and show that if K_1 and K_2 are affinely equivalent, then they must be similar. Next, we make a choice of A ensuring that K_1 and K_2 can be obtained by rotating planar bodies about the x_n-axis, and use this fact to prove that K_1 and K_2 cannot be similar. Finally, we verify that for this K_1 and K_2, the preceding expressions do define support functions.

Let $S \in \mathcal{G}(n, 2)$. For $i = 1, 2$, the support function in S of the planar convex body $K_i|S$ is h_{K_i} restricted to S, by (0.21). Let v and w be orthogonal unit vectors in S, and suppose that $u = u(\theta) = v \cos \theta + w \sin \theta$ is any unit vector in S. Then

$$h_{K_1}(u(\theta)) = \exp((v \cos \theta + w \sin \theta) \cdot A(v \cos \theta + w \sin \theta))$$

and

$$h_{K_2}\left(u\left(\theta + \frac{\pi}{2}\right)\right) = \exp(-(-v \sin \theta + w \cos \theta) \cdot A(-v \sin \theta + w \cos \theta)).$$

It follows that

$$h_{K_1}(u(\theta))/h_{K_2}\left(u\left(\theta + \frac{\pi}{2}\right)\right) = \exp(v \cdot Av + w \cdot Aw).$$

Since the last expression is independent of θ, the property (0.23) of support functions implies that $K_2|S$ is a dilatate of a rotation of $K_1|S$ by $\pi/2$ in S about the origin.

Let us take A to be a diagonal matrix, with eigenvalues a_i, $1 \leq i \leq n$, satisfying

$$a_1 \geq a_2 \geq \cdots \geq a_n.$$

Suppose that $\phi K_1 = K_2$, where ϕ is an affine map. In Section 0.2, we show that since K_i is centered, $i = 1, 2$, ϕ must actually be a linear map. By (0.27), we have $h_{K_1}(\phi^t u) = h_{K_2}(u)$, so

$$\|\phi^t u\|^2 = \exp\left(-2 \sum_{i=1}^n \frac{a_i(\phi^t u)_i^2}{\|\phi^t u\|^2} - 2 \sum_{i=1}^n a_i u_i^2\right), \tag{3.1}$$

for all $u \in S^{n-1}$. Suppose first that $u \in S^{n-1}$ is such that $u_i = 0$ for $3 \leq i \leq n$, in which case we can set $u_1 = 2z/(1+z^2)$ and $u_2 = (1-z^2)/(1+z^2)$. We claim that both sides of (3.1) are then constant. The substitution yields an equation of the form $p(z)/(1+z^2)^2 = e^{f(z)}$, where $p(z)$ is a polynomial of degree at most four. Note that $p(z)$ is nonzero for real z, since $\|\phi'(u)\|$ cannot be zero. We rewrite this in the form

$$p(z) = (1+z^2)^2 e^{f(z)},$$

which then holds for all real z. Further, $f(z)$ is a rational function whose denominator $p(z)(1+z^2)^2$ is nonzero for each real z. Therefore both sides of the equation represent functions analytic for z in a domain in the complex plane $\mathbb{C}$ containing the real axis. Since $p(z)$ is a polynomial, we may take its domain to be the whole of $\mathbb{C}$, and then a standard uniqueness theorem (see, for example, [167, Theorem 1, p. 261]) implies that the last equation holds for all $z \in \mathbb{C}$. The exponential function has no zeros in $\mathbb{C}$, so the only zeros of the right-hand side are double zeros at $z = \pm i$. These must then be precisely the zeros of the left-hand side, implying that $p(z)$ is a constant multiple of $(1+z^2)^2$, and hence that $e^{f(z)}$ is constant. Therefore both sides of (3.1) are constant, under the assumption that $u_i = 0$ for $3 \leq i \leq n$. This implies that the first two columns of the matrix ϕ' are orthogonal, and the sum of the squares of the entries in each of these columns is the same.

The same conclusion can now be drawn for any pair of columns by replacing u_1 and u_2 with the appropriate pair of coordinates of u. It follows that ϕ' is an orthogonal matrix $C = (c_{ij})$ multiplied by a constant, b^{-1} say. This means that ϕ must be a similarity (see Section 0.2).

Let us take the eigenvalues of A to be $a_i = 1$, $1 \leq i \leq n-1$, and $a_n = 1/2$. This gives

$$h_{K_1}(x) = \|x\| \exp \left(\left(\sum_{i=1}^{n-1} x_i^2 + \frac{1}{2} x_n^2 \right) / \|x\|^2 \right)$$

and

$$h_{K_2}(x) = \|x\| \exp \left(\left(-\sum_{i=1}^{n-1} x_i^2 - \frac{1}{2} x_n^2 \right) / \|x\|^2 \right),$$

respectively. Let H_1 and H_2 be the planar convex bodies whose support functions are given for nonzero $x = (x_1, x_n)$ in the $\{x_1, x_n\}$-plane by

$$h_{H_1}(x) = \|x\| \exp \left(\left(x_1^2 + \frac{1}{2} x_n^2 \right) / \|x\|^2 \right)$$

and

$$h_{H_2}(x) = \|x\| \exp \left(\left(-x_1^2 - \frac{1}{2} x_n^2 \right) / \|x\|^2 \right),$$

respectively. Then K_j is obtained from H_j by rotation about the x_n-axis, for $j = 1, 2$.

We claim that K_1 and K_2 cannot be similar. To prove this, we set $x_1 = \cos\theta$ and $x_n = \sin\theta$ in the previous expression for h_{H_1}, and then use (0.40) to obtain an equation for the boundary of H_1 in terms of the parameter θ. Differentiating with respect to θ, we see that the radial function (cf. Section 0.7) of H_1 is strictly decreasing as θ increases from 0 to $\pi/2$. The intersection of K_1 with the subspace $x_n = 0$ is an $(n-1)$-dimensional ball of radius e. If the intersection of K_1 with any other $(n-1)$-dimensional subspace is an $(n-1)$-dimensional ball, it must also have radius e, and this is impossible. Therefore the x_n-axis is the only axis of revolution of K_1, and K_1 and K_2 can only be similar if H_1 is a dilatate of H_2. This is not the case, so the claim is proved.

Finally, we shall prove that h_{K_1} is indeed a support function; the proof for h_{K_2} is similar. Since K_1 is obtained by rotating H_1 about the x_n-axis, it suffices to show that h_{H_1} is really a support function. To this end, consider the planar body L whose radial function is given in polar coordinates by

$$\rho_L(\theta) = h_{H_1}(\theta)^{-1} = \exp\left(-\frac{1}{2} - \frac{1}{2}\cos^2\theta\right).$$

A straightforward calculation shows that the numerator of the expression (0.41) for the curvature of L is nonnegative, so L must be convex. Now the expression (0.36) for the radial function of a polar body shows that $h_{H_1} = h_{L^*}$, the support function of the polar body L^* of L. ∎

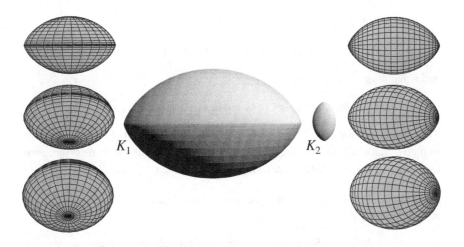

Figure 3.2. Nonsimilar convex bodies with similar projections.

The bodies K_1 and K_2 for the case $n = 3$ of the previous theorem are shown in Figure 3.2. The bodies themselves are depicted in the middle of the picture. On the left and right are pairs of projections of K_1 and K_2; these are scaled and, in the case of K_2, rotated by $\pi/2$.

3.2. The width function and central symmetral

In Theorem 3.1.3, we excluded the case $k = 1$, corresponding to projections on lines through the origin. The theorem is false in this case, since if K is any compact convex set, the reflection $-K$ of K in the origin projects in each 1-dimensional subspace onto a segment of the same length as that of K. By Lemma 3.1.4, $-K$ is a translate of K if and only if K is centrally symmetric, so the two sets K, $-K$ for any compact convex set (or polytope) K that is not centrally symmetric will provide counterexamples.

The length of the projection of K on a 1-dimensional subspace is just the width of K in a direction parallel to the subspace, defined as follows.

Definition 3.2.1. Let K be a compact convex set in $\mathbb{E}^n$. Then K has two (possibly identical) supporting hyperplanes orthogonal to a unit vector u, namely, H_u and H_{-u}, where $H_u = \{x : x \cdot u = h_K(u)\}$. (See Figure 0.2.) The distance between these hyperplanes is the *width* $w_K = v_{1,K}$ of K in the direction u. In other words,

$$w_K(u) = h_K(u) + h_K(-u),$$

for $u \in S^{n-1}$. The *minimal width* of K is

$$\text{minw}\, K = \min_{u \in S^{n-1}} w_K(u).$$

If $0 \le t \le 1$, then by (0.23),

$$h_{(1-t)K+t(-K)} = (1-t)h_K + th_{-K},$$

and since $w_{-K} = w_K$, we have

$$w_{(1-t)K+t(-K)} = (1-t)w_K + tw_{-K} = w_K.$$

This means that for each K that is not centrally symmetric, there is a whole continuum of generally noncongruent compact convex sets with the same projections as K on 1-dimensional subspaces. Of all of these, one is of special interest.

Definition 3.2.2. Let K be a compact convex set. The *central symmetral* $\triangle K$ of K is defined by $\triangle K = \frac{1}{2}(K + (-K))$.

The term *difference body* has also been used for $\triangle K$, but it makes more sense to reserve this for $2\triangle K = K + (-K)$.

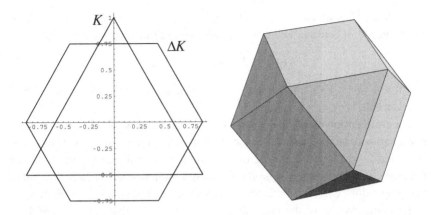

Figure 3.3. Central symmetrals of an equilateral triangle and a regular tetrahedron.

Clearly $\triangle K$ is a centered compact convex set, and $\triangle K$ is a translate of K if and only if K is centrally symmetric. Therefore the only interesting examples occur when K is not centrally symmetric. Suppose that $o \in \operatorname{int} K$. To construct $\triangle K$, one takes the union of K and all translates of $-K$ placed so that the corresponding translate of o lies on the boundary of K, and then dilatates by a factor of $1/2$. Figure 3.3 depicts the central symmetrals of an equilateral triangle and a regular tetrahedron; they are a regular hexagon and cuboctahedron, respectively. Generally, $\triangle K$ is the unique centered compact convex set such that

$$w_{\triangle K}(u) = w_K(u),$$

for each $u \in S^{n-1}$.

Let us note the following inequality concerning the volume of $\triangle K$.

Theorem 3.2.3. *Let K be a convex body in $\mathbb{E}^n$. Then*

$$\lambda_n(\triangle K) \geq \lambda_n(K),$$

with equality if and only if K is centrally symmetric.

Proof. This is an immediate consequence of the Brunn–Minkowski inequality (B.10) (with $t = 1/2$). ∎

There still remains the possibility that Theorem 3.1.3 (for translations) works if $k = 1$ and K_1 is some fixed special compact convex set. This is, in fact, true for parallelotopes, for example; see Note 3.5. But it even fails if K_1 is a ball, and this serves as an introduction to a much-studied family of convex bodies.

Definition 3.2.4. A convex body K is of *constant width* if $w_K(u)$ is constant for all $u \in S^{n-1}$.

In view of Definition 3.2.1, an alternative, and useful, characterization of convex bodies K of constant width is that their support functions should satisfy

$$h_K(u) + h_K(-u) = \text{constant},$$

for all $u \in S^{n-1}$. Evidently, a convex body K is of constant width if and only if $\triangle K$ is a ball.

An n-dimensional ball in $\mathbb{E}^n$ is obviously of constant width. The most famous nonspherical examples are the Reuleaux polygons in the plane; their shape has been borrowed for some coins currently used in Britain (see Figure 3.4). The boundary of a Reuleaux polygon is composed of finitely many circular arcs, all of the same radius, each of whose centers is located at the end of one of the arcs.

It is perhaps one of the great surprises of convexity that not only are there myriad nonspherical convex bodies of constant width, but that in addition a full understanding of some of their basic properties still eludes us.

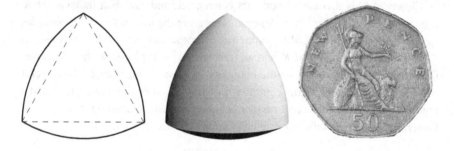

Figure 3.4. Convex bodies of constant width.

Theorem 3.2.5. *Nonspherical convex bodies of constant width exist in $\mathbb{E}^n$ for all $n \geq 2$.*

Geometric proof. Let T be a Reuleaux triangle, placed in $\mathbb{E}^n$ so that it is contained in the $\{x_1, x_n\}$-plane and is symmetric with respect to the x_n-axis. Let K be the convex body formed by rotating T about this axis, so the intersection of K with a hyperplane H orthogonal to the axis is an $(n-1)$-dimensional ball whose diameter has length equal to that of the intersection of T and H. We shall prove that K is of constant width in $\mathbb{E}^n$.

Let u be any direction, and suppose that H_i, $i = 1, 2$, are the hyperplanes orthogonal to u supporting K at points p_i, $i = 1, 2$. Denote by J_i the hyperplane orthogonal to the x_n-axis passing through p_i, and let q_i be the intersection of J_i with the x_n-axis, $i = 1, 2$. Now $K \cap J_i$ is an $(n-1)$-dimensional ball. Therefore $L_i = H_i \cap J_i$ is an $(n-2)$-dimensional plane tangent to this ball, and so orthogonal to the line segment $s_i = [p_i, q_i]$. By their definition, L_1 and L_2 are parallel, so s_1

and s_2 (both orthogonal to the x_n-axis) are also parallel. Consequently, s_1 and s_2 lie in the same 2-dimensional plane F containing the x_n-axis; F is parallel to u, so if $t_i = F \cap H_i, i = 1, 2$, then the distance between H_1 and H_2 is the same as that between t_1 and t_2. Since the lines t_i support $F \cap K$, and the latter is a rotation of T about the x_n-axis, the proof is complete. ∎

Analytic proof. Let $u = (u_1, \ldots, u_n)$ be any direction. Then

$$h_K(u) = h_T\left((u_1^2 + \cdots + u_{n-1}^2)^{1/2}, u_n\right),$$

so

$$
\begin{aligned}
h_K(u) + h_K(-u) &= h_T\left((u_1^2 + \cdots + u_{n-1}^2)^{1/2}, u_n\right) \\
&\quad + h_T\left((u_1^2 + \cdots + u_{n-1}^2)^{1/2}, -u_n\right) \\
&= h_T\left((u_1^2 + \cdots + u_{n-1}^2)^{1/2}, u_n\right) \\
&\quad + h_T\left(-(u_1^2 + \cdots + u_{n-1}^2)^{1/2}, -u_n\right),
\end{aligned}
$$

a constant independent of u. ∎

If we let K_1 be the unit ball and K_2 be a nonspherical convex body of constant width 2, we see again that Theorem 3.1.3 is false if $k = 1$.

Next, we prove for convex bodies of constant width an analogue of our earlier results. A geometric proof is not difficult to construct, but we shall only give a more transparent and much shorter analytic proof.

Theorem 3.2.6. *Suppose that $2 \leq k \leq n - 1$ and that all projections $K|S$, $S \in \mathcal{G}(n, k)$, of a compact convex set in $\mathbb{E}^n$ are of constant width (in S). Then K is a convex body of constant width.*

Proof. By (0.21) and the definition of w_K, we have

$$w_K(u) = w_{K|S}(u) = c_S,$$

for all $u \in S^{n-1} \cap S$, where c_S is a constant. However, c_S is actually independent of S, since there is a $u \in S^{n-1}$ common to any two members of $\mathcal{G}(n, k)$, $k \geq 2$. ∎

In Theorem 1.2.4 we proved that, up to a translation, no less information is available from a set of at least two X-rays, if these are also only known up to a translation. If $2 \leq k \leq n - 1$, and the projections of a compact convex set K on all k-dimensional subspaces are known only up to a translation, then Theorem 3.1.3 says that K is determined, up to a translation; so the first part of Theorem 3.1.1 still holds, in this sense. Moreover, by the proof of Theorem 3.1.3, this applies also to the second part of Theorem 3.1.1. However, we have already seen that if the projections of K on all 1-dimensional subspaces are only known up to a translation, then K is no longer determined, even up to a translation and even if all these projections are of equal length.

To circumvent these difficulties, we shall consider a special subclass of the class of compact convex sets for which they do not occur, namely, those that are centrally symmetric. Note that if K is centrally symmetric with center c, then $-K = K - 2c$, by Lemma 3.1.4. Therefore if $0 \le t \le 1$,

$$(1 - t)K + t(-K) = (1 - t)K + t(K - 2c) = K - 2ct,$$

and in particular $\triangle K = K - c$, is a translate of K. This eliminates the examples mentioned just before Definition 3.2.2; the next theorem does the same for those involving convex bodies of constant width.

Theorem 3.2.7. *If K is a centrally symmetric convex body of constant width in $\mathbb{E}^n$, then K is a ball.*

Proof. Central symmetry implies that $h_K(u) = h_K(-u)$, and constant width c implies that $h_K(u) + h_K(-u) = c$, for all $u \in S^{n-1}$. Therefore $h_K(u) = c/2$, for all $u \in S^{n-1}$. Since convex bodies are determined by their support functions (see Section 0.6), K must be a ball. ∎

It is easily seen that a centrally symmetric compact convex set is determined, up to a translation, if all its projections on 1-dimensional subspaces are known, up to translation. We shall soon see that much more can be proved, by embarking upon a trip into the heart of the classical theory of convexity.

3.3. Projection functions and the Blaschke body

If we know the projection of a compact convex set on a 1-dimensional subspace only up to a translation, then we simply know the length of this projection; in other words, the width function. More generally, we could measure the k-dimensional volume of the projection of the set on k-dimensional subspaces. Alternatively, it might be possible to measure the surface area, or the mean width of the projections. Minkowski's theory of mixed volumes, and in particular intrinsic volumes, provides a common generalization of all of these possibilities. (An introduction to mixed volumes and intrinsic volumes is given in Appendix A.) If K is a compact convex set in $\mathbb{E}^n$, and $1 \le i \le k \le n - 1$, we can consider $V_i(K|S)$, where $S \in \mathcal{G}(n, k)$ is a k-dimensional subspace. The function $V_i(K|\cdot)$ is called a *projection function*. The special case $i = k = 1$ corresponds to the width function w_K, and when $i = k = n - 1$ we usually refer to the *brightness function* of K and denote it by v_K. Note that when $i = k$, $V_i(K|S)$ is the λ_i-measure of $K|S$; in this case we speak of the *ith projection function*. If $i = k - 1$, or $i = 1$, $V_i(K|S)$ is a constant multiple of the surface area, or the mean width, respectively, of $K|S$ (measured in S). Also of special significance is the case $k = n - 1$, where we refer to $V_i(K|\cdot)$ as the *i*th *girth function* of K. In this section we shall investigate the extent to which a compact convex set is determined by its projection functions.

If K is a compact convex set in $\mathbb{E}^n$, we can define, for each i, $1 \leq i \leq n - 1$, a finite Borel measure $S_i(K, \cdot)$ in S^{n-1}, called the *i*th *area measure* of K. See Section A.1 for an introductory example and Section A.2 for several properties of these measures used in the sequel.

The first important result says that an area measure essentially determines the associated set. It is not necessary to know the constant $c_{n,i}$, but it is actually defined by (A.27).

Theorem 3.3.1 (Aleksandrov's uniqueness theorem). *Let $1 \leq i \leq n - 1$, and let K_1, K_2 be compact convex sets, of dimension at least $i + 1$, in $\mathbb{E}^n$. If $S_i(K_1, \cdot) = S_i(K_2, \cdot)$, then K_1 is a translate of K_2.*

Proof. By (A.32) and (A.26),

$$
\begin{aligned}
V(K_1; K_2, i; B, n - i - 1) &= \frac{1}{n} \int_{S^{n-1}} h_{K_1}(u) \, dS_i(K_2, u) \\
&= \frac{1}{n} \int_{S^{n-1}} h_{K_1}(u) \, dS_i(K_1, u) \\
&= c_{n,i+1} V_{i+1}(K_1).
\end{aligned}
$$

The special case (B.19) of the Aleksandrov–Fenchel inequality gives

$$
V(K_1; K_2, i; B, n - i - 1)^{i+1} \geq c_{n,i+1}^{i+1} V_{i+1}(K_2)^i V_{i+1}(K_1).
$$

Substituting, we obtain $V_{i+1}(K_1) \geq V_{i+1}(K_2)$. Interchanging K_1 and K_2 yields the opposite inequality, so $V_{i+1}(K_1) = V_{i+1}(K_2)$. This implies that equality holds in (B.19), and hence that K_1 and K_2 are homothetic. If we put $K_2 = r K_1 + x$ in the first set of equalities, we get

$$
\begin{aligned}
c_{n,i+1} V_{i+1}(K_1) &= V(K_1; K_2, i; B, n - i - 1) \\
&= V(K_1; r K_1 + x, i; B, n - i - 1) \\
&= r^i V(K_1, i + 1; B, n - i - 1) \\
&= r^i c_{n,i+1} V_{i+1}(K_1),
\end{aligned}
$$

where we have used the linearity property (A.16) and translation invariance (A.19) of mixed volumes. Therefore $r = 1$, and K_1 is a translate of K_2. ■

Our next theorem employs mixed volumes and area measures to produce necessary and sufficient conditions that the projections of two compact convex sets on each $(n - 1)$-dimensional subspace have the same *i*th intrinsic volumes; in other words, that they have the same *i*th girth function.

Theorem 3.3.2. *Let* $1 \le i \le n - 1$, *and let* K_1, K_2 *be compact convex sets in* $\mathbb{E}^n$. *The following conditions are equivalent:*

$$V_i(K_1|u^\perp) = V_i(K_2|u^\perp), \tag{3.2}$$

for all $u \in S^{n-1}$;

$$S_i(K_1, \cdot) + S_i(-K_1, \cdot) = S_i(K_2, \cdot) + S_i(-K_2, \cdot); \tag{3.3}$$

$$V(K_1, i; B, n-i-1; K) = V(K_2, i; B, n-i-1; K), \tag{3.4}$$

for all centrally symmetric compact convex sets K.

If K_1 *and* K_2 *are convex bodies of class* C_+^2, *these conditions are also equivalent to*

$$F_{K_1}^{(i)} + F_{-K_1}^{(i)} = F_{K_2}^{(i)} + F_{-K_2}^{(i)}. \tag{3.5}$$

Proof. Suppose that (3.2) holds. By the generalized Cauchy projection formula (A.45),

$$c_{n-1,i} V_i(K_j|u^\perp) = \frac{1}{2} \int_{S^{n-1}} |u \cdot v| \, dS_i(K_j, v)$$

$$= \frac{1}{2} \int_{S^{n-1}} |u \cdot v| \, dS_i(-K_j, v),$$

for $j = 1, 2$ and for all $u \in S^{n-1}$. Set

$$\mu = S_i(K_1, \cdot) + S_i(-K_1, \cdot) - S_i(K_2, \cdot) - S_i(-K_2, \cdot).$$

Then μ is a signed finite even Borel measure in S^{n-1}, and

$$\int_{S^{n-1}} |u \cdot v| \, d\mu(v) = 0,$$

for all $u \in S^{n-1}$. By Theorem C.2.1, this implies that μ is identically zero, so (3.3) is proved.

Assume (3.3). Let K be any centrally symmetric compact convex set. Note that $h_K = h_{-K}$. Using this and (A.32), we have for $j = 1, 2$,

$$V(K_j, i; B, n-i-1; K) = \frac{1}{n} \int_{S^{n-1}} h_K(u) \, dS_i(K_j, u)$$

$$= \frac{1}{2n} \int_{S^{n-1}} h_K(u) \, d\big(S_i(K_j, u) + S_i(-K_j, u)\big),$$

giving (3.4).

Suppose that (3.4) is true. Let K be the line segment $[-u, u]$, $u \in S^{n-1}$. By (0.25), $h_K(v) = |u \cdot v|$ for $v \in S^{n-1}$, so

$$V(K_j, i; B, n - i - 1; [-u, u]) = \frac{1}{n} \int_{S^{n-1}} |u \cdot v| \, dS_i(K_j, v)$$

$$= \frac{2c_{n-1,i}}{n} V_i(K_j | u^{\perp}),$$

for $j = 1, 2$, by (A.32) and the generalized Cauchy projection formula (A.45). This proves (3.2).

The definitions of C_+^2 and $F_K^{(i)}$ may be found in Section 0.9 (see especially (0.42)). Equation (A.7) shows that for convex bodies of class C_+^2, (3.5) is equivalent to (3.3). ∎

For a generalization of the last theorem, see Note 4.9.

Remark 3.3.3. (i) The equivalence (3.2)⇔(3.4) of Theorem 3.3.2, with $i = n-1$, implies that if a convex body K is centrally symmetric, then

$$V(K_1, n - 1; K) = V(K_2, n - 1; K)$$

for all compact convex K_1, K_2 such that $\lambda_{n-1}(K_1 | u^{\perp}) = \lambda_{n-1}(K_2 | u^{\perp})$ for all $u \in S^{n-1}$. This is actually a characterization of central symmetry. To see this, suppose that this condition holds for some arbitrary convex body K. Choosing $K_1 = K$ and $K_2 = -K$, and applying Minkowski's first inequality (B.13), we see that

$$\lambda_n(K)^n = V(K, n - 1; K)^n = V(-K, n - 1; K)^n \geq \lambda_n(-K)^{n-1} \lambda_n(K).$$

But since $\lambda_n(K) = \lambda_n(-K)$, equality holds throughout, implying that $-K$ is homothetic to, and hence a translate of, K. By Lemma 3.1.4, K is centrally symmetric.

(ii) It can be shown that (3.2)–(3.4) are equivalent to (3.5) holding for almost all $u \in S^{n-1}$, under the weaker condition that K_1 and K_2 both slide freely inside a ball. (This means that for $j = 1, 2$, there is a ball $R_j B$ such that if $x \in \text{bd} \, R_j B$, then there is a $y \in \mathbb{E}^n$ such that $x \in K_j + y \subset R_j B$.) We shall need this extra generality in the proof of Theorem 3.3.15 and to achieve it, one can argue as follows. If K slides freely in the ball RB, then K is a summand of RB, that is, there is a convex body K' such that $RB = K + K'$; see [737, Theorem 3.2.2]. Since K is a summand of a ball, it follows easily from the positive multilinearity of mixed area measures (see [737, p. 279, eq. (5.1.25)]) that $S_i(K, \cdot)$ is bounded above by a multiple of spherical Lebesgue measure. This implies that $S_i(K, \cdot)$ is absolutely continuous with respect to spherical Lebesgue measure, so (A.7) holds, as we noted just after that equation. The equivalence of (3.3) and (3.5) follows immediately.

We would like to consider intrinsic volumes of projections on lower-dimensional, as well as $(n-1)$-dimensional, subspaces. The next lemma provides a link between the two.

Lemma 3.3.4. *Let $1 \leq i \leq k \leq n-1$, and let K_1, K_2 be compact convex sets in $\mathbb{E}^n$. If $V_i(K_1|S) = V_i(K_2|S)$, for all $S \in \mathcal{G}(n,k)$, then $V_i(K_1|u^\perp) = V_i(K_2|u^\perp)$, for all $u \in S^{n-1}$.*

Proof. By Kubota's integral recursion (A.46), for any compact convex set K in $\mathbb{E}^n$ we have

$$V_i(K|u^\perp) = c \int V_i(K|S) \, dS,$$

where the integration is with respect to normalized Haar measure on the set of $S \in \mathcal{G}(n,k)$ such that $S \subset u^\perp$, identified with $\mathcal{G}(n-1,k)$, and the constant c depends only on i, k, and n. Applying this to K_1 and K_2 proves the lemma. ∎

It seems to be unknown whether the converse of the previous lemma is true (see Problem 3.8 and Note 3.5). We can only prove this when $i = 1$.

Theorem 3.3.5. *Let K_1, K_2 be compact convex sets in $\mathbb{E}^n$. Then*

$$V_1(K_1|u^\perp) = V_1(K_2|u^\perp),$$

for all $u \in S^{n-1}$, if and only if

$$w_{K_1}(u) = w_{K_2}(u),$$

for all $u \in S^{n-1}$.

Proof. Suppose that the first equality holds. If $\dim K_1 = 0$, then clearly $\dim K_2 = 0$, and the second equality is trivially true. If $\dim K_1 = 1$, then by considering u orthogonal and parallel to K_1, we see that K_1 and K_2 must be parallel line segments. The first equality then implies that the lengths of these segments must be equal, so we are finished in this case also.

Assume that both K_1 and K_2 are of dimension at least two. For any compact convex set K, $V_1(K|u^\perp) = V_1(-K|u^\perp)$, for all $u \in S^{n-1}$. Also,

$$V_1(\triangle K|u^\perp) = \frac{1}{2}\left(V_1(K|u^\perp) + V_1(-K|u^\perp)\right),$$

for all $u \in S^{n-1}$, by the linearity property (A.16) of mixed volumes. Therefore $V_1(\triangle K_1|u^\perp) = V_1(\triangle K_2|u^\perp)$, for all $u \in S^{n-1}$. Since $\triangle K_1$ and $\triangle K_2$ are centrally symmetric, we have $S_1(\triangle K_j, \cdot) = S_1(-\triangle K_j, \cdot)$ for $j = 1, 2$. By Theorem 3.3.2, $S_1(\triangle K_1, \cdot) = S_1(\triangle K_2, \cdot)$. Now Aleksandrov's uniqueness theorem, Theorem 3.3.1, implies that $\triangle K_1$ and $\triangle K_2$ are translates, and so, being centered, they are equal. The fact that $w_{\triangle K}(u) = w_K(u)$ for each K and $u \in S^{n-1}$ yields the second equality of the theorem. The other direction was proved in Lemma 3.3.4. ∎

Theorem 3.3.6 (Aleksandrov's projection theorem). *Let* $1 \leq i \leq k \leq n - 1$, *and let* K_1, K_2 *be centrally symmetric compact convex sets, of dimension at least* $i + 1$, *in* $\mathbb{E}^n$. *If* $V_i(K_1|S) = V_i(K_2|S)$ *for all* $S \in \mathcal{G}(n, k)$, *then* K_1 *is a translate of* K_2.

Proof. For $j = 1, 2$, $S_i(K_j, \cdot) = S_i(-K_j, \cdot)$, because K_j is centrally symmetric. By Lemma 3.3.4 and Theorem 3.3.2, we see that $S_i(K_1, \cdot) = S_i(K_2, \cdot)$. The proof is completed by Aleksandrov's uniqueness theorem, Theorem 3.3.1. ∎

Note that the restriction to sets of dimension at least $i + 1$ is necessary. For example, suppose that K_1 and K_2 are 2-dimensional compact convex sets contained in the same plane in $\mathbb{E}^3$ and of equal areas. Then their projections on each 2-dimensional subspace also have equal areas, but of course K_1 and K_2 need not be translates, or even congruent.

A powerful corollary follows easily from Theorem 3.3.6. It may be regarded as a deep generalization or analogue of the elementary Theorem 3.2.7. We state it for convex bodies, although Theorem 3.3.6 shows that it holds for compact convex sets of appropriately high dimension.

Corollary 3.3.7. *Let* K *be a centered convex body in* $\mathbb{E}^n$. *Then* K *is determined among all centered convex bodies in* $\mathbb{E}^n$ *by any of the following measurements of all its projections on k-dimensional subspaces:*

(i) *(for* $1 \leq k \leq n - 1$) *the* λ_k-*measures;*

(ii) *(for* $2 \leq k \leq n - 1$) *the surface areas (i.e., the* λ_{k-1}-*measures of the boundaries);*

(iii) *(for* $1 \leq k \leq n - 1$) *the mean widths.*

Proof. With Aleksandrov's projection theorem, Theorem 3.3.6, in hand, we merely have to note that the quantities in (i)–(iii) are constant multiples of V_k, V_{k-1}, and V_1, respectively (see (A.28), (A.35), and (A.50)). ∎

Aleksandrov's projection theorem is a beautiful and fundamental result that has been the starting point of several lines of research (see Notes 3.5 and 4.9). The theorem is not true if K_1 is not centrally symmetric, since in this case setting $K_2 = -K_1$ provides a counterexample. (Moreover, in the absence of central symmetry, one cannot generally even conclude that K_1 and K_2 are congruent; see Theorems 3.3.17 and 3.3.18.) For $k = 1$, we have seen that if K is not centrally symmetric, then the centered set $\triangle K$ has the same width function. We now introduce an analogous set for the brightness function.

Suppose that K is a convex body in $\mathbb{E}^n$ and that $0 \leq t \leq 1$. Recall (see Section A.2) that $S(K, \cdot) = S_{n-1}(K, \cdot)$ denotes the surface area measure of K. By the consequence (A.23) of Minkowski's existence theorem, Theorem A.3.2, there is a convex body whose surface area measure is $(1 - t)S(K, \cdot) + tS(-K, \cdot)$.

Using the equivalence (3.2)⇔(3.3) of Theorem 3.3.2, with $i = n - 1$, we see that each such set has the same brightness function as K, so there will be a continuum of generally noncongruent sets with this property. The most important member of this family corresponds to $t = 1/2$.

Definition 3.3.8. If K is a convex body in $\mathbb{E}^n$, the *Blaschke body* ∇K is the unique centered convex body such that

$$S(\nabla K, \cdot) = \frac{1}{2}S(K, \cdot) + \frac{1}{2}S(-K, \cdot).$$

It is a major obstacle that when $1 < i < n - 1$, there is no corresponding object for the ith projection function; see Note 3.4.

Aleksandrov's projection theorem implies that ∇K is the unique centered convex body with the same brightness function as K. If K is also centrally symmetric, then ∇K is a translate of K, so the only interesting examples are for those K that are not centrally symmetric. In $\mathbb{E}^2$, the width and brightness functions, and therefore ∇K and $\triangle K$, are equal up to a rotation by $\pi/2$ about the origin, but this is not true in higher dimensions. As is shown in Figure 3.5, the Blaschke body of a regular tetrahedron is a centered octahedron. This is because the surface area measure of a regular tetrahedron K consists of four equal point masses located on S^2 at the outer unit normal vectors to its facets (cf. Section A.2); reflecting these masses in the origin, one obtains the surface area measure of $-K$. Adding, we get eight equal point masses situated at the outer unit normal vectors to the facets of an octahedron.

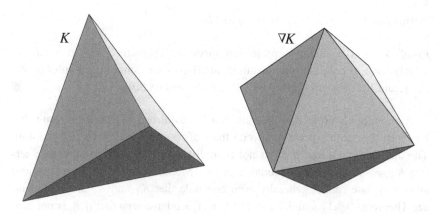

Figure 3.5. The Blaschke body of a tetrahedron.

The following interesting fact about ∇K should be compared with the inequality of Theorem 3.2.3 for the central symmetral; see Note 4.9 for a common generalization of these two results.

Theorem 3.3.9. *Let K be a convex body in $\mathbb{E}^n$. Then*

$$\lambda_n(\nabla K) \geq \lambda_n(K),$$

with equality if and only if K is centrally symmetric.

Proof. Since K and ∇K have the same brightness function, equation (3.4) holds with $i = n - 1$, K_1 and K replaced by ∇K, and K_2 replaced by K. By (A.13), this yields

$$\lambda_n(\nabla K) = V(K, n - 1; \nabla K).$$

Applying Minkowski's first inequality (B.13), we obtain

$$\lambda_n(\nabla K)^n = V(K, n - 1; \nabla K)^n \geq \lambda_n(K)^{n-1}\lambda_n(\nabla K),$$

and hence $\lambda_n(\nabla K) \geq \lambda_n(K)$. Equality holds if and only if K is a translate of ∇K, and therefore if and only if K is centrally symmetric. ∎

It is known that certain centrally symmetric convex bodies – parallelotopes, for example – are determined, among *all* convex bodies, by their width function. In other words, Aleksandrov's projection theorem for $i = 1$ also holds for certain fixed centrally symmetric bodies K_1, even when K_2 is not necessarily centrally symmetric. For the brightness function ($i = n - 1$), however, only parallelotopes have this property. (These facts are discussed in more detail in Note 3.5.) In particular, we shall soon see that the theorem is not true for $i = n - 1$, without the central symmetry of K_2, when K_1 is a ball.

Definition 3.3.10. Let $1 \leq i \leq n - 1$. If K is a convex body in $\mathbb{E}^n$ such that $V_i(K|S)$ $(= \lambda_i(K|S))$ has the same value for each $S \in \mathcal{G}(n, i)$, we say K is of *constant i-brightness*. If $i = 1$, this is equivalent to constant width (cf. Definition 3.2.4). If $i = n - 1$, we simply say K has *constant brightness*. If $V_i(K|u^{\perp})$ has the same value for all $u \in S^{n-1}$, we say that K is of *constant i-girth*. If $i = n - 2$, we simply say K has *constant girth*.

A convex body in $\mathbb{E}^3$ has constant brightness if its shadow on any plane always has the same area, and has constant girth if the length of the perimeter of the shadow is always the same. (Note that by (A.49), $V_{n-2}(K|u^{\perp})$ is half the surface area of $K|u^{\perp}$ in $u^{\perp}$.) For $i = n - 1$, the definitions of constant i-brightness and constant i-girth coincide.

We now collect some results following easily from the previous ones.

Theorem 3.3.11. *Let $1 \leq i \leq k \leq n - 1$, and let K be a centrally symmetric convex body in $\mathbb{E}^n$. If $V_i(K|S)$ has the same value for all $S \in \mathcal{G}(n, k)$, then K is a ball. In particular, the only centrally symmetric convex bodies of constant i-brightness or constant i-girth are balls.*

Proof. The first statement follows from Aleksandrov's projection theorem, Theorem 3.3.6, on taking $K_1 = K$ and K_2 to be a ball of the appropriate radius. To obtain the second, we put $i = k$ or $k = n - 1$.   ■

Theorem 3.3.12. *Let K be a convex body in $\mathbb{E}^n$, and suppose that $1 \leq i \leq n - 1$. If K has constant i-brightness, then K has constant i-girth.*

Proof. In Lemma 3.3.4, take $K_1 = K$ and $K_2 = rB$ for suitable $r > 0$.   ■

It is not known if the converse to the previous theorem is true (see Problem 3.7). However, constant 1-girth is actually just constant width in disguise:

Theorem 3.3.13. *Let K be a convex body in $\mathbb{E}^n$. Then K is of constant 1-girth if and only if K has constant width.*

Proof. Simply take $K_1 = K$ and $K_2 = rB$ for a suitable $r > 0$, and apply Theorem 3.3.5.   ■

Next, we give a characterization of constant i-girth in terms of surface area measures.

Theorem 3.3.14. *Let K be a convex body in $\mathbb{E}^n$, and suppose that $1 \leq i \leq n - 1$. Then K has constant i-girth if and only if $S_i(K, \cdot) + S_i(-K, \cdot)$ is a constant multiple of λ_{n-1}. Further, if K is a convex body of class C_+^2 (or if K slides freely inside a ball) then K has constant i-girth if and only if $F_K^{(i)}(u) + F_K^{(i)}(-u)$ is constant for all (or almost all, respectively) $u \in S^{n-1}$.*

Proof. This follows immediately from Theorem 3.3.2 and Remark 3.3.3(ii) on taking $K_1 = K$ and K_2 to be a ball of suitable radius.   ■

Notice that the case $i = n - 1$ in the previous theorem applies when K has constant brightness (i.e., constant $(n - 1)$-brightness), since this is equivalent to constant $(n - 1)$-girth. By Theorem 3.3.2, these properties are equivalent to saying that ∇K is a ball.

Theorem 3.3.14 and others like it would be pointless if every convex body of constant i-girth were a ball. Our next task is to show that this is not the case.

Theorem 3.3.15. *For each $n \geq 2$ and $1 \leq i \leq n - 1$, there is a nonspherical convex body in $\mathbb{E}^n$ of constant i-girth.*

Proof. Suppose that $1 \leq i \leq n-1$. Our examples will be solids of revolution, with the x_n-axis as axis of revolution. We shall describe the 2-dimensional meridian section of our example K lying in the $\{x_1, x_n\}$-plane. For each $u \in S^{n-1}$, let

x_u be the point on the boundary of K where the outer normal vector is u. Let $u_1, \ldots, u_{n-1}$ be the principal curvature directions corresponding to the radii of curvature $R_1(u), \ldots, R_{n-1}(u)$ at x_u. We shall take u_1 in the plane of u and the x_n-axis and orthogonal to u. The directions $u_2, \ldots, u_{n-1}$ are any orthonormal set of directions in the $(n-2)$-dimensional plane orthogonal to the plane spanned by u and u_1. Then, for each j, $2 \le j \le n-1$, $R_j(u)$ is the distance from x_u to the x_n-axis measured along the line through x_u parallel to u. Let $R(u)$ be the common value of $R_2(u), \ldots, R_{n-1}(u)$. See Figure 3.6.

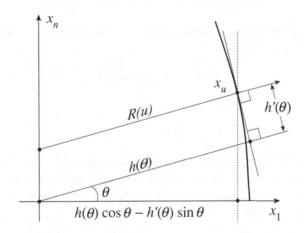

Figure 3.6. Part of the meridian section of K.

The body K will be constructed so that the principal radii of curvature satisfy

$$R(u)^i + R(-u)^i = c \qquad (3.6)$$

and

$$R_1(u)R(u)^{i-1} + R_1(-u)R(-u)^{i-1} = c, \qquad (3.7)$$

where c is a constant. By (0.42), this implies that

$$\binom{n-1}{i}\left(F_K^{(i)}(u) + F_K^{(i)}(-u)\right) = c.$$

Since it will be also be clear from the construction that K slides freely in a ball (see Remark 3.3.3(ii)), it will follow from Theorem 3.3.14 that K is of constant i-girth.

We may now concentrate on the slice of K lying in the $\{x_1, x_n\}$-plane; denote by $h(\theta)$ the support function of this 2-dimensional convex body, where $-\pi/2 \le \theta \le \pi/2$ is the angle between u and the x_1-axis. Then (see (0.40), (0.43), and Figure 3.6)

$$R(u) = h(\theta) - h'(\theta)\tan\theta$$

and

$$R_1(u) = h(\theta) + h''(\theta).$$

Letting $f(\theta) = \big(h(\theta)\cos\theta - h'(\theta)\sin\theta\big)^i / c$, we see that

$$R(u)^i = c f(\theta) \sec^i \theta$$

and

$$R_1(u)R(u)^{i-1} = -c f'(\theta) \sec^{i-1}\theta \csc\theta / i.$$

In terms of f, the two conditions (3.6) and (3.7) we wish to impose on the radii of curvature of K become

$$f(\theta) + f(-\theta) = \cos^i \theta \tag{3.8}$$

and

$$f'(\theta) - f'(-\theta) = (\cos^i \theta)'. \tag{3.9}$$

(Of course (3.9) also follows from (3.8) on differentiating.) These equations are independent of n. This means that if K is a convex body of revolution of constant i-girth in $\mathbb{E}^{i+1}$, the meridian section and axis of K may be used to construct a convex body of revolution in $\mathbb{E}^n$, for any $n \geq i+1$, also of constant i-girth. This is the same phenomenon that we saw in Theorem 3.2.5 in the case $i = 1$.

To construct the meridian section of K so that (3.8) is satisfied, we begin by setting $\theta_0 = \arccos(2^{-1/i})$. For $-\theta_0 < \theta \leq 0$, we let $h(\theta) = \cos\theta$. This implies that for these values of θ, the corresponding support lines all pass through $(1, 0)$, at which the meridian section will have a corner point (see Figure 3.7). It also implies that $f(0) = 1/c$, which together with the case $\theta = 0$ of (3.8) yields $c = 2$.

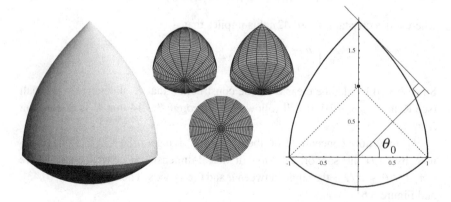

Figure 3.7. A convex body of constant brightness, three of its projections, and a meridian section.

Next, let $g(\theta) = (2\cos^i \theta - 1)^{1/i}$. Then (3.8) and our choice of $h(\theta)$ for $-\theta_0 < \theta \leq 0$ imply that for $0 \leq \theta < \theta_0$,

$$h'(\theta) \sin \theta - h(\theta) \cos \theta = -g(\theta).$$

The last equation is a first-order linear differential equation. The solution

$$h(\theta) = g(\theta) \cos \theta + 2 \sin \theta \int_0^\theta g(t)^{1-i} \cos^i t \, dt \qquad (3.10)$$

may be found by means of an integrating factor followed by integration by parts. The coefficients of the cosine and sine in this expression for $h(\theta)$ are the x_1- and x_n-coordinates of the point x_u with outer normal vector in the direction θ (cf. (0.40)); as θ tends to θ_0, these converge to zero and a finite limit $x_n(\theta_0)$, say, respectively. This means that support lines orthogonal to θ for $\theta_0 \leq \theta \leq \pi/2$ all pass through another corner point at $(0, x_n(\theta_0))$. Equation (3.10) determines the meridian section of K in the first quadrant.

At the point $(0, x_n(\theta_0))$, the principal radii of curvature are zero. Equations (3.6) and (3.7) then give $R_1(-u) = R(-u) = 2^{1/i}$, where u is in the direction θ for $\theta_0 \leq \theta \leq \pi/2$. So in the fourth quadrant the meridian section of K is the part of the circle with center on the x_n-axis, radius $2^{1/i}$, and containing $(1,0)$. The construction of K is now complete. ∎

Remark 3.3.16. The body of constant i-girth constructed in the previous example is actually of constant i-brightness; see Note 3.3.

The obvious question brought up by Theorem 3.3.15 is: What additional conditions (other than central symmetry) must be placed on a convex body of constant i-girth to force it to be a ball? This in turn raises the question of whether it is possible to obtain some version of Aleksandrov's projection theorem applying to arbitrary convex bodies for which the ith intrinsic volumes of projections are known for at least two values of i. We shall now construct some examples to show that even this is not enough, without extra assumptions.

Theorem 3.3.17. *For $n \geq 2$, there are noncongruent convex polytopes P_1 and P_2 in $\mathbb{E}^n$ such that*

$$V_i(P_1|S) = V_i(P_2|S),$$

for each $S \in \mathcal{G}(n,k)$, where either $i = 1$ and $1 \leq k \leq n-1$, or $i = k = n-1$.

Proof. We shall actually construct noncongruent polytopes P_1 and P_2 in $\mathbb{E}^n$ with the property that

$$\{h_{P_1}(u), h_{P_1}(-u)\} = \{h_{P_2}(u), h_{P_2}(-u)\},$$

for all $u \in S^{n-1}$. It will follow immediately from Definition 3.2.1 that $w_{P_1} = w_{P_2}$, proving the theorem for $i = k = 1$, and then Kubota's integral recursion (A.46) takes care of the cases $i = 1$ and $i \leq k \leq n - 1$. Our construction will also make it clear that

$$S(P_1, \cdot) + S(-P_1, \cdot) = S(P_2, \cdot) + S(-P_2, \cdot),$$

yielding the case $i = k = n - 1$ by Theorem 3.3.2.

We first deal with the case $n = 2$. Let

$$P_1 = \text{conv}\, \{e^{m\pi i/6} \in \mathbb{C} : m = 0, 1, 2, 4, 5, 6, 8, 9, 10\}$$

and

$$P_2 = \text{conv}\, \{e^{m\pi i/6} \in \mathbb{C} : m = 0, 1, 2, 3, 4, 5, 6, 8, 10\}.$$

Then P_1 and P_2 are nine-sided polygons (see Figure 3.8); it is useful to think of them as the union of a hexagon and three triangles, so that one of the polygons is obtained from the other by reflecting one of these triangles in the origin.

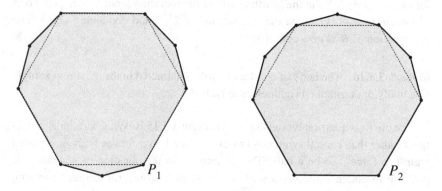

Figure 3.8. Noncongruent polygons with equal width functions.

We can now complete the proof by induction on n. Suppose that Q_1 and Q_2 are noncongruent $(n - 1)$-dimensional polytopes such that

$$\{h_{Q_1}(u), h_{Q_1}(-u)\} = \{h_{Q_2}(u), h_{Q_2}(-u)\},$$

for all $u \in S^{n-2}$. Let us regard Q_1 and Q_2 as sitting in the hyperplane $x_n = 0$ in $\mathbb{E}^n$, and define $P_j = Q_j \times [-1, 1]$, $j = 1, 2$. Let $u \in S^{n-1}$, and put $u = (v, \varphi)$ in spherical polar coordinates. Let F be the two-dimensional plane containing the vector u and the unit vector in the direction of the x_n-axis. Then $v \in F$, and $P_j \cap F = (Q_j \cap l_v) \times [-1, 1]$, for $j = 1, 2$, where l_v is the line through o parallel to v. Since $Q_1 \cap l_v = \pm Q_2 \cap l_v$, we have $P_1 \cap F = \pm P_2 \cap F$. It follows that P_1 and P_2 have the property set out in the first paragraph of the proof. ∎

The restriction on i can be removed; this is also true for the next example, producing pairs of very smooth bodies exhibiting the same behavior. See Note 3.6 for references and comments.

Theorem 3.3.18. *For $n \geq 2$, there are noncongruent convex bodies of revolution K_1 and K_2, of class C_+^∞ in $\mathbb{E}^n$, such that*

$$V_i(K_1|S) = V_i(K_2|S),$$

for each $S \in \mathcal{G}(n, k)$, where either $i = 1$ and $1 \leq k \leq n-1$, or $i = k = n-1$.

Proof. Proceeding as in the previous theorem, we shall construct noncongruent convex bodies of revolution K_1 and K_2, of class C_+^∞ in $\mathbb{E}^n$ such that

$$\{h_{K_1}(u), h_{K_1}(-u)\} = \{h_{K_2}(u), h_{K_2}(-u)\},$$

for all $u \in S^{n-1}$. It will also be clear from the construction that

$$F_{K_1}^{(n-1)} + F_{-K_1}^{(n-1)} = F_{K_2}^{(n-1)} + F_{-K_2}^{(n-1)}.$$

The theorem will then follow by the same arguments that we used at the beginning of the proof of Theorem 3.3.17.

We require very smooth versions of the polygons constructed in that example. To this end, we define a function $g \in C^\infty(\mathbb{R})$ as follows. Choose $g \in C^\infty([0, \pi/3])$ such that $g^{(m)}(0) = g^{(m)}(\pi/3) = 0$ for all nonnegative integers m, $g(\theta) > 0$ for $0 < \theta < \pi/3$, and $g(\pi/6 - \theta) = g(\pi/6 + \theta)$ for $0 \leq \theta \leq \pi/6$. Extend g by taking it to be of period $\pi/3$.

Define a function $f_1 \in C^\infty(\mathbb{R})$ of period 2π by $f_1(\theta) = g(\theta)$ if $0 \leq \theta \leq \pi$ and $f_1(\theta) = -g(\theta)$ if $\pi \leq \theta \leq 2\pi$. Now define another function $f_2 \in C^\infty(\mathbb{R})$, of period $2\pi/3$, by letting $f_2(\theta) = g(\theta)$ if $0 \leq \theta \leq \pi/3$ and $f_2(\theta) = -g(\theta)$ for $\pi/3 \leq \theta \leq 2\pi/3$. Then for $j = 1, 2$, the function f_j is odd and $f_j(\theta - \pi/2)$ is even. Using these properties, we see that

$$f_j\left(\theta - \frac{\pi}{2}\right) = f_j\left(-\theta - \frac{\pi}{2}\right) = -f_j\left(\theta + \frac{\pi}{2}\right),$$

for $0 \leq \theta \leq 2\pi$ and $j = 1, 2$. Consequently

$$f_j(\theta) = -f_j(\theta + \pi), \qquad (3.11)$$

for all θ and $j = 1, 2$.

Let $c > 0$ be such that $c + f_j(\theta) > 0$, $j = 1, 2$. If c is sufficiently large, then since f_j is bounded, the positively homogeneous extension of $c + f_j$ to $\mathbb{E}^n$ is sublinear (cf. Section 0.6). Let K_j' be the planar convex body, symmetric about the y-axis in $\mathbb{E}^2$, whose support function is given by $h_{K_j'}(\theta) = c + f_j(\theta)$ for all θ. Using (3.11) and the definitions of f_1 and f_2, we have

$$\{h_{K_1'}(\theta), h_{K_1'}(\theta + \pi)\} = \{c + g(\theta), c - g(\theta)\}$$

$$= \{h_{K_2'}(\theta), h_{K_2'}(\theta + \pi)\},$$

for all θ. Note that K_1' and K_2' are not congruent, since their support functions have different periods. Moreover, when c is sufficiently large, (0.43) ensures that the radius of curvature of each body is positive everywhere, so each body is of class C_+^∞ (cf. Section 0.9). This completes the proof for $n = 2$.

For $n > 2$, our examples are the noncongruent convex bodies K_j, $j = 1, 2$, obtained by rotating copies of K_j', $j = 1, 2$, placed in the $\{x_1, x_n\}$-plane, about the x_n-axis. If u is any direction, the vectors u and $-u$ belong to the same 2-dimensional subspace containing the x_n-axis. It follows that K_1 and K_2 have all the required properties. ∎

In the spirit of the question inspiring these examples, we shall now consider convex bodies of constant i-girth for two or more different values of i. We need the following lemma.

Lemma 3.3.19. *Let K be a convex body in $\mathbb{E}^n$ of class C_+^2 and of constant width w. For $1 \le i \le n - 1$, the principal radii of curvature of K satisfy*

$$R_i(u) + R_{n-i}(-u) = w,$$

for each $u \in S^{n-1}$.

Proof. For each $u \in S^{n-1}$, let x_u be the point in bd K with outer normal vector u. Note that the chord $[x_u, x_{-u}]$ of K is orthogonal to the supporting hyperplanes to K at its endpoints, since otherwise the diameter of K is greater than w, which is impossible. Therefore $x_{-u} = x_u - wu$.

The principal radii of curvature $R_i(u)$ are the eigenvalues of the reverse Weingarten map $\overline{W}_u = dx_u$ from $u^\perp$ to itself (see Section 0.9). Therefore we have $dx_u = R_i(u)du$ for a principal curvature direction du with corresponding radius of curvature $R_i(u)$. Now

$$dx_{-u} = dx_u - wdu = \big(R_i(u) - w\big)du = \big(w - R_i(u)\big)d(-u),$$

so $d(-u)$ is a principal curvature direction with corresponding radius of curvature $w - R_i(u)$. Since the principal radii of curvature are ranked by magnitude, we obtain

$$R_{n-i}(-u) = w - R_i(u),$$

and the lemma is proved. ∎

Theorem 3.3.20. *Let K be a convex body in $\mathbb{E}^3$ of class C_+^2 and of constant width and constant brightness. Then K is a ball.*

Proof. Since K has constant brightness b, say, Theorem 3.3.2 and (0.42) imply that

$$F_K^{(2)}(u) + F_K^{(2)}(-u) = 2b/\pi.$$

Using the fact that K has constant width w, say, and Lemma 3.3.19, we obtain

$$\begin{aligned} 2b/\pi &= R_1(u)R_2(u) + R_1(-u)R_2(-u) \\ &= R_1(u)R_2(u) + (w - R_2(u))(w - R_1(u)) \\ &= 2((R_1(u) - w/2)(R_2(u) - w/2) + (w/2)^2). \end{aligned}$$

Now bd K has an umbilic point, that is, a point x_{u_0} such that $R_1(u_0) = R_2(u_0)$. (Figure 3.9 depicts an umbilic point of an ellipsoid.) To see this, suppose that $R_1(u) \neq R_2(u)$ for each $u \in S^2$. Then one can associate with each $u \in S^2$ the principal curvature direction corresponding to $R_1(u)$, in an unambiguous way, thereby defining a continuous tangent vector field on S^2 which is nonzero everywhere, and contradicting a well-known theorem in topology (see, for example, [163, Theorem 34.1, p. 123]). (The latter basically says that one cannot comb a hairy ball without producing a cowlick.) Substituting u_0 into the previous displayed equation, we see that $2b/\pi \geq w^2/2$. If $u \in S^2$, then $K|u^\perp$ has constant width w in $u^\perp$, so Cauchy's surface area formula (A.49) and the isoperimetric inequality (B.14) for $n = 2$ imply that $w^2/4 \geq b/\pi$. Therefore equality holds in this inequality, so $K|u^\perp$ is a disk. It follows from Corollary 3.1.6 that K is a ball. ∎

The smoothness assumption in the previous theorem can be removed; see Notes 3.3 and 3.6.

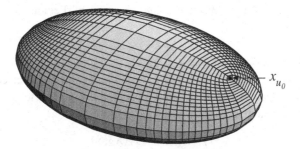

Figure 3.9. Principal curves and an umbilic point of an ellipsoid.

Open problems

Problem 3.1. (See Theorems 3.1.3 and 3.1.8 and Note 3.1.) Suppose that $2 < k \leq n - 1$, and that K_1, K_2 are convex bodies in $\mathbb{E}^n$ with $K_1|S$ similar to $K_2|S$, for all $S \in \mathcal{G}(n, k)$. Is K_1 homothetic to $\pm K_2$?

Problem 3.2. (See Theorems 3.1.3 and 3.1.8 and Note 3.1.) Suppose that $2 \leq k \leq n - 1$ and that K_1 and K_2 are arbitrary convex bodies in $\mathbb{E}^n$ such that $K_1|S$ is congruent to $K_2|S$, for all $S \in \mathcal{G}(n, k)$. Is K_1 a translate of $\pm K_2$?

Problem 3.3. (See Note 3.2.) Let $2 \leq k \leq n - 1$, and let K be a convex body in $\mathbb{E}^n$ such that all the projections $K|S$, $S \in \mathcal{G}(n, k)$, are affinely equivalent. Must K be an ellipsoid?

Problem 3.4. (See Note 3.4.) Let K be a convex body in $\mathbb{E}^n$. Is there an inequality of the form

$$\lambda_n(\nabla K) \leq c_n \lambda_n(K),$$

where c_n is a constant independent of K, with equality if and only if K is a simplex?

Problem 3.5. (See Notes 3.4 and 6.1.) Suppose that the Blaschke body ∇K of a convex body K in $\mathbb{E}^n$ is homothetic to its central symmetral $\triangle K$. Is K centrally symmetric?

Problem 3.6. (See Notes 3.8 and 4.12 and Section 4.4.) Let $1 \leq i < n - 1$. Find an algorithm by which a centered compact convex set K in $\mathbb{E}^n$ can be constructed from its ith projection function.

Problem 3.7. (See Note 3.3.) For $1 < i < n - 1$, are there convex bodies of constant i-girth but not of constant i-brightness?

Problem 3.8. (See Note 3.6.) Does the converse of Lemma 3.3.4 hold for $1 < i \leq k < n - 1$?

Problem 3.9. (See Notes 3.6 and 4.9.) Let $1 \leq k \leq n - 1$, let i and j be distinct integers between 1 and k, and suppose that K_1, K_2 are compact convex sets, of dimension at least $k + 1$, in $\mathbb{E}^n$. If K_2 is centrally symmetric and there exist constants a and b such that

$$V_i(K_1|S) = aV_i(K_2|S)$$

and

$$V_j(K_1|S) = bV_j(K_2|S),$$

for all $S \in \mathcal{G}(n, k)$, then is K_1 homothetic to K_2?

Notes

3.1. *Homothetic and similar projections.* Theorem 3.1.3 was first published by Süss [796] and Nakajima (= Matsumura) [644]. These authors considered only the case $n = 3$, but their proofs can be generalized. Thirty years later, Groemer [346] gave a short proof for the case when one of the sets is centrally symmetric. Hadwiger [372] established the general result, and actually showed that it is not necessary to consider projections on all $(n - 1)$-dimensional subspaces; the hypothesis need only be true for one fixed subspace S, together with all subspaces containing a line orthogonal to S. In other words, one requires only a "ground" projection and all corresponding "side" projections. In a sense, this is best possible. (In $\mathbb{E}^2$, let H_1 be the unit disk and H_2 a nonspherical convex body of constant width 2. For $j = 1, 2$, let $K_j = H_j \times [0, 1]$ be convex bodies in $\mathbb{E}^3$. If S is any plane containing the z-axis, then $K_1|S$ is a translate of $K_2|S$, yet K_1 is not a translate of K_2.) Hadwiger also noted, however, that in $\mathbb{E}^n$ for $n \geq 4$, the ground projection may be dispensed with. Another proof of Theorem 3.1.3 is given by Székely [799], but ours is taken from the paper [693] of Rogers. It is rather simpler than that of Hadwiger, and again, not all the projections are used, but only those on subspaces containing a line parallel to a diameter of one of the bodies. The reader who finds Hadwiger's approach more useful or appealing may consult the stability version Theorem 4.3.4 of Groemer (cf. Remark 4.3.5(ii)). This also uses the ground and side projections.

Corollary 3.1.5 was originally proved by Blaschke and Hessenberg [73], and Corollary 3.1.6 for $n = 3$ appears to go back at least to Fujiwara [250] and Kubota [480]. In [633], Montejano considers *linear subfamilies* of $\mathcal{G}(n, k)$, those with the property that every 1-dimensional subspace is contained in some member of the subfamily. He shows that projections on subspaces in a linear subfamily will suffice for these two results, when in addition $k > 2n/3$ and (in the case of Corollary 3.1.5) the body is strictly convex.

Groemer [357] calls a set of hyperplanes in $\mathbb{E}^n$ a *hyperplane bundle* of direction u if every member contains the line through the origin parallel to u and every point in $\mathbb{E}^n$ is contained in some member. He proves that if one of the sets K_1 or K_2 is strictly convex, then projections onto hyperplanes in a hyperplane bundle suffice in Theorem 3.1.3. Groemer proceeds to obtain corresponding versions of Corollaries 3.1.5 and 3.1.6.

For Theorem 3.1.7, we have followed the argument given by Chakerian [151, Lemma 2]. This in turn is an adaptation of a proof of Süss [798] for the 3-dimensional case. Again, the theorem seems to have first appeared (with a quite different proof and in three dimensions) in [73]. A proof using some simple Banach space theory is given by Grinberg [334, Theorem 3].

The remarkable Theorem 3.1.8 was discovered by Petty and McKinney [667]. They actually show that the pairs of bodies given by their construction are the only pairs with the property that the projection on any 2-dimensional subspace of one body is a dilatate of a rotation by $\pi/2$ about the origin of the projection of the other. The authors also show that these are precisely the centered convex bodies in $\mathbb{E}^n$, $n \geq 3$, with the property that every circumscribed box has the same volume (any convex body of constant width has this property). The fact that there are such pairs that are not affinely equivalent is proved by Gardner and Volčič [282].

Golubyatnikov has obtained some partial answers to Problem 3.2. In [294], he states that the answer is positive when $k = n - 1$ and K_1 and K_2 are convex polytopes, and in [296] (see also [297]) he proves that the answer is positive when $k = 2$ and none of the projections $K_1|S$ and $K_2|S$ have extra symmetries with respect to rotations. (In other words, the only direct rigid motion taking $K_j|S$ onto itself is the identity, for $j = 1, 2$.) He also shows that the same assumptions allow Problem 3.1 to be answered positively, for the excluded case $k = 2$ of that question. Golubyatnikov finds further partial results concerning Problem 3.2 when $k = 3$ in [300] and for convex bodies in n-dimensional complex space $\mathbb{C}^n$ in [301].

In some of these papers, Golubyatnikov calls a body C in $\mathbb{E}^n$ k-*convex* if $1 \leq k \leq n - 1$ and for each point not in C, there is a k-dimensional plane containing the point and not intersecting C. (The concept goes back at least to Meisters and Ulam [598, Corollary 1].) Golubyatnikov proves that the results of the previous paragraph extend to $(n - 2)$-convex bodies in $\mathbb{E}^n$, and in this case the extra assumption concerning symmetry is necessary (see [297, Figure 1]; of course, we already know from Theorem 3.1.8 that Problem 3.1 has a negative answer in the excluded case $k = 2$, even for convex bodies).

Barov, Cobb, and Dijkstra [35] (see also [36] and [186]) also find versions of some of these results for non-convex bodies. For example, [35, Theorem 1] states that if $0 < k < n$ and a convex body K and compact set C in $\mathbb{E}^n$ satisfy $C|S = K|S$ for all $S \in \mathcal{G}(n, k)$, then C contains a topological $(k - 1)$-sphere.

Fix a $u \in S^{n-1}$. Kuz'minykh [489] calls a convex body K in $\mathbb{E}^n$ *recoverable* if it is determined by the family of all its noncongruent projections of the form $(\phi K)|u^\perp$, where $\phi \in SO_n$. No planar convex body is recoverable, since for $0 < a \le b$, there are noncongruent planar convex bodies with minimal width a and diameter b. In [489], it is shown that for $n \ge 3$, there is a convex body in $\mathbb{E}^n$ that is not recoverable, and that the class of recoverable bodies is dense in $\mathcal{K}_0^n$.

3.2. *Bodies with congruent or affinely equivalent projections.* Corollary 3.1.6 provokes the question of whether a convex body in $\mathbb{E}^n$ must be a ball if all its projections on k-dimensional subspaces (for some k with $2 \le k \le n - 1$) are all congruent to each other. (For the closely related results on the corresponding question for sections, see Theorem 7.2.17 and Note 7.2.) Süss [797] provided a positive answer for $n = 3$, using Brouwer's fixed-point theorem. In [371], Hadwiger introduced the following ideas. If C is an $(n - 1)$-dimensional compact convex set in $\mathbb{E}^n$, a *complete turning* of C is an even function $C(u)$ on S^{n-1}, where $C(u) \subset u^\perp$ is congruent to C and the function $C(u)$ is continuous in the Hausdorff metric. The point is that if all projections $K|S$ of a convex body K in $\mathbb{E}^n$ on $(n - 1)$-dimensional subspaces S are congruent, then there exists a complete turning of $C = K|S$ for any such S. Using topological methods, Mani [567] proved that if n is odd and C admits a complete turning, then C must be an $(n - 1)$-dimensional ball. A positive answer to our question for n odd and $k = n - 1$ follows from Corollary 3.1.6. Generalizing an earlier result of Burton, Montejano [631] shows that if C admits a complete turning, then C must be centrally symmetric. If K has congruent projections, the central symmetry of K follows from Corollary 3.1.5, and Theorem 3.3.11 then implies that K is a ball. This yields an unconditional positive answer to our question. In [633, Theorem 3.7], Montejano uses the weaker hypothesis that all projections of K are similar to derive the same conclusion, that K is a ball. In fact, he shows that this follows quite easily from the analogous problem for sections; see Note 7.2.

Kuz'minykh [488] generalizes the result of Süss in the previous paragraph in a different way, by showing that an arbitrary compact set in $\mathbb{E}^3$, all of whose projections on planes are congruent to each other, must be the union of a sphere and a set enclosed by it. He also gives examples to show that analogous statements are false without the compactness assumption.

Problem 3.3, in which the hypothesis is that all projections are merely affinely equivalent, can be disposed of when n is odd, by the following argument of Burton [119]. Let K be a convex body in $\mathbb{E}^n$ with affinely equivalent projections $K|u^\perp$, $u \in S^{n-1}$. It is known that for each u there is a unique $(n-1)$-dimensional ellipsoid $E(u)$ of least λ_{n-1}-measure containing $K|u^\perp$. (This is a version of the Löwner ellipsoid in $u^\perp$; see Theorem 9.2.1.) Let ϕ_u be the affine map taking $E(u)$ to $B \cap u^\perp$. Now all the images $\phi_u E(u)$ are congruent, so, using the easily proved fact that the Löwner ellipsoid of an affine image of a body is the same affine image of its Löwner ellipsoid, we see that the bodies $\phi_u(K|u^\perp)$ are also congruent. Therefore the function $C(u) = \phi_u(K|u^\perp)$ is a complete turning of any one of them. By Mani's result, each $\phi_u(K|u^\perp)$ is an $(n - 1)$-dimensional ball, so each $K|u^\perp$ is an ellipsoid. Theorem 3.1.7 implies that K is an ellipsoid. Note that by using Corollary 3.1.6 or Theorem 3.1.7, we also obtain an answer to Problem 3.3 for arbitrary n and even k.

The applicability of these topological methods when n is odd stems from the fact that for such n, there is no continuous tangent vector field on S^{n-1} that is nonzero everywhere. (We used this fact for $n = 3$ in the proof of Theorem 3.3.20.

For general odd n, see, for example, [191, Theorem 3.3, p. 343].) For n even, however, such a field is obtained by assigning to $x = (x_j) \in S^{n-1}$ the vector $(-x_2, x_1, -x_4, x_3, \ldots, -x_n, x_{n-1})$.

3.3. *Sets of constant width and brightness.* Sets of convex width have a long and rich history. Planar examples were discovered by Euler, who called them orbiforms. The terms constant breadth, equiwide, and (in three dimensions) spheroform have also been used. Many references can be found in [83, Chapter 15] and especially in the survey papers of Chakerian and Groemer [154] and Heil and Martini [386]. A connection to *rotors* in polytopes, convex bodies that can rotate inside a polytope without losing contact with each of its facets, has led to applications in cinematography and rotary engines; see [53, Figure 12.10.5.2], [154, Section 10], [353, Section 3.6], [356, Section 5.7], [386, Section 5], [737, p. 196], and [845, p. 72].

Theorem 3.2.6 is noted by Blaschke and Hessenberg [73]. (Montejano [633] shows that projections on subspaces in a linear subfamily of $\mathcal{G}(n, k)$ with $k > n/2$ suffice, and Groemer [357] does the same for projections in a hyperplane bundle (see Note 3.1).) The deeper Theorem 3.3.13 is due to Minkowski. His proof, in [625], employs spherical harmonics; two more proofs, by Funk and McShane, are given in [83, Section 67].

A planar convex body K of constant width has the property that bd K is a *P-curve*. This means that for each chord $[p_1, q_1]$ of K such that there are parallel supporting lines to K at p_1 and q_1, there is another such chord $[p_2, q_2]$ parallel to these lines, and the supporting lines to K at p_2 and q_2 are parallel to $[p_1, q_1]$. See [154, p. 83]. P-curves were introduced by Blaschke [72], who showed that a centrally symmetric convex body K in $\mathbb{E}^n$, $n \geq 3$, is an ellipsoid if and only if all its projections on 2-dimensional planes are bounded by P-curves. (A centrally symmetric P-curve is called a *Radon curve*.) Krautwald [475] proved that a convex body K in $\mathbb{E}^n$, $n \geq 3$, is an affine image of a body of constant width if and only if all its projections on 2-dimensional planes are bounded by P-curves. Blaschke's characterization of ellipsoids follows from this and Theorem 3.2.7. Another consequence is an earlier result of Chakerian [151]: A convex body K in $\mathbb{E}^n$, $n \geq 3$, is an affine image of a body of constant width if and only if all its projections on hyperplanes are affine images of bodies of constant width.

The concept of convex width extends to certain surfaces called hedgehogs that do not necessarily form the boundary of a convex body; see the article of Martinez-Maure [571].

Inspired by work of G. Herglotz, Blaschke constructed the first example of a nonspherical convex body in $\mathbb{E}^3$ of constant brightness. This is given in his famous book [71, pp. 151–4], where the illustration on the right in Figure 3.7 also appears. Much later, Chakerian [152] defined sets of constant i-girth, and proved Theorem 3.3.14, at least when K is of class C_+^2. Klee [447] raised the question of whether there are nonspherical convex bodies in $\mathbb{E}^n$ of constant i-brightness. Firey [240] answered this with the result in Remark 3.3.16, by generalizing Blaschke's example to obtain Theorem 3.3.15 and showing that constant i-brightness and constant i-girth are equivalent for convex bodies of revolution; he also proved Theorem 3.3.12 and posed Problem 3.7. Our proof of Theorem 3.3.15 uses Remark 3.3.3(ii), an observation of W. Weil (private communication).

Schneider [731, pp. 60–2] extends a result of L. Berwald by showing that a convex body is of constant brightness if and only if the integral of the surface areas of its sections by hyperplanes orthogonal to a direction u is independent of u. Schneider's result is re-proved and generalized by Goodey [307]; see Note 4.6.

Theorem 3.3.20 is due to Nakajima [642]. The proof we give is similar to that of a generalization due to Chakerian [152]; see Note 3.6. (The existence of an umbilic point is noted in [83, Section 68] and proved explicitly in [783, pp. 288–90].) Howard [400] has recently proved that any convex body in $\mathbb{E}^3$ of constant width and constant brightness is a ball, thereby settling a major problem dating back to 1926; see Note 3.6 for more details.

3.4. *Blaschke bodies and Blaschke sums.* In denoting the Blaschke body by ∇K we follow Lutwak [537]. The inequality of Theorem 3.3.9 was proved by Petty [662] (or see [537, eq. (5.5)] for a quite different proof, using the Kneser–Süss inequality (B.32)). Firey [239] conjectured that Problem 3.4, concerning the complementary inequality, has a positive answer.

The interesting Problem 3.5 was posed by R. Schneider at the Vienna Conference on Convexity in 1981. If the answer is positive, then a body of constant width and constant brightness must be centrally symmetric, and so, by Theorem 3.2.7, a ball; consequently, such a result would generalize Howard's theorem from [400] (see Notes 3.3 and 3.6).

Formation of the Blaschke body is a special case of a more general process. Suppose that K_1 and K_2 are convex bodies in $\mathbb{E}^n$ and that $a_j \geq 0$, $j = 1, 2$. By Minkowski's existence theorem, Theorem A.3.2, there is a convex body K, unique up to translation, such that

$$S(K, \cdot) = a_1 S(K_1, \cdot) + a_2 S(K_2, \cdot)$$

(cf. (A.23)). This can be used to define a new kind of addition and scalar multiplication, called *Blaschke addition* and *Blaschke scalar multiplication*. These are denoted by + and · ($\sharp$ and $\times$ are also commonly used), so the previous equation becomes

$$K = a_1 \cdot K_1 + a_2 \cdot K_2.$$

When $a_1 = a_2 = 1$, K is called the *Blaschke sum* of K_1 and K_2. One can also use this notation (convenient in certain contexts, but generally avoided in this book) to express the Blaschke body as

$$\nabla K = \frac{1}{2} \cdot K + \frac{1}{2} \cdot (-K).$$

Blaschke scalar multiplication is related to the usual scalar multiplication by $r \cdot K = r^{1/(n-1)}K$, but Blaschke addition is quite different from Minkowski addition.

The process of forming a Blaschke sum is mentioned by Blaschke [71, p. 112]. It was apparently first considered by Minkowski [623, p. 117] for polytopes; see Grünbaum's book [367, Section 15.3] for a readable introduction in this setting. Other places in the literature where Blaschke sums occur are listed in the survey [733, p. 48] and book [737, p. 395] of Schneider. More recently, Grinberg and Zhang [337] have shown that every centered convex body in $\mathbb{E}^n$ is a Hausdorff limit of finite Blaschke sums of n-dimensional ellipsoids; compare Corollary 4.1.12, which can be deduced by applying the cosine transform (cf. Section C.2) and using Definition 4.1.1.

The existence of additions, related to the corresponding area measures and suitable for working with projections on 1-dimensional subspaces (Minkowski addition) and $(n-1)$-dimensional subspaces (Blaschke addition), raises the question of others suitable for projections on i-dimensional subspaces when $2 \leq i \leq n-2$. To be more specific, suppose that $2 \leq i \leq n - 2$, and let K_1 and K_2 be convex bodies in $\mathbb{E}^n$. Is there always a convex body K such that

$$S_i(K, \cdot) = S_i(K_1, \cdot) + S_i(K_2, \cdot)?$$

Firey [241] showed that this is true when K_1 and K_2 are sufficiently smooth coaxial convex bodies of revolution, but in general the answer, unfortunately, is no. This was proved independently by Fedotov [226] (see also [227]) and by Goodey and Schneider [313]. Fedotov showed that this fails to hold when K_1 is a regular simplex in $\mathbb{E}^4$ and $K_2 = -K_1$. In fact, the answer to the question is negative for most (in the sense of Baire category) pairs (K_1, K_2) of convex bodies, as Schneider [738] demonstrates.

3.5. *Determination by one projection function.* As we have seen in this chapter, the problem of determining a compact convex set from natural metrical properties of its projections involves a good part of the considerable machinery developed by the pioneers. A full historical account could be another book in itself.

The potent Theorems 3.3.1 and 3.3.6 of Aleksandrov were developed in a series of papers (see [2], [3], [4], [5], especially the first two). (Aleksandrov confined his arguments to the case $k = n - 1$.) When one considers the length of these papers, it may seem that the proofs we present in this chapter are extraordinarily short. In fact they essentially follow Aleksandrov's original argument, but most of the hard work must still be done in proving the results referred to in the appendixes. A proof rather different in some ways and apparently more concise than Aleksandrov's was provided independently by Fenchel and Jessen [230]. It is our impression, however, that many pages of details lie behind this elegant proof; see [737, p. 332] for some interesting comments on this issue. In any case, both proofs involve area measures, mixed volumes, and the Aleksandrov–Fenchel inequality.

Chakerian [152] proves Theorem 3.3.2, at least when K is of class C_+^2. The equivalence (3.2)⇔(3.4) is also established by Petty [662]. There is a generalization of this equivalence, yielding Aleksandrov's projection theorem (Theorem 3.3.6), Theorem 3.2.3, and Theorem 3.3.9 as special cases; see Note 4.9.

As an aside, we mention the following complementary inequality to that of Theorem 3.2.3, usually phrased in terms of the difference body $2\triangle K$: If K is a convex body in $\mathbb{E}^n$, then

$$\lambda_n(2\triangle K) \leq \binom{2n}{n} \lambda_n(K),$$

with equality if and only if K is a simplex. This appeared in [696], and is called the *Rogers–Shephard inequality* after the authors. For more information, see [366, p. 258], [737, Section 7.3], and the references given there, as well as [284], [859], and Note 9.5. We have noted already that the complementary inequality for Theorem 3.3.9 has not yet been established.

Schneider [730] proves the following result: Let K_1 be a centrally symmetric convex polytope in $\mathbb{E}^n$ and $1 \leq i \leq n - 1$; then there are subsets A of S^{n-1} of arbitrarily small λ_{n-1}-measure such that if K_2 is a centrally symmetric convex body and $V_i(K_1|u^\perp) = V_i(K_2|u^\perp)$ for all $u \in A$, then K_1 is a translate of K_2. (Here, A can be any open set containing, for each i-dimensional face of K_1, a unit vector parallel to that face.) Another related observation is due to Schneider and Weil [744]. Let K_1 be a centrally symmetric convex body in $\mathbb{E}^n$, with $h_{K_1} \in C^{n+3}(S^{n-1})$. Suppose that A is a subset of S^{n-1}, centered but not dense in S^{n-1}. Then there is a centrally symmetric convex body K_2 such that $\lambda_{n-1}(K_1|u^\perp) = \lambda_{n-1}(K_2|u^\perp)$ for all $u \in A$, but K_1 is not a translate of K_2.

The fact that even the ball is not determined, among all convex bodies, by the volumes of its projections does not discount the possibility that some other centrally symmetric convex bodies may be. There are, in fact, several results of this type. A centered convex body K is called *irreducible* if it is not the central symmetral of a convex body that is not centrally symmetric, that is, if $K = \triangle C$ implies that C is a translate of K. It is easy to see that a centrally symmetric convex body K is determined, among all convex bodies, by the lengths of its projections on 1-dimensional subspaces (or, equivalently, by its width function) if and only if its centered translate is irreducible. It is known that centered parallelotopes in $\mathbb{E}^n$ are irreducible, and that in $\mathbb{E}^2$ these are the only irreducible convex bodies (see the papers of Groemer [344] and Grünbaum [365]). This is not true in higher dimensions; for $n > 2$, every centered convex polytope with no more than $4n - 2$ vertices is irreducible. This and many other facts, including a necessary and sufficient condition for a polytope to be irreducible, can be found in the article [852] by Yost.

Goodey, Schneider, and Weil [314] not only show that for $1 \leq i < n - 1$ a centered parallelotope in $\mathbb{E}^n$ is determined, among all convex bodies, by the i-dimensional volumes of its projections on i-dimensional subspaces (in short, by its ith projection function), but also prove that other centered polytopes share this property; for example, they show that the Cartesian product of a centrally symmetric polygon and an $(n - 2)$-dimensional cube is determined in this sense. Another example is any centrally symmetric polytope whose $(i + 1)$-dimensional faces all are simplices. Schneider [738] proves this and draws the remarkable conclusion that for $1 \leq i < n - 1$, most (in the sense of Baire category) centrally symmetric convex bodies in $\mathbb{E}^n$ are determined, among all convex bodies, by their ith projection function.

For determination by brightness functions, the situation is much simpler. Gardner and Volčič [281] prove that a convex body in $\mathbb{E}^n$ is determined, among all convex bodies, by its brightness function if and only if it is a parallelotope. (Martini [573] established this result for polytopes.) The method used in [281] owes something to work of Vincensini [816] and Firey [237]. In a generalization, Bauer [48] characterizes those convex bodies K in $\mathbb{E}^n$ that are *almost determined* by their brightness functions. This means that the only convex bodies with the same brightness function as K are of the form $(1 - t) \cdot K + t \cdot (-K)$, for $0 \leq t \leq 1$, in the notation of Note 3.4. She also proves that if $2 \leq i \leq n-2$, a convex polytope in $\mathbb{E}^n$ is determined by its ith projection function if all its $(i + 1)$-dimensional faces are parallelotopes.

In related work, Campi, Colesanti, and Gronchi [144] give a characterization of those convex bodies K in $\mathbb{E}^n$ with the property that if K is the Blaschke sum of convex bodies K_1 and K_2, each with the same brightness function as K, then K_1 and K_2 are translates of K. (Compare the notion of an irreducible body defined above.)

We have seen that the centered body $\triangle K$ has the same width function as K, and the centered body ∇K has the same brightness function. Thus, if a convex body in $\mathbb{E}^n$ is not centrally symmetric, it is not even determined up to translation and reflection in the origin by its ith projection function, when $i = 1$ or $n - 1$. For $2 \leq i \leq n - 2$, however, there are convex bodies in $\mathbb{E}^n$ that are not centrally symmetric but that are so determined. This surprising result is demonstrated by Goodey, Schneider, and Weil [314], an example being the Cartesian product of a triangle and an $(n-2)$-dimensional cube. In fact, as Bauer [47] proves, most (in the sense of Baire category) convex bodies have this property. This is refined by Schneider [741] who proves that if $2 \leq i \leq j \leq n-1$ and $i < n-1$, then most convex bodies in $\mathbb{E}^n$ are determined by the values $V_i(K|S)$, where S lies in an open neighborhood of the subset of $\mathcal{G}(n, j)$ consisting of those j-dimensional subspaces containing a fixed direction.

Goodey, Schneider, and Weil [314] also contribute to Problem 3.8, by proving that the converse to Lemma 3.3.4 is always true if either K_1 or K_2 is a polytope. This converse is trivially true for $i = n - 1$, and we observed earlier that it is also true for $i = 1$, by Theorem 3.3.5 (a result of Nakajima [643] when $n = 3$). In view of these results, it seems reasonable to hope that Problem 3.8 has a positive answer, and this is one reason the proof of the result in Remark 3.3.16 has been omitted.

All this is an indication that for $1 < i < n - 1$ the ith projection functions behave somewhat differently from the width or brightness functions. This seems to be partly due to the loss of injectivity properties in higher-order cosine and spherical Radon transforms. We direct the reader who wishes to learn more to the work of Goodey and Weil [320, Section 6] and the references given there.

3.6. *Determination by more than one projection function.* As was mentioned in Note 3.3, an old problem dating back to Nakajima's 1926 paper [642] asks whether any convex body in $\mathbb{E}^3$ of constant width and constant brightness must be a ball. Over the years this question has been the motivation of much work. For example, Goodey, Schneider, and Weil [315] demonstrate that most (in the sense of Baire category) convex bodies in $\mathbb{E}^n$ are determined, up to translation and reflection in the origin, by their width and brightness functions.

At last, this problem has been solved by Howard [400]. In view of the importance of this major achievement, we shall sketch his solution.

Howard begins by quoting the known result that a convex body K in $\mathbb{E}^n$ has boundary in $C^{1,1}$ if and only if some ball slides freely inside K. (A convex body K_1 in $\mathbb{E}^n$ slides freely inside a convex body K_2 if for each $x \in \mathrm{bd}\, K_2$ there is a $y \in \mathbb{E}^n$ such that $x \in K_1 + y \subset K_2$.) This has a dual form, due to D. Hug: $h_K \in C^{1,1}$ if and only if K slides freely inside some ball. Using this, Howard proves that if $h_{\triangle K} \in C^{1,1}$, then $h_K \in C^{1,1}$. In particular, if K has constant width, then $h_{\triangle K} = c$ for some constant c, so $h_K \in C^{1,1}$, an extra smoothness that allows the rest of the proof to work. Writing

$$h_K(u) = \frac{1}{2}\left(h_K(u) + h_K(-u)\right) + \frac{1}{2}\left(h_K(u) - h_K(-u)\right) = c + f(u),$$

say, we see that the even part of h_K is constant on S^{n-1} and the odd part $f \in C^{1,1}$. Next, Howard uses the fact that $h_K \in C^{1,1}$ to establish the formula

$$\lambda_{n-1}(K|u^{\perp}) = \frac{1}{2}\int_{S^{n-1}} \det\left(h_K(u)I + \nabla^2 h_K(u)\right)|u \cdot v|\, dv,$$

for the brightness function of K at $u \in S^{n-1}$, where I is the identity matrix. The previous integral is the cosine transform (see (C.10)) of $\det(h_K I + \nabla^2 h_K)$, so if K has constant brightness, Theorem C.2.1 implies that the even part of $\det(h_K I + \nabla^2 h_K)$ is constant. Substitution of $h_K = c + f$ and some further arguments reveal that

$$\det\left(fI + \nabla^2 f\right) = -c' < 0,$$

where c' is a constant and f is the odd function defined above. But a rather long and technical argument, involving known properties of the so-called functions of bounded distortion, shows that this is impossible for any odd $f \in C^{1,1}$. This contradiction completes Howard's proof.

Howard's result is actually more general. To describe how, we shall first recall some earlier work with a similar flavor.

In proofs such as that of Aleksandrov's uniqueness theorem, Theorem 3.3.1, one can sometimes replace the unit ball B by an arbitrary centrally symmetric convex body E of class C_+^2. In this manner, one derives the corresponding facts about convex bodies whose width, brightness, and so forth are now measured relative to that of E. For example, Chakerian [152] obtains the following generalization of Theorem 3.3.20: If K is a convex body in $\mathbb{E}^3$ of class C_+^2 such that $\triangle K$ is homothetic to $\triangle E$, and $V_2(K|u^{\perp})$ is a constant multiple of $V_2(E|u^{\perp})$ for all $u \in S^2$, then K is homothetic to E. Rephrased, this says that two convex bodies in $\mathbb{E}^3$ of class C_+^2, with one centrally symmetric, must be homothetic if the perimeters of their shadows on any plane are in a fixed ratio and the areas of their shadows on any plane are also in a fixed (but possibly different) ratio. (This does not contradict Theorem 3.1.8, since there the dilatation factor of the projections of K_1 and K_2 on different planes is generally different.) For convex bodies in $\mathbb{E}^3$ of class C_+^2, this represents a considerable improvement on Theorem 3.1.3. Howard [400] proves that the smoothness assumption on K can be removed; thus Problem 3.9 has a positive answer when $i = 1$, $j = 2$, and $K_2 \in C_+^2$ is centrally symmetric. His result on constant width and constant brightness described above is the special case when $K_2 = B$.

Problem 3.9 is still open, but many other partial results are known. The answer is positive if $a = b = 1$; see Note 4.9. Howard and Hug [401] prove that the answer is positive when $K_1 \in C^2$, $K_2 \in C_+^2$ is centrally symmetric, and (i, j) is not $(1, n - 1)$ or $(n - 2, n - 1)$. They also show that the answer is positive in the two exceptional cases when K_1 and K_2 are coaxial bodies of revolution. (Howard and Hug acknowledge that some of their techniques and proofs owe much to important insights of Haab, whose earlier paper [369] unfortunately contains some errors.) Finally, Howard and Hug [402] obtain a positive answer for $K_2 \in C_+^2$ (and arbitrary K_1) when $i = 1$ and either $2 \le j < (n + 1)/2$ or $j = 3, n = 5$. Despite all these results, it is still unknown whether a convex body in $\mathbb{E}^n$ of constant width and constant brightness must be a ball when $n \ge 4$.

The following theorem, applicable when both the areas and the perimeters of the projections of a 3-dimensional convex body are known, is stated by Anikonov [10]: If K_1 and K_2 are convex bodies in $\mathbb{E}^3$, such that for $i = 1, 2$, $V_i(K_1|u^\perp) = V_i(K_2|u^\perp)$ for all $u \in S^2$, and if the support functions of K_1 and K_2 are real analytic, then K_1 is a translate of $\pm K_2$. (According to Goodey, Schneider, and Weil [316], at least one essential detail of the proof in [10] is not carried out.) An example similar to that of Theorem 3.3.18, showing that one cannot expect to weaken the smoothness assumption in this statement, was found by S. Campi in the case $n = 3$ (private communication). Theorem 3.3.18 itself and the polytopes of Theorem 3.3.17 appear in a slightly different guise in work of Gardner and Volčǐ c [283]. Goodey, Schneider and Weil [315] prove that Theorems 3.3.17 and 3.3.18 hold for all i with $1 \le i \le n - 1$.

3.7. *Determination by directed projection functions, etc.* Suppose that K is a compact convex set and u is a direction. The part of K *illuminated* in the direction u is the set of all points x in K so that the ray emanating from x in the direction $-u$ does not contain any other point of K. Let K be a compact convex set in $\mathbb{E}^n$, let $1 \le i < k \le n - 1$, and let $S \in \mathcal{G}(n, k)$. Goodey and Weil [324] introduce the *directed projection function* $v_{K,i,k}(S, u)$, giving the ith intrinsic volume of the part of $K|S$ that is illuminated in the direction $u \in S^{n-1} \cap S$. They prove that $v_{K,i,k}$ determines an arbitrary compact convex set K of dimension at least $i + 1$, up to translation. Thus the extra information allows the symmetry assumption in Aleksandrov's projection theorem, Theorem 3.3.6, to be dropped. Related stability results can be found in [324] as well.

Let $1 \le i < k \le n - 1$, let K be a compact convex set in $\mathbb{E}^n$, and consider the function of $u \in S^{n-1}$ giving the average of $V_i(K|S)$ over all $S \in \mathcal{G}(n, k)$ with $u \in S$. That this function determines centrally symmetric compact convex sets of dimension at least $i + 1$, up to translation, is a consequence of the injectivity of certain higher-order Radon transforms on projection functions. More details can be found in [324], where Goodey and Weil also introduce the *average directed projection function* $\overline{v}_{K,i,k}(u)$ giving the average of $v_{K,i,k}(S, u)$ over all $S \in \mathcal{G}(n, k)$ with $u \in S$. They prove that $\overline{v}_{K,i,k}$ determines an arbitrary compact convex set K of dimension at least $i + 1$, up to translation, when $2 \le k < (2n - 3)/5$ and $(n - 2)/2 \le k \le n - 1$, but not when $k = (2n - 3)/5$ is an integer. In a remarkable analysis, they find that other sporadic cases of non-determination when $(2n - 3)/5 < k < (n - 2)/2$ are linked to the existence of integer points on certain nonsingular algebraic curves.

In a sequel, Goodey and Weil [325] define a different type of directed projection function in such a way that the case $i = j$ can be included and moreover both the shape and location of convex bodies can be determined.

These results inspired Schneider [743] to study another type of data where symmetry is not needed. He proves that two convex bodies whose projections on each hyperplane have the same mean width and the same Steiner point are equal, and provides a stability version of this result. (The *Steiner point* $st(K)$ of a convex body in $\mathbb{E}^n$ is given by

$$st(K) = \frac{1}{\kappa_n} \int_{S^{n-1}} h_K(u)u \, du.)$$

Also related to the above work of Goodey and Weil is that of astrophysicists on inverse problems concerning various projection data that take into account shadowing and light scattering effects of the body. See, for example, the paper of Kaasalainen and Lamberg [413] and the references given there.

3.8. *Reconstruction.* Every uniqueness theorem in tomography raises the important question of reconstruction. Reconstruction from brightness functions is treated in detail in Section 4.4. If the measurements are exact, the case $i = 1$ of Problem 3.6 is trivial: For each measurement $w_K(u)$ of an unknown centered convex body K, consider the closed half-spaces whose bounding hyperplanes have outer unit normal vectors u and $-u$ and have distance $h_K(u) = w_K(u)/2$ from the origin, and take the intersection of all such half-spaces. Nevertheless this case is of interest in robotics, where width function measurements can be obtained using an instrumented parallel-jaw gripper. For example, Rao and Goldberg [684] and Arkin et al. [12] use this model in attempting to recognize a shape among a known finite set of shapes. Li [506] finds the precise number of exact width function measurements which together with location information allow one to reconstruct a polygon with a known number of edges.

   Of course, if K is an unknown arbitrary convex body, the problem of reconstructing an approximation to K from finitely many exact support function measurements can be solved in the same way. Such data has also featured in robotics via the notion of a *line probe* or *hyperplane probe*. However, the focus here has been on algorithmic complexity issues; see, for example, the work of Lindenbaum and Bruckstein [509] and Richardson [692].

   The simple approach above for exact data fails when the measurements are noisy. Algorithms for the approximate reconstruction of planar convex bodies from a finite set of noisy support function measurements arose from the electrical engineering program of Willsky (see Note 1.5), and can immediately be applied to the case $n = 2$ and $i = 1$ of Problem 3.6. Prince [678], who mentions applications to tactile sensing, robot vision, and chemical component analysis, was the first to propose such algorithms. Prince and Willsky [679] use support function algorithms in computerized tomography as a prior to improve performance, particularly when only limited data are available. Theirs is the special case, in which $n = 2$ and the vectors u_i, $1 \le i \le k$, are equally spaced in S^1, of the following natural extension. Let u_i, $1 \le i \le k$, be fixed vectors in S^{n-1} whose *positive hull* (the set of all convex combinations with nonnegative coefficients) is $\mathbb{E}^n$. We say that the nonnegative real numbers h_i, $1 \le i \le k$, are *consistent* if there is a compact convex set C in $\mathbb{E}^n$ such that $h_C(u_i) = h_i$, $1 \le i \le k$. If h_i, $1 \le i \le k$, are consistent, there will be many such sets C; let $P(h_1, \ldots, h_k)$ denote the one that is the polytope defined by

$$P(h_1, \ldots, h_k) = \cap_{i=1}^k \{x \in \mathbb{E}^n : x \cdot u_i \le h_i\}. \qquad (3.12)$$

Algorithm NoisySupportLSQ

Input: Natural numbers $n \ge 2$ and $k \ge n + 1$; vectors $u_i \in S^{n-1}$, $1 \le i \le k$, whose positive hull is $\mathbb{E}^n$; noisy support function measurements

$$y_i = h_K(u_i) + N_i,$$

$1 \le i \le k$, of an unknown convex body K in $\mathbb{E}^n$, where the N_i's are independent normal random variables with zero mean and variance σ^2.

Task: Construct a convex polytope P_k in $\mathbb{E}^n$ that approximates K, with facet outer unit normal vectors belonging to the set $\{u_i : 1 \le i \le k\}$.

Action: Solve the following constrained linear least squares problem:

$$\min_{h_1,\ldots,h_k} \sum_{i=1}^{k}(y_i - h_i)^2,$$

subject to $h_i, 1 \le i \le k$, are consistent.   (3.13)

Let $\hat{h}_1, \ldots, \hat{h}_k$ be a solution and let $P_k = P(\hat{h}_1, \ldots, \hat{h}_k)$.   ∎

Naturally any implementation of Algorithm NoisySupportLSQ involves making explicit the constraint (3.13). For $n = 2$, this was done by Prince and Willsky [679] for vectors $u_i, 1 \le i \le k$, equally spaced in S^1, and by Lele, Kulkarni, and Willsky [504] for arbitrary vectors $u_i, 1 \le i \le k$, by means of an inequality constraint of the form $Ah \le 0$, where $h = (h_1, \ldots, h_k)$ and A is a certain matrix. (The latter paper also applies support function measurements to target reconstruction from range-resolved and Doppler-resolved laser-radar data and considers the reconstruction of the convex hull of nonconvex objects.) Poonawala, Milanfar, and Gardner [677] give further variations of the algorithm for $n = 2$ and provide a statistical analysis in this case.

For general n, it is more difficult to deal with the constraint (3.13). This was studied by Karl et al. [417], but the authors did not implement the algorithm for $n \ge 3$. An implementation for $n = 3$ and certain special sets of directions was carried out by Gregor and Rannou [332], who apply their algorithm to projection magnetic resonance imaging.

If the positive hull of $u_i, 1 \le i \le k$, is not $\mathbb{E}^n$, then (3.12) could still be considered as output of Algorithm NoisySupportLSQ, if consistency of $h_i, 1 \le i \le k$, is extended to closed convex sets which may be unbounded. However, if $\{u_i\}$ is a dense sequence of vectors in S^{n-1}, then for sufficiently large k, the positive hull of $u_i, 1 \le i \le k$, is $\mathbb{E}^n$ and in this case, Algorithm NoisySupportLSQ produces a polytope P_k as output.

We call a sequence $\{u_i\}$ in S^{n-1} *evenly spread* if for all $0 < t < 2$, there is a constant $c = c(t) > 0$ and an $N = N(t)$ such that

$$|\{u_1, \ldots, u_k\} \cap C_t(u)| \ge ck,$$

for all $u \in S^{n-1}$ and $k \ge N$. It is easy to show that an evenly spread sequence is dense in S^{n-1} and to construct specific evenly spread sequences. Gardner, Kiderlen, and Milanfar [274] employ techniques from the theory of empirical processes to show that if $\{u_i\}$ is evenly spread, then $\delta(P_k, K) \to 0$ almost surely as $k \to \infty$. Moreover, rates of convergence depending on k are estimated. Despite this, it seems that there is at present no completely satisfactory implementation of the algorithm for $n \ge 3$.

A different approach to the case $n = 2$ is taken by Fisher et al. [245], who use spline interpolation and the so-called von Mises kernel to fit a smooth curve to the data. This method was taken up by Hall and Turlach [378] and Mammen et al. [566], the former dealing with convex bodies with corners and the latter giving an example to show that the algorithm in [245] may fail for a given data set.

Karl [416] (see also Karl and Verghese [418]) obtains algorithms for the approximate reconstruction of smooth convex sets from curvature measurements of their projections. He notes that cardiac cineangiography and computer tracking of moving objects involve essentially 4-dimensional reconstruction problems.

The preliminary Theorem 3.1.2 shows that a finite set of projections provides insufficient data for verification of convex polytopes. Despite this, the problem of "shape-from-silhouette," in which an approximation to a shape is to be reconstructed from some of its orthogonal projections given as sets, is a popular one in the pattern recognition community. This finds many applications such as non-invasive 3-dimensional model acquisition, obstacle avoidance, and human motion tracking and analysis. The paper [86] by Bottino and Laurentini is an example, with some references that will aid the interested reader. See also Note 3.10. A closely related problem is that of reconstructing a compact smooth closed hypersurface M in $\mathbb{E}^n$ from its apparent contours in sufficiently large sets of directions, where the *apparent contour* of M in the direction $u \in S^{n-1}$ is the set of all $x \in u^\perp$ such that the line $x + l_u$ is tangent to M. See, for example, the articles of Golubyatnikov et al. [302] and Pointet [675], and the references given there.

3.9. *Mean projection bodies.* If μ is a finite Borel measure in S^{n-1}, and K is a convex body in $\mathbb{E}^n$, the function

$$h_{K_\mu}(u) = \int_{S^{n-1}} h_{K|v^\perp}(u)\, d\mu(v)$$

is the support function of a convex body K_μ. Schneider [732] calls K μ-*reconstructible* if $K_\mu = K$, and *reconstructible* if it is μ-reconstructible for some μ. He notes that boxes, 2-dimensional centrally symmetric compact convex sets, and balls are reconstructible in this sense, and proves that if K is μ-reconstructible when μ is a multiple of Lebesgue measure λ_{n-1}, then K is a ball.

Related to this, and prompted by earlier work on mean section bodies (see Note 7.13), are results of Goodey, Kiderlen, and Weil [311]. They call the convex body $P_k K$ whose support function is the integral of the support functions of $K|F$ over all k-dimensional planes F the kth *mean projection body*. They derive various relations between this body, the Blaschke section body, and the projection body (see Note 7.13). Spriestersbach [784] proves that K is determined by $P_{n-1}K$, a result improved by Goodey [307], who shows that shows that K is determined by $P_k K$ for $k \geq n/2$. Goodey further discovers the quite amazing fact that this is also true for $k = 2$ in all dimensions with the single exception of $n = 14$! Goodey and Jiang [310] prove that K is determined by $P_3 K$ for $n \geq 4$. In [430], Kiderlen proves that $P_k K$ determines K among centrally symmetric convex bodies in $\mathbb{E}^n$ if $2 \leq k \leq n - 1$ is even and among convex bodies of constant width if $3 \leq k \leq n - 1$ is odd. He also obtains corresponding stability results and shows that a convex body K in $\mathbb{E}^n$ is determined by $P_k K$ and $P_{k'}K$ if $2 \leq k \neq k' \leq n - 1$.

3.10. *Projections of convex polytopes.* The natural question of whether a set in $\mathbb{E}^n$ must be a convex polytope if all its projections on k-dimensional subspaces (for some k with $2 \leq k \leq n - 1$) are also convex polytopes was answered positively by Klee [444]. (The case $n = 3$ is due to Bol [78].) This paper contains several related results, including an example of a nonpolyhedral 3-cell in $\mathbb{E}^3$ whose projections on planes are all polygonal (but not necessarily convex).

Klee [445] also proved that if P is an n-dimensional centrally symmetric convex polytope, $n \geq 3$, then there is a $u \in S^{n-1}$ such that $P|u^\perp$ has at least $2n$ vertices. So, of course, if every projection of a polytope in $\mathbb{E}^n$, $n \geq 3$, has less than $2n$ vertices, it cannot be centrally symmetric. For a related open problem and references to other work of this sort, see [173, Problem B10] and [367, Section 5.1]. Knox [451] finds best possible inequalities relating the number of facets of a convex polytope to the number of facets of one of its projections on a hyperplane. Also, results exist concerning the expected number of j-dimensional faces of a random projection of a regular convex polytope onto a k-dimensional subspace. See the paper of Böröczky, Jr. and Henk [84], and the references given there.

Projections are one type of probe used in geometric probing (cf. Note 1.6); the terms *projection probe*, *silhouette*, and *silhouette probe* have also been used. Extra a priori information must be assumed. For example, Li [506] proves that $3n - 2$ projections are necessary and sufficient to successively determine a convex n-gon in $\mathbb{E}^2$, among all convex polygons. Let $f_k(P)$ denote the number of k-dimensional faces of a polytope. Dobkin, Edelsbrunner, and Yap [189] demonstrate that $5 f_0(P) + f_2(P)$ projections suffice, and $[f_2(P)/2] + 1$ are necessary, to successively determine the convex polyhedron P in $\mathbb{E}^3$, among all convex polyhedra.

3.11. *Critical projections.* Filliman [231] demonstrated that for $1 \leq i \leq n - 1$ the ith projection function of a polytope in $\mathbb{E}^n$ is piecewise linear on $\mathcal{G}(n, i)$, and used this fact in connection with the location of *critical projections* of a polytope, those with maximal or minimal volume. The problem is surprisingly difficult, and is not fully solved even for regular polytopes, despite work of Filliman, Chakerian, McMullen, and others, listed in the survey [581, Section 3] of Martini.

Suppose that $1 \leq p \leq \infty$ and that

$$B_p^n = \{x \in \mathbb{E}^n : \sum_{i=1}^{n} \|x_i\|^p \leq 1\}$$

is the unit ball in l_p^n ($\mathbb{E}^n$ with the l^p norm, or sup norm if $p = \infty$). Note that B_2^n and B_∞^n are the usual unit ball and centered cube of side length 2 in $\mathbb{E}^n$. Let $u \in S^{n-1}$. Barthe and Naor [41] prove that the ratio

$$\frac{\lambda_{n-1}\left(B_p^n | u^\perp\right)}{\lambda_{n-1}\left(B_p^{n-1}\right)}$$

increases with $p \geq 1$. This ratio is 1 when $p = 2$, so it follows that the minimal projection of B_p^n for $p \geq 2$ and the maximal projection of B_p^n for $1 \leq p \leq 2$ both occur when $u = (1, 0, \ldots, 0)$. They also find the maximal projections of B_p^n onto hyperplanes for $p \geq 2$ and the minimal such projections for $p = 1$. This leaves open only the problem of finding the minimal projections for $1 < p < 2$. Barthe and Naor use probabilistic methods, but later Koldobksy, Ryabogin, and Zvavitch [467] find a more direct approach using the Fourier transform (see Note 4.2), also presented in Koldobsky's book [465, Chapter 8]. Compare Note 7.8, dealing with the corresponding results for sections; in fact, the latter came earlier and inspired the work outlined in this paragraph.

Some results on characterizations of the shape of minimal projections of a cube are obtained by Ostrovskii; see [653] and the references given there.

Burger and Gritzmann [114] (see also Gritzmann and Klee's survey [341, Section 8.1]) study the computational complexity of finding critical projections of polytopes. They give, for example, a randomized algorithm for maximizing the volume of projections of a polytope in $\mathbb{E}^n$ (described in terms of its facets) on hyperplanes with an error bound of essentially $O(\sqrt{n/\log n})$.

3.12. *Almost-spherical or almost-ellipsoidal projections, and related results.* Groemer [347] concludes that every convex body in $\mathbb{E}^3$ has a reasonably circular projection. Specifically, he shows that for each compact convex set K in $\mathbb{E}^3$, there is a $u \in S^2$ such that

$$\frac{r(K|u^\perp)}{R(K|u^\perp)} \geq \frac{1}{2},$$

where r and R denote inradius and circumradius in $u^\perp$, and another $u \in S^2$ for which

$$\frac{\text{minw}(K|u^\perp)}{\text{diam}(K|u^\perp)} \geq \frac{1}{\sqrt{2}}.$$

(Quantities such as minimal width are taken in the plane $u^\perp$, and the left-hand sides are defined to be 1 when $K | u^\perp$ is a single point. The first inequality was originally proved by F. John.) See [173, Problem A10] for a related open problem of L. Santaló.

Dvoretzky's theorem and its embellishments show that each centered convex body of sufficiently high dimension has a k-dimensional projection that is almost spherical (in a different sense), and many that are almost ellipsoidal. Related to this topic are results on the distance between projections of a convex body. See the discussion in Note 7.9.

3.13. *Aleksander Danilovich Aleksandrov (1912–1999).* (Sources: [204], [765].) Aleksander Danilovich Aleksandrov was born in Volyn', near Ryazan' in Russia, but from early childhood lived in Leningrad (now St. Petersburg), where his parents were high school teachers. In 1933 he graduated in physics at the University of Leningrad, where he studied and taught until 1964. By 1930 he had already begun research in optics and theoretical physics, as well as in geometry: One of his teachers, B. N. Delone, was interested in the geometry of numbers and the structure of crystals. Aleksandrov continued this research until 1936, simultaneously teaching in the Faculty of Mathematics and Mechanics. In 1937 he obtained his PhD and became acting Professor of Geometry, and from 1938 to 1952 was also a Senior Scientific Worker at the Leningrad branch of the Steklov Institute. In 1944 Aleksandrov became Professor of Geometry, and from 1952 to 1964 served as rector of the university. From 1964 to 1986 he headed the Geometry Department at the Mathematical Institute in Novosibirsk, Siberia. After 1987 he again worked at the Leningrad branch of the Steklov Institute.

Figure 3.10. Aleksander Danilovich Aleksandrov.

Aleksandrov's long career brought him many honors and awards. In 1942 his work in geometry won a State prize, and in 1951 he received the Lobachevsky prize, given every four years by the Academy of Sciences of the USSR for work in geometry. (Previous recipients include Hilbert, Poincaré, and Klein.) From 1946 he was a corresponding member of the Academy of Sciences of the USSR, becoming a full member in 1964. In 1975 he was elected a member of the Italian National Academy, and in 1990 was the first recipient of the Euler gold medal awarded by the Russian Academy of Sciences.

This text is a testament to some of Aleksandrov's fundamental contributions to convex geometry, in particular to the theory of area measures (see Section A.2),

which relieved many earlier results from implicit and often unnecessary differentiability assumptions, and the discovery of the Aleksandrov–Fenchel inequality
(see Section B.3). He also developed a theory of 2-dimensional manifolds of
bounded curvature and found new geometric methods for studying elliptic partial
differential equations. In later work, Aleksandrov proved the "fundamental theorem
of chronogeometry," which says that any bijection of $\mathbb{E}^n$, $n \geq 3$, onto itself taking
light-cones onto light-cones is affine; this initiated a vast fresh area of research.
Comprehensive summaries of Aleksandrov's work and a long list of his publications
can be found in [204], [556], and [765].

   Married twice, Aleksandrov had a son and a daughter. He enjoyed mountaineer-
ing and in 1949 became a USSR master of this sport, a title requiring expertise of
the highest caliber.

4

Projection bodies and volume inequalities

In the first section of this chapter, we meet projection bodies. A projection body is a centered convex body whose support function gives the volumes of projections on hyperplanes of another convex body. In Theorem 4.1.11, we see that projection bodies are just centered zonoids. Every ellipsoid is a zonoid; moreover, Corollary 4.1.12 characterizes zonoids as Hausdorff limits of finite Minkowski sums of ellipsoids. Zonoids are "supersymmetric" convex bodies, of proven and wide applicability in mathematics and other subjects. Particularly susceptible to analytical methods, zonoids have found a prominent place not only in geometry, but also in the local theory of Banach spaces. Generalized zonoids, which constitute a stepping-stone from zonoids to general centrally symmetric convex bodies, are also introduced. Here some properties of the cosine transform, expounded in Appendix C, come into play.

Section 4.2 deals with a question, sometimes called Shephard's problem, of comparative information: If the area of the shadow on a plane of one convex body is always less than that of another, is the volume of the one body less than that of the other? The solution employs the Brunn–Minkowski theory and zonoids. Theorem 4.2.4 exhibits a pair of centrally symmetric convex bodies for which the answer is negative. If the body whose shadows have larger area is a zonoid, however, the answer is positive, as Corollary 4.2.7 demonstrates. Furthermore, Theorem 4.2.13 shows that for arbitrary centrally symmetric convex bodies in $\mathbb{E}^n$, the hypotheses of Shephard's problem imply that the volume of one body is no more than $\sqrt{n}$ times the volume of the other. This follows readily from a basic property of the John ellipsoid of a convex body, the ellipsoid of maximal volume contained in the body, proved in Theorem 4.2.12.

A taste of stability results is offered in Section 4.3. One of these, stated in Remark 4.3.13, is crucial in Section 4.4, for proving the convergence of Algorithm BrightLSQ, an algorithm for reconstructing an approximation to an unknown centered convex body from finitely many measurements of its brightness function.

There should be no extra demands on the reader who has found the previous chapter manageable.

4.1. Projection bodies and related concepts

Definition 4.1.1. Let K be a convex body in $\mathbb{E}^n$, $n \geq 2$. The *projection body* ΠK of K is the centered convex body such that

$$h_{\Pi K}(u) = \lambda_{n-1}(K|u^{\perp}) = \frac{1}{2} \int_{S^{n-1}} |u \cdot v| \, dS(K, v),$$

for all $u \in S^{n-1}$.

Note the useful fact that $\Pi(-K) = \Pi K$.

The last equality in the previous definition is from Cauchy's projection formula ((A.45) with $i = n-1$; recall (see Section A.2) that $S(K, \cdot) = S_{n-1}(K, \cdot)$ denotes the surface area measure of K). To put the definition in other words, the projection body ΠK has width in any direction equal to twice the $(n-1)$-dimensional volume of the projection of K orthogonal to that direction. It is obvious that ΠK is centered, and it follows directly from the characterization of support functions in Section 0.6 that $h_{\Pi K}$ is indeed a support function; but we shall prove a much stronger result in Theorem 4.1.11.

As we shall see, it turns out that projection bodies in $\mathbb{E}^n$ have another, apparently quite different, geometrical description; they are precisely the centered n-dimensional zonoids.

Definition 4.1.1 makes sense when K is an arbitrary compact convex set, and this is sometimes useful, though ΠK is then not a body. Indeed, suppose that K is a k-dimensional compact convex set in $\mathbb{E}^n$. If $k < n - 1$, then ΠK reduces to the single point $\{o\}$. If $k = n - 1$, then we would get $h_{\Pi K}(u) = \lambda_{n-1}(K)|u \cdot v|$, for all $u \in S^{n-1}$, where $v \in S^{n-1}$ is orthogonal to K; by (0.25), this means that ΠK is simply a centered line segment of length $2\lambda_{n-1}(K)$ parallel to v.

The map Π taking a convex body to its projection body is sometimes called the *Minkowski map*, and it provides a convenient notation for discussing projections. For example, the case $i = k = n - 1$ of Aleksandrov's projection theorem, Theorem 3.3.6, becomes: If K_1 and K_2 are centered convex bodies in $\mathbb{E}^n$, then

$$\Pi K_1 = \Pi K_2 \Rightarrow K_1 = K_2;$$

in other words, the Minkowski map is injective when restricted to centered bodies.

If K is an arbitrary convex body in $\mathbb{E}^n$, there will generally be many other convex bodies whose projection bodies are also ΠK (for example, any convex body of constant brightness equal to one has the unit ball as its projection body). Among these there is a unique centered convex body, namely, ∇K; we highlight in the next theorem the relationship between the projection body and the Blaschke body.

Theorem 4.1.2. *If K is a convex body in $\mathbb{E}^n$, then ∇K is the unique centered convex body such that*

$$\Pi(\nabla K) = \Pi K.$$

Proof. We observed in Chapter 3 that by the equivalence $(3.2) \Leftrightarrow (3.3)$ of Theorem 3.3.2, with $i = n - 1$, ∇K has the same brightness function as K, and that the uniqueness follows from Aleksandrov's projection theorem, Theorem 3.3.6, again with $i = n - 1$. ■

We also know from Chapter 3 that equality of projection bodies can be stated in terms of Blaschke bodies or mixed volumes, as follows.

Theorem 4.1.3. *Suppose that K_1 and K_2 are convex bodies in $\mathbb{E}^n$. Then the following are equivalent:*

$$\Pi K_1 = \Pi K_2;$$

$$\nabla K_1 = \nabla K_2;$$

$$V(K_1, n - 1; K) = V(K_2; n - 1; K),$$

for each centrally symmetric compact convex set K.

Proof. This is just a reformulation of Theorem 3.3.2, with $i = n - 1$. ■

The term *projection class* of K is sometimes used for the class of all convex bodies K' for which $\Pi K' = \Pi K$. For any such K',

$$\lambda_n(\nabla K) = \lambda_n(\nabla K') \geq \lambda_n(K'),$$

by the previous theorem and Theorem 3.3.9, so each projection class has precisely one centered member, and this is the unique body in the class with the largest volume. For a significant generalization, see Note 4.9.

Let us begin to look at some simple examples of projection bodies. If K is the unit disk in $\mathbb{E}^2$, then ΠK is the centered disk of radius 2, and if K is the centered unit square, then $\Pi K = 2K$. This follows from the next theorem, which is false in higher dimensions; see Figure 4.1 and Remark 4.1.19(i).

Theorem 4.1.4. *Let K be a convex body in $\mathbb{E}^2$. Then ΠK is a rotation by $\pi/2$ about the origin of the difference body $2\triangle K$. Consequently, every centered convex body in $\mathbb{E}^2$ is a projection body.*

Proof. If K is any convex body in $\mathbb{E}^2$, $u \in S^1$, and $v \in S^1$ is orthogonal to u, then

$$h_{\Pi K}(u) = \lambda_1(K|u^\perp) = w_K(v)$$
$$= w_{\triangle K}(v) = h_{2\triangle K}(v).$$

This establishes the first statement of the theorem. If K is centered, then $\triangle K = K$, and it follows that $\Pi K_0 = K$, where K_0 is $\frac{1}{2}K$ rotated by $\pi/2$ about the origin. ∎

In $\mathbb{E}^n$, the projection body of the unit ball is the centered ball of radius κ_{n-1}. In order to examine ellipsoids, the following lemma is useful.

Theorem 4.1.5. *The projection bodies of affinely equivalent convex bodies are also affinely (in fact, linearly) equivalent. Specifically, if $\phi \in GL_n$, then*

$$\Pi(\phi K) = |\det \phi|\phi^{-t}(\Pi K).$$

Proof. The first statement follows from the second, since translating a convex body leaves its projection body unchanged. Let $\phi \in GL_n$. If K is a convex body in $\mathbb{E}^n$, $u \in S^{n-1}$, and $\phi w = u$, then

$$h_{\Pi(\phi K)}(u) = \frac{1}{2}\int_{S^{n-1}} |u \cdot v|\, dS(\phi K, v)$$
$$= \frac{n}{2}V(\phi K, n-1; [-u, u])$$
$$= \frac{n}{2}V(\phi K, n-1; \phi[-w, w])$$
$$= \frac{n}{2}|\det \phi|V(K, n-1; [-w, w])$$
$$= |\det \phi|h_{\Pi K}(w) = |\det \phi|h_{\Pi K}(\phi^{-1}u),$$

where we have used Cauchy's projection formula ((A.45) with $i = n - 1$), equation (A.37), and the invariance of mixed volumes under volume-preserving linear transformations (A.17). We now apply the formula (0.27) for the change in support functions under linear transformations to get

$$h_{\Pi K}(\phi^{-1}u) = h_{\phi^{-t}(\Pi K)}(u),$$

which, together with the previous equation, completes the proof. ∎

In the next theorem, E^* denotes the polar body of E; see Section 0.8.

Corollary 4.1.6. *The projection body of an ellipsoid is an ellipsoid. Specifically, if E is a centered n-dimensional ellipsoid in $\mathbb{E}^n$, then*

$$\Pi E = \frac{\kappa_{n-1}\lambda_n(E)}{\kappa_n}E^*.$$

Conversely, every centered n-dimensional ellipsoid is a projection body.

Proof. Let E be an n-dimensional ellipsoid in $\mathbb{E}^n$. We may assume that E is centered, so $E = \phi B$, where $\phi \in GL_n$. Then, by Theorem 4.1.5,

$$\Pi E = \Pi(\phi B) = |\det \phi| \phi^{-t}(\Pi B) = |\det \phi| \phi^{-t}(\kappa_{n-1} B) = |\det \phi| \kappa_{n-1} \phi^{-t} B.$$

Now

$$\phi^{-t} B = \phi^{-t} B^* = (\phi B)^* = E^*,$$

by (0.37). Also, $|\det \phi| = \lambda_n(E)/\kappa_n$, by (0.7); the first statement of the theorem follows.

For the second part, observe first that if K is a convex body in $\mathbb{E}^n$ and $r > 0$, then $\Pi(rK) = r^{n-1} \Pi K$, by (0.8). Suppose that E is a centered n-dimensional ellipsoid in $\mathbb{E}^n$. Let $E_0 = rE^*$, where

$$r = \left(\frac{\kappa_{n-1} \lambda_n(E^*)}{\kappa_n} \right)^{-1/(n-1)}.$$

Then (cf. Section 0.8)

$$\Pi E_0 = \frac{\kappa_n}{\kappa_{n-1} \lambda_n(E^*)} \Pi E^* = (E^*)^* = E. \qquad \blacksquare$$

If K is the centered unit cube in $\mathbb{E}^n$, $\Pi K = 2K$. To see this, we use the formula

$$h_{\Pi K}(u) = \frac{1}{2} \int_{S^{n-1}} |u \cdot v| \, dS(K, v).$$

When K is the centered unit cube, $S(K, \cdot)$ is a sum of point masses of weight 1 placed at the intersections of the coordinate axes with S^{n-1}; see Section A.2. Let e_i be the unit vector in the ith coordinate direction. The integral then reduces to a sum of the n terms $|u \cdot e_i|$, $1 \le i \le n$. By (0.25), each term is the support function of the line segment $[-e_i, e_i]$, and so by (0.23), ΠK is the vector sum of these segments, the centered unit cube expanded by a factor of 2.

The same argument will also show that if K is any polytope, then ΠK is a vector sum of line segments orthogonal to the facets of K, each of length equal to the $(n-1)$-dimensional volume of the corresponding facet. For example, if K is a regular tetrahedron, ΠK is a rhombic dodecahedron (see Figure 4.1).

(Figure 4.1 also depicts the projection body of a convex double bounded cone C. By approximating C by double pyramids over regular m-gons, and letting $m \to \infty$, it can be shown that the boundary of the spindle-shaped ΠC is actually formed by revolving a cosine curve around the vertical axis; Chilton and Coxeter [162] give a proof. As a simple but instructive exercise, we invite the reader to construct the projection body of the cylinder in $\mathbb{E}^3$ obtained by rotating a centered unit square (side length 1) about the z-axis. The answer is given in Note 4.3.)

The rhombic dodecahedron is a member of an important class of polytopes we are now led to investigate.

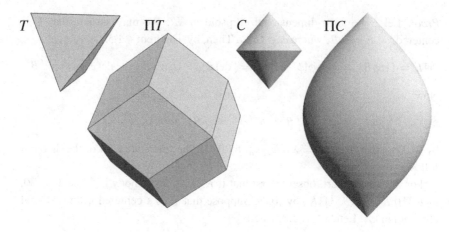

Figure 4.1. Projection bodies.

Definition 4.1.7. A *zonotope* is a finite vector sum of line segments.

The name "zonotope" derives from the fact that there are bands or zones of facets in the boundary, with the property that adjacent facets in the zone share edges all parallel to one of the line segments in the defining vector sum. Such zones can be seen in the zonotope ΠT depicted in Figure 4.1.

Lemma 4.1.8. *Let Z be a centered zonotope in $\mathbb{E}^n$. The support function of Z has the form*

$$h_Z(u) = \sum_{i=1}^m a_i |u \cdot v_i|,$$

for $u \in S^{n-1}$, where $a_i > 0$ and $v_i \in S^{n-1}$, $1 \le i \le m$.

Proof. If Z is an arbitrary zonotope, then $Z = \sum_{i=1}^m (a_i[-v_i, v_i] + w_i)$, where $a_i > 0$, $v_i \in S^{n-1}$, and $w_i \in \mathbb{E}^n$, $1 \le i \le m$. By (0.23), (0.24), and (0.25), the support function of Z is given for $u \in S^{n-1}$ by

$$h_Z(u) = \sum_{i=1}^m (a_i |u \cdot v_i| + u \cdot w_i).$$

Now if Z is centered, we have $h_Z(u) = h_Z(-u)$, implying that $\sum_{i=1}^m u \cdot w_i = 0$ and yielding the required representation of h_Z. ∎

Definition 4.1.9. A compact convex set in $\mathbb{E}^n$ is a *zonoid* if it is the Hausdorff limit of a sequence of zonotopes.

The class of zonotopes is closed under Minkowski linear combinations and affine transformations. It follows that the zonoids form the smallest nontrivial class of compact convex sets closed under these two operations and also under Hausdorff limits. In particular, the projection of a zonotope (or zonoid) is also a zonotope (or zonoid, respectively).

Zonotopes and zonoids are centrally symmetric. The proof of Lemma 4.1.8 shows that every zonotope is a translation of a centered zonotope and must therefore be centrally symmetric. Since central symmetry is preserved under limits in the Hausdorff metric, every zonoid is also centrally symmetric. See Remark 4.1.19 in this regard.

Other special properties of zonotopes and zonoids make them important in surprisingly diverse ways. We shall say more about this later and in the notes for this chapter. It is convenient to begin by establishing the characterization of centered zonoids most relevant for our purposes.

Theorem 4.1.10. *A compact convex set K in $\mathbb{E}^n$ is a centered zonoid if and only if*

$$h_K(u) = \int_{S^{n-1}} |u \cdot v| \, d\mu(v),$$

for all $u \in S^{n-1}$, where μ is a finite even Borel measure in S^{n-1}.

Proof. Suppose first that the support function h_K of K is given by the previous equation; note that K is centered, because $h_K(u) = h_K(-u)$ for all $u \in S^{n-1}$. Let $\varepsilon > 0$ be given. Choose $\delta = \varepsilon/\mu(S^{n-1})$. Let $\{E_1, \ldots, E_m\}$ be a partition of S^{n-1} into nonempty disjoint Borel sets of diameters less than δ. Choose $v_i \in E_i$, $1 \le i \le m$. Then

$$\left| \int_{S^{n-1}} |u \cdot v| \, d\mu(v) - \sum_{i=1}^{m} |u \cdot v_i| \mu(E_i) \right| = \left| \sum_{i=1}^{m} \int_{E_i} (|u \cdot v| - |u \cdot v_i|) \, d\mu(v) \right|$$

$$\le \sum_{i=1}^{m} \int_{E_i} \left| |u \cdot v| - |u \cdot v_i| \right| d\mu(v)$$

$$\le \sum_{i=1}^{m} \int_{E_i} |u \cdot (v - v_i)| \, d\mu(v)$$

$$\le \sum_{i=1}^{m} \int_{E_i} |v - v_i| \, d\mu(v)$$

$$< \delta \sum_{i=1}^{m} \int_{E_i} d\mu(v)$$

$$= \delta \, \mu(S^{n-1}) = \varepsilon.$$

Let Z be the zonotope with support function $h_Z(u) = \sum_{i=1}^{m} \mu(E_i)|u \cdot v_i|$, for $u \in S^{n-1}$. Then we have shown that

$$|h_K(u) - h_Z(u)| < \varepsilon,$$

for all $u \in S^{n-1}$. By (0.22), the distance between K and Z in the Hausdorff metric is less than ε, and consequently K is a zonoid.

Conversely, let K be a centered zonoid. Then K is the limit of a sequence $\{Z_j\}$ of zonotopes. Furthermore K, since it is centered, must also be the limit of the sequence $\{\triangle Z_j\}$, where each central symmetral $\triangle Z_j$ is a centered zonotope. By Lemma 4.1.8, a centered zonotope Z has support function

$$h_Z(u) = \sum_{i=1}^{m} a_i |u \cdot v_i| = \int_{S^{n-1}} |u \cdot v| \, d\mu(v),$$

for $u \in S^{n-1}$, where μ is the sum of point masses $a_i/2$ at the points $\pm v_i$. Therefore it suffices to show that the set of compact convex sets whose support functions have the desired form is closed in $\mathcal{K}^n$.

To this end, let $\{K_j\}$ be a sequence of compact convex sets with

$$h_{K_j}(u) = \int_{S^{n-1}} |u \cdot v| \, d\mu_j(v),$$

for all $u \in S^{n-1}$, where μ_j is a finite even Borel measure in S^{n-1}, and suppose that $K_j \to K$ in the Hausdorff metric (so $h_{K_j} \to h_K$ uniformly on S^{n-1}; cf. (0.22)). Clearly there is an $r > 0$ such that $K_j \subset rB$ for sufficiently large j.

Let $\|\mu_j\| = \mu_j(S^{n-1})$. Using Fubini's theorem, we obtain

$$2\kappa_{n-1} \|\mu_j\| = \int_{S^{n-1}} 2h_{\Pi B}(v) \, d\mu_j(v)$$

$$= \int_{S^{n-1}} \int_{S^{n-1}} |u \cdot v| \, du \, d\mu_j(v)$$

$$= \int_{S^{n-1}} \int_{S^{n-1}} |u \cdot v| \, d\mu_j(v) \, du$$

$$= \int_{S^{n-1}} h_{K_j}(u) \, du$$

$$\leq r \int_{S^{n-1}} h_B(u) \, du = r\omega_n.$$

Therefore $\{\|\mu_j\|\}$ is bounded. This means that there is a subsequence $\{\mu_{j_k}\}$ and a finite even Borel measure μ in S^{n-1} such that μ_{j_k} converges weakly to μ; see (0.16) and, for example, [199, p. 424]. Since the scalar product $|u \cdot v|$ is continuous, we see that

$$h_{K_{j_k}}(u) \to \int_{S^{n-1}} |u \cdot v| \, d\mu(v),$$

for each $u \in S^{n-1}$. At u, these support functions also converge to $h_K(u)$, completing the proof. ∎

The measure μ of the previous theorem is called the *generating measure* of the zonoid K. It is essentially unique, by Theorem C.2.1.

Theorem 4.1.11. *A projection body is a centered zonoid. Conversely, every centered n-dimensional zonoid in $\mathbb{E}^n$ is the projection body of a unique centered convex body.*

Proof. Let ΠK be the projection body of a convex body K in $\mathbb{E}^n$. Then by Cauchy's projection formula ((A.45) with $i = n - 1$),

$$
\begin{aligned}
h_{\Pi K}(u) &= \frac{1}{2} \int_{S^{n-1}} |u \cdot v| \, dS(K, v) \\
&= \frac{1}{4} \left(\int_{S^{n-1}} |u \cdot v| \, dS(K, v) + \int_{S^{n-1}} |u \cdot v| \, dS(-K, v) \right) \\
&= \int_{S^{n-1}} |u \cdot v| \, d\mu(v),
\end{aligned}
$$

where $\mu = \frac{1}{4}\big(S(K, \cdot) + S(-K, \cdot)\big)$ is even. So ΠK is a centered zonoid, by Theorem 4.1.10.

If K is a centered zonoid, we have, by Theorem 4.1.10,

$$
h_K(u) = \int_{S^{n-1}} |u \cdot v| \, d\mu(v),
$$

for all $u \in S^{n-1}$, where μ is a finite even Borel measure in S^{n-1}. If K is also n-dimensional, then μ cannot be concentrated on a great subsphere of S^{n-1}, so Minkowski's existence theorem, Theorem A.3.2, provides a centered convex body K_0 such that $2\mu = S(K_0, \cdot)$. Therefore

$$
h_K(u) = \frac{1}{2} \int_{S^{n-1}} |u \cdot v| \, dS(K_0, v),
$$

for all $u \in S^{n-1}$. This implies that $K = \Pi K_0$, and the uniqueness follows from Aleksandrov's projection theorem, Theorem 3.3.6. ∎

Corollary 4.1.12. *A compact convex set in $\mathbb{E}^n$ is a zonoid if and only if it is a Hausdorff limit of finite Minkowski sums of n-dimensional ellipsoids.*

Proof. A zonoid is a Hausdorff limit of finite vector sums of line segments. Each such line segment may be approximated arbitrarily closely in the Hausdorff metric by n-dimensional ellipsoids, so zonoids have the stated property.

Conversely, each n-dimensional ellipsoid is a zonoid, by Corollary 4.1.6 and Theorem 4.1.11. A Hausdorff limit of finite Minkowski sums of zonoids is again a zonoid, so the proof is finished. ∎

For an analogous result involving Blaschke sums, see Note 3.4.

By Theorem 4.1.11, a convex body in $\mathbb{E}^n$ is a projection body if and only if it is a centered zonoid. Consequently, the information we have gathered about zonoids can be transferred to projection bodies. For example, the class of projection bodies is closed under Minkowski linear combinations (see Note 4.3). Further, this class is also closed under full-dimensional Hausdorff limits and nonsingular linear transformations (the latter also follows from Theorem 4.1.5). Now if K is a k-dimensional compact convex set in $\mathbb{E}^n$, $k < n$, we could define its projection body by identifying its affine hull with $\mathbb{E}^k$ (so that n is replaced by k in Definition 4.1.1). If we did this, we could say that the class of projection bodies is closed under Hausdorff limits and linear transformations, and infer the feasible-sounding statement that the projection of a projection body is again a projection body. We choose not to adopt this approach, however.

The following result provides more information about how a zonoid may be approximated by a zonotope.

Theorem 4.1.13. *Let K be a centered zonoid in $\mathbb{E}^n$ and let $u_i \in S^{n-1}$, $1 \leq i \leq k$. There is a centered zonotope Z in $\mathbb{E}^n$, the sum of at most $k+1$ line segments, such that $h_Z(\pm u_i) = h_K(\pm u_i)$ for $1 \leq i \leq k$.*

Proof. Let μ be the generating measure of K, and assume without loss of generality that μ is normalized so that $\mu(S^{n-1}) = 1$. Let $C \subset \mathbb{E}^k$ be the set of all vectors $(h_Z(u_1), \ldots, h_Z(u_k))$ such that Z is a centered zonoid in $\mathbb{E}^n$ with normalized generating measure. Using the support function representation in Theorem 4.1.10, it is straightforward to check that C is compact and convex. If $v \in S^{n-1}$, the line segment $[-v, v]$ has by (0.25) support function $|u \cdot v|$, $u \in S^{n-1}$, and normalized generating measure (the sum of point masses of weight 1/2 at $\pm v$). Since the map that takes v to $(|u_1 \cdot v|, \ldots, |u_k \cdot v|)$ is a continuous map from S^{n-1} into C, the subset D of C arising from centered line segments with normalized generating measures is compact. Each centered zonoid is the Hausdorff limit of a sequence of centered zonotopes, so $C = \mathrm{cl}\,\mathrm{conv}\,D$, and since D is closed, this yields $C = \mathrm{conv}\,D$. By Carathéodory's theorem (see [737, Theorem 1.1.4]), each point in C is a convex combination of at most $k + 1$ points in D. Therefore there are $v_1, \ldots, v_{k+1} \in S^{n-1}$ and reals $a_j \geq 0$, $1 \leq j \leq k+1$, summing to one, such that

$$h_K(u_i) = \sum_{j=1}^{k+1} a_j |u_i \cdot v_j|,$$

for $1 \leq i \leq k$. Consequently, the zonotope $Z = \sum \{a_j[-v_j, v_j] : 1 \leq j \leq k+1\}$ has the desired property. ∎

The previous theorem will find application in Section 4.4, though its full strength is not needed there. In fact, the number $k + 1$ of line segments can be reduced to k; see Note 4.1.

We now turn to an important relaxation of the definition of projection bodies and zonoids.

Definition 4.1.14. Any translation of a centered compact convex set K in $\mathbb{E}^n$, $n \geq 2$, for which there is a *signed* finite even Borel measure μ in S^{n-1} such that

$$h_K(u) = \int_{S^{n-1}} |u \cdot v| \, d\mu(v),$$

for all $u \in S^{n-1}$, is called a *generalized zonoid*. The measure μ is called the *generating signed measure* of K. An n-dimensional centered generalized zonoid is also called a *generalized projection body*.

The term "the generating signed measure" can be used unambiguously because of Theorem C.2.1, which implies that this measure is essentially unique. Theorem 4.1.10 tells us that a projection body (or zonoid) is a generalized projection body (or generalized zonoid, respectively). In fact, a zonoid is just a generalized zonoid whose generating signed measure happens to be a measure.

Remark 4.1.15. It is not at all obvious from this definition that there are generalized zonoids that are not zonoids, but such bodies do exist, and Figure 4.2 shows a 3-dimensional example. The construction is as follows. On the left of Figure 4.2 are shown six planar members of a one-parameter family $\{K(\alpha), -2/5 \leq \alpha \leq 1/2\}$ whose support functions are given for unit vectors u by

$$h_{K(\alpha)}(u) = 1 + \frac{\alpha}{2}\big(3(u \cdot e)^2 - 1\big),$$

where e is a unit vector in the vertical direction. (The parameter α is confined to the interval $-2/5 \leq \alpha \leq 1/2$ in order to ensure that $h_{K(\alpha)}$ is sublinear.) With the same definition in $\mathbb{E}^3$, $K(\alpha)$ is a body of revolution about the vertical axis. It can be shown that

$$h_{K(\alpha)}(u) = \frac{1}{2\pi} \int_{S^2} |u \cdot v| \left(1 + 2\alpha\big(3(v \cdot e)^2 - 1\big) \right) dv,$$

for all $u \in S^2$. This representation shows that $K(\alpha)$ is a generalized zonoid for $-2/5 \leq \alpha \leq 1/2$. The function $1 + 2\alpha(3t^2 - 1)$ is nonnegative for $-1 \leq t \leq 1$ if and only if $-1/4 \leq \alpha \leq 1/2$, so $K(\alpha)$ is a zonoid if and only if $-1/4 \leq \alpha \leq 1/2$. (In particular, $K(0)$ is a disk.) The 3-dimensional example in Figure 4.2 corresponds to $\alpha = -1/3$.

Generalized zonoids can be viewed as "differences" of zonoids, in the sense made precise by the following theorem.

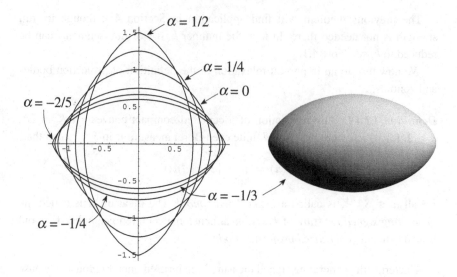

Figure 4.2. A generalized zonoid that is not a zonoid.

Theorem 4.1.16. *Let K be a generalized zonoid. There exist zonoids Z_1, Z_2 such that*

$$Z_1 = K + Z_2.$$

Proof. By translating, if necessary, we may assume that K is centered. Let μ be the generating signed measure of K. By the Jordan decomposition theorem (see, for example, [700, p. 274]), there are Borel measures μ_1 and μ_2 such that $\mu = \mu_1 - \mu_2$. For $j = 1, 2$, define

$$\bar{\mu}_j(E) = \frac{1}{2}\mu_j(E) + \frac{1}{2}\mu_j(-E),$$

for each Borel subset E of S^{n-1}. Then $\bar{\mu}_j$ is even, $j = 1, 2$, and $\mu = \bar{\mu}_1 - \bar{\mu}_2$, since μ is even. Let Z_j be the centered zonoid whose generating measure is $\bar{\mu}_j$, for $j = 1, 2$. Then $h_K = h_{Z_1} - h_{Z_2}$, so $Z_1 = K + Z_2$, by (0.23) and the uniqueness of support functions. ∎

We also have the following result (cf. Remark 4.1.19(ii)).

Theorem 4.1.17. *Every centrally symmetric convex body K in $\mathbb{E}^n$ with $h_K \in C^\infty(S^{n-1})$ is a generalized zonoid.*

Proof. We may assume that K is centered. By Theorem C.2.2, the cosine transform C (see Section C.2) is surjective as a map from $C_e^\infty(S^{n-1})$ to itself. Therefore there is an $f \in C_e^\infty(S^{n-1})$ such that $h_K = Cf$. We can then take μ in

Definition 4.1.14 to be defined by

$$\mu(E) = \int_E f(v)\,dv,$$

for each Borel subset E of S^{n-1}. ∎

We do not actually need h_K to be infinitely differentiable; from known results (see Note 4.4) it follows that $h_K \in C^{(n+5)/2}(S^{n-1})$ is enough. The next corollary follows directly from the previous theorem and the fact that the class of convex bodies with C^∞ support functions is dense in $\mathcal{K}^n$ (see Section 0.9).

Corollary 4.1.18. *The class of generalized zonoids is dense in the class of centrally symmetric compact convex sets.*

Theorem 4.1.16 and Corollary 4.1.18 provide a useful passage from zonoids, via generalized zonoids, to centrally symmetric compact convex sets; see Note 4.9.

The next remark collects some other basic facts about zonotopes, zonoids, and generalized zonoids to aid in their visualization (see also the notes for this chapter).

Remark 4.1.19. (i) As we have seen, zonotopes are centrally symmetric. It can be shown that every face of a zonotope is a zonotope, so zonotopes have centrally symmetric faces. A zonotope is therefore much more symmetric than the typical centrally symmetric polytope. An octahedron is an example of a centrally symmetric polyhedron that is not a zonotope, since its facets are not centrally symmetric. Another example of a centrally symmetric convex body that is not a zonoid is a convex double bounded cone. By Theorem 4.2.4 and Corollary 4.2.7, the body K_2 illustrated in Figure 4.4 is not a zonoid; neither is any nonsingular affine image of it, by Theorem 4.1.5.

(ii) A polytopal generalized zonoid must be a zonotope. Therefore, by (i), an octahedron is also an example of a centrally symmetric convex body that is not a generalized zonoid. This and Corollary 4.1.18 show that the class of generalized zonoids is not closed in $\mathcal{K}^n$. On the other hand, by its very definition, the class of zonoids is closed in $\mathcal{K}^n$. This implies that there exists a generalized zonoid that is not a zonoid; an explicit example was given in Remark 4.1.15.

(iii) A convex body (or polygon) in $\mathbb{E}^2$ is a zonoid (or zonotope, respectively) if and only if it is centrally symmetric. This can be deduced from Theorems 4.1.4 and 4.1.11 and remark (ii), but is also quite easily proved directly.

(iv) A convex polytope in $\mathbb{E}^n$ is a zonotope if all its 2-dimensional faces are centrally symmetric. (By (i) and (iii), this property characterizes zonotopes.)

4.2. Smaller bodies with larger projections

Aleksandrov's projection theorem, Theorem 3.3.6, implies that two centered convex bodies in $\mathbb{E}^n$ must be equal if the i-dimensional volumes of their projections on any i-dimensional subspace are always equal. In particular, of course, the volumes of the bodies must also be equal. This prompts the following question.

Question 4.2.1. *Let* $1 \leq i \leq n - 1$, *and let* K_1 *and* K_2 *be compact convex sets in* $\mathbb{E}^n$. *If* $\lambda_i(K_1|S) \leq \lambda_i(K_2|S)$ *for all* $S \in \mathcal{G}(n, i)$, *is it true that* $\lambda_n(K_1) \leq \lambda_n(K_2)$?

One quickly sees that this question must be modified if a positive answer is desired. Consider the easiest case, when $n = 2$ and $i = 1$. Suppose that the first body K_1 is rB, where $r < 1$ is to be chosen later, and that the second, K_2, is nonspherical and of constant width 2. The perimeter of a convex body of constant width 2 is 2π; this fact, sometimes called Barbier's theorem, follows from the case $n = 2$ of Cauchy's surface area formula (A.49). By the case $n = 2$ of the isoperimetric inequality (B.14) and its equality condition, the area of K_2 must be less than π, the area of B. Therefore we can choose $r < 1$ so that the area of K_2 is still less than the area of rB. But now the width of K_1 in any direction is $2r$, less than that of K_2. See Figure 4.3.

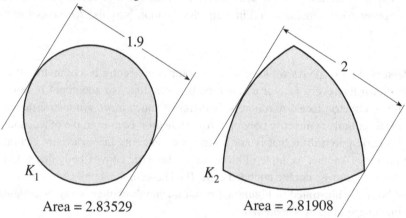

Area = 2.83529   Area = 2.81908

Figure 4.3. A smaller convex body with larger projections.

This is hardly surprising, since we saw in Chapter 3 that Aleksandrov's projection theorem fails for bodies that are not centrally symmetric. In fact, it is possible to construct examples for which the answer to Question 4.2.1 is negative, by using a result from the previous section.

Theorem 4.2.2. *Let* K_2 *be any convex body in* $\mathbb{E}^n$ *that is not centrally symmetric. There exists a centrally symmetric convex body* K_1 *such that* $\lambda_{n-1}(K_1|u^\perp) < \lambda_{n-1}(K_2|u^\perp)$ *for all* $u \in S^{n-1}$, *but* $\lambda_n(K_1) > \lambda_n(K_2)$.

Proof. In Theorem 3.3.9, replace K by K_2. By that theorem, $\lambda_n(\nabla K_2) > \lambda_n(K_2)$, since K_2 is not centrally symmetric. Choose $r < 1$ so that the latter inequality still holds with ∇K_2 replaced by $K_1 = r\nabla K_2$. Using Theorem 4.1.2, we obtain

$$\Pi K_1 = \Pi(r\nabla K_2) = r^{n-1}\Pi(\nabla K_2) = r^{n-1}\Pi K_2.$$

By Definition 4.1.1, this completes the proof. ∎

As we have already seen in the 2-dimensional case, it does not even help if K_1 is a ball. The next theorem extends this observation to n dimensions.

Theorem 4.2.3. *Let $1 \le i \le n - 1$, and suppose that K_2 is any nonspherical convex body of the same constant i-brightness as the unit ball B. There is an $r < 1$ such that with $K_1 = rB$, $\lambda_i(K_1|S) < \lambda_i(K_2|S)$ for all $S \in \mathcal{G}(n, i)$, but $\lambda_n(K_1) > \lambda_n(K_2)$.*

Proof. Recall that K_2 exists by Theorem 3.3.15 and Remark 3.3.16. Then $\lambda_i(K_2|S) = \lambda_i(B|S)$ for all $S \in \mathcal{G}(n, i)$. By the special case (A.47) of Kubota's integral recursion, this implies that $V_i(K_2) = V_i(B)$. Applying the isoperimetric inequality (B.14) and its equality condition, we obtain

$$\lambda_n(B) > \lambda_n(K_2).$$

Therefore it is possible to choose an $r < 1$ so that the conclusion of the theorem is true when $K_1 = rB$. ∎

The only hope, then, is to impose some symmetry condition on the other body, K_2. Generally speaking, however, even the central symmetry of K_2 is not enough. In fact, the next surprising result shows that Question 4.2.1 has a negative answer for $n = 3$ and $i = 2$, even when K_1 is a ball and K_2 is a centrally symmetric convex body of revolution. See also Remark 4.2.10 for a different approach.

Theorem 4.2.4. *There is a ball K_1 and a centrally symmetric convex body of revolution K_2 in $\mathbb{E}^3$ such that $\lambda_2(K_1|u^\perp) < \lambda_2(K_2|u^\perp)$ for all $u \in S^2$, but $\lambda_3(K_1) > \lambda_3(K_2)$.*

Proof. The body K_2 will be the convex double bounded cone obtained by rotating the triangle with vertices $(0, 0, \pm k)$ and $(1, 0, 0)$ around the z-axis, where $k = (\sqrt{2}\cos\alpha)^{-1}$ and α is given by

$$2\alpha + \tan\alpha = \pi.$$

($\alpha = 0.9175\ldots$ radians and $k = 1.1634\ldots$.) See Figure 4.4.

It is enough, by symmetry, to consider those $u \in S^2$ for which $u = (\sin\varphi, 0, \cos\varphi)$, $0 \le \varphi \le \pi/2$, where φ is the usual spherical polar coordinate denoting the angle between u and the z-axis. For $\cot\varphi \ge k$, that is, for u close to the north

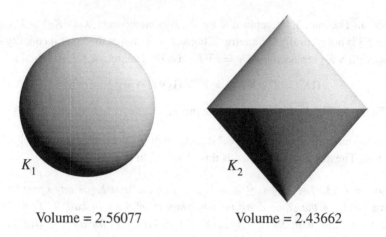

Volume = 2.56077   Volume = 2.43662

Figure 4.4. A smaller centered convex body with larger projections.

pole, the projection of K_2 on $u^\perp$ is the same as that of its intersection with the xy-plane, the unit disk. Let us set up x- and y-axes in $u^\perp$ that are projections of those in $\mathbb{E}^3$. Then the projection of the unit disk is the ellipse $E(\varphi)$ meeting the x-axis and y-axis in $u^\perp$ at $(\pm\cos\varphi, 0)$ and $(0, \pm1)$, respectively. The area of $E(\varphi)$ is $\pi\cos\varphi$. For $\cot\varphi \leq k$, the projection of K_2 on $u^\perp$ is the convex hull of $E(\varphi)$ and the points $(\pm k\sin\varphi, 0)$. See Figure 4.5.

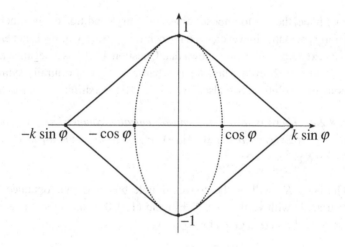

Figure 4.5. A projection of K_2.

The area of this projection can be computed as follows. First, find the coordinates (x_0, y_0) of the point where the tangent line from $(k\sin\varphi, 0)$ meets the ellipse $E(\varphi)$ in the first quadrant. On setting $\cos \psi = k^{-1} \cot \varphi$, we obtain

$$x_0 = \frac{\cos^2 \varphi}{k \sin \varphi} = \cos \psi \cos \varphi$$

and

$$y_0 = \left(1 - \frac{\cot^2 \varphi}{k^2}\right)^{1/2} = \sin \psi.$$

Now divide the projection into the triangle with vertices (x_0, y_0), $(x_0, -y_0)$, and $(k \sin \varphi, 0)$, its reflection in the y-axis, and the part of the ellipse between the lines $x = -x_0$ and $x = x_0$. Using integration and summing these the three areas, we find that

$$\lambda_2(K_2|u^\perp) = f(\varphi) = \begin{cases} (\pi - 2\psi) \cos \varphi + 2k \sin \varphi \sin \psi & \text{if } \cot \varphi \le k, \\ \pi \cos \varphi & \text{if } \cot \varphi \ge k. \end{cases}$$

The next task is to find the minimum of this function. After some manipulation, we obtain

$$\frac{df}{d\varphi} = \begin{cases} (-\pi + 2\psi + 2k^2 \sin \psi \cos \psi) \sin \varphi & \text{if } \cot \varphi < k, \\ -\pi \sin \varphi & \text{if } \cot \varphi > k. \end{cases}$$

The minimum must occur when $\cot \varphi < k$, since f is decreasing for $\cot \varphi > k$. Therefore it occurs when

$$2\psi + 2k^2 \sin \psi \cos \psi = \pi.$$

This equation is satisfied when $\psi = \alpha$, by the earlier choice of α and k. Substituting $\psi = \alpha$ into f gives, after a little calculation, that the minimum value of f is $\sqrt{6k} \sin \alpha$.

If $r > 0$, then $\lambda_2(rB|u^\perp) = \pi r^2$, for all $u \in S^2$. So we have $\lambda_2(r_0 B|u^\perp) \le \lambda_2(K_2|u^\perp)$, for all $u \in S^2$, if

$$r_0 = (\sqrt{6k} \sin \alpha / \pi)^{1/2}.$$

Furthermore,

$$\lambda_3(r_0 B) = \frac{4}{3}\pi r_0^3 = 2.5607\ldots$$

and

$$\lambda_3(K_2) = \frac{2}{3}\pi k = 2.4366\ldots.$$

Consequently, we can choose an $r < r_0$ so that the ball $K_1 = rB$ and K_2 have all the required properties.   ∎

Since the central symmetry of K_2 is still not enough to ensure a positive answer to Question 4.2.1, something stronger is required. It transpires that K_2 should be a zonoid, at least for $i > 1$. We shall use the following lemma.

Lemma 4.2.5. *Let* $1 \leq i \leq n - 1$, *and let* K *be a zonoid in* $\mathbb{E}^n$. *Then*

$$V(K_1, i; B, n - i - 1; K) \leq V(K_2, i; B, n - i - 1; K),$$

for all compact convex K_1, K_2 *such that*

$$V_i(K_1 | u^\perp) \leq V_i(K_2 | u^\perp),$$

for all $u \in S^{n-1}$.

Proof. By the translation invariance (A.19) of mixed volumes, it will suffice to prove the lemma when K is centered. This implies, by Theorem 4.1.10, that

$$h_K(u) = \int_{S^{n-1}} |u \cdot v| \, d\mu(v),$$

for all $u \in S^{n-1}$, where μ is a finite even Borel measure in S^{n-1}. For $j = 1, 2$,

$$V(K_j, i; B, n - i - 1; K) = \frac{1}{n} \int_{S^{n-1}} h_K(u) \, dS_i(K_j, u)$$

$$= \frac{1}{n} \int_{S^{n-1}} \int_{S^{n-1}} |u \cdot v| \, d\mu(v) \, dS_i(K_j, u)$$

$$= \frac{1}{n} \int_{S^{n-1}} \int_{S^{n-1}} |u \cdot v| \, dS_i(K_j, v) \, d\mu(u),$$

where we have used (A.32), and Fubini's theorem to change the order of integration. Since $V_i(K_1 | u^\perp) \leq V_i(K_2 | u^\perp)$, we have

$$\int_{S^{n-1}} |u \cdot v| \, dS_i(K_1, v) \leq \int_{S^{n-1}} |u \cdot v| \, dS_i(K_2, v),$$

for all $u \in S^{n-1}$, by the generalized Cauchy projection formula (A.45). Combining this inequality and the previous equation completes the proof. ∎

Theorem 4.2.6. *Let* $1 \leq i \leq k \leq n - 1$, *and let* K_1 *and* K_2 *be compact convex sets in* $\mathbb{E}^n$. *Suppose that* $V_i(K_1 | S) \leq V_i(K_2 | S)$ *for all* $S \in \mathcal{G}(n, k)$. *If* $K_2 | T$ *is a zonoid for each* $T \in \mathcal{G}(n, k + 1)$, *then* $V_{i+1}(K_1 | T) \leq V_{i+1}(K_2 | T)$, *for all* $T \in \mathcal{G}(n, k + 1)$. *If* K_1 *and* K_2 *are convex bodies, then equality holds for all* $T \in \mathcal{G}(n, k + 1)$ *if and only if* K_1 *is a translate of* K_2.

Proof. Choose $T \in \mathcal{G}(n, k + 1)$, and for $j = 1, 2$, let $K'_j = K_j | T$. Then for all $S \in \mathcal{G}(n, k)$ with $S \subset T$, $K'_j | S = K_j | S$. Moreover, K'_2 is a zonoid. We identify T

with $\mathbb{E}^{k+1}$ and apply Lemma 4.2.5 with $n = k+1$, $K_1 = K_1'$, and $K = K_2 = K_2'$. Using (A.29), we obtain

$$V(K_1', i; B \cap T, k - i; K_2') \leq V(K_2', i + 1; B \cap T, k - i) = c_{k+1,i+1}V_{i+1}(K_2').$$

By the special case (B.19) of the Aleksandrov–Fenchel inequality,

$$V(K_1', i; B \cap T, k - i; K_2')^{i+1} \geq c_{k+1,i+1}^{i+1}V_{i+1}(K_1')^i V_{i+1}(K_2').$$

Combining these facts, we get $V_{i+1}(K_1') \leq V_{i+1}(K_2')$, as required.

Now suppose that $V_{i+1}(K_1|T) = V_{i+1}(K_2|T)$ for all $T \in \mathcal{G}(n, k+1)$ and that K_1 and K_2 are convex bodies. Then

$$\dim K_j' = \dim(K_j|T) = k + 1 \geq i + 1,$$

for $j = 1, 2$. With this restriction on dimension, equality in the special case of the Aleksandrov–Fenchel inequality employed earlier holds if and only if K_1' and K_2' are homothetic. Since their $(i + 1)$th intrinsic volumes are also equal, K_1' and K_2' must be translates. Therefore equality holds for all $T \in \mathcal{G}(n, k + 1)$ if and only if $K_1|T$ and $K_2|T$ are translates for all $T \in \mathcal{G}(n, k + 1)$. By Theorem 3.1.3, this means that K_1 must be a translate of K_2. ∎

The following corollary can now be obtained from Theorem 4.2.6 by induction; note, however, that we need the fact that the projection of a zonoid is a zonoid. The special case of this corollary for which $k = i$ gives the promised positive answer to Question 4.2.1 when K_2 is a zonoid.

Corollary 4.2.7. *Let $1 \leq i \leq k \leq n - 1$, and let K_1 and K_2 be compact convex sets in $\mathbb{E}^n$. Suppose also that K_2 is a zonoid. If $V_i(K_1|S) \leq V_i(K_2|S)$ for all $S \in \mathcal{G}(n, k)$, then $V_{n+i-k}(K_1) \leq V_{n+i-k}(K_2)$. If K_1 and K_2 are convex bodies, then equality holds if and only if K_1 is a translate of K_2.*

It turns out that when $i = 1$, we only need the central symmetry of K_2. Note that the assumption in the next theorem merely says that the width of K_1 in any direction is no larger than that of K_2.

Theorem 4.2.8. *Suppose that K_1 and K_2 are convex bodies in $\mathbb{E}^n$, with K_2 centrally symmetric. If $\lambda_1(K_1|S) \leq \lambda_1(K_2|S)$ for all $S \in \mathcal{G}(n, 1)$, then $\lambda_n(K_1) \leq \lambda_n(K_2)$, with equality if and only if K_1 is a translate of K_2.*

Proof. As we observed in Chapter 3, the width function of a convex body equals that of its central symmetral. Therefore $\lambda_1(\Delta K_j|S) = \lambda_1(K_j|S)$, for each $S \in \mathcal{G}(n, 1)$ and $j = 1, 2$, so

$$\lambda_1(\Delta K_1|S) \leq \lambda_1(\Delta K_2|S),$$

for all $S \in \mathcal{G}(n, 1)$. Since $\triangle K_1$ and $\triangle K_2$ are centered, this implies that $\triangle K_1 \subset \triangle K_2 = K_2$. Therefore

$$\lambda_n(K_1) \leq \lambda_n(\triangle K_1) \leq \lambda_n(K_2),$$

by Theorem 3.2.3. Equality holds here if and only if K_1 is centrally symmetric and $\triangle K_1 = \triangle K_2$, and this is true precisely when K_1 is a translate of K_2.   ∎

In Question 4.2.1, we make the assumption $\lambda_i(K_1|S) \leq \lambda_i(K_2|S)$ for all $S \in \mathcal{G}(n, i)$, rather than take $i \leq k$ and assume that $V_i(K_1|S) \leq V_i(K_2|S)$ for all $S \in \mathcal{G}(n, k)$. The reason for this is illustrated by the following example, corresponding to the latter hypothesis for $i = 1$, $k = 2$, and $n = 3$.

Theorem 4.2.9. There are boxes K_1, K_2 in $\mathbb{E}^3$ such that

$$V_1(K_1|u^\perp) < V_1(K_2|u^\perp),$$

for all $u \in S^2$, yet

$$\lambda_3(K_1) > \lambda_3(K_2).$$

Proof. Let J_1, J_2 be centered squares in the xy-plane in $\mathbb{E}^3$, with J_1 contained in the relative interior of J_2. Then there is an $\varepsilon > 0$ such that

$$V_1(J_1|u^\perp) < V_1(J_2|u^\perp) - \varepsilon,$$

for all $u \in S^2$. Our boxes will be defined by $K_j = J_j \times [0, c_j]$, for suitable $c_j > 0$, $j = 1, 2$. First choose $c_1 > 0$ small enough that

$$V_1(K_1|u^\perp) < V_1(J_2|u^\perp) - \varepsilon/2,$$

for all $u \in S^2$. Then $\lambda_3(J_2) = 0 < \lambda_3(K_1)$, so we can now choose $c_2 > 0$ small enough to ensure that $\lambda_3(K_2) < \lambda_3(K_1)$. Since $V_1(J_2|u^\perp) \leq V_1(K_2|u^\perp)$ for all $u \in S^2$, the proof is complete.   ∎

Remark 4.2.10. Suppose that $1 \leq i \leq n - 1$, and that K_1 is a smooth centrally symmetric convex body in $\mathbb{E}^n$ that is not a zonoid. Then it can be shown that there is a centrally symmetric convex body K_2 (which is also not a zonoid by Corollary 4.2.7 with $k = n - 1$) such that $V_i(K_1|u^\perp) < V_i(K_2|u^\perp)$ for all $u \in S^{n-1}$, but $V_{i+1}(K_1) > V_{i+1}(K_2)$. For $i = n - 1$, this gives a negative answer to Question 4.2.1 when K_1 and K_2 are centrally symmetric. Moreover, this process can be generalized: Given any i with $2 \leq i \leq n - 1$, it can be shown that there are centrally symmetric convex bodies K_1 and K_2 in $\mathbb{E}^n$ such that the answer to Question 4.2.1 is negative. See Note 4.10 for more details.

Theorem 4.2.11. *Suppose that K_1 and K_2 are convex bodies in $\mathbb{E}^n$. Then*

$$\lambda_{n-1}(K_1|u^{\perp}) \leq \lambda_{n-1}(K_2|u^{\perp}),$$

for all $u \in S^{n-1}$, if and only if

$$\lambda_{n-1}(\operatorname{bd}\phi K_1) \leq \lambda_{n-1}(\operatorname{bd}\phi K_2),$$

for every $\phi \in GA_n$.

Proof. Suppose that the first condition holds, and let ϕ be a nonsingular affine transformation. By Lemma 4.2.5 with $i = n - 1$, we know that

$$V(K_1, n - 1; K) \leq V(K_2, n - 1; K),$$

whenever K is a zonoid. Since ellipsoids are zonoids by Corollary 4.1.12, we have

$$V(K_1, n - 1; \phi^{-1}B) \leq V(K_2, n - 1; \phi^{-1}B).$$

Now for $j = 1, 2$, (A.35), (A.26), and (A.17) tell us that

$$\lambda_{n-1}(\operatorname{bd}\phi K_j) = nV(\phi K_j, n - 1; B) = n|\det\phi|V(K_j, n - 1; \phi^{-1}B),$$

and this yields the second condition.

Conversely, assume the second condition, and suppose that $u \in S^{n-1}$. From the previous two equations we know that

$$V(K_1, n - 1; E) \leq V(K_2, n - 1; E),$$

for all n-dimensional ellipsoids E. Applying this inequality to each of a sequence of n-dimensional ellipsoids converging to the line segment $[o, u]$, and using the continuity of mixed volumes (see Section A.3), we obtain

$$V(K_1, n - 1; [o, u]) \leq V(K_2, n - 1; [o, u]).$$

The first condition follows immediately from this and (A.37). ∎

We have seen in Remark 4.2.10 that Question 4.2.1 has a negative answer, even for centrally symmetric K_1 and K_2, when $2 \leq i \leq n - 1$. One can still attempt to find some sort of relationship between the volumes of K_1 and K_2 from the given information concerning their brightness functions. The following theorem, which has many applications, will allow us to do this when K_2 is centrally symmetric. The proof constructs an ellipsoid E of maximal volume contained in a given centrally symmetric convex body K in $\mathbb{E}^n$ and with the same center as K. The ellipsoid E, which by an argument similar to that of Theorem 9.2.1 can be shown to be unique, is called the *John ellipsoid* of K.

Theorem 4.2.12. *For each centrally symmetric convex body K in $\mathbb{E}^n$, there is an ellipsoid E with the same center as K such that*

$$E \subset K \subset \sqrt{n}E.$$

Proof. We may assume that K is centered. We first show that there is a centered ellipsoid of maximal volume contained in K. Let $\mathcal{E}$ be the class of centered n-dimensional ellipsoids contained in K, and suppose that $a = \sup\{\lambda_n(E) : E \in \mathcal{E}\}$. Then $a > 0$. Choose a sequence of sets E_m in $\mathcal{E}$ whose volumes converge to a. For each m, there is a $\phi_m \in GL_n$ with $E_m = \phi_m B$. Now $\phi_m x = A_m x$, where A_m is a nonsingular $n \times n$ matrix; so we can choose a subsequence of $\mathbb{N}$ such that A_m converges to an $n \times n$ matrix A as $m \to \infty$ through this subsequence. (Note that since there is an $R > 0$ such that $K \subset RB$, the entries in each A_m are bounded by R.) Since $\lambda_n(E_m) = |\det A_m|\kappa_n$, by (0.7), $|\det A_m|$ converges to a/κ_n as $m \to \infty$ through this subsequence, so $\det A \neq 0$ and A is nonsingular. Then if $\phi x = Ax$, we have $\phi \in GL_n$, and $E = \phi B$ is an n-dimensional centered ellipsoid with $E \subset K$ and $\lambda_n(E) = a$.

By the previous construction, the unit ball B is a centered ellipsoid of maximal volume contained in $K' = \phi^{-1}K$. If we can show that $K' \subset \sqrt{n}B$, then $K \subset \sqrt{n}E$. Let $x_0 \in \operatorname{bd} K'$. We may assume that $x_0 = (b, 0, \ldots, 0)$ lies on the x_1-axis with $b \geq 1$, so that $\operatorname{conv}\{\pm x_0, B\} \subset K'$. For $0 < c \leq 1$, let ϕ_c be the linear transformation defined by

$$\phi_c(x_1, x_2, \ldots, x_n) = (cx_1, x_2, \ldots, x_n).$$

Let S be any 2-dimensional subspace containing the x_1-axis, and set up Cartesian coordinates (x, y) in S so that $x = x_1$. Then the point where the tangent line from x_0 meets $B \cap S$ is $p = (b^{-1}, b^{-1}\sqrt{b^2 - 1})$. The transformed body $\phi_c K'$ contains $\phi_c B$ and $\pm cx_0$, and the point where the tangent line from cx_0 meets $\phi_c B \cap S$ is $\phi_c p = (cb^{-1}, b^{-1}\sqrt{b^2 - 1})$. A calculation shows that the distance $r = r(c)$ from o to the line through cx_0 and $\phi_c p$ is

$$r = \frac{bc}{\sqrt{1 + c^2(b^2 - 1)}}.$$

Because $c \leq 1$, the centered ball rB is contained in $\operatorname{conv}\{\pm cx_0, \phi_c B\}$, and hence in $\phi_c K'$; so K' contains the ellipsoid $E_c = \phi_c^{-1}(rB)$ given by

$$c^2 x_1^2 + x_2^2 + \cdots + x_n^2 \leq r^2.$$

Using (0.11), we find that

$$\lambda_n(E_c) = \frac{b^n c^{n-1} \kappa_n}{\left(1 + c^2(b^2 - 1)\right)^{n/2}}.$$

Suppose that $b \geq \sqrt{n}$. Then we can choose c so that $cb = \sqrt{n}$. Substituting for c in the expression for $\lambda_n(E_c)$, and differentiating the resulting function of b, we find that $\lambda_n(E_c)$ has its minimum value of κ_n when $b = \sqrt{n}$. If $b > \sqrt{n}$, the ellipsoid E_c with $c = \sqrt{n}/b$ is of larger volume than B, a contradiction. So $b \leq \sqrt{n}$, which means that $x_0 \in \sqrt{n}B$, as required.   ∎

Theorem 4.2.13. *Suppose that K_1 and K_2 are convex bodies in $\mathbb{E}^n$ such that*

$$\lambda_{n-1}(K_1|u^\perp) \le \lambda_{n-1}(K_2|u^\perp),$$

for all $u \in S^{n-1}$. If K_2 is centrally symmetric, then

$$\lambda_n(K_1) \le \sqrt{n}\lambda_n(K_2).$$

Proof. Let E be the John ellipsoid of K_2, and put $E_0 = \sqrt{n}E$, so that

$$K_2 \subset E_0 \subset \sqrt{n}K_2,$$

by Theorem 4.2.12. Since ellipsoids are zonoids by Corollary 4.1.12, we can apply Lemma 4.2.5, with $i = n - 1$ and $K = E_0$. Using these facts, (A.18), (A.28), and Minkowski's first inequality (B.13), we obtain

$$
\begin{aligned}
\lambda_n(K_1)^{(n-1)/n}\lambda_n(E_0)^{1/n} &\le V(K_1, n-1; E_0) \\
&\le V(K_2, n-1; E_0) \\
&= V(K_2, n-1; E_0)^{1/n} V(K_2, n-1; E_0)^{(n-1)/n} \\
&\le \lambda_n(E_0)^{1/n}\left(\sqrt{n}\lambda_n(K_2)\right)^{(n-1)/n},
\end{aligned}
$$

yielding the required inequality. ∎

Remark 4.2.14. It can be shown that the factor $\sqrt{n}$ in the inequality of the previous theorem is of the correct order. However, the estimate can be significantly improved asymptotically, and, moreover, this improvement applies to arbitrary convex bodies K_1, K_2 satisfying the hypotheses of Theorem 4.2.13. The main ideas are as follows. An argument similar to that in the first paragraph of the proof of Theorem 4.2.12 shows that there is an ellipsoid E (also called the John ellipsoid) of maximal volume contained in an arbitrary convex body K in $\mathbb{E}^n$. The *volume ratio* of K is then defined by

$$vr(K) = \left(\frac{\lambda_n(K)}{\lambda_n(E)}\right)^{1/n}.$$

It can be proved that

$$vr(K) \le \left(\frac{n^{n/2}(n+1)^{(n+1)/2}}{n!\kappa_n}\right)^{1/n},$$

with equality when K is a simplex; see Note 4.10 for references. Without using central symmetry, the argument of Theorem 4.2.13 shows that

$$\lambda_n(K_1)^{(n-1)/n}\lambda_n(E)^{1/n} \le \lambda_n(K_2),$$

where E is the John ellipsoid of K_2, or

$$\lambda_n(K_1) \le vr(K_2)^{n/(n-1)}\lambda_n(K_2).$$

Using the previous bound for the volume ratio, one obtains

$$\frac{vr(K_2)^{n/(n-1)}}{\sqrt{n}} \rightarrow \sqrt{\frac{e}{2\pi}} = 0.6577\ldots$$

as $n \rightarrow \infty$. In fact, the expression on the left is less than 1 for $n \geq 4$, and always no larger than

$$\frac{3}{\pi}\sqrt{\frac{3}{2}} = 1.16955\ldots,$$

its value when $n = 2$.

4.3. Stability

In this section we take another look at some results on projections, from the point of view of stability. For example, consider the trivial Theorem 3.1.1, stating that when $1 \leq k \leq n - 1$, a convex body in E^n is determined by its projections on k-dimensional subspaces. If all these projections are only approximately known, is the convex body approximately determined? Such questions are of practical significance, and some of the known answers are presented in the sequel.

For precise formulation, some notion of distance between convex bodies is necessary. One difficulty in presenting stability results is that many such metrics are available, the most natural one depending on the situation. Furthermore, by their very nature, stability theorems tend to be rather messy to state and hard to prove. (This difficulty is not surprising when one considers that a uniqueness theorem can usually be retrieved from a stability version of it.) For these reasons we shall not always attempt to prove the most general or powerful theorem known.

Recall the definitions (0.4), (0.6), and (0.22) of the Hausdorff metric δ. As a warm-up, let us answer the easy question we just posed.

Theorem 4.3.1. *Let $1 \leq k \leq n - 1$, and let K_1, K_2 be compact convex sets in E^n. If $\delta(K_1|S, K_2|S) \leq \varepsilon$ for all $S \in \mathcal{G}(n, k)$, then $\delta(K_1, K_2) \leq \varepsilon$.*

Proof. Note that if E_1, E_2 are sets with $\delta(E_1, E_2) \leq \varepsilon$, and S is a subspace, then $\delta(E_1|S, E_2|S) \leq \varepsilon$, since projections cannot increase distances. Therefore it suffices to take $k = 1$. Suppose that $\delta(K_1, K_2) > \varepsilon$. Without loss of generality, we may assume that there is a point $x_1 \in K_1$ such that $\|x_1 - x_2\| > \varepsilon$, where x_2 is the point in K_2 nearest to x_1. Let t be the line through o parallel to the line through x_1 and x_2. Then $\delta(K_1|t, K_2|t) > \varepsilon$, as required. ∎

The next goal is to state a stability version of Theorem 3.1.3. To do this, we first consider the case of translations, and here it is convenient to introduce a new metric.

Definition 4.3.2. The *translative Hausdorff distance* between two compact convex sets K_1, K_2 in $\mathbb{E}^n$ is defined by

$$\delta_t(K_1, K_2) = \inf_{x \in \mathbb{E}^n} \delta(K_1, K_2 + x).$$

Lemma 4.3.3. *The function δ_t satisfies the triangle inequality.*

Proof. If K_1, K_2 are compact convex sets, there is a point x_2 such that $\delta_t(K_1, K_2) = \delta(K_1, K_2 + x_2)$. If K_3 is a third compact convex set, there is an x_3 with $\delta_t(K_2, K_3) = \delta(K_2, K_3 + x_3)$. Then, using the fact that δ is a metric, we obtain

$$
\begin{aligned}
\delta_t(K_1, K_3) &\leq \delta(K_1, K_3 + x_2 + x_3) \\
&= \delta(K_1 - x_2, K_3 + x_3) \\
&\leq \delta(K_1 - x_2, K_2) + \delta(K_2, K_3 + x_3) \\
&= \delta(K_1, K_2 + x_2) + \delta(K_2, K_3 + x_3) \\
&= \delta_t(K_1, K_2) + \delta_t(K_2, K_3).
\end{aligned}
$$
∎

The other conditions for a metric are easily checked. Of course, δ_t is not really a metric on $\mathcal{K}^n$, but rather on the set of equivalence classes formed by identifying translates of a compact convex set.

Theorem 4.3.4. *Let K_1, K_2 be compact convex sets in $\mathbb{E}^n$, $n \geq 3$, such that for all $S \in \mathcal{G}(n, n-1)$, $\delta_t(K_1|S, K_2|S) \leq \varepsilon$. Then*

$$\delta_t(K_1, K_2) \leq (1 + 2\sqrt{2})\varepsilon.$$

Proof. Let $S_0 \in \mathcal{G}(n, n-1)$. By translating K_2 (first parallel to S_0 and then parallel to $S_0^\perp$), if necessary, we may assume that

$$\delta(K_1|S_0, K_2|S_0) \leq \varepsilon$$

and

$$\delta(K_1|S_0^\perp, K_2|S_0^\perp) \leq \varepsilon.$$

For $j = 1, 2$, let H_j be a supporting hyperplane to K_j such that H_1 and H_2 are parallel and their distance apart is equal to $\delta(K_1, K_2)$. Suppose that $S \in \mathcal{G}(n, n-1)$ contains both $S_0^\perp$ and a line orthogonal to H_j. Since the hyperplanes H_j are orthogonal to S,

$$\delta(K_1, K_2) = \delta(K_1|S, K_2|S).$$

By hypothesis, there is a point $x \in \mathbb{E}^n$ such that $\delta\big(K_1|S, (K_2|S) + x\big) \leq \varepsilon$. Note that we may assume that $x \in S$.

Let y, z denote the projections of x on $T = S_0 \cap S$ and $S_0^{\perp}$, respectively. Because T is contained in both S_0 and S,

$$
\begin{aligned}
\|y\| &= \delta\big(K_2|T, (K_2|T) + y\big) \\
&\leq \delta(K_1|T, K_2|T) + \delta\big(K_1|T, (K_2|T) + y\big) \\
&\leq \delta(K_1|S_0, K_2|S_0) + \delta\big(K_1|S, (K_2|S) + x\big) \\
&\leq \varepsilon + \varepsilon = 2\varepsilon.
\end{aligned}
$$

Similarly, $\|z\| \leq 2\varepsilon$, and it follows that $\|x\| \leq 2\sqrt{2}\varepsilon$. Consequently,

$$
\begin{aligned}
\delta_t(K_1, K_2) &\leq \delta(K_1, K_2) \\
&= \delta(K_1|S, K_2|S) \\
&\leq \delta\big(K_1|S, (K_2|S) + x\big) + \delta\big(K_2|S, (K_2|S) + x\big) \\
&\leq \varepsilon + 2\sqrt{2}\varepsilon,
\end{aligned}
$$

completing the proof.   ∎

Remark 4.3.5. (i) The factor $(1 + 2\sqrt{2})$ in the previous theorem cannot be removed. This can be seen by considering the unit ball and an inscribed tetrahedron in $\mathbb{E}^3$; we shall not give the details here. This factor may be removed in the special case where K_1 and K_2 are centrally symmetric. This is shown by translating K_1 and K_2 so that they are centered. Following the proof of Theorem 4.3.4, we see that since $K_j|S$ is also centered, $j = 1, 2$, $\delta\big(K_1|S, (K_2|S) + x\big)$ is a minimum when $x = o$. This means that in this case

$$
\delta(K_1, K_2) = \delta(K_1|S, K_2|S) = \delta_t(K_1|S, K_2|S) \leq \varepsilon.
$$

(ii) The proof of Theorem 4.3.4 shows that it is not necessary to assume that $\delta_t(K_1|S, K_2|S) \leq \varepsilon$ for *all* $S \in \mathcal{G}(n, n-1)$; it is enough that this holds for one $S_1 \in \mathcal{G}(n, n-1)$ together with all those $S \in \mathcal{G}(n, n-1)$ such that $S_1^{\perp} \subset S$ (or even, in fact, those containing a fixed $(n-2)$-dimensional subspace orthogonal to S_1).

For homothets, the introduction of a suitable metric is not so straightforward. In fact, we shall work with a function that is not symmetric.

Definition 4.3.6. If K_1, K_2 are compact convex sets in $\mathbb{E}^n$, we define

$$
\delta_1(K_1, K_2) = \inf_{a \geq 0} \delta_t(K_1, aK_2).
$$

Notice that δ_1 is invariant under homotheties of K_2. Recall that $r(K)$ and $R(K)$ denote the inradius and circumradius of K, respectively.

Theorem 4.3.7. *Let K_1, K_2 be compact convex sets in $\mathbb{E}^n$, $n \geq 3$, such that for all $S \in \mathcal{G}(n, n-1)$, $\delta_1(K_1|S, K_2|S) \leq \varepsilon$. Then*

$$\delta_1(K_1, K_2) \leq \left(1 + 2R(K_2)/r(K_2)\right)(1 + 2\sqrt{2})\varepsilon.$$

Proof. Let $S_1 \in \mathcal{G}(n, n-1)$. By applying a homothety to K_2, if necessary, we may assume that $\delta(K_1|S_1, K_2|S_1) \leq \varepsilon$. If $S \in \mathcal{G}(n, n-1)$, then there is an $a \geq 0$ such that $\delta_t(K_1|S, aK_2|S) \leq \varepsilon$. Therefore if $S \in \mathcal{G}(n, n-1)$ contains $S_1^\perp$, and $T = S \cap S_1$, we also have $\delta(K_1|T, K_2|T) \leq \varepsilon$ and $\delta_t(K_1|T, aK_2|T) \leq \varepsilon$. Let us fix such an S. Using Lemma 4.3.3, we obtain

$$\delta_t(aK_2|T, K_2|T) \leq \delta_t(K_1|T, K_2|T) + \delta_t(K_1|T, aK_2|T)$$

$$\leq \delta(K_1|T, K_2|T) + \delta_t(K_1|T, aK_2|T)$$

$$\leq \varepsilon + \varepsilon = 2\varepsilon.$$

Since $K_2|T$ and $aK_2|T$ are homothetic, this implies that

$$|\text{diam}\,(aK_2|T) - \text{diam}\,(K_2|T)| \leq 4\varepsilon,$$

so

$$|a - 1| \leq 4\varepsilon/\text{diam}\,(K_2|T) \leq 2\varepsilon/r(K_2).$$

Choose $x \in S$ (the same S as before) so that the circumscribed ball of $C = (K_2|S) + x$ is centered. Then

$$\delta_t(aK_2|S, K_2|S) \leq \delta(aC, C)$$

$$= \sup_{u \in S^{n-1}} |h_{aC}(u) - h_C(u)|$$

$$= |a - 1| \sup_{u \in S^{n-1}} h_C(u)$$

$$\leq |a - 1| R(K_2)$$

$$\leq 2\varepsilon R(K_2)/r(K_2).$$

Then, applying Lemma 4.3.3, we have

$$\delta_t(K_1|S, K_2|S) \leq \delta_t(K_1|S, aK_2|S) + \delta_t(aK_2|S, K_2|S)$$

$$\leq \varepsilon + 2\varepsilon R(K_2)/r(K_2).$$

This is true for all $S \in \mathcal{G}(n, n-1)$ containing $S_1^\perp$, and we also have the inequality $\delta(K_1|S_1, K_2|S_1) \leq \varepsilon$. Consequently, the result follows from Theorem 4.3.4 and Remark 4.3.5(ii). ∎

Corollary 4.3.8. *Let K be a convex body in $\mathbb{E}^n$, $n \geq 3$, such that for each $S \in \mathcal{G}(n, n - 1)$, $\delta(K|S, B_S) \leq \varepsilon$ for some $(n - 1)$-dimensional ball B_S in S. Then there is an $a \geq 0$ such that $\delta(K, aB) \leq 3(1 + 2\sqrt{2})\varepsilon$.*

Proof. Just apply Theorem 4.3.7, with $K_1 = K$ and $K_2 = B$. ∎

Corollary 4.3.9. *Let K be a convex body in $\mathbb{E}^n$, $n \geq 3$, such that for each $S \in \mathcal{G}(n, n - 1)$, $R(K|S) - r(K|S) \leq \varepsilon$. Then*

$$R(K) - r(K) \leq 6(1 + 2\sqrt{2})\varepsilon.$$

Proof. For each $S \in \mathcal{G}(n, n - 1)$, let B_S be either the inscribed or circumscribed $(n - 1)$-dimensional ball of $K|S$. Then $\delta(K|S, B_S) \leq \varepsilon$. By Corollary 4.3.8, $\delta_1(K, B) \leq 3(1 + 2\sqrt{2})\varepsilon$, from which it follows easily that $R(K) - r(K) \leq 6(1 + 2\sqrt{2})\varepsilon$. ∎

Theorem 4.3.10. *Let K be a convex body in $\mathbb{E}^3$, and let $\varepsilon \geq 0$. If*

$$V_1(K|u^\perp)^2 - \pi\lambda_2(K|u^\perp) \leq \varepsilon,$$

for all $u \in S^2$, then

$$V_2(K)^{1/2} - \left(\frac{9\pi}{2}\right)^{1/6} \lambda_3(K)^{1/3} \leq 6(4 + \sqrt{2})\sqrt{\varepsilon}.$$

Proof. Bonnesen's inequality (B.31) implies that if C is a convex body in $\mathbb{E}^2$, then

$$\pi\left(R(C) - r(C)\right)^2 \leq V_1(C)^2 - \pi\lambda_2(C).$$

Therefore

$$R(K|u^\perp) - r(K|u^\perp) \leq \sqrt{\varepsilon/\pi},$$

for all $u \in S^2$. By Corollary 4.3.9,

$$R(K) - r(K) \leq 6(1 + 2\sqrt{2})\sqrt{\varepsilon/\pi}.$$

By (A.35), the quantity $V_2(K)$ is half the surface area of K. The latter is no larger than the surface area of the circumscribed ball of K; this is a direct consequence of Cauchy's surface area formula (A.49). Using this, we obtain

$$V_2(K)^{1/2} - \left(\frac{9\pi}{2}\right)^{1/6} \lambda_3(K)^{1/3} \leq \left(2\pi R(K)^2\right)^{1/2} - \left(\frac{9\pi}{2}\right)^{1/6} \left(\frac{4}{3}\pi r(K)^3\right)^{1/3}$$

$$= \sqrt{2\pi}\left(R(K) - r(K)\right)$$

$$\leq 6(4 + \sqrt{2})\sqrt{\varepsilon}. \quad ∎$$

The isoperimetric inequality in the plane implies that for any convex body C in $\mathbb{E}^2$, $V_1(C)^2 \geq \pi \lambda_2(C)$. (See (B.14) or (B.21). Recall by (A.35) that in $\mathbb{E}^n$ the intrinsic volume V_{n-1} equals half the surface area.) The standard analogue of this in $\mathbb{E}^3$, again obtained by (B.14) or (B.21), is that any convex body K satisfies $V_2(K)^3 \geq 9\pi \lambda_3(K)^2/2$. So Theorem 4.3.10 measures an isoperimetric deficit of K in terms of the isoperimetric deficit of its projections. Note that we say *an* isoperimetric deficit; the particular deficit used here is in terms of the last inequality raised to the power $1/6$. There are many other ways one might show that if all projections of a convex body are nearly spherical, then the body itself is nearly spherical. Another approach is taken in the next theorem, in which, by the isoperimetric inequality, the maximal value of the constant c is 1, and this characterizes the ball.

Theorem 4.3.11. *Suppose that K is a convex body in $\mathbb{E}^3$ such that*

$$\frac{\pi \lambda_2(K|u^\perp)}{V_1(K|u^\perp)^2} \geq c,$$

for all $u \in S^2$. Then

$$\frac{8 V_2(K)}{\pi V_1(K)^2} \geq c.$$

Proof. The special case (A.48) of Kubota's integral recursion implies that

$$V_1(K) = \frac{1}{\pi^2} \int_{S^2} V_1(K|u^\perp) \, du$$

and Cauchy's surface area formula (A.49) gives

$$V_2(K) = \frac{1}{2\pi} \int_{S^2} \lambda_2(K|u^\perp) \, du.$$

Therefore, using the Cauchy–Schwarz inequality ((B.9) with $p = q = 2$), we have

$$
\begin{aligned}
V_1(K)^2 &= \frac{1}{\pi^4} \left(\int_{S^2} \frac{V_1(K|u^\perp)}{\lambda_2(K|u^\perp)^{1/2}} \lambda_2(K|u^\perp)^{1/2} \, du \right)^2 \\
&\leq \frac{1}{\pi^4} \int_{S^2} \frac{V_1(K|u^\perp)^2}{\lambda_2(K|u^\perp)} \, du \int_{S^2} \lambda_2(K|u^\perp) \, du \\
&\leq \frac{1}{\pi^4} \left(\int_{S^2} \frac{\pi}{c} \, du \right) 2\pi V_2(K) \\
&= \frac{8}{\pi c} V_2(K),
\end{aligned}
$$

giving the required inequality. ∎

We now turn to the question of the stability of Aleksandrov's projection theorem, Theorem 3.3.6. Let us consider the case $i = k = n - 1$, where, as we have seen, Aleksandrov's theorem shows that the map $\Pi: \mathcal{K}_0^n \to \mathcal{K}_0^n$, taking a convex body K to its projection body ΠK, is injective when restricted to centered bodies. The most natural stability question here is whether given $\varepsilon > 0$, there is an $\eta > 0$ such that if K_1, K_2 are centered convex bodies with $\delta(\Pi K_1, \Pi K_2) < \eta$, then $\delta(K_1, K_2) < \varepsilon$. This is the same as asking if the map Π^{-1} taking a projection body ΠK onto ∇K is uniformly continuous. This is true for $n = 2$. For $n \geq 3$, however, we can expect a negative answer, for the following reason. Suppose that $n = 3$ and that C_1, C_2 are centered 2-dimensional compact convex sets, of equal areas, contained in the same plane. Then the areas of their projections on any plane are also equal, though C_1 and C_2 may be far apart in the Hausdorff metric. If we now fatten C_1 and C_2 slightly, we might expect to produce two centered convex bodies whose projection bodies are close, but that are not themselves close. The next theorem makes this precise.

Theorem 4.3.12. *For $n \geq 3$, the map Π^{-1} is not uniformly continuous.*

Proof. We shall exhibit two sequences $\{K_m\}$ and $\{K'_m\}$ of centrally symmetric convex bodies in $\mathbb{E}^n$, $n \geq 3$, such that $\delta(\Pi K_m, \Pi K'_m)$ tends to zero and yet $\delta(K_m, K'_m)$ is bounded away from zero.

Let $K = \sum_{i=1}^n c_i[-e_i/2, e_i/2]$, where $c_i > 0$ and e_i is the unit vector in the ith coordinate direction, $1 \leq i \leq n$. By Definition 4.1.1, the support function of ΠK is given for $u \in S^{n-1}$ by

$$h_{\Pi K}(u) = \sum_{i=1}^n |u \cdot e_i| \prod_{j \neq i} c_j.$$

For each $m \in \mathbb{N}$, let

$$K_m = 2m^{n-3}\left[-\frac{1}{2}e_1, \frac{1}{2}e_1\right] + m^{n-3}\left[-\frac{1}{2}e_2, \frac{1}{2}e_2\right] + m^{-2}\sum_{i=3}^n\left[-\frac{1}{2}e_i, \frac{1}{2}e_i\right]$$

and

$$K'_m = m^{n-3}\left[-\frac{1}{2}e_1, \frac{1}{2}e_1\right] + 2m^{n-3}\left[-\frac{1}{2}e_2, \frac{1}{2}e_2\right] + m^{-2}\sum_{i=3}^n\left[-\frac{1}{2}e_i, \frac{1}{2}e_i\right].$$

Then

$$h_{\Pi K_m}(u) = |u \cdot e_1|m^{1-n} + 2|u \cdot e_2|m^{1-n} + 2\sum_{i=3}^n |u \cdot e_i|$$

and

$$h_{\Pi K'_m}(u) = 2|u \cdot e_1|m^{1-n} + |u \cdot e_2|m^{1-n} + 2\sum_{i=3}^n |u \cdot e_i|.$$

Since $(h_{\Pi K_m} - h_{\Pi K'_m}) \to 0$ uniformly on S^{n-1}, $\delta(\Pi K_m, \Pi K'_m) \to 0$. On the other hand, the projections of K_m and K'_m on the first coordinate axis are centered line segments of lengths $2m^{n-3}$ and m^{n-3}, respectively. Therefore $\delta(K_m, K'_m) \geq \frac{1}{2}m^{n-3}$, and this gives the desired conclusion. ∎

Remark 4.3.13. Despite Theorem 4.3.12, there is still the possibility of proving a stability result for bodies not too flat or too large in diameter. It can be proved that if $n \geq 3$, $0 < a < 2/n(n+4)$, and K_1 and K_2 are centered convex bodies with $rB \subset K_1, K_2 \subset RB$, then there is a constant $c = c(a, n, r, R)$ such that

$$\delta(K_1, K_2) \leq c\delta(\Pi K_1, \Pi K_2)^a. \tag{4.1}$$

See Note 4.11 for related results and references.

4.4. Reconstruction from brightness functions

Aleksandrov's projection theorem, Theorem 3.3.6, with $i = k = n - 1$, states that a centered convex body K in $\mathbb{E}^n$ is determined by its brightness function $v_K(u) = \lambda_{n-1}(K|u^\perp) = h_{\Pi K}(u)$, $u \in S^{n-1}$. In this section, we provide an algorithm for reconstructing an approximation to K from a finite number of measurements of v_K. We shall need the following definition.

Definition 4.4.1. Let $U = \{u_i \in S^{n-1} : 1 \leq i \leq k\}$ be a finite set of directions that span $\mathbb{E}^n$. The *nodes* corresponding to U are defined as follows. The hyperplanes $u_i^\perp$, $1 \leq i \leq k$, partition $\mathbb{E}^n$ into a finite set of convex polyhedral cones, which intersect S^{n-1} in a finite set of regions. The nodes $\pm v_j \in S^{n-1}$, $1 \leq j \leq l$, are the vertices of these regions. Thus, when $n = 2$, the nodes are simply the $2k$ unit vectors each of which is orthogonal to some u_i, $1 \leq i \leq k$. When $n = 3$, each v_j is of the form $(u_i \times u_{i'})/\|u_i \times u_{i'}\|$, where $1 \leq i < i' \leq k$, and $l \leq k(k-1)/2$. (In general, $l = O(k^{n-1})$.)

The following result will be crucial for the reconstruction algorithm to be presented later.

Theorem 4.4.2. *Let K be a convex body in $\mathbb{E}^n$ and let $U = \{u_i \in S^{n-1} : 1 \leq i \leq k\}$ span $\mathbb{E}^n$. There is a centered convex polytope P, with each of its facets orthogonal to one of the nodes corresponding to U, such that $v_P(u_i) = v_K(u_i)$ for $1 \leq i \leq k$.*

Proof. The projection body ΠK of K is a centered zonoid, by Theorem 4.1.11. Extend the finite set U to an infinite set $\{u_i \in S^{n-1} : i \in \mathbb{N}\}$ that is dense in S^{n-1}. For each $m \geq k$, apply Theorem 4.1.13 with K replaced by ΠK to obtain a centered zonotope Z_m such that $h_{Z_m}(u_i) = h_{\Pi K}(u_i)$ for $1 \leq i \leq m$. Choose m_0

large enough to ensure that dim $Z_{m_0} = n$, let $Z = Z_{m_0}$, and let Q be the centered n-dimensional convex polytope such that $Z = \Pi Q$ (cf. Theorem 4.1.11). Then

$$v_Q(u_i) = h_Z(u_i) = h_{\Pi K}(u_i) = v_K(u_i),$$

for $1 \leq i \leq k$. Therefore it suffices to prove the theorem when K is a centered n-dimensional convex polytope.

Suppose, then, that K has facets with volumes a_p and outer unit normal vectors w_p, $1 \leq p \leq r$. Comparing (A.22), we observe that

$$\sum_{p=1}^{r} a_p w_p = o. \tag{4.2}$$

Let V be the set of nodes corresponding to U, as in Definition 4.4.1. By relabeling we may assume that the nodes are v_j, $1 \leq j \leq 2l$, that $w_j = v_j$ for $j \in I$, and that w_p is not a node for $p \notin I$. Let q be the smallest integer such that $q \notin I$ and let E be the region in S^{n-1} as in Definition 4.4.1 to which w_q belongs. The fact that U spans $\mathbb{E}^n$ implies that E is the intersection of S^{n-1} with a convex polyhedral cone that is a proper subset of a half-space. Therefore the vertices of E span $\mathbb{E}^n$, so there is a subset of the vertices of E, nodes v_j for $j \in J$, say, and corresponding reals $b_j > 0$, such that

$$a_q w_q = \sum_{j \in J} b_j v_j. \tag{4.3}$$

Then

$$
\begin{aligned}
o = \sum_{p=1}^{r} a_p w_p &= \sum_{j \in I} a_j w_j + a_q w_q + \sum_{p \notin I \cup \{q\}} a_p w_p \\
&= \sum_{j \in I} a_j v_j + \sum_{j \in J} b_j v_j + \sum_{p \notin I \cup \{q\}} a_p w_p \\
&= \sum_{j \in I \cap J} (a_j + b_j) v_j + \sum_{j \in I \setminus J} a_j v_j + \sum_{j \in J \setminus I} b_j v_j + \sum_{p \notin I \cup \{q\}} a_p w_p. \tag{4.4}
\end{aligned}
$$

The vectors $\{w_p : 1 \leq p \neq q \leq r\}$ in (4.4) cannot all lie in a hyperplane, since otherwise (4.2) implies that w_q also lies in this hyperplane, contradicting $\dim K = n$. By Minkowski's existence theorem, Theorem A.3.2, the set of unit vectors and corresponding positive reals in (4.4) are the outer unit normal vectors and volumes of the facets of some convex polytope P_1. Note that the set of outer unit normal vectors of P_1 does not contain w_q.

The way that the nodes are defined guarantees that for $1 \leq i \leq k$, $u_i \cdot w_q$ has the same sign as $u_i \cdot v_j$ for each $j \in J$. Consequently, by (4.3),

$$a_q |u_i \cdot w_q| = \sum_{j \in J} b_j |u_i \cdot v_j|$$

and hence, by Cauchy's projection formula ((A.45) with $i = n - 1$),

$$2v_K(u_i) = \sum_{p=1}^{r} a_p |u_i \cdot w_p|$$

$$= \sum_{j \in I} a_j |u_i \cdot v_j| + \sum_{j \in J} b_j |u_i \cdot v_j| + \sum_{p \notin I \cup \{q\}} a_p |u_i \cdot w_p| = 2v_{P_1}(u_i),$$

for $1 \leq i \leq k$.

Applying the above argument inductively, we arrive after a finite number of steps at a polytope P' that satisfies the statement of the theorem, except that it may not be centered. In this case, note that the outer unit normal vector to any facet of $P = \nabla P'$ is orthogonal to some facet of P', so each facet of P is orthogonal to one of the nodes in U. By Theorem 4.1.2, $P = \nabla P'$ is the required polytope. ∎

The following algorithm is one of several for the purpose of reconstructing centered convex bodies from their brightness functions. There are also versions that apply to noisy brightness function measurements; see Figure 4.8 and the remarks just before, and Note 4.12.

Algorithm BrightLSQ

Input: Natural numbers $n \geq 2$ and $k \geq n$; mutually nonparallel directions $u_i \in S^{n-1}$, $1 \leq i \leq k$ that span $\mathbb{E}^n$; brightness function values $v_K(u_1), \ldots, v_K(u_k)$ of an unknown centered convex body K in $\mathbb{E}^n$.

Task: Construct a centered convex polytope P in $\mathbb{E}^n$ such that $v_P(u_i) = v_K(u_i)$, $1 \leq i \leq k$.

Action:

Phase I:

1. Calculate the nodes (see Definition 4.1.1) $\pm v_j \in S^{n-1}$, $1 \leq j \leq l$ corresponding to $u_i \in S^{n-1}$, $1 \leq i \leq k$.

2. Write $\alpha = (\alpha_1, \ldots, \alpha_l)$ when $\alpha_j \geq 0$, $1 \leq j \leq l$. Solve the following constrained linear least squares problem (LLS):

$$\min_{\alpha} \sum_{i=1}^{k} \left(v_K(u_i) - \sum_{j=1}^{l} \alpha_j |u_i \cdot v_j| \right)^2, \tag{4.5}$$

$$\text{subject to} \quad \alpha_j \geq 0, \quad 1 \leq j \leq l.$$

Let $\widehat{\alpha}_j$, $1 \leq j \leq l'$, $l' \leq l$, be a suitably relabeled solution to (LLS) in which any $\widehat{\alpha}_j$ with $\widehat{\alpha}_j = 0$ has been discarded. Let $m = 2l'$, and let $a_j = a_{j+m/2} = \widehat{\alpha}_j$ and $v_{j+m/2} = -v_j$ for $1 \leq j \leq m/2$. Then (A.22) is satisfied. If the vectors v_j, $1 \leq j \leq m$ span $\mathbb{E}^n$, Minkowski's existence theorem, Theorem A.3.2, implies that there is a unique centered convex polytope P with facets of volumes a_j and outer unit normal vectors v_j, $1 \leq j \leq m$.

Phase II:

3. Use Algorithm MinkData (see Section A.4) to reconstruct P from a_j and v_j, $1 \leq j \leq m$. ∎

As stated, Algorithm BrightLSQ may not succeed in its task, since Phase II cannot be performed if the output vectors v_j from Phase I lie in a proper subspace of $\mathbb{E}^n$. In practice, of course, one can run the algorithm and simply discard the output of Phase I in the case of such a degeneracy. However, some reasonable extra conditions on the input can be imposed which guarantee that the algorithm will succeed. To prove this, some technical lemmas are needed.

Let $\varepsilon > 0$. A set $U \subset S^{n-1}$ is called an *ε-net* if each point in S^{n-1} is within a distance ε of some point in U.

Lemma 4.4.3. *Let $R > 0$, let $0 < \varepsilon < 1$, and let U be an ε-net in S^{n-1}. Suppose that K_1 and K_2 are compact convex sets in $\mathbb{E}^n$ such that $h_{K_1}(u) = h_{K_2}(u) \leq R$ for each $u \in U$. Then*

$$\delta(K_1, K_2) \leq \frac{2\varepsilon R}{1 - \varepsilon}.$$

Proof. Let u^* be the point in S^{n-1} where h_{K_1} attains its maximum. Let u_0 be the point in U nearest to u^* and let $w_0 = (u^* - u_0)/\|u^* - u_0\|$. Then the positive homogeneity (0.19) and subadditivity (0.20) of h_{K_1} give

$$h_{K_1}(u^*) \leq h_{K_1}(u_0) + h_{K_1}(u^* - u_0) = h_{K_1}(u_0) + \|u^* - u_0\| h_{K_1}(w_0)$$
$$\leq R + \varepsilon h_{K_1}(u^*).$$

Therefore, for all $u \in S^{n-1}$,

$$h_{K_1}(u) \leq h_{K_1}(u^*) \leq \frac{R}{1 - \varepsilon}.$$

The same argument shows that $h_{K_2}(u) \leq R/(1 - \varepsilon)$ for each $u \in U$.

Let $u \in S^{n-1}$, let u_1 be the point in U nearest to u, and let $w_1 = (u - u_1)/\|u - u_1\|$. Using the subadditivity and positive homogeneity of h_{K_1} and h_{K_2}, we obtain

$$h_{K_1}(u) - h_{K_2}(u) \leq h_{K_1}(u_1) + h_{K_1}(u - u_1) - \left(h_{K_2}(u_1) - h_{K_2}(u_1 - u) \right)$$
$$= h_{K_1}(u - u_1) + h_{K_2}(u_1 - u)$$
$$= \|u - u_1\| \left(h_{K_1}(w_1) + h_{K_2}(-w_1) \right) \leq \frac{2\varepsilon R}{1 - \varepsilon}.$$

The same bound applies to $h_{K_2}(u) - h_{K_1}(u)$, and this proves the lemma. ∎

Lemma 4.4.4. *Let $0 < r < R$, let $0 < \varepsilon < r^{n-1}/(5R^{n-1})$, and let U be an ε-net in S^{n-1}. Let K be a centered convex body in $\mathbb{E}^n$ such that $rB \subset K \subset RB$,* *let $m \in \mathbb{N}$ be even, and let $a_j > 0$ and $v_j \in S^{n-1}$, $1 \le j \le m$, be such that $a_j = a_{j+m/2}$ and $v_{j+m/2} = -v_j$ for $1 \le j \le m/2$, and*

$$v_K(u) = \sum_{j=1}^{m} a_j |u \cdot v_j|, \qquad (4.6)$$

for each $u \in U$. Then the vectors v_j, $1 \le j \le m$, span $\mathbb{R}^n$.

Proof. Suppose that the assumptions of the lemma hold but that the span of the vectors v_j, $1 \le j \le m$, is a q-dimensional subspace S in $\mathbb{E}^n$ for some $1 \le q \le n - 1$. Since $rB \subset K \subset RB$, we have

$$s = \kappa_{n-1} r^{n-1} \le h_{\Pi K}(u) = h_Z(u) \le \kappa_{n-1} R^{n-1} = t,$$

for $u \in U$, where

$$h_Z(u) = \sum_{j=1}^{m} a_j |u \cdot v_j|,$$

for $u \in S^{n-1}$, is the support function of a q-dimensional centered zonotope Z in S (see Lemma 4.1.8). Lemma 4.4.3 applied with K, L, and R replaced by ΠK, Z, and t, respectively, gives

$$\delta(\Pi K, Z) \le \frac{2\varepsilon t}{1 - \varepsilon} < \frac{r^{n-1} t}{2R^{n-1}} = \frac{s}{2}.$$

Since $sB \subset \Pi K$, we have $(s/2)B \subset Z$, which is impossible. ∎

Theorem 4.4.5. *Suppose that $0 < r < R$, that $U = \{u_i \in S^{n-1} : 1 \le i \le k\}$ is a set of mutually nonparallel directions such that $U \cup -U$ is an ε-net with $0 < \varepsilon < r^{n-1}/(5R^{n-1})$, and that K is a centered convex body in $\mathbb{E}^n$ with $rB \subset K \subset RB$. Then Algorithm BrightLSQ succeeds in its task.*

Proof. By Theorem 4.4.2, there is a centered convex polytope with the same brightness function measurements as K in the input directions, each of whose facets has outer unit normal vector equal to some node $\pm v_j$, $1 \le j \le l$, corresponding to U. Since the output of Phase I of Algorithm BrightLSQ could be the volumes and outer unit normal vectors of the facets of this polytope, the optimal value of the objective function in (4.5) is zero. It follows that (4.6) holds. Lemma 4.4.4 implies that the vectors v_j, $1 \le j \le m$, span $\mathbb{E}^n$, and Phase II can be performed to construct the output polytope P. By Cauchy's projection formula ((A.45) with $i = n - 1$) and (4.6), $v_P(u_i) = v_K(u_i)$ for $1 \le i \le k$. ∎

We now show that Algorithm BrightLSQ can provide approximations to K that converge to K as the number of measurements increases. A further technical lemma is required.

Lemma 4.4.6. *Let* $0 < r < R$, *let* $0 < \varepsilon < r^{n-1}/(5R^{n-1})$, *and let* U *be an* ε-*net in* S^{n-1}. *Let* K_1 *and* K_2 *be centered convex bodies in* $\mathbb{E}^n$ *such that* $rB \subset K_1 \subset RB$ *and* $v_{K_1}(u) = v_{K_2}(u)$ *for each* $u \in U$. *Then* $r_0 B \subset K_2 \subset R_0 B$, *where*

$$R_0 = \frac{3n\kappa_n}{\kappa_{n-1}} \left(\frac{3}{2}\right)^{1/(n-1)} \frac{R^n}{r^{n-1}} \quad \text{and} \quad r_0 = \frac{\kappa_{n-1} r^{n-1}}{2^n R_0^{n-2}}. \tag{4.7}$$

Proof. The assumptions $rB \subset K_1 \subset RB$ and $v_{K_1}(u) = v_{K_2}(u)$ for each $u \in U$ imply that

$$s = \kappa_{n-1} r^{n-1} \le h_{\Pi K_1}(u) = h_{\Pi K_2}(u) \le \kappa_{n-1} R^{n-1} = t,$$

for all $u \in U$. Since $\varepsilon < s/(5t)$, Lemma 4.4.3 applied with K_1, K_2, and R replaced by ΠK_1, ΠK_2, and t, respectively, gives

$$\delta(\Pi K_1, \Pi K_2) \le \frac{2\varepsilon t}{1 - \varepsilon} < \frac{s}{2}.$$

From this and $sB \subset \Pi K_1 \subset tB$ we obtain

$$\frac{s}{2} B \subset \Pi K_2 \subset \frac{3t}{2} B. \tag{4.8}$$

Cauchy's surface area formula (A.49) now yields

$$S(K_2) = \frac{1}{\kappa_{n-1}} \int_{S^{n-1}} v_{K_2}(u) \, du \le \frac{3n\kappa_n t}{2\kappa_{n-1}}.$$

From the isoperimetric inequality (B.14), it follows that

$$V(K_2) \le \kappa_n \left(\frac{3t}{2\kappa_{n-1}}\right)^{n/(n-1)}.$$

Let $x \in \mathrm{bd}\, K_2$, and let $u = x/\|x\|$. The monotonicity (A.18) and invariance under translation (A.19) of mixed volumes yield

$$V(K_2, n-1; [o, x]) \le V(K_2, n-1; K_2) = V(K_2).$$

Using this, (A.37), (4.8), and the positive multilinearity (A.16) of mixed volumes, we obtain

$$\frac{s}{2}\|x\| \le \|x\| v_{K_2}(u) = nV(K_2, n-1; [o, x]) \le nV(K_2).$$

Consequently,

$$\|x\| \le \frac{2n}{s} V(K_2) \le R_0,$$

where R_0 is given by (4.7), and hence $K_2 \subset R_0 B$.

Let c be the largest number such that $cB \subset K_2$. There are common parallel supporting hyperplanes to cB and K_2 at contact points $z, -z \in \mathrm{bd}\,(cB) \cap \mathrm{bd}\, K_2$. Let S be the closed slab bounded by these hyperplanes, and note that

$K_2 \subset S \cap R_0 B$. If u is any direction orthogonal to z, then $(S \cap R_0 B)|u^{\perp} \subset E$, where E is an $(n-1)$-dimensional box, the product of $n-1$ mutually orthogonal edges, with one edge parallel to z of length $2c$ and $n-2$ edges orthogonal to z, each of length $2R_0$. Therefore

$$\frac{s}{2} \le v_{K_2}(u) \le 2^{n-1} c R_0^{n-2},$$

so $r_0 B \subset K_2$, where r_0 is as in (4.7).   ∎

Theorem 4.4.7. *Let $n \ge 2$, let $0 < r < R$, and let K be a centered convex body in $\mathbb{E}^n$ such that $r B \subset K \subset R B$. Let $u_i, i \in \mathbb{N}$, be a sequence of mutually nonparallel directions such that $\{\pm u_i\}$ is dense in S^{n-1}. Let $P_k, k \ge n$, be an output, if any, to Algorithm BrightLSQ corresponding to input directions $u_i, 1 \le i \le k$. Then $\delta(K, P_k) \to 0$ as $k \to \infty$.*

Proof. For each k, let

$$\Delta_k = \max_{u \in S^{n-1}} \min_{1 \le i \le k} \{\|u - u_i\|, \|u - (-u_i)\|\}. \tag{4.9}$$

Then the set $U_k = \{u_i : 1 \le i \le k\}$ is such that $U_k \cup -U_k$ is an ε-net in S^{n-1}, where $\varepsilon = \Delta_k$. Since $\{\pm u_i\}$ is dense in S^{n-1}, $\Delta_k \to 0$ as $k \to \infty$, so we can choose an N_0 so that $\Delta_k < r^{n-1}/(5R^{n-1})$ for $k \ge N_0$. By Theorem 4.4.5, for $k \ge N_0$, Algorithm BrightLSQ produces a convex polytope P_k such that

$$h_{\Pi K}(\pm u) = v_K(\pm u) = v_{P_k}(\pm u) = h_{\Pi P_k}(\pm u), \tag{4.10}$$

for $u \in U_k$.

By assumption $K \subset R B$, so $\Pi K \subset \kappa_{n-1} R^{n-1} B$. Using (4.10), we apply Lemma 4.4.3 with $U = U_k$ and K, L, and R replaced by ΠK, ΠP_k, and $\kappa_{n-1} R^{n-1}$, respectively, to obtain

$$\delta(\Pi K, \Pi P_k) \le \frac{2\Delta_k \kappa_{n-1} R^{n-1}}{1 - \Delta_k} < \frac{5}{2} \Delta_k \kappa_{n-1} R^{n-1}, \tag{4.11}$$

for $k \ge N_0$.

Suppose that $n = 2$. By Theorem 4.1.4, the projection body of any centered convex body is just its dilatate by a factor of 2, rotated by $\pi/2$ about o. From (4.11) it then follows directly that

$$\delta(K, P_k) = \frac{1}{2}\delta(\Pi K, \Pi P_k) \le \frac{5}{2}\Delta_k R, \tag{4.12}$$

for $k \ge N_0$.

Now suppose that $n \ge 3$. From Lemma 4.4.6 with $\varepsilon = \Delta_k$, $U = U_k \cup -U_k$, and $K_2 = P_k$, we conclude that for $k \ge N_0$,

$$r_0 B \subset P_k \subset R_0 B, \tag{4.13}$$

where R_0 and r_0 are given by (4.7), and these containments also hold for K since $r_0 \leq r$ and $R \leq R_0$. By (4.1), (4.7), (4.11), and (4.13), for each $0 < a < 2/(n(n+4))$ there is a $c' = c'(a, n, r, R) > 0$ such that

$$\delta(K, P_k) \leq c'\Delta_k^a, \tag{4.14}$$

for $k \geq N_0$.

The estimates (4.12) and (4.14) imply that for $n \geq 2$, $\delta(K, P_k) \to 0$ as $k \to \infty$. ∎

The proof above of Theorem 4.4.7 can be shortened somewhat, but we have included estimates that are useful in estimating the speed of convergence. These estimates show that for a fixed k it is desirable to choose the input directions u_i, $1 \leq i \leq k$, in order to minimize Δ_k. When $n = 2$, one can take these directions with equally spaced polar angles in the interval $[0, \pi)$, so that $\Delta_k = \pi/k$ and

$$\delta(K, P_k) \leq \frac{5\pi R}{2k}$$

when $k > 5\pi R/r$. For $n \geq 3$ and general k, it is not known how to minimize Δ_k, but special cases of this problem are solved; see Note 4.12.

Algorithm BrightLSQ has been successfully implemented. Some sample reconstructions follow.

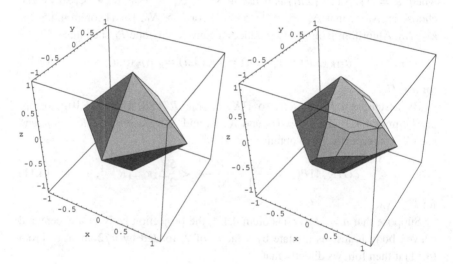

Figure 4.6. Reconstruction of an octahedron from its brightness function.

Figure 4.6 (right) shows a reconstruction produced by Algorithm BrightLSQ of the regular octahedron on the left. The input consisted of exact brightness function measurements in the 21 directions corresponding to the spherical polar angles

$$\{(2r\pi/5, s\pi/8) : r = 0, \ldots, 4, \ s = 0, \ldots, 4\}.$$

The reconstructed polyhedron has 20 facets but the same brightness function as the octahedron in the above 21 directions.

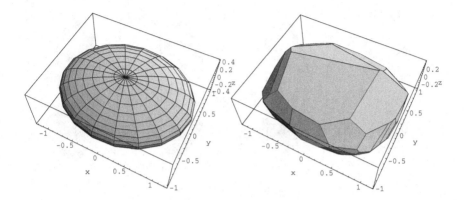

Figure 4.7. Reconstruction of an ellipsoid from its brightness function.

Figure 4.7 (left) displays the standard ellipsoid

$$\left\{ (x, y, x) : \frac{x^2}{a^2} + \frac{y^2}{b^2} + \frac{z^2}{c^2} \le 1 \right\}$$

with parameters $a = 1.2$, $b = 1$, and $c = 0.4$. Figure 4.7 (right) is a reconstruction produced by Algorithm BrightLSQ. This convex polytope with 72 facets resulted from a set of 50 exact brightness function measurements taken in randomly generated directions. The corresponding 50 brightness function values of the reconstructed polyhedron agree with those of the ellipsoid to within 2%.

We shall call Algorithm NoisyBrightLSQ the modification of Algorithm BrightLSQ in which nothing is changed except that the input brightness function measurements are noisy; specifically, they are of the form

$$y_i = v_K(u_i) + N_i, \tag{4.15}$$

for $1 \le i \le k$, where the N_i's are independent normal random variables with zero mean and variance σ^2. Convergence proofs can be constructed for Algorithm NoisyBrightLSQ, and moreover, rates of convergence can be estimated; see Note 4.12.

In Figure 4.8 (left), a polyhedron P is depicted that has 18 facets; it is the convex hull of the points $(0, 0, \pm 3/2)$, H, and $-H$, where H is the hexagon with vertices $(\cos(k\pi/6), \sin(k\pi/6), 1/2)$, $k = 1, 3, 5, 7, 9, 11$. To the right, reconstructions of P using Algorithm NoisyBrightLSQ, corresponding to noise level $\sigma = 0$ (no noise), $\sigma = 0.01$, and $\sigma = 0.05$, are shown in that order. The polyhedra have 58, 80, and 66 facets, respectively. The input consisted of noisy brightness function measurements in 55 optimally spaced directions (see Note 4.12).

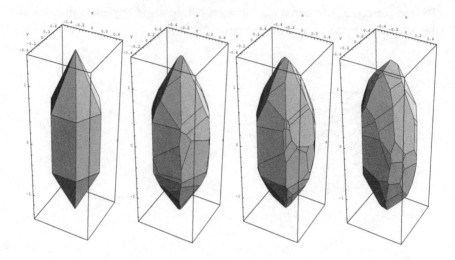

Figure 4.8. Reconstructions from noisy brightness function measurements.

Open problems

Problem 4.1. (See Notes 4.1, 9.1, and 9.10.) Which convex bodies in $\mathbb{E}^n$, $n \geq 3$, have the property that $\Pi^* K (= (\Pi K)^*)$ and $\triangle K$ are homothetic?

Problem 4.2. (See Note 4.6.) Which convex bodies K in $\mathbb{E}^n$ are such that for each $m \in \mathbb{N}$, $K = \Pi^m K_m$ for some convex body K_m?

Problem 4.3. (See Theorem 4.3.11 and Notes 4.11 and 7.11.) Suppose that K is a convex body in $\mathbb{E}^3$ such that

$$\frac{\pi \lambda_2(K|u^\perp)}{V_1(K|u^\perp)^2} \geq c,$$

for all $u \in S^2$. Is it true that

$$\frac{9\pi \lambda_3(K)^2}{2V_2(K)^3} \geq c\,?$$

Problem 4.4. (See Notes 4.6 and 9.4.) Suppose that $1 \leq i \leq n-1$. Which convex bodies K in $\mathbb{E}^n$ are such that $\Pi_i^2 K$ is homothetic to K?

Problem 4.5. (See Note 4.6.) Suppose that $1 \leq i \leq n-1$. Which convex bodies K in $\mathbb{E}^n$ are such that $\Pi_i K$ is homothetic to K?

Notes

4.1. *Projection bodies and zonoids.* Projection bodies, zonoids, and zonotopes have a long and complicated history. Two extensive articles that detail this are by Bolker [79] and Schneider and Weil [745]; also, the surveys of Goodey and Weil [320] and

Martini [581] cover more recent results. It seems that the story of zonotopes begins with Kepler, but they were first considered as a class by the crystallographer Fedorov [224]. Seven equivalent definitions of them are given by Bolker [79, Theorem 3.3], and this theorem contains the proofs of the statements in Remark 4.1.19. (However, for the reader who has absorbed the point of view of this chapter, we would rather recommend the proofs given in Schneider's paper [731, Section 6] and book [737, Section 3.5]. The historical remarks in [79] about these statements should be augmented by those in [745, Section 1].) For example, they are precisely the affine images in $\mathbb{E}^n$ of the unit cube in $\mathbb{E}^k$ for some k. Zonotopes are important in geometry. Every convex polyhedron tiling $\mathbb{E}^3$ by translation is a zonotope, and zonotopes in $\mathbb{E}^3$ are exactly those convex polyhedra that are equidissectable with the unit cube by translations. For references to these facts, and a wealth of additional information, consult [737, Notes to Section 3.5]. Other good general references for zonotopes are provided by Burger, Gritzmann, and Klee [115] and Martini [574], and some pictures of zonotopes generated by Mathematica can be found on the website of Eppstein [210].

Projection bodies go back to Minkowski (see [83, Section 30, p. 50]). Petty [662] proved Theorem 4.1.5, but the proof we use here was found by Lutwak [538]. The fact (Theorem 4.1.11) that a convex body whose support function has the integral representation of Theorem 4.1.10 is a projection body was noted by Firey [237]. The link between projection bodies and zonoids was completed by Lindquist [512] (see also [513]). We have followed Lindquist in the first part of the proof of Theorem 4.1.10, but have based the second part on the more direct proof in [79, Theorems 4.2 and 4.3]. These results are also to be found in [731].

Campi, Haas, and Weil [150] provide two proofs of Theorem 4.1.13, ascribing the one we give to J. Bourgain, and cite several other papers on approximating zonoids by zonotopes. They note that since the set D in the proof of Theorem 4.1.13 is connected, a refinement of Carathéodory's theorem allows the number $k + 1$ of line segments to be reduced to k. The other proof of Theorem 4.1.13 in [150] is constructive and is applied to an estimation problem for fiber processes. See also Note 4.12.

Zonoids, like their special polytopal forms the zonotopes, have many equivalent definitions. Some of these are quite surprising. For example, the centered zonoids are precisely the ranges of nonatomic $\mathbb{E}^n$-valued vector measures. This is a slight strengthening of the usual statement of Liapounoff's theorem, that such ranges are compact and convex. (We again refer the reader to the survey paper [745] for references.) Now let K be a centered convex body in $\mathbb{E}^n$. Then (see Section 0.8) the support function h_K of K defines a norm on $\mathbb{E}^n$, with the polar body K^* of K acting as the unit ball in the corresponding finite-dimensional Banach space or *Minkowski space* $(\mathbb{E}^n, K^*)$. It turns out that $(\mathbb{E}^n, K^*)$ is isometric to a subspace of $L^1([0, 1])$ if and only if K is a zonoid. It is natural, then, that functional analysts have become interested in this part of geometry; see, for example, the paper [91] of Bourgain and Lindenstrauss. References to articles connecting the study of zonoids to combinatorics, stochastic geometry, random determinants, Hilbert's fourth problem, mathematical economics, and physics (light field theory) can be found in the works of Goodey and Weil [320, Section 8], Lutwak [541], Martini [577], and Schneider [737, p. 195].

Given any centered convex body K in $\mathbb{E}^n$, there is a Minkowski space $(\mathbb{E}^n, K)$, with norm defined by the gauge function of K, for which K is the unit ball; this is just the dual of the Minkowski space considered in the previous paragraph. Length in $(\mathbb{E}^n, K)$ is determined by the norm, and it transpires that the volume of a body in $(\mathbb{E}^n, K)$ may, without loss of generality, be defined as the Euclidean volume of the body. On the other hand, there are several natural but quite different ways to define surface area in $(\mathbb{E}^n, K)$. Under the Holmes–Thompson definition (see [397] and [803]), the Minkowski surface area of a face F of a polytope in $(\mathbb{E}^n, K)$ is

$$\frac{\lambda_{n-1}(F)\lambda_{n-1}(K^*|S)}{\kappa_{n-1}},$$

where F is contained in a translate of the $(n-1)$-dimensional subspace S of $\mathbb{E}^n$. One can now pose the corresponding isoperimetric problem: Which convex bodies have maximal volume among all with a given Minkowski surface area? With the Holmes–Thompson definition, Johnson and Thompson [409] proved that convex bodies homothetic to $\Pi(K^*)$ provide the answer.

The natural question arises whether K must be an ellipsoid (which would imply that $(\mathbb{E}^n, K)$ is Euclidean – just a linear image of $(\mathbb{E}^n, B) = \mathbb{E}^n)$ if K itself is a solution to the isoperimetric problem for $(\mathbb{E}^n, K)$. With the Holmes–Thompson definition of surface area, this is equivalent to asking whether K must be an ellipsoid if K is centered and ΠK is homothetic to K^*. For $n = 2$ this is not true; it is known that K has this property if and only if bd K is a Radon curve, so a nonelliptical example is provided by an affinely regular hexagon. For $n \geq 3$, the question is implicit in work of Petty [663, p. 35], who gave a formulation involving minimal circumscribed cylinders, and was also posed by Holmes and Thompson [397, Problem 7.5] and Gruber [362]. Martini [579] provides a partial answer to the more general Problem 4.1, by showing that a convex polytope with this property must be a simplex. See also the works of Lutwak [542, Section 12], Martini [581, Section 4], Schneider [737, p. 416], and Thompson [803] for other comments and references.

In a long series of papers summarized in [131], Busemann, Ewald, and Shephard define a notion of convexity for functions on $\mathcal{G}(n, i)$, $1 \leq i \leq n - 1$. They show that although the ith projection function of a zonoid is convex in their sense of the term, this is not true for arbitrary convex bodies (or even convex polytopes) when $1 < i < n - 1$. See also the papers of Phadke [671] and Silverman [767].

Martinez-Maure [572] defines projection bodies for a class of surfaces called hedgehogs that do not necessarily form the boundary of a convex body.

4.2. *The Fourier transform approach I: The brightness function and projection bodies.* A convex body K in $\mathbb{E}^n$ is said to have a *curvature function* $f_K : S^{n-1} \to \mathbb{R}$ if its surface area measure $S(K, \cdot)$ is absolutely continuous with respect to spherical Lebesgue measure in S^{n-1} and the Radon–Nikodym derivative of $S(K, \cdot)$ with respect to spherical Lebesgue measure in S^{n-1} is f_K. In this case

$$\lambda_{n-1}\left(K|u^{\perp}\right) = \frac{1}{2}\int_{S^{n-1}}|u \cdot v|\, dS(K, v) = \frac{1}{2}\int_{S^{n-1}}|u \cdot v| f_K(v)\, dv = \frac{1}{2}Cf_K(u),$$

for all $u \in S^{n-1}$, where C is the cosine transform, by Cauchy's projection formula ((A.45) with $i = n - 1$). (If K is of class C_+^2, then f_K is just the reciprocal Gauss curvature, viewed as a function of the outer unit normal vector.) Extend f_K to a positively homogeneous function of degree $-n - 1$ on $\mathbb{E}^n$ by defining $f_K(x) = \|x\|^{-n-1} f_K(x/\|x\|)$ for all $x \in \mathbb{E}^n$. If K is centrally symmetric, then f_K is also even, so by (C.24), we have

$$h_{\Pi K}(u) = \lambda_{n-1}\left(K|u^{\perp}\right) = -\frac{1}{\pi}\widehat{f_K}(u),$$

for all $u \in S^{n-1}$. Note that since $h_{\Pi K}$ is positively homogeneous of degree one on $\mathbb{E}^n$ by (0.19), and the Fourier transform of a positively homogeneous function of degree p on $\mathbb{E}^n$ is positively homogeneous of degree $-n - p$ (see, for example, [465, Lemma 2.21]), our extension of f_K was the natural one.

The previous displayed formula is useful, but requires K to have a curvature function. For general convex bodies, there is a corresponding formula,

$$h_{\Pi K}(u) = \lambda_{n-1}\left(K|u^{\perp}\right) = -\frac{1}{\pi}\widehat{S(K,\cdot)}_e(u),$$

for $u \in S^{n-1}$, where $S(K,\cdot)_e$ is the *extended measure* of the surface area measure $S(K,\cdot)$ of K. The definition of the extended measure μ_e of a finite Borel measure μ in S^{n-1} requires the Fourier transform of distributions; for background, see the book of Strichartz [791], for example. Details are provided by Koldobsky, Ryabogin, and Zvavitch [467], [468] and Koldobksy [465, Chapter 8], who ascribe this formula for the brightness function of K to Semyanistyi [761]. This formula, the Fourier inversion formula, and Minkowski's existence theorem, Theorem A.3.2, yield the following characterization of projection bodies (see [467] or [465, Theorem 8.6]).

A centered convex body K in $\mathbb{E}^n$ is the projection body of some centered convex body K' in $\mathbb{E}^n$ if and only if there is a finite Borel measure μ in S^{n-1} such that $\widehat{h_K} = -(2\pi)^{n-1}\mu_e$, in which case $\mu = S(K',\cdot)$.

Koldobsky, Ryabogin, and Zvavitch [467] (see also [465, Theorem 8.7]) also obtain a necessary and sufficient condition for a centered convex body K in $\mathbb{E}^n$ with $h_K \in C^{\infty}(S^{n-1})$ to be a projection body. The condition is the nonnegativity of expressions involving the parallel section function A_{K^*} of the polar body K^* of K, very similar to the characterization of intersection bodies via formulas (8.14) and (8.15) discussed in Note 8.1.

4.3. *The Minkowski map and Minkowski linear combinations of projection bodies.* Theorem 4.1.5, which says that affinely equivalent convex bodies have affinely equivalent projection bodies, makes the Minkowski map Π especially interesting in Banach space theory. Ludwig [524], [526] observes that with the operation of Minkowski addition, $\mathcal{K}^n$ and the class $\mathcal{P}^n$ of convex polytopes in $\mathbb{E}^n$ are Abelian semigroups and Π is a valuation. She proves in [526] that if $Z : \mathcal{P}^n \to \mathcal{K}^n$ is a translation-invariant valuation that satisfies the affine invariance property of Theorem 4.1.5 (when Π is replaced by Z), then $Z = c\Pi$ for some constant $c \geq 0$.

We noted after Corollary 4.1.12 that the class of projection bodies is closed under Minkowski linear combinations. This means that if K_1 and K_2 are convex bodies and $a_j \geq 0$, $j = 1, 2$, then there is a convex body K such that

$$\Pi K = a_1 \Pi K_1 + a_2 \Pi K_2,$$

or, equivalently, by Definition 4.1.1 and Theorem C.2.1,

$$S(K,\cdot) = a_1 S(K_1,\cdot) + a_2 S(K_2,\cdot).$$

Using the Blaschke addition defined in Note 3.4, and taking $a_1 = a_2 = 1$, we arrive at the formula (which actually holds for compact convex sets)

$$\Pi(K_1 \# K_2) = \Pi K_1 + \Pi K_2.$$

This can be illustrated by the following example. Let K be the cylinder in $\mathbb{E}^3$ obtained by rotating the square $[-1, 1]^2$ about the z-axis. The support function of K is given by

$$h_K(\varphi) = \sin\varphi + \cos\varphi,$$

where φ is the usual vertical angle in spherical polar coordinates. One can use this to calculate directly from Definition 4.1.1 that ΠK is a centered cylinder of radius 4 and height 2π, but this lacks insight. Alternatively, one can argue as follows. Let K_n be a centered polyhedral cylinder of height 2, the Cartesian product of a regular $2n$-gon F_0 inscribed in the unit circle and $[-1, 1]$. Then $S_2(K_n,\cdot)$ is the sum of

$(2n+2)$ point masses located at the outward normals of the facets of K_n. Separating these masses into $(n+1)$ antipodal pairs, we see that

$$K_n = F_0 + F_1 + \cdots + F_n,$$

where F_j, $1 \le j \le n$, are consecutive vertical facets of K_n. As we noted after Definition 4.1.1, ΠF_j is a centered line segment of length $2\lambda_2(F_j)$ orthogonal to F_j, $0 \le j \le n$. Therefore

$$\Pi K_n = \Pi F_0 + \Pi F_1 + \cdots + \Pi F_n$$

is also a centered polyhedral cylinder whose cross-section is a regular $2n$-gon in the xy-plane. Letting $n \to \infty$, we see that ΠK is indeed a centered cylinder.

4.4. *Generalized zonoids.* Generalized zonoids are implicit in the work of Blaschke. The proof of Remark 4.1.19(ii), showing that for polytopes we get nothing new, can be found in work of Schneider [731], [737, Corollary 3.5.6]. The first examples of generalized zonoids that are not zonoids are those of Remark 4.1.15, discovered by Schneider. The details of the proof are given in [731, p. 69]. The support functions of these convex bodies satisfy

$$h_K(u) = \int_{S^{n-1}} |u \cdot v| f(v)\, dv,$$

for all $u \in S^{n-1}$, where $f \in C_e(S^{n-1})$. (Other examples with this property are produced by Weil [833], [835], who also coined the term "generalized zonoid" in [832].) Theorem 4.1.17 implies that the support function of a sufficiently smooth centered convex body has such an integral representation. This fact is quite often used in the literature; despite its convenience, we have avoided it in our proofs, since its demonstration makes heavy use of spherical harmonics. It was known to Blaschke [71, p. 154], but the result was first made precise by Schneider [727] (see also [737, Theorem 3.5.3] and Groemer's book [356, Proposition 3.6.4]), who proved Theorem 4.1.17 under the weaker assumption that $h_K \in C^{n+2}(S^{n-1})$. An improvement in which it is only assumed that $h_K \in C^{(n+5)/2}(S^{n-1})$ has been found by Goodey and Weil; see [318] for the precise statement of their result.

The existence of centered convex bodies that are not generalized zonoids shows that not all centered convex bodies have support functions with the foregoing integral representation. This contradicts the mistaken assertion in Bonnesen and Fenchel's book (unfortunately not corrected in the English translation [83, Section 19, p. 32]).

It can be shown that each centered convex body K in $\mathbb{E}^n$ has a *generating distribution* ϱ_K, defined in terms of the cosine transform C by

$$\varrho_K(f) = \int_{S^{n-1}} h_K(u)(C^{-1}f)(u)\, du,$$

for $f \in C_e^\infty(S^{n-1})$. This important result is due to Weil [831]. From this, one obtains a representation of the support function of K. The distributional approach and the link (C.12) between the cosine and spherical Radon transforms allow projection bodies (or generalized projection bodies) to be characterized as those convex bodies whose first area measures are Radon transforms of finite Borel measures (or finite signed Borel measures, respectively). We refer the reader to articles of Goodey and Weil [317], [318], [320], Lonke [518], and Weil [831], [834].

Suppose that K is a centered generalized zonoid. Just as h_K, and therefore w_K, allows an integral representation as in Definition 4.1.14, the other projection functions $\lambda_i(K|\cdot)$, $1 < i \le n - 1$, also allow an integral representation; the measure μ must be replaced by a suitable *ith projection generating measure*, and the integrand $|u \cdot v|$ by an appropriate generalization. However, the uniqueness result corresponding to Theorem C.2.1 fails when $1 < i < n - 1$. See [320, Section 6].

4.5. *Bodies whose projections are zonoids.* We noted in the text that the projection of a
zonoid is a zonoid. For $2 \leq i \leq n - 1$, consider the class $\mathcal{K}_i^n$ of centered convex
bodies in $\mathbb{E}^n$ such that $K|S$ is a zonoid for each $S \in \mathcal{G}(n, i)$. Is each member of
$\mathcal{K}_i^n$ a zonoid? The answer is negative, since this is false for an octahedron in $\mathbb{E}^3$, by
Remark 4.1.19. Nevertheless, the question has provoked some interesting material.

Suppose that K is a centered polytope in $\mathcal{K}_3^n$ and that F is a 2-dimensional face
of K. Then we can choose an $S \in \mathcal{G}(n, 3)$ so that F is also a 2-dimensional face
of the zonotope $K|S$. Now F is centrally symmetric, so K is a zonotope; compare
Remark 4.1.19(i) and (iv). This was noted by Witsenhausen [843]. On the other
hand, Weil [835] shows that there are arbitrarily smooth members of $\mathcal{K}_{n-1}^n$ that are
not zonoids, and further that $\mathcal{K}_{i+1}^n$ is a proper subset of $\mathcal{K}_i^n$ for $2 \leq i \leq n - 1$.

4.6. *Projection bodies of order i.* Let $1 \leq i \leq n - 1$. A *projection body of order i* is a
convex body $\Pi_i K$ whose support function is of the form

$$h_{\Pi_i K}(u) = c_{n-1,i} V_i(K|u^\perp) = \frac{1}{2} \int_{S^{n-1}} |u \cdot v| \, dS_i(K, v),$$

for all $u \in S^{n-1}$ and some convex body K in $\mathbb{E}^n$. By this definition and Theorem 4.1.10, each such body is a centered zonoid; however, no characterization
of projection bodies of order i seems to be known except in the case $i = n - 1$,
which of course reduces to that of the ordinary projection bodies. The equivalence
$(3.2) \Leftrightarrow (3.4)$ of Theorem 3.3.2 gives a necessary and sufficient condition for equality
of projection bodies of order i. The special case $i = k$ of Aleksandrov's projection
theorem, Theorem 3.3.6, says that the map Π_i, when restricted to the class of centered convex bodies, is injective. If $K_1 \subset K_2$, then $\Pi_i K_1 \subset \Pi_i K_2$, but the results
of Section 4.2 show, in a dramatic way, that the converse of this is false.

Suppose $1 \leq i \leq n - 2$ and K is a convex body in $\mathbb{E}^n$. Then $\Pi_{i+1} K$ is a ball
if and only if the integral of the ith intrinsic volumes of its sections by hyperplanes
orthogonal to a direction u is independent of u. This result is proved by Goodey
[308]; the Berwald–Schneider characterization of constant brightness mentioned in
Note 3.3 is the special case $i = n - 2$.

Iterates of the maps Π_i (including $\Pi_{n-1} = \Pi$) have also been considered. Suppose, for example, that a convex body K has the property that for every m there is a
convex body K_m such that K is a translate of $\Pi^m K_m$. Evidently K must be a special kind of zonoid. A geometric characterization of such K is not known (cf. Problem 4.2), but Weil [830] characterizes those that are polytopes as Cartesian products
of (possibly degenerate) centrally symmetric polygons. It turns out that these are
also exactly the polytopes for which $\Pi^2 K$ is homothetic to K (see Problem 4.4).
In the same paper, Weil solves Problem 4.5 when $i = n - 1$ and K is a polytope.
Schneider [732] deals with the case $i = 1$ of these questions, by showing that balls
are the only convex bodies for which there is an m such that $\Pi_1^m K$ is homothetic to
K; Grinberg and Zhang [337] find another route to this fact.

In [592], McMullen shows that the unit cube C in $\mathbb{E}^n$ has the property that for
each $S \in \mathcal{G}(n, i)$, $1 \leq i \leq n - 1$, $\lambda_i(C|S) = \lambda_{n-i}(C|S^\perp)$. (Schnell [748] finds
that a suitable generalization holds for parallelotopes.) In a sequel [593], he calls
this property of the unit cube *property (VP)*, and shows that for $n \geq 3$ the centered
polytopes P with property (VP) are precisely those for which $\Pi P = 2P$. Via work
of Weil [830], this allows a characterization of centrally symmetric polytopes with
these properties in terms of Cartesian products.

Goodey [306] produces a family of centrally symmetric convex bodies of revolu-
tion K in $\mathbb{E}^4$ that have the property that $\lambda_2(K|S) = \lambda_2(K|S^\perp)$ for each $S \in \mathcal{G}(4, 2)$.

Extending the notion of property (VP), Schneider [739] calls a pair (P, Q) of
convex polytopes in $\mathbb{E}^n$ a *(VP)-pair* if $\lambda_i(P|S) = \lambda_{n-i}(Q|S^\perp)$ for each $S \in \mathcal{G}(n, i)$
and $1 \leq i \leq n - 1$. He characterizes such VP-pairs and applies the result to
Minkowski geometry.

4.7. *The L^p-Brunn–Minkowski theory and L^p-projection bodies.* The L^p-Brunn–Minkowski theory has its roots in work of Firey [235] and has also been called the Brunn–Minkowski–Firey theory. If $p \geq 1$ and K_1 and K_2 are convex bodies in $\mathbb{E}^n$ containing the origin in their interiors, a convex body $K_1 +_p K_2$ can be defined by

$$h_{K_1+_pK_2}(u)^p = h_{K_1}(u)^p + h_{K_2}(u)^p,$$

for $u \in S^{n-1}$. The operation $+_p$, called *p-Minkowski addition*, was studied by Firey [235], [236], who proved a generalization of the Brunn–Minkowski inequality (B.10) in which Minkowski addition is replaced by p-Minkowski addition. Note the connection with the ith radial addition defined in Note 6.1; via (0.36), we have $K_1 +_p K_2 = K_1^* \tilde{+}_{-p} K_2^*$.

The new theory really gained steam with the results of Lutwak [543], who found the appropriate L^p-analog of the surface area measure $S(K, \cdot)$ of a convex body K in $\mathbb{E}^n$ containing the origin in its interior, the differential of which turns out to be $h_K(u)^{1-p}dS(K, u)$. He also generalized Firey's inequality and proved the L^p versions of Minkowski's existence theorem (Theorem A.3.2) and the Kneser–Süss inequality (B.32). For $p = 1$, there is of course no need for the bodies to contain the origin in their interiors, and the whole theory reduces to the usual Brunn–Minkowski theory. For $p > 1$, however, translation invariance is lost.

The L^p-*projection body* P_pK, $p \geq 1$, of a convex body K in $\mathbb{E}^n$ containing origin in its interior is defined by

$$h_{P_pK}(u)^p = \frac{1}{2} \int_{S^{n-1}} |u \cdot v|^p h_K(v)^{1-p} \, dS(K, v),$$

for all $u \in S^{n-1}$. From Definition 4.1.1 we have $\Pi K = P_1 K$. The notation $\Pi_p K$ has also been used, and sometimes different constants appear before the integral. These bodies were introduced by Lutwak [544] (see also the papers of Lutwak and Oliker [545] and Lutwak, Yang, and Zhang [547]), whose definition was slightly different. From known properties of the p-cosine transform (see Section C.2), it follows that a centered convex body is determined by its L^p-projection body when p is not an even integer, a remarkable extension of Aleksandrov's projection theorem, Theorem 3.3.6. Ryabogin and Zvavitch [711] obtain a Fourier transform characterization of L^p-projection bodies analogous to that of projection bodies in Note 4.2, and note that an earlier result of A. Koldobsky shows that a centered convex body in $\mathbb{E}^n$ is an L^p-projection body if and only if its polar body K^* is the unit ball of a subspace of $L^p([0, 1])$. Ludwig [526] obtains a characterization of certain valuations in terms of the operator P_p in the spirit of that for Π described in Note 4.3.

Gardner [264, Section 18.3] provides further references to the L^p-Brunn–Minkowski theory. See also Notes 4.11 and 9.5.

4.8. *Characterizations in terms of mixed volumes.* Zonoids and generalized zonoids have characterizations analogous to that of centrally symmetric compact convex sets given in Remark 3.3.3(i). Suppose that K_1 is a fixed smooth centrally symmetric convex body in $\mathbb{E}^n$ and that $1 \leq i \leq n - 1$. Goodey and Zhang [326] prove that a centrally symmetric compact convex set K in $\mathbb{E}^n$ is a zonoid if and only if

$$V(K_1, i; B, n-i-1; K) \leq V(K_2, i; B, n-i-1; K),$$

for all centrally symmetric compact convex K_2 such that $\Pi_i K_1 \subset \Pi_i K_2$. Lemma 4.2.5 gives one direction, but the other direction is more difficult to establish. The result strengthens a theorem of Weil [831] (see also the article [304] of Goodey), and is based on a functional-analytic approach due to Weil and mentioned by Schneider and Weil [745, Theorems 6.1 and 9.8].

Suppose, again, that K_1 is a fixed smooth centrally symmetric convex body in $\mathbb{E}^n$ and that $1 \leq i \leq n - 1$. Goodey and Zhang [326] also prove that a centrally symmetric compact convex set K in $\mathbb{E}^n$ is a generalized zonoid if and only if there is a constant c, depending only on K, such that

$$|V(K_1, i; B, n-i-1; K) - V(K_2, i; B, n-i-1; K)| \le c\,\delta(\Pi_i K_1, \Pi_i K_2),$$

for all compact convex K_2. This strengthens a theorem of Goodey [304] (see also Weil's paper [834]).

A functional-analytic setting for these and other results is given by Goodey, Lutwak, and Weil [312]; see Note 8.8 for commentary.

4.9. *Results related to Aleksandrov's projection theorem.* It was pointed out in Note 3.4 that there is no equivalent of the central symmetral or Blaschke body corresponding to $1 < i < n-1$. This deficiency is one reason for the comparative lack of results for the intermediate values of i. The paper [157] of Chakerian and Lutwak shows that in at least one situation, this can be rectified by the use of zonoids and generalized zonoids. Theorem 4.1.16 and Corollary 4.1.18 can be combined with (A.43) to yield the following result. Suppose that $1 \le i \le k \le n-1$ and that K_1 and K_2 are compact convex sets in $\mathbb{E}^n$. Then

$$V_i(K_1|S) = V_i(K_2|S),$$

for all $S \in \mathcal{G}(n,k)$, if and only if

$$V(K_1, i; B, k-i; C_1, \dots, C_{n-k}) = V(K_2, i; B, k-i; C_1, \dots, C_{n-k}),$$

whenever $C_1, \dots, C_{n-k}$ are centrally symmetric compact convex sets in $\mathbb{E}^n$. (The case $k = n-1$ is the equivalence $(3.2)\Leftrightarrow(3.4)$ of Theorem 3.3.2.) The following result in [157] can be inferred from this and the special case (B.18) of the Aleksandrov–Fenchel inequality. Suppose that $1 \le i \le k \le n-1$ and that K_1 and K_2 are compact convex sets in $\mathbb{E}^n$, of dimension at least $i + 1$, with K_2 centrally symmetric. If

$$V_i(K_1|S) = V_i(K_2|S),$$

for all $S \in \mathcal{G}(n,k)$, then for any l with $1 \le l \le n-k$, we have

$$V_{i+l}(K_1) \le V_{i+l}(K_2),$$

with equality if and only if K_1 is a translate of K_2.

The previous result subsumes several others we have seen. The special case $K_2 = \triangle K_1, i = k = 1, l = n-1$, is Theorem 3.2.3. If $K_2 = \nabla K_1$ and $i = k = n-1$, $l = 1$, we retrieve Theorem 3.3.9. Moreover, Aleksandrov's projection theorem, Theorem 3.3.6, is also implied, since if K_1 is also centrally symmetric, we may reverse the inequality in the conclusion to get equality. It is also interesting to compare Corollary 4.2.7 with the result for $l = n-k$, for which the conclusion $V_{n+i-k}(K_1) \le V_{n+i-k}(K_2)$ is the same. We see that the stronger hypothesis here allows the assumption in Corollary 4.2.7 that K_2 is a zonoid to be weakened considerably.

In addition, as Goodey, Schneider, and Weil [316] observe, we can obtain a positive answer to Problem 3.9 if $a = b = 1$. To see this, let $S \in \mathcal{G}(n,j)$ and apply the above result to obtain $V_j(K_1|S) \le V_j(K_2|S)$. Since equality holds by the hypotheses of Problem 3.9, $K_1|S$ is a translate of $K_2|S$. Then K_1 is a translate of K_2, by Theorem 3.1.3.

4.10. *Smaller bodies with larger projections.* The most interesting special case of the motivating Question 4.2.1 of Section 4.2 was asked by Shephard [762]. The negative answer to *Shephard's problem*, in which $i = n-1$ and K_1 and K_2 are centrally symmetric, was provided independently by Schneider and Petty. Petty's example in Theorem 4.2.4 was published in [662]. (It is interesting that the projected area of a convex double bounded cone is useful in the design of interceptor missile systems;

see the article of Pennell and Deignan [659].) The existence of Schneider's example follows from his original version [727, Satz 3] of the result in Remark 4.2.10, in which $i = n - 1$ and K_1 is of class C_+^{n+2}. (See also Groemer's book [356, Theorem 5.5.14]. Goodey and Zhang [326] prove the stronger form in Remark 4.2.10, by means of their characterization of zonoids mentioned in Note 4.8. In any case, the result requires the surjectivity of the cosine transform; cf. Theorem C.2.2.) The papers of Petty and Schneider both gave the positive answer in the case that K_2 is a zonoid, with Schneider's also proving explicitly Theorem 4.2.6 in its general form. (In fact, further generalizations are indicated in [727].) Theorem 4.2.9 was communicated to the author by P. R. Goodey.

The article [537] and an unpublished manuscript of Lutwak were very useful in the preparation of Section 4.2. In particular, Theorems 4.2.2, 4.2.3, and 4.2.8 are to be found in them. The interesting Theorem 4.2.11 is due to Chakerian and Lutwak [156], and answers a question of Fáry and Makai (see [173, p. 22]).

The existence of the John ellipsoid, Theorem 4.2.12, was proved by John [407], but our proof is taken from Amir's book [9, pp. 46–7]. The cube shows that the factor $\sqrt{n}$ is the best possible. John [407] also observed that for an arbitrary convex body K in $\mathbb{E}^n$, the ellipsoid E of maximal volume contained in K satisfies

$$E \subset K \subset nE.$$

(A proof is also given by Leichtweiss [502]. Palmon [655] proves that the factor n cannot be reduced when, and only when, K is a simplex.) Interesting comments about applications of the John ellipsoid can be found in Berger's survey article [54]. For example, approximation by ellipsoids is now a standard method in linear programming (Khachiyan's polynomial time algorithm) and nonlinear programming (Schor's algorithm); see the references in the works of Gruber [364, p. 338] and Schrijver [749]. The John ellipsoid is also of importance in the local theory of Banach spaces, where distance between two Banach spaces X and Y is often measured by the *Banach–Mazur distance*

$$d(X, Y) = \inf\{\|\phi\|\,\|\phi^{-1}\| : \phi \text{ is a linear isomorphism from } X \text{ to } Y\}.$$

(This is actually a multiplicative distance, so

$$d(X, Z) \le d(X, Y)d(Y, Z),$$

for Banach spaces X, Y, and Z; also, $d(X, Y) = 1$ if and only if X and Y are isometric.) Suppose that $X = (\mathbb{E}^n, K_X)$ and $Y = (\mathbb{E}^n, K_Y)$, where K_X and K_Y are centered convex bodies in $\mathbb{E}^n$, the unit balls in X and Y, respectively. Then $d(X, Y) \le d$ if and only if there is a $\phi \in GL_n$ such that

$$K_Y \subset \phi K_X \subset dK_Y.$$

The existence of the John ellipsoid is therefore equivalent to the statement that if $\dim X = n$, then $d(X, l_2^n) \le \sqrt{n}$ (where $l_2^n = \mathbb{E}^n$, of course). See the surveys of Giannopoulos and Milman [287] and Lindenstrauss and Milman [511] for a wealth of information and references.

The proof of Theorem 4.2.13 is due to R. Schneider (private communication), though the theorem can be found, with an extra factor of $3/2$ on the right-hand side, in the paper [24] of Ball. However, the volume ratio estimate of Remark 4.2.14, showing that this constant can be replaced by $1.16955\ldots$ (and by a number less than one, for $n \ge 4$), even for arbitrary pairs of convex bodies, is also due to Ball [25]. (Note that without the assumption of central symmetry, the proof of Theorem 4.2.13 would result in the much larger factor of n on the right-hand side of the inequality.) The article [27] of Ball includes an excellent introduction to the strong form of John's Theorem 4.2.12 and Ball's own volume-ratio results of Remark 4.2.14 that depend on it. Barthe [38] uses techniques similar to those employed by Ball to prove the following result: Among all convex bodies in $\mathbb{E}^n$ whose John ellipsoid is B, a regular simplex has the maximum mean width.

Ball [24] also shows that there is a $c > 0$ such that for each $n \in \mathbb{N}$, there is a centered convex body K in $\mathbb{E}^n$ (the unit ball of a "random" n-dimensional subspace of l_∞^{2n}, which is $\mathbb{E}^{2n}$ with the sup norm) such that

$$\lambda_{n-1}(K|u^\perp) \geq c\sqrt{n}\lambda_n(K)^{(n-1)/n},$$

for all $u \in S^{n-1}$. Let K_1 be the centered ball of volume one, and let K_2 be a dilatate of K such that K_2 also has volume one. Using (0.8), (0.9), and Stirling's formula, one finds that for large n, a shadow of a ball in $\mathbb{E}^n$ of volume one on a hyperplane has $(n-1)$-dimensional volume close to $\sqrt{e}$. Also, it follows from (0.8) that $\lambda_{n-1}(K_2|u^\perp) \geq c\sqrt{n}$, for all $u \in S^{n-1}$. Consequently, as Ball remarks in [24], this choice of K_1 (expanded slightly) and K_2 provides a very strong negative answer to Shephard's question, for large n.

In Petty's example of Theorem 4.2.4, both K_1 and K_2 are polar bodies of projection bodies, in short, *polar projection bodies*. Zhang [861] notes that his methods yield the existence of polar projection bodies K_1 and K_2 in $\mathbb{E}^n$, $n \geq 3$, for which Shephard's problem has a negative answer. On the other hand, he uses Ball's volume-ratio estimate to show that when K_2 is a polar projection body, the factor $\sqrt{n}$ in Theorem 4.2.13 can be replaced by the constant $4/3$. See Note 9.1 for more on polar projection bodies.

By using techniques from harmonic analysis to study the kernel of the cosine transform on a Grassmann manifold, Goodey and Zhang [327] find a complete answer to Question 4.2.1 for centrally symmetric convex bodies K_1 and K_2. It is shown in [327] that the answer is positive when $K_2 \in \mathcal{K}(n-i)$, where $\mathcal{K}(i)$ is the class of centrally symmetric convex bodies with a positive ith projection generating measure (see Note 4.4). (This is consistent with the case $i = n-1$, since $\mathcal{K}(1)$ is the class of zonoids.) Goodey and Zhang then generalize Remark 4.2.10 by showing that if there is a centrally symmetric convex body of revolution of class C^∞, with $K_1 \notin \mathcal{K}(n-i)$, $2 \leq i \leq n-1$, then there is a centrally symmetric body of revolution K_2 for which the answer to Question 4.2.1 is negative. They show that such a body K_1 can be obtained by an approximation to a convex double bounded cone, so the question under consideration generally has a negative answer for all i with $2 \leq i \leq n-1$.

Koldobsky, Ryabogin, and Zvavitch [467] (see also [468] and Koldobksy's book [465, Chapter 8]) give a treatment of Shephard's problem using the Fourier transform approach (see Note 4.2). Ryabogin and Zvavitch [712] ask whether if $p > 1$ and K_1 and K_2 are centrally symmetric convex bodies in $\mathbb{E}^n$ such that $P_p K_1 \subset P_p K_2$, then $\lambda_n(K_1) \leq \lambda_n(K_2)$ (if $1 < p < n$) and $\lambda_n(K_1) \geq \lambda_n(K_2)$ (if $p > n$). They show that the answer is negative for all $n \geq 2$ but positive if K_2 is an L^p-projection body.

4.11. *Stability results.* Interest in stability results concerning projections is a relatively new phenomenon. In the 1970s V. I. Diskant established some useful theorems, but most of the work we quote directly here began around 1980. A survey of stability in geometric inequalities has been made by Groemer [354] (see especially Section 7 of this article). Several of the results mentioned in this note are also presented in Groemer's book [356].

All the results in Section 4.3 up to Corollary 4.3.9 are taken from the paper [348] of Groemer. As we noted in the text, Theorem 4.3.4 is the stability version of Theorem 3.1.3; historical comments on the latter are given in Note 3.1. (Groemer elaborates on Theorem 4.3.4 in [352], where he obtains upper bounds for $\delta_t(K_1, K_2)$ in terms of the pth mean of the function $\delta_t(K_1|u^\perp, K_2|u^\perp)$, $u \in S^{n-1}$, for $p > 0$.) In Corollaries 4.3.8 and 4.3.9, the constant can be slightly improved for $n > 3$

(cf. [348, Theorem 2 and Corollary 3]). Groemer [351] also published Theorem 4.3.10, inspired by the Problem 4.3 of Chakerian. This question appeared in [153], where Theorem 4.3.11 is also to be found.

Using his approach via hyperplane bundles mentioned in Note 3.1, Groemer [357] finds corresponding stronger forms of the stability results in Section 4.3 up to Corollary 4.3.9. He goes on to obtain estimates of the deviation of a convex body in $\mathbb{E}^3$ from a ball or body of constant width from measurements of the corresponding deviations of their projections on a *finite* set of planes.

Golubyatnikov [298], [299] has obtained stability versions of his results on congruent projections mentioned in Note 3.1.

The instability Theorem 4.3.12 is due to Goodey [305]. In fact, he shows that all the maps Π_i, Π_i^{-1}, $1 \leq i \leq n - 1$, are continuous, but only Π_1 is uniformly continuous. The important and deep stability theorem of Remark 4.3.13, which is naturally rather difficult to prove, is due to Bourgain and Lindenstrauss [91] and Campi [139] (for $n = 3$). Campi's article followed his earlier paper [138] dealing with bodies of revolution, and, like [91], exploited the work of V. I. Diskant. The result in [91] is actually stated in terms of a different metric. For this reason we direct the reader to Groemer's book [356, Theorem 5.5.7], where a slightly stronger version of the result in Remark 4.3.13 is proved in detail.

Another stability theorem, due to Goodey and Groemer [309], can be loosely rephrased to say that if K_1 and K_2 are centered convex bodies in $\mathbb{E}^n$, with controlled inradii and diameters, and $\Pi_1 K_1$ is close to $\Pi_1 K_2$, then K_1 is close to K_2. A similar result was obtained previously for $n = 3$ by Campi [138]. From this it follows that if K_1 and K_2 are arbitrary convex bodies of nearly equal 1-girth in every direction, then K_1 and K_2 have nearly equal width in every direction. Taking one body to be a ball, we see that if the 1-girth of a convex body is nearly constant, then it is of nearly constant width, a stability version of Theorem 3.3.13. This raises the question of whether a convex body of almost constant width must actually be close, in the Hausdorff metric, say, to a convex body of constant width. This is true and is also proved in [309] (Groemer [350] supplies further results of this type).

If $1 < i < n$, the proof of Bourgain and Lindenstrauss [91] can be modified to produce a stability result very similar to that of Remark 4.3.13 where the map Π is replaced by the map Π_i (see Note 4.6). This observation is made by Hug and Schneider [404], who give details and also obtain many other stability results, including some pertaining to the L^p-Brunn–Minkowski theory (see Note 4.7).

4.12. *Reconstruction from brightness functions.* Algorithm BrightLSQ was developed by Gardner and Milanfar in [278], where the proof of convergence, Theorem 4.4.7, can also be found. By combining the supporting analysis with the polynomial-time algorithm of Gritzmann and Hufnagel [340] for reconstruction from surface area measures (see Section A.4) and a refinement of the stability result in Remark 4.3.13, Gardner and Milanfar also describe an oracle-polynomial-time algorithm for reconstruction from brightness functions.

Phase I of Algorithm BrightLSQ was considered earlier, though in a quite different context. Kiderlen [428] (see also [432]) needs to invert the cosine transform in order to estimate the directional distribution of a fiber process from its rose of intersection, a function on the unit sphere giving certain averages of intersections of the process with hyperplanes through the origin. One of his algorithms is a linear program based on Theorem 4.1.13, and this can also be applied to reconstruction from brightness functions. In [278], Kiderlen's algorithm is called Algorithm BrightLP. The input and task for this algorithm are as in Algorithm BrightLSQ. It proceeds in the same way except that in Step 2, (LLS) is replaced by the following linear program (LP):

$$\min_{\alpha} \; \sum_{i=1}^{k} \left(v_K(u_i) - \sum_{j=1}^{l} \alpha_j |u_i \cdot v_j| \right),$$

$$\text{subject to} \quad \sum_{j=1}^{l} \alpha_j |u_i \cdot v_j| \le v_K(u_i), \quad i = 1, \ldots, k$$

$$\text{and} \quad \alpha_j \ge 0, \quad j = 1, \ldots, l.$$

Gardner and Milanfar [278] also propose an algorithm, Algorithm BrightNFacets, designed to reconstruct from brightness-function data an origin-symmetric convex polytope with less than or equal to a prescribed even number N of facets. However, this has not yet been implemented, except in two dimensions, since it involves a constrained *nonlinear* least squares problem. The output of Algorithm BrightLSQ has a number of facets that increases with the number of measurements, but what makes it so effective is that problem (LLS) requires optimizing only over the volumes of the facets and not their outer unit normal vectors. The underlying result that makes this possible is Theorem 4.4.2, due to Campi, Colesanti, and Gronchi [142] (see also Note 9.6), which allows the outer unit normal vectors of the output to be fixed in advance.

Algorithm BrightLSQ was implemented with the assistance of Chris Street, then a Western Washington University (WWU) student. The optimization problem (LLS) is solved by the function "lsqnonneg" from Matlab's Optimization Toolbox. With improvements made by a team of WWU students, Chris Eastman, Thomas Riehle, and Greg Richardson, reconstructions even more ambitious than those shown in Section 4.4 take only a few minutes.

For $n \ge 3$ and general k, it is not known how to minimize the spread Δ_k of directions u_i, $1 \le i \le k$, defined by (4.9), though special cases of this problem, of interest in many applications, are solved. The web page of N. Sloane contains a library of best-known chordal packings of lines at http://www.research .att.com/˜njas/grass/dim3, from which the set of 55 directions used for the reconstruction in Figure 4.8 was taken.

It was pointed out in [278] that all the above algorithms can be run with noisy brightness function measurements as input. However, Algorithm NoisyBrightLSQ is more suitable for this purpose than the corresponding noisy version of Algorithm BrightLP, because although (LP) can in principle be solved faster than (LLS), the first constraint in (LP) may then cause solutions to degenerate, even to a single point. In [274], Gardner, Kiderlen, and Milanfar employ techniques from the theory of empirical processes and new entropy estimates to prove a version of Theorem 4.4.7 for Algorithm NoisyBrightLSQ. They show that if the sequence $\{u_i\}$ of directions is evenly spread (see Note 3.8), then the outputs from Algorithm NoisyBrightLSQ corresponding to input directions u_i, $1 \le i \le k$, converge almost surely to K in the Hausdorff metric as $k \to \infty$. Moreover, rates of convergence depending on k are estimated.

Algorithm BrightLSQ shows that the process is divided into two phases: In Phase I, the surface area measure is obtained from the brightness function, and in Phase II, a suitable representation of the body, essentially equivalent to the support function, is obtained from the surface area measure. Ryabogin and Zvavitch [712] find a formula for the support function of a suitably smooth convex body of revolution in terms of its surface area measure.

4.13. *Hermann Minkowski (1864–1909)*. (Sources: [185], [687].) Hermann Minkowski was born in Alexotas, Russia (now Lithuania), to German parents. The family had to leave, since Jewish children were not allowed to attend Russian schools. On arriving in Königsberg (now Kaliningrad, Russia) when Minkowski was eight years old, his father became a rag merchant, but ensured a good education for his gifted children. (One of Minkowski's brothers, Oskar, became a famous pathologist, best known for his discovery and exploration of the role of the pancreas in diabetes.) Minkowski received most of his education in Königsberg, where he obtained his doctorate in 1885, and where he became a friend of fellow student David Hilbert. He held positions at Bonn and Zurich before taking up the professorship at Göttingen created for him by Hilbert in 1902. In 1897 he married and had two children. His untimely death followed a sudden attack of appendicitis.

Minkowski's mathematical genius is beyond question. At the age of 18, he wrote a 140-page paper on quadratic forms, which jointly with a paper of H. J. S. Smith won the Grand Prix des Sciences Mathématiques of 1883, awarded by the Paris Academy of Sciences. (Smith, a professor at Oxford, had actually solved the problem

Figure 4.9. Hermann Minkowski

set by the Academy 30 years earlier. Minkowski shared the prize despite not having had time to have his paper translated into French.) Quadratic forms continued to inspire much of Minkowski's later work. In particular, it was the study of a polyhedral fundamental region associated with such forms which led him to found the geometry of numbers as a distinct branch of number theory. He proved that a centered convex body in $\mathbb{E}^n$ of volume at least 2^n must contain a pair $\pm x \neq o$ of integer lattice points, a result of enormous importance in Diophantine approximation. Becoming interested in convexity for its own sake, he was led to emphasize the significance of supporting planes, to show that convex bodies in $\mathbb{E}^3$ are the convex hulls of their extreme points, to introduce support functions, and to discover the theory of mixed volumes (cf. Appendix A) and associated inequalities (cf. Appendix B). Even before metric spaces were introduced, Minkowski understood that each centered convex body can be used to define a new distance, with its corresponding Minkowski geometry (see Section 0.8 and Note 4.1). The construction of a coherent theory of convexity was not only important for geometry, but also cleared the way for the development of functional analysis. Minkowski was always interested in physics, and published papers on capillarity and electrodynamics. In the last years of his life, Minkowski realized that the principles of relativity formulated by H. Lorentz and Einstein result in a union of space and time in 4-dimensional space-time, which he also defined mathematically, creating the geometric setting for Einstein's subsequent general theory of relativity.

Considerable insight into Minkowski's personality can be gained through Reid's book [687] on Hilbert, which contains a moving portrayal of the deep and constant friendship between Minkowski and Hilbert as they progressed from students to leading mathematicians of their day. Minkowski's friendly charm and humor shine through a basic shyness, which caused Einstein to give up attending Minkowski's class in Zurich. (In 1905, hearing of Einstein's work, Minkowski said, "Oh, that Einstein, always missing lectures – I really would not have believed him capable of it!") Announcing before a class that he could solve the four color problem, where "mathematicians of the third class" had failed, Minkowski admitted to the class some weeks later, after a thunderclap on a rainy day, "Heaven is angered by my arrogance; my proof of the four color theorem is also defective." The four color problem, of course, had to wait until 1976 for a solution which made unprecedented use of computer calculation.

5

Point X-rays

In Chapters 1 and 2 we considered parallel X-rays, that is, X-rays taken from infinity, of bounded Lebesgue measurable sets. In this chapter and the next one we turn our attention to X-rays emanating from finite points. This corresponds to the "fan-beam" X-rays of great importance in medicine; in fact, modern CAT scanners use only this type of X-ray.

It seems that even basic results on point X-rays of planar convex bodies require some knowledge of measure theory, so in this chapter, this is assumed from the outset. Two types of point X-ray and the related chordal symmetrals are introduced in the first section, in the context of measurable sets in $\mathbb{E}^n$. The second section contains some lemmas also useful in the next chapter. These could be skipped at first, and read when encountered again in Section 5.3, which deals only with point X-rays of planar convex bodies.

An X-ray (or directed X-ray) of a planar convex body at a point gives the lengths of all the intersections of the body with lines through the point (or rays issuing from the point, respectively). The main result here is Theorem 5.3.8, stating that any set of four points in general position in the plane have the property that the X-rays at these points will distinguish between any two convex bodies. For directed X-rays, three noncollinear points will suffice for this purpose, as Theorem 5.3.6 demonstrates. It is unknown whether two points are enough, except in special circumstances requiring a priori the position or nature of the body.

The last section of the chapter concerns point X-rays of convex bodies in the projective plane. To avoid too much repetition, the proofs here are not written in such great detail, and two are even omitted, since the techniques are much the same as those in Chapter 1 and Section 5.3.

Throughout, the phrase "measurable set in $\mathbb{E}^n$" always refers to a bounded λ_n-measurable set, and in the last two sections, "measurable set" means a bounded λ_2-measurable subset of $\mathbb{E}^2$.

5.1. Point X-rays and chordal symmetrals

Suppose that we take a point X-ray of a set E in $\mathbb{E}^n$, $n \geq 2$. Each beam of the X-ray issues from a point p and, ideally, the information we collect is the total length of the intersection with E of each ray originating from p. We can think of this as a function on the unit sphere S^{n-1}, formally defined as follows.

Definition 5.1.1. Let p be a point and let E be a bounded measurable set in $\mathbb{E}^n$, $n \geq 2$. The *directed X-ray of E at p* is defined for λ_{n-1}-almost all $u \in S^{n-1}$ by

$$D_p E(u) = D_p 1_E(u) = \lambda_1 \big(E \cap (r_u + p) \big),$$

where r_u is the ray emanating from o in the direction u.

In the definition, we are equating the directed X-ray of E with the divergent beam transform (cf. (C.5)) of the characteristic function 1_E of E, but this will not be used directly here. The fact that $D_p E$ is defined for λ_{n-1}-almost all $u \in S^{n-1}$ is a consequence of Fubini's theorem; when E is a Borel set, $D_p E$ is defined everywhere on S^{n-1}. Usually we shall avoid technicalities by assuming that E is a body star-shaped at p, so that each ray emanating from p that meets E does so in a (possibly degenerate) line segment. In this case, the directed X-ray simply gives the length of each such line segment.

It turns out that the concept of a directed X-ray at a point, although useful, is not as interesting mathematically as the related notion defined in the following way.

Definition 5.1.2. Let p be a point and let E be a bounded measurable set in $\mathbb{E}^n$, $n \geq 2$. The *X-ray of E at p* is defined for λ_{n-1}-almost all $u \in S^{n-1}$ by

$$X_p E(u) = X_p 1_E(u) = \lambda_1 \big(E \cap (l_u + p) \big),$$

where l_u is the line through o parallel to u.

The definition relates to a well-known integral transform, the line transform (cf. (C.6)), but again, this fact will not be important for the sequel. As with the directed X-ray, the X-ray of a Borel set at a point p is defined everywhere on S^{n-1}. With the X-ray, however, we regard the rays issuing from p in both the directions u and $-u$ as a single beam, so when E is star-shaped at p, $X_p E$ is the function on S^{n-1} giving the length of the line segment in which each line through p intersects E.

We sometimes use the term *point X-ray* as a convenient way of referring to an X-ray or directed X-ray at a point. In the planar case, it is often convenient to think of these as functions of the polar coordinate angle θ, where $0 \leq \theta < 2\pi$.

The X-ray carries less information than the directed X-ray. For example, the congruent disks shown in Figure 5.1 have the same X-rays at the point p, but

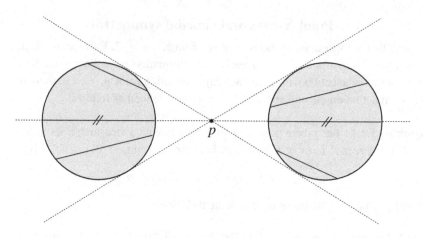

Figure 5.1. Disks with equal X-rays.

different directed X-rays. Note that in general $X_pE(u) = X_pE(-u)$, for each $u \in S^{n-1}$, so X_pE is even, whereas D_pE is not necessarily an even function.

In Chapters 1 and 2 we saw that in geometric tomography the parallel X-ray of a convex body can be identified with its corresponding Steiner symmetral. A similar identification is possible for point X-rays, with symmetrals defined with respect to a point, rather than a hyperplane.

Definition 5.1.3. Suppose that p is a point and that E is a set in $\mathbb{E}^n$, $n \geq 2$. For each $u \in S^{n-1}$, define $c(u)$ as follows. If $E \cap (l_u + p) = \emptyset$, or if it is not λ_1-measurable, let $c(u) = \emptyset$. Otherwise, let $c(u)$ be the (possibly degenerate) closed line segment parallel to u with center p and length $\lambda_1(E \cap (l_u + p))$. The union of all the line segments $c(u)$ will be denoted by $\tilde{\triangle}_pE$ and called the *chordal symmetral of E at p*. If $p = o$, we drop the suffix and simply refer to $\tilde{\triangle}E$ as the *chordal symmetral of E*.

By essentially the same argument as for the Steiner symmetral, the centrally symmetric set $\tilde{\triangle}_pE$ can be shown to be compact if E is, and hence, by approximation, λ_n-measurable if E is. However, we are really only interested in the special case when E is a body star-shaped at p; $\tilde{\triangle}_pE$ is the set whose intersection with each line through p is a centered line segment congruent to that in which the line intersects E.

For a body E star-shaped at p, the X-ray X_pE determines the chordal symmetral $\tilde{\triangle}_pE$, and vice versa. When E is a measurable set, $\tilde{\triangle}_pE$ still determines X_pE, but knowledge of X_pE only specifies $\tilde{\triangle}_pE$ up to the singleton $\{o\}$.

It is not generally true that $\tilde{\triangle}_pE$ is a body whenever E is a body star-shaped at p. For example, let K be the closed semidisk in $\mathbb{E}^2$ formed by intersecting the unit disk with the upper closed half-plane. Then $\tilde{\triangle}K$ is the union of the centered disk of radius $1/2$ and the closed line segment with endpoints $(\pm 1, 0)$. This is not a regular set, so $\tilde{\triangle} K$ is not a body. When K is a convex body such that $p \in \operatorname{int} K$ or $p \notin K$, however, $\tilde{\triangle} K$ is a body.

Suppose that E is a body star-shaped at o. To construct $\tilde{\triangle} E$, one slides each chord of E lying on a line through o until it is bisected by o, and takes the union of the new chords obtained in this way. Note that $\tilde{\triangle} E = E$ if and only if E is centered, so the only interesting examples are for noncentered bodies. Figure 5.2 depicts the chordal symmetrals of an equilateral triangle and a regular tetrahedron with centroid at the origin, showing that even when K is convex and $o \in \operatorname{int} K$, $\tilde{\triangle} K$ may not be convex. (For the tetrahedron, the chordal symmetral is a curved version of the rhombic dodecahedron (cf. Figure 4.1).)

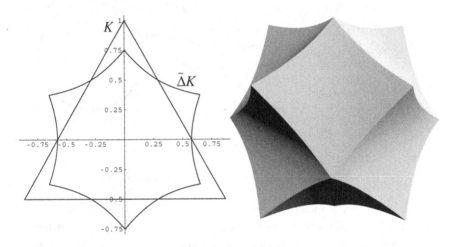

Figure 5.2. Chordal symmetrals of an equilateral triangle and a regular tetrahedron.

The chordal symmetral is the concept dual to that of the central symmetral of Definition 3.2.2. To explain this, we direct the reader's attention to Section 0.7, where the radial function is defined. The radial function ρ_E of a body E star-shaped at o, when evaluated at a unit vector u, gives the signed distance along the line l_u from o to the boundary of E (see Figure 0.3). Then

$$\rho_{\tilde{\triangle} E}(u) = \frac{1}{2}\left(\rho_E(u) + \rho_{-E}(u)\right),$$

for each $u \in S^{n-1}$, whereas

$$h_{\triangle K}(u) = \frac{1}{2}\left(h_K(u) + h_{-K}(u)\right),$$

for each $u \in S^{n-1}$, gives the central symmetral $\triangle K$ of a convex body K in terms of its support function. Alternatively, one can compare Definition 3.2.2 with the observation that if $o \in \text{int } E$, then

$$\tilde{\triangle} E = \frac{1}{2}(E \tilde{+} (-E)),$$

where $\tilde{+}$ denotes radial addition (see (0.30)).

For a convex body K, $\tilde{\triangle}_p K$ is never convex when $p \notin K$, but rather (in the plane, at least) figure-eight shaped, consisting of the union of a body and its reflection in p, as in Figure 5.3. It is interesting, however, that each half of $\tilde{\triangle}_p K$ is convex, as we shall show.

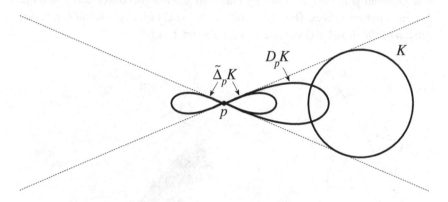

Figure 5.3. Chordal symmetral and directed chordal symmetral of a disk.

This motivates a modified version of $\tilde{\triangle}_p E$, in which we follow Definition 5.1.3, but replace lines through p by rays with endpoint p. Let us also denote the resulting set by $D_p E$, justifying this by observing that this set has the same directed X-ray as E at p. We call the set $D_p E$ the *directed chordal symmetral* of E at p. When E is star-shaped at p and $p \in E$, then $D_p E = E$. If K is convex and $p \notin K$, $D_p K$ is just one of the halves of $\tilde{\triangle}_p K$, magnified by a factor of 2, and it is also convex (see Figure 5.3). The proof of this fact requires the following lemma, referred to again in subsequent chapters.

Lemma 5.1.4. *Let E be a body in $\mathbb{E}^n$ star-shaped at o. Then E is convex if and only if*

$$\rho_E(u + v)^{-1} \le \rho_E(u)^{-1} + \rho_E(v)^{-1},$$

for all $u, v \in S^{n-1}$ such that $\rho_E(u)$ and $\rho_E(v)$ are either both positive or both negative.

Proof. Note that E is convex if all its sections by 2-dimensional subspaces are convex. Indeed, given points $x, y \in E$, we can find a 2-dimensional subspace S containing them; then, if $E \cap S$ is convex, we have $[x, y] \subset E \cap S \subset E$. This means that we can assume that $n = 2$.

As is explained in Section 0.7, the radial function of a body can be defined for all $x \in \mathbb{E}^n \setminus \{o\}$ by virtue of (0.29), from which it is positively homogeneous of degree -1. Note that we are using this extended radial function in the statement of the lemma, since $u + v$ is generally not a unit vector.

By our assumption that E is a body star-shaped at o, E is convex if and only if for all x, $y \in \mathrm{bd}\, E$ such that $\rho_E(x)$ and $\rho_E(y)$ are either both positive or both negative, the line segment $[x, y]$ is contained in E. This is equivalent to the condition that for all u, $v \in S^1$ such that $\rho_E(u)$ and $\rho_E(v)$ are either both positive or both negative, the point $z \in [\rho_E(u)u, \rho_E(v)v]$ with direction $w = (u + v)/\|u + v\|$ is contained in E. We shall assume that $\rho_E(u)$ and $\rho_E(v)$ are both positive; the case when they are both negative can be dealt with in a similar fashion.

For some t with $0 \le t \le 1$, we have

$$\frac{\|z\|(u+v)}{\|u+v\|} = z = (1-t)\rho_E(u)u + t\rho_E(v)v,$$

giving

$$t = \frac{\rho_E(u)}{\rho_E(u) + \rho_E(v)} \quad \text{and} \quad \|z\| = \frac{\rho_E(u)\rho_E(v)\|u+v\|}{\rho_E(u) + \rho_E(v)}.$$

The point z is contained in E if and only if it lies on the line segment $[-\rho_E(-w)w, \rho_E(w)w]$. Since our assumptions imply that $-\rho_E(-w) \le \|z\|$, this occurs if and only if

$$\|z\| \le \rho_E(w) = \rho_E\left(\frac{u+v}{\|u+v\|}\right) = \|u+v\|\rho_E(u+v).$$

Substituting the expression for $\|z\|$ just obtained, we get

$$\frac{\rho_E(u)\rho_E(v)}{\rho_E(u) + \rho_E(v)} \le \rho_E(u+v),$$

as required. ∎

Theorem 5.1.5. *If K is a convex body in $\mathbb{E}^n$, then $D_p K$ is also a convex body.*

Proof. We may assume that $p = o$. If $o \in K$, then $D_o K = K$, so suppose that $o \notin K$. Then

$$\rho_{D_o K}(u) = \rho_K(u) - |\rho_K(-u)| = \rho_K(u) - |\rho_{-K}(u)|,$$

for each $u \in S^{n-1}$ such that $\rho_K(u) > 0$. Suppose that $\rho_K(u) > 0$ and that $\rho_K(v) > 0$. By Lemma 5.1.4,

$$\rho_K(u+v)^{-1} \le \rho_K(u)^{-1} + \rho_K(v)^{-1}$$

and

$$|\rho_{-K}(u+v)^{-1}| \ge |\rho_{-K}(u)^{-1}| + |\rho_{-K}(v)^{-1}|.$$

The inequality

$$(c_1^{-1} + d_1^{-1})^{-1} - (c_2^{-1} + d_2^{-1})^{-1} \geq ((c_1 - c_2)^{-1} + (d_1 - d_2)^{-1})^{-1}$$

holds whenever c_j and d_j are positive real numbers with $c_1 \geq c_2$ and $d_1 \geq d_2$; this is just a rearrangement of Minkowski's inequality (B.4) with $p = -1$. Using this, we obtain

$$\rho_{D_oK}(u + v) = \rho_K(u + v) - |\rho_{-K}(u + v)|$$
$$\geq \left(\rho_K(u)^{-1} + \rho_K(v)^{-1}\right)^{-1} - \left(|\rho_{-K}(u)^{-1}| + |\rho_{-K}(v)^{-1}|\right)^{-1}$$
$$\geq \left((\rho_K(u) - |\rho_{-K}(u)|)^{-1} + (\rho_K(v) - |\rho_{-K}(v)|)^{-1}\right)^{-1}$$
$$= \left(\rho_{D_oK}(u)^{-1} + \rho_{D_oK}(v)^{-1}\right)^{-1}.$$

In view of Lemma 5.1.4, this completes the proof.   ∎

We observed before that the X-ray X_pE and chordal symmetral $\tilde{\Delta}_pE$ of a measurable set E determine each other up to the singleton $\{o\}$. A similar relationship holds between the directed X-ray and directed chordal symmetral. Since they carry essentially the same information, each point X-ray and its corresponding chordal symmetral can be identified for our purposes, and we make this identification in the definitions to follow. The symbol $\simeq$ denotes, as usual, equivalence up to a set of measure zero; so, for example, $X_pE \simeq X_pE'$ means that $X_pE(u) = X_pE'(u)$ for λ_{n-1}-almost all $u \in S^{n-1}$.

Definition 5.1.6. Let $\mathcal{E}$ be a class of bounded measurable sets and P a fixed set of points in $\mathbb{E}^n$. We say that $E \in \mathcal{E}$ is *determined* by the X-rays (or directed X-rays) at the points in P if whenever $E' \in \mathcal{E}$ and $X_pE \simeq X_pE'$ (or $D_pE \simeq D_pE'$, respectively) for all $p \in P$, then $E \simeq E'$.

As in Chapter 2, we augment this definition by saying that sets in $\mathcal{E}$ are *determined by m point X-rays* if there is a set P of m points such that each set in $\mathcal{E}$ is determined by the X-rays at the points in P.

Definition 5.1.7. Let $\mathcal{E}$ be a class of bounded measurable subsets of $\mathbb{E}^n$. We say that $E \in \mathcal{E}$ can be *verified* by the X-rays (or directed X-rays) at a set P of points if P can be chosen (depending on E) such that if $E' \in \mathcal{E}$ and $X_pE \simeq X_pE'$ (or $D_pE \simeq D_pE'$, respectively) for all $p \in P$, then $E \simeq E'$.

As with parallel X-rays, then, the process of determination involves a set of points specified in advance, whereas verification allows the points to vary with the particular set being X-rayed.

5.2. The X-ray of order i

The definition of the X-ray of a body E star-shaped at a point p can be reformulated in terms of its radial function ρ_E. We have, for each $u \in S^{n-1}$,

$$
X_p E(u) = \begin{cases} \rho_{E-p}(u) + \rho_{E-p}(-u) & \text{if } p \in E, \\ \big| |\rho_{E-p}(u)| - |\rho_{E-p}(-u)| \big| & \text{if } p \notin E. \end{cases}
$$

In the next chapter, we shall define the i-chord function of E, in which the terms in this alternative definition of $X_p E$ (for $p = o$) are raised to the power of a real number i. The corresponding generalizations of certain preliminary lemmas for point X-rays are also required, and to avoid duplication, we shall derive them in the general form in this section. Only the case $i = 1$ is necessary for the next two sections of this chapter, however.

As a tool for proving some of these lemmas, it is convenient to have a corresponding generalization of the X-ray of a bounded measurable set at a point. This runs as follows.

Definition 5.2.1. Let p be a point and E a bounded measurable set in $\mathbb{E}^n$, $n \geq 2$, and suppose that $i \in \mathbb{R}$. The *X-ray of order i of E at p* is defined by

$$
X_{i,p} E(u) = X_{i,p} 1_E(u) = \int_{-\infty}^{\infty} 1_E(p + tu)|t|^{i-1}\, dt,
$$

for $u \in S^{n-1}$ for which the integral exists.

Note that $X_{i,p}$ is also the line transform of order i (see (C.9)); here it is just being applied to characteristic functions of measurable sets. When $i = 1$, we retrieve Definition 5.1.2 of the X-ray of E at p.

The next lemma provides an opportunity to understand this definition. We are taking $p = o$ for convenience of notation.

Lemma 5.2.2. Let E be a body in $\mathbb{E}^n$ star-shaped at o. Suppose either that $o \notin E$ or that $i \in \mathbb{R}$ is positive. If $i \neq 0$, then for each $u \in S^{n-1}$,

$$
X_{i,o} E(u) = \begin{cases} \dfrac{1}{i}\left(\rho_E(u)^i + \rho_E(-u)^i\right) & \text{if } o \in E, \\[2mm] \dfrac{1}{i}\big| |\rho_E(u)|^i - |\rho_E(-u)|^i \big| & \text{if } o \notin E. \end{cases}
$$

If $i = 0$ and $o \notin E$, then for each $u \in S^{n-1}$,

$$
X_{0,o} E(u) = \big| \log|\rho_E(u)/\rho_E(-u)| \big|.
$$

Proof. Suppose that $o \notin E$. Assume that $\rho_E(u) \geq -\rho_E(-u) > 0$ (the case when $\rho_E(-u) \geq -\rho_E(u) > 0$ is handled similarly). Then

$$X_{i,o}E(u) = \int_0^\infty 1_E(tu)t^{i-1}\,dt$$

$$= \int_{|\rho_E(-u)|}^{\rho_E(u)} t^{i-1}\,dt$$

$$= \begin{cases} \frac{1}{i}\left||\rho_E(u)|^i - |\rho_E(-u)|^i\right| & \text{if } i \neq 0, \\ \left|\log|\rho_E(u)/\rho_E(-u)|\right| & \text{if } i = 0. \end{cases}$$

Now suppose that $i > 0$ and that $o \in E$. For each $u \in S^{n-1}$,

$$X_{i,o}E(u) = \int_0^\infty 1_E(tu)t^{i-1}\,dt + \int_0^\infty 1_E(-su)s^{i-1}\,ds$$

$$= \int_0^{\rho_E(u)} t^{i-1}\,dt + \int_0^{\rho_E(-u)} s^{i-1}\,ds$$

$$= \frac{1}{i}\left(\rho_E(u)^i + \rho_E(-u)^i\right). \qquad \blacksquare$$

The reader may recall from Chapters 1 and 2 that Lebesgue measure, via the Cavalieri principle (Lemma 1.2.2, or, in more general form, Lemma 2.1.5), has the appropriate invariance properties for application to parallel X-rays. This invariance is unfortunately lost when the X-rays issue from a finite point, and one has to substitute a new measure; in fact, for X-rays of order i, a different measure for each i. We shall confine ourselves to two dimensions, since this is all we need here.

Definition 5.2.3. Let $i \in \mathbb{R}$. For a bounded measurable set E in $\mathbb{E}^2$ and a line t chosen as the x-axis of a Cartesian coordinate system, let

$$v_i(E) = \iint_E |y|^{i-2}\,dx\,dy.$$

The line t will be called the *base line* of v_i.

Observe that $v_2 = \lambda_2$. If $i > 1$, then $v_i(E)$ is always finite, but if $i \leq 1$, then v_i is a σ-finite measure in $\mathbb{E}^2$, finite on bounded sets bounded away from t. In fact, even for $i \leq 1$, v_i is sometimes finite on sets not bounded away from t, as the next lemma shows.

Lemma 5.2.4. *Let $i \in \mathbb{R}$ be positive. The triangle $T = \{(x, y) : a|x - x_0| \leq y \leq b\}$, for $a > 0$, has finite v_i-measure.*

Proof. This is easily seen by computing the integral in polar coordinates centered at the point $(x_0, 0)$. $\qquad \blacksquare$

The next two useful lemmas are analogues of Lemmas 1.2.2 and 1.2.3.

Lemma 5.2.5. *Let $i \in \mathbb{R}$. Suppose that E_1 and E_2 are measurable subsets of $\mathbb{E}^2$ with finite v_i-measure and the same X-rays of order i at $p = (x_0, 0)$. Then $v_i(E_1) = v_i(E_2)$.*

Proof. For $j = 1, 2$, we have, using polar coordinates centered at p,

$$
\begin{aligned}
v_i(E_j) &= \iint_{E_j} |y|^{i-2} \, dx \, dy \\
&= \int_0^\pi \int_{E_j \cap (l_\theta + p)} |r|^{i-1} |\sin \theta|^{i-2} \, dr \, d\theta \\
&= \int_0^\pi X_{i,p} E_j(\theta) |\sin \theta|^{i-2} \, d\theta,
\end{aligned}
$$

where $l_\theta + p$ is the line through p at angle θ. Since $X_{i,p} E_1 = X_{i,p} E_2$, the result follows. ∎

Lemma 5.2.6. *Let $i \in \mathbb{R}$. Suppose that E_1 and E_2 are measurable subsets of $\mathbb{E}^2$, contained in the upper open half-plane and of finite positive v_{i-1}-measure, with equal X-rays of order i at the point $p = (x_0, 0)$. Then the centroids of E_1 and E_2, with respect to the measure v_{i-1}, lie on the same line through p.*

Proof. Let c_j be the centroid of E_j, $j = 1, 2$, with respect to the measure v_{i-1}. Then, by (0.15), $c_j = (x_j, y_j)$, where

$$
x_j = \frac{1}{v_{i-1}(E_j)} \iint_{E_j} x y^{i-3} \, dx \, dy
$$

and

$$
y_j = \frac{1}{v_{i-1}(E_j)} \iint_{E_j} y^{i-2} \, dx \, dy = \frac{v_i(E_j)}{v_{i-1}(E_j)},
$$

for $j = 1, 2$. (The assumption that $v_{i-1}(E_j)$ is finite, $j = 1, 2$, ensures that the integrals are finite, and in particular that $v_i(E_j)$ is finite, $j = 1, 2$.) Using polar coordinates centered at p as in the previous lemma, we get

$$
\frac{y_j}{x_j} = v_i(E_j) \Big/ \int_0^\pi X_{i,p} E_j(\theta) \cos \theta \sin^{i-3} \theta \, d\theta,
$$

for $j = 1, 2$. Using Lemma 5.2.5, the equality of the X-rays of order i implies that $(y_1/x_1) = (y_2/x_2)$, completing the proof. ∎

The next lemma shows how the position of p allows us to derive, from the assumption that E_1 and E_2 have equal X-rays of order i at p, information about their v_i-measures. We refer the reader to Figure 5.4.

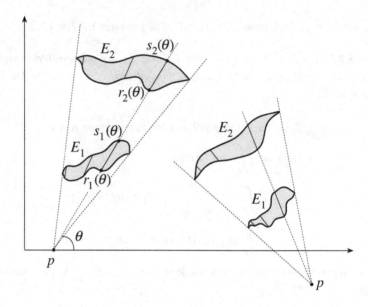

Figure 5.4. Star-shaped sets.

Lemma 5.2.7. *Let $i \in \mathbb{R}$. Suppose that $p = (x_0, y_0)$, and let (r, θ) denote polar coordinates centered at p. Let $0 \le \alpha < \beta \le \pi$, and let*

$$E_j = \{(r, \theta) : 0 < r_j(\theta) \le r \le s_j(\theta), \ \alpha \le \theta \le \beta\},$$

$j = 1, 2$, be bodies in $\mathbb{E}^2$ star-shaped at p, with equal X-rays of order i at p. Suppose also that $r_1(\theta) \le r_2(\theta)$ for $\alpha \le \theta \le \beta$. Then

(i) *if $y_0 = 0$ and $k > \max\{i, 1\}$, then $v_k(E_1) < v_k(E_2)$, if $i \ge 0$, and $v_k(E_1) > v_k(E_2)$, if $i < 0$;*

(ii) *if $y_0 < 0$, and E_1 has finite positive v_i-measure, then $v_i(E_1) < v_i(E_2)$, if $i > 2$, and $v_i(E_1) > v_i(E_2)$, if $i < 2$; and if $y_0 > 0$, these inequalities are reversed.*

Proof. Using polar coordinates centered at p, we have, for $j = 1, 2$,

$$v_i(E_j) = \int_\alpha^\beta \int_{r_j(\theta)}^{s_j(\theta)} (r \sin \theta + y_0)^{i-2} r \, dr \, d\theta. \tag{5.1}$$

By Lemma 5.2.2, we see that since E_1 and E_2 have the same X-rays of order i at p, and do not contain p, either

$$s_1(\theta)^i - r_1(\theta)^i = s_2(\theta)^i - r_2(\theta)^i$$

or

$$s_1(\theta)/r_1(\theta) = s_2(\theta)/r_2(\theta)$$

holds for $\alpha \le \theta \le \beta$, according as $i \ne 0$ or $i = 0$, respectively. With these expressions in hand we attack each part of the lemma in turn.

(i) If $i \ne 0$, we make the substitution $t = r^i$ into the equation resulting from (5.1) on replacing i by k, to obtain

$$v_k(E_j) = \int_\alpha^\beta \int_{r_j(\theta)^i}^{s_j(\theta)^i} \frac{1}{i} t^{(k-i)/i} \sin^{k-2}\theta \, dt \, d\theta.$$

The integrand increases with t, since $k - i > 0$, and $v_k(E_1)$ is finite, because $k \ge 2$.

If $i = 0$, we instead substitute $t = \log r$ into the equation resulting from (5.1) on replacing i by k. This gives

$$v_k(E_j) = \int_\alpha^\beta \int_{\log r_j(\theta)}^{\log s_j(\theta)} e^{kt} \sin^{k-2}\theta \, dt \, d\theta,$$

and again the integrand increases with t.

In both cases the range of the inner integrals is of the same length for $j = 1, 2$, so if $i \ge 0$, $v_k(E_1) < v_k(E_2)$, as required. If $i < 0$, however, we have $s_j(\theta)^i \le r_j(\theta)^i$, for $j = 1, 2$, so by interchanging the limits of the inner integral we see that in this case $v_k(E_1) > v_k(E_2)$.

(ii) If $i \ne 0$, the same substitutions in (5.1) yield

$$v_i(E_j) = \int_\alpha^\beta \int_{r_j(\theta)^i}^{s_j(\theta)^i} \frac{1}{i} (t^{1/i} \sin\theta + y_0)^{i-2} t^{(2-i)/i} \, dt \, d\theta, \qquad (5.2)$$

for $j = 1, 2$. If $i = 0$, we substitute $t = \log r$ in (5.1) instead, giving

$$v_0(E_j) = \int_\alpha^\beta \int_{\log r_j(\theta)}^{\log s_j(\theta)} (e^t \sin\theta + y_0)^{-2} e^{2t} \, dt \, d\theta, \qquad (5.3)$$

for $j = 1, 2$.

The derivative with respect to t of the integrand in (5.2) is

$$-\frac{(i-2)}{i^2} y_0 (t^{1/i} \sin\theta + y_0)^{i-3} t^{(2-2i)/i}.$$

Suppose that $y_0 < 0$. If $i > 2$, the integrand increases with t, so using the equality of the X-rays of order i as before yields $v_i(E_1) < v_i(E_2)$. If $0 < i < 2$, the integrand decreases with t, so $v_i(E_1) > v_i(E_2)$. If $i < 0$, we have $s_j(\theta)^i \le r_j(\theta)^i$, for $j = 1, 2$, so by interchanging the limits of the inner integral, we see that $v_i(E_1) > v_i(E_2)$ remains true. The latter also holds when $i = 0$, since the derivative of the integrand in (5.3) is

$$2 y_0 e^{2t} (e^t \sin\theta + y_0)^{-3},$$

implying that the integrand decreases with t. The case when $y_0 > 0$ is dealt with similarly. ∎

5.3. Point X-rays of planar convex bodies

All sets in this section are subsets of the plane $\mathbb{E}^2$.

A uniqueness theorem for the divergent beam and line transforms, Theorem C.1.1, states that (under a mild restriction) any object, even with varying density, is uniquely determined by an infinite set of point X-rays. On the other hand, no finite set of point X-rays suffice; this is a consequence of Theorem 7.3.1. The successful use of the fan-beam X-ray in computerized tomography depends on the reconstruction of approximate images. Our goal is to obtain, for certain sets of uniform density, some uniqueness theorems for finite sets of point X-rays.

In this section the main program is to find finite sets of points in the plane such that the corresponding X-rays determine convex bodies, in the sense of Definition 5.1.6. We already know from Section 5.1 that at least two point X-rays will be required, since if p is a point not in a convex body K, the different convex body D_pK has the same directed X-ray as K at p.

In Chapter 2, we declared that two measurable sets are regarded as having equal parallel X-rays if these agree almost everywhere, and we noted that if the sets are bodies, then "almost everywhere" is equivalent to "everywhere." For point X-rays, unfortunately, this convenient state of affairs is upset by an example mentioned in the previous section. The upper half of the unit disk and a centered disk of radius $1/2$ are convex bodies with X-rays at the origin equal almost everywhere, but not everywhere.

In most situations, however, this cannot occur. Let us agree to write $X_pK_1 \simeq X_pK_2$ when $X_pK_1(u) = X_pK_2(u)$ for λ_1-almost all $u \in S^1$, as in Section 5.1. If K_1 and K_2 are convex bodies such that $p \in \text{int}\, K_j$, $j = 1, 2$, or $p \notin K_j$, $j = 1, 2$, and if $X_pK_1 \simeq X_pK_2$, then $X_pK_1 = X_pK_2$. Observe also that if two convex bodies have directed X-rays that are equal almost everywhere then they must be equal everywhere. In these cases, therefore, we can replace equality almost everywhere by exact equality, and we shall do this without further comment. Nevertheless, one must pay attention to the difference between "equal" and "equal almost everywhere" when working with X-rays at a point in the boundary of a body.

We need some notation. Suppose that K and K' are convex bodies with equal X-rays at $p \notin K \cup K'$ and that C is a component of $\text{int}\,(K \triangle K')$. Let

$$C' = \cup\{t \cap \text{int}\,(K \triangle K')\} \setminus C,$$

where the union is taken over all rays t issuing from p such that $t \cap C \neq \emptyset$. If C is nearer to p than C', we shall write $pC = C'$, whereas if C' is nearer to p, we write $p^{-1}C = C'$; in this way, either pC or $p^{-1}C$ is defined, but not both. Note that if K and K' have the same X-rays at p, then so do the components C and C'. The same is true for directed X-rays, provided p does not belong to K or K'.

The first theorem concerns directed X-rays. In the proof, we use the measure from Definition 5.2.3 and some of the lemmas from the previous section. Here, and for the remainder of this chapter, we only need these for $i = 1$, however.

Theorem 5.3.1. *Suppose that K is a convex body and that p_1, p_2 are distinct points such that the line l through p_1 and p_2 intersects K. If K' is a convex body with the same directed X-rays as K at p_1 and p_2, then $K = K'$.*

Proof. Assume that $K' \neq K$. Since K and K' have the same directed X-rays at p_1 and p_2, neither p_1 nor p_2 belongs to K or K', and K' must meet the same component of $l \setminus \{p_1, p_2\}$ as K.

Let us first show that no component C of int $(K \triangle K')$ has its closure meeting l. Suppose that C is such a component and that $C' = p_1C$ is defined. If K intersects l between p_1 and p_2, then the components p_2C' and C intersect, and therefore $p_2C' = C$. By Lemma 5.2.7(i) with $i = 1$ and $k = 2$ (any $k > 1$ would do), equality of the directed X-rays at p_1 gives $\lambda_2(C) < \lambda_2(C')$, while the equality of the directed X-rays at p_2 gives the opposite inequality, a contradiction. We now suppose without loss of generality that p_1, p_2, and $K \cap l$ are in that order on l. Then p_2C and $C' = p_1C$ intersect, so they coincide. Let v_1 be the measure of Definition 5.2.3 with $i = 1$, corresponding to a base line t through p_2 separating p_1 from $K \cup K'$. Since $p_2C = C'$, Lemma 5.2.5 implies that $v_1(C) = v_1(C')$; but $p_1C = C'$ also, giving $v_1(C) > v_1(C')$, by Lemma 5.2.7(ii), a contradiction.

We now consider the various ways that the line l can meet K. Suppose initially that $l \cap$ int $K \neq \emptyset$. There is a nonempty component C_1 of int $(K \triangle K')$, and this must be bounded away from l. A sequence of disjoint components is generated in the following way (see Figure 5.5). Assume that p_1, p_2, and $K \cap l$ are in that order on l and that $p_2^{-1}C_1 = C_2$ is defined (the other cases can be treated similarly). Then $p_1C_2 = C_3$ and $p_2^{-1}C_3 = C_4$ are also defined. Continuing in this fashion, we let $C_{2n+1} = p_1C_{2n}$ and $C_{2n+2} = p_2^{-1}C_{2n+1}$ for all $n \in \mathbb{N}$. By Lemma 5.2.5, $v_1(C_n) = v_1(C_1)$ for each n. The convexity of K ensures that the components $C_{2n}, n \in \mathbb{N}$, which are all disjoint, lie in a triangle T such that $T \cap l$ is a vertex of T. Since $v_1(T) < \infty$, by Lemma 5.2.4, we get a contradiction.

We are left with the case when l supports K. Because all components of int $(K \triangle K')$ are at a positive distance from l, $K \cap l = K' \cap l$. We have to deal with two possibilities: Either there is a line t distinct from l supporting K at some $q \in$ bd $K \cap l$, or else l is the only line supporting K at every point in bd $K \cap l$. We claim that in both cases l (and t) have the same property for K', too. To see this, suppose that l is the only supporting line at $q \in$ bd $K \cap l$, but that there is a line t through q, distinct from l, supporting K'. Then $t \cap$ int $K \neq \emptyset$, so there is a component C of int $(K \setminus K')$ such that cl $C \cap l \neq \emptyset$; but this was excluded at the beginning of the proof.

Assume that a line $t \neq l$ supports K at $q \in$ bd $K \cap l$. Let $\{C_n\}$ be a sequence of components of int $(K \triangle K')$ defined as before. Then the same argument can be

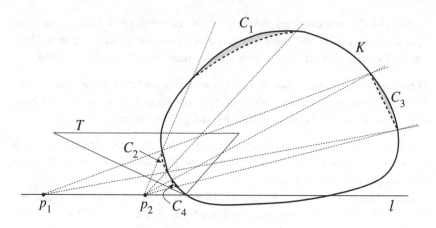

Figure 5.5. A sequence of components.

applied to reach a contradiction; the triangle T may be chosen so that one of its
sides lies on t. Suppose, on the other hand, that l is the only line supporting K
at all points of $K \cap l$. We can use the same approach, working with the measure
ν_1 with base line l, if $\nu_1(K \triangle K') < \infty$. In this case it may also happen, however,
that $\nu_1(K \triangle K') = \infty$. For this reason, we need a different argument here, which
could be described as "chord chasing."

Suppose that bd $K \cap l = [q_1, q_2]$ (where possibly $q_1 = q_2$). Let p_2, p_1, q_1,
and q_2 be in that order on l. (The other cases can be treated similarly.) If $K \neq K'$,
there exists a line l', parallel to l, such that $K \cap l' \neq K' \cap l'$ and such that every
supporting line or chord of K or K' between l and l' makes an angle with l smaller
than $\pi/4$. We may also assume that for any $x \in$ bd $K \cup$ bd K' between l and l', the
lines through x and p_j, $j = 1, 2$, make an angle with l smaller than $\pi/4$. Select
$y_1 \in$ bd $K' \cap l'$ such that $[p_1, y_1] \cap K \neq \emptyset$ and $y_1 \notin K$. See Figure 5.6.

Let p_1, x_2, x_1, and y_1 be in that order on the line through p_1 and y_1, with
$x_j \in$ bd K, $j = 1, 2$, and let y_2 be the other endpoint of the chord of K' on that
line. Then $\|x_1 - y_1\| = \|x_2 - y_2\|$. Let z_2, z_3 belong to bd K and to the line through
y_2 and p_2, and let $y_3 \neq y_2$ belong to the same line and to bd K'. If z_2, y_2, z_3, y_3
are in that order, then $\|y_2 - z_2\| = \|y_3 - z_3\|$. Now take $x_3, x_4 \in$ bd K on the
line through p_1 and y_3, with x_4, x_3, y_3 in that order. Let $y_4 \neq y_3$ be the other
endpoint of the chord of K' on that line.

Consider a line m supporting K at z_2. Let T be the triangle determined by
m, l, and the line through p_1 and x_2. From the assumptions it follows that the
angle at the vertex v of T not belonging to l is larger than $\pi/2$. The same angle
is opposite to $[z_2, y_2]$ in the triangle with vertices z_2, y_2, and v, and therefore
$\|y_2 - z_2\| > \|v - y_2\|$. From the convexity of K we have $\|v - y_2\| \geq \|x_2 - y_2\|$,

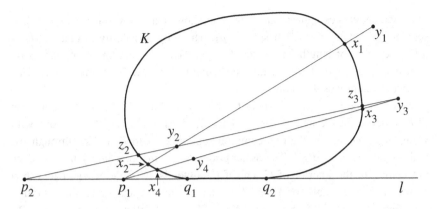

Figure 5.6. Chord chasing.

and hence $\|y_2 - z_2\| > \|x_2 - y_2\|$. Similarly we see that $\|x_3 - y_3\| > \|y_3 - z_3\|$, which then implies that $\|x_3 - y_3\| > \|x_1 - y_1\|$.

Iterating this construction, we obtain two sequences $\{x_{2n+1}\}$ and $\{y_{2n+1}\}$ such that

$$\|x_{2n+1} - y_{2n+1}\| > \|x_{2n-1} - y_{2n-1}\|,$$

for every n, and

$$\lim_{n\to\infty} x_{2n+1} = \lim_{n\to\infty} y_{2n+1} = q_2,$$

which is impossible.   ∎

It is natural to wonder if the assumption about the position of the line joining the points p_1 and p_2 is necessary, but this question is unresolved at present; see Problem 5.1.

The next lemma says that X-rays at two points sometimes yield the same information as directed X-rays at those points.

Lemma 5.3.2. *Suppose that K is a convex body and that l is a line through two points p_1 and p_2. If $l \cap \text{int } K = \emptyset$ and K' is a convex body such that $X_{p_j} K \simeq X_{p_j} K'$, $j = 1, 2$, then $D_{p_j} K = D_{p_j} K'$, $j = 1, 2$.*

Proof. We claim that $l \cap \text{int } K' = \emptyset$. If this is not true, then $l \cap K'$ is a line segment, of length $a > 0$, say. Since the X-rays of K and K' at p_1 are equal almost everywhere, there are sequences $\{l_n\}$, $n \in \mathbb{N}$, of lines through p_1, approaching l, and such that $l_n \cap K'$ has the same length as $l_n \cap K$, for all n. Since $l \cap \text{int } K = \emptyset$, $l_n \cap K$ lies on one side of p_1, for every line l_n in each such sequence. By choosing two such sequences appropriately, we can show that $l \cap K$ must contain a line segment of length a on each side of p_1. The only possibility is that $l \cap K$ is a line segment of length $2a$, centered at p_1. The same argument applied to p_2 instead of p_1 implies that p_2 is also the center of $l \cap K$, which is impossible. This contradiction proves the claim.

Since $l \cap \text{int } K' = \emptyset$, it suffices to show that K and K' lie on the same side of l. Suppose that K and K' lie on opposite sides of l. Let C be the cone with vertex p_1, bounded by the two common support lines to K and K' through p_1, and containing K and K'. The assumptions of the theorem imply that p_2 does not belong to the interior of C, and if C subtends an angle at p_1 of less than π, then it is impossible for K and K' to have common supporting lines through p_2. Therefore $p_1 \in K$. Since $X_{p_1} K \simeq X_{p_1} K'$, it follows that $p_1 \in K'$. By considering lines through p_1 approaching l, we see that p_1 must be the midpoint of the (possibly degenerate) line segment $l \cap \text{bd } K \cap \text{bd } K'$. The same argument, with p_1 and p_2 interchanged, will then imply that $p_1 = p_2$, which is not the case.   ∎

For X-rays, the result corresponding to Theorem 5.3.1 becomes more difficult to state. This is partly because nontrivial questions arise when X-rays are used at points interior to a body.

Theorem 5.3.3. *A convex body K is determined by X-rays at distinct points p_1 and p_2 in each of the following situations:*

(i) *The line l through p_1 and p_2 meets $\text{int } K$, p_1 and p_2 do not belong to $\text{int } K$, and it is specified whether or not K meets $[p_1, p_2]$;*

(ii) *l supports K;*

(iii) *p_1, p_2 belong to $\text{int } K$.*

Proof. (i) If K does not meet $[p_1, p_2]$, then any convex body K' with the same X-rays as K at p_1 and p_2 must meet the same component of $l \setminus \{p_1, p_2\}$ as K, for otherwise K and K' could not have the same supporting lines through p_1 and p_2. Then we know that p_1 and p_2 do not belong to $\text{int } K$, and also which component of $l \setminus \{p_1, p_2\}$ intersects K, so the distinction between X-rays and directed X-rays disappears. Therefore we argue in this case exactly as in the first part of the proof of Theorem 5.3.1.

(ii) In this case Lemma 5.3.2 shows that we can replace X-rays by directed X-rays, and Theorem 5.3.1 does the rest.

(iii) Suppose that p_j is interior to K, $j = 1, 2$. Let K' be a convex body distinct from K with the same X-rays as K at the points p_1 and p_2. We first show that $\text{bd } K$ and $\text{bd } K'$ intersect in some point not belonging to the line l through p_1 and p_2. Suppose the contrary. Then $\text{int}(K \triangle K')$ has exactly two components, C and C'. Consider the lines t_j through p_j orthogonal to l. Denote by E_j, E_j' the open half-planes determined by these lines, with $p_1 \in E_2$, $p_2 \in E_1'$; see Figure 5.7.

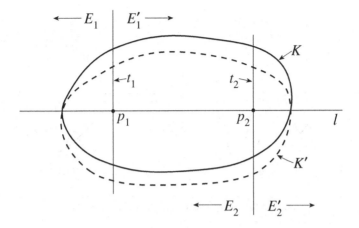

Figure 5.7. Case (iii) of Theorem 5.3.3.

Let ν_1 be the measure with the base line l. Lemma 5.2.4 and convexity imply that C and C' both have finite ν_1-measure. Also, $E_j \cap C$ and $E'_j \cap C'$ have the same X-rays at p_j, $j = 1, 2$, and therefore, by Lemma 5.2.5, $\nu_1(E_1 \cap C) = \nu_1(E'_1 \cap C')$ and $\nu_1(E_2 \cap C) = \nu_1(E'_2 \cap C')$. But on the other hand, by inclusion, $\nu_1(E_2 \cap C) > \nu_1(E_1 \cap C)$ and $\nu_1(E'_2 \cap C') < \nu_1(E'_1 \cap C')$, a contradiction.

We have shown that there is a component C of int $(K \triangle K')$ with one endpoint, x say, not lying on l. We claim that cl $C \cap l = \emptyset$. If this is not true, the other endpoint, y, of C belongs to l, and we may suppose that p_1, p_2 and y lie in that order on l. Let $x_1 \neq x$ belong to bd K and the line through x and p_1, and let $x_2 \neq x_1$ belong to bd K and the line through x_1 and p_2. Then x_1 and x_2 belong to bd $K \cap$ bd K', but x_2 must lie strictly between x and y in bd K, which is impossible.

From the component C we can construct, as in the proof of Theorem 5.3.1, a sequence of disjoint components. These all have the same ν_1-measure and are contained in int $(K \triangle K')$, which is of finite ν_1-measure. This final contradiction completes the proof. ∎

The last theorem allows only two possibilities when the line through p_1 and p_2 meets K, but it is unknown whether these can actually occur. See Figure 5.8, Problem 5.2, and Note 5.2.

The following corollary characterizes centrally symmetric bodies.

Corollary 5.3.4. *If the X-rays of a convex body K at distinct points p_1, p_2 are equal, then K is centrally symmetric about the midpoint of the segment $[p_1, p_2]$.*

Proof. The hypotheses mean that each chord of K on a line through p_1 has the same length as the chord of K on the parallel line through p_2. Without loss of

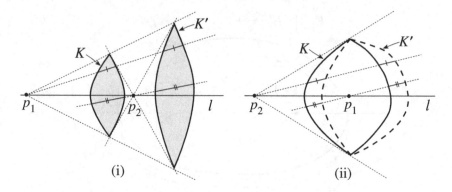

Figure 5.8. Possible examples for X-rays at two points.

generality, suppose that the midpoint of $[p_1, p_2]$ is o, and let $K' = -K$. Then K and K' have the same X-rays at p_1 and p_2. Since the hypotheses imply that int K meets the line segment $[p_1, p_2]$, K' does also. If neither p_1 nor p_2 belongs to int K, we can apply Theorem 5.3.3(i) to deduce that $K' = K$. If one of the points belongs to int K, the other must also, and then Theorem 5.3.3(iii) implies that $K' = K$. Therefore $-K = K$, so K is centrally symmetric about o. ∎

Theorems 5.3.1 and 5.3.3 are not powerful enough to give an unrestricted uniqueness theorem. As a first step to such a result, we now consider adding another X-ray at a point not collinear with the first two.

Theorem 5.3.5. *X-rays at three noncollinear points determine a convex body K in the interior of the triangle formed by the points.*

Proof. Let K be contained in the interior of the triangle T with vertices at $p_j, 1 \le j \le 3$, and suppose that $K' \ne K$ is another convex body with the same X-rays as K at each p_j. Then K' must also lie in the interior of T. Suppose that C is a component of int $(K \triangle K')$ of maximal area. Denote by E the open half-plane containing C and bounded by the line through the endpoints of C in bd $K \cap$ bd K'; see Figure 5.9.

For some j, $p_j \in E$. But then $p_j C$ is defined, and by Lemma 5.2.7(i) with $i = 1$ and $k = 2$, it has area strictly larger than C, a contradiction. ∎

So far we have had to take advantage of some a priori knowledge about the position of the body K relative to the points at which the X-rays are taken. The methods introduced in the first part of the chapter and the idea of the last theorem can now be combined to eliminate this defect.

Theorem 5.3.6. *Convex bodies are determined by directed X-rays at any set of three noncollinear points.*

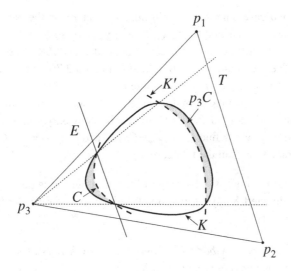

Figure 5.9. Illustration for Theorem 5.3.5.

Proof. Let p_j, $1 \leq j \leq 3$, be noncollinear points, and let K be a convex body. If K contains one of the points in its interior, the result is trivial. Suppose, then, that K and K' are convex bodies with the same directed X-rays at these points and not containing them in their interiors. If a line through two of the points intersects K, or if K is contained in the interior of the triangle T whose vertices are the points p_j, $1 \leq j \leq 3$, then we know from Theorem 5.3.1 and Theorem 5.3.5, respectively, that $K' = K$. Then by permuting the indices if necessary, we may assume that K is contained in a cone determined by the lines through p_1 and p_j, $j = 2, 3$, and containing the triangle T. We have only two cases to consider: Either p_1 separates K from T, or $[p_2, p_3]$ does. In each case the directed X-rays tell us that K' intersects K; we shall assume that $K' \neq K$ and derive a contradiction.

Consider the first case. Take the line through p_1 and p_2 as the base line for the measure ν_1. We assume that $K \subset \{(x, y) : y > 0\}$ so that the point p_3 has negative y-coordinate. Let C_1 be a component of int $(K \triangle K')$ having maximal ν_1-measure. Then $p_3 C_1$ is defined, since otherwise $p_3^{-1} C_1$ would have larger ν_1-measure, according to Lemma 5.2.7(ii). We may suppose without loss of generality that $C_2 = p_2^{-1} C_1$ is defined; by Lemma 5.2.5, $\nu_1(C_2) = \nu_1(C_1)$. As before, $p_3 C_2$ must be defined. This means, by convexity, that $C_3 = p_1 C_2$ is also defined, and by Lemma 5.2.5, $\nu_1(C_3) = \nu_1(C_2)$. Again $p_3 C_3$ has to be defined, by the maximality of $\nu_1(C_1)$, and consequently, by convexity, $C_4 = p_2^{-1} C_3$ is defined. Iterating this process, we define $C_{2n+1} = p_1 C_{2n}$ and $C_{2n+2} = p_2^{-1} C_{2n+1}$. These sets are disjoint, have the same ν_1-measure, and are contained in the set $K \cup K'$ of finite ν_1-measure, a contradiction.

In the second case, take the line through p_2 and p_3 as the base line for the measure ν_1. We assume that $K \subset \{(x, y) : y > 0\}$ so that the point p_1 has negative y-coordinate. Let C_1 be a component of int $(K \triangle K')$ having maximal ν_1-measure. The maximality of $\nu_1(C_1)$ and Lemma 5.2.7(ii) imply that $p_1 C_1$ is defined, so either $p_2 C_1$ or $p_3 C_1$ is defined, say the former. Put $C_2 = p_2 C_1$, and note that $\nu_1(C_2) = \nu_1(C_1)$ by Lemma 5.2.5. From the maximality of $\nu_1(C_1)$ it follows that $p_1 C_2$ is defined and therefore, by convexity, that $C_3 = p_3 C_2$ is defined. Iteratively, we define $C_{2n} = p_2 C_{2n-1}$ and $C_{2n+1} = p_3 C_{2n}$, and this yields a contradiction as in the first case. ∎

The previous theorem is the strongest known, though it is possible (cf. Problem 5.1) that directed X-rays at two points suffice. For X-rays, we have the following weaker version.

Theorem 5.3.7. *A convex body K is determined by X-rays at any set of three noncollinear points not contained in* int K.

Proof. If no line through two of the points meets the interior of K, then Lemma 5.3.2 allows us to replace X-rays with directed X-rays, and the result follows from Theorem 5.3.6. Suppose, then, that the line l through two of the points, p_1 and p_2, say, meets int K. If $K' \neq K$ is a convex body such that $X_{p_j} K \simeq X_{p_j} K'$, $j = 1, 2$, then, by Theorem 5.3.3(i) and (ii), l also meets int K' and the bodies K and K' must be separated by p_1 or p_2, as in Figure 5.8(i). It is now impossible for K and K' to have common supporting lines through the third point, and this contradiction proves the theorem. ∎

The last theorem might be false if one of the three points belongs to int K; see Problem 5.3. Figure 5.10 shows that in this case the knowledge of three X-rays does not assist in any obvious way.

A uniqueness theorem for X-rays can now be proved, but to compensate for the weaker information, a fourth point seems necessary.

Theorem 5.3.8. *Convex bodies are determined by X-rays at any set of four points, with no three collinear.*

Proof. Suppose that P is a set of four points, with no three collinear, and that K is a convex body. If two of the points in P belong to int K, the conclusion follows from Theorem 5.3.3(iii). Therefore at least three points do not belong to int K, and we may apply Theorem 5.3.7 to complete the proof. ∎

The following result is analogous to Theorem 1.2.11, but, in contrast to the case of parallel X-rays, it is not known if it is the best possible (cf. Problem 5.1). See Section 0.2 for the definition of "projectively equivalent."

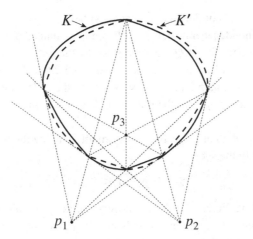

Figure 5.10. Possible example for X-rays at three points.

Theorem 5.3.9. *Convex bodies are determined by X-rays (and therefore also by directed X-rays) at any set of collinear points not projectively equivalent to a subset of directions of the edges of a regular polygon.*

Proof. Let K and K' be distinct convex bodies, and p_j, $1 \leq j \leq n$, distinct points on a line l, such that $X_{p_j}K \simeq X_{p_j}K'$ for $1 \leq j \leq n$. We know (cf. Section 0.2) that any three directions are affinely equivalent to a subset of the directions of the edges of an equilateral triangle. Therefore any three points on a line are projectively equivalent to such a set of directions, and from this we see that $n \geq 4$.

Suppose that the line l meets K. By Theorem 5.3.3(ii), we can assume that $l \cap \mathrm{int}\, K \neq \emptyset$. If two of the points belong to $\mathrm{int}\, K$, then Theorem 5.3.3(iii) applies. Therefore at least three of the points do not belong to $\mathrm{int}\, K$. The fact that K and K' have equal X-rays almost everywhere at two of these points means, by Theorem 5.3.3(i), that they must be separated by one of the points, as in Figure 5.8(i); but then K and K' cannot have equal supporting lines through the third point.

Consequently, we may assume that $K \cap l = \emptyset$ and hence $K' \cap l = \emptyset$. Let ν_1 be the measure with base line l, and let C be a component of $\mathrm{int}(K \triangle K')$. Then, for each j, either $p_j C$ or $p_j^{-1}C$ is defined, and these components have the same ν_1-measure, by Lemma 5.2.5. As in Chapter 1, we may associate with C a system of components; here, it is the family $\mathcal{C}$ of distinct sets of the form

$$\mathcal{C} = \{q_{j_m} \cdots q_{j_1}C : m \in \mathbb{N}, 1 \leq j_k \leq n\},$$

where q_j stands for p_j or p_j^{-1}. Since these are disjoint, have the same ν_1-measure, and are contained in the set $\mathrm{int}(K \triangle K')$ of finite ν_1-measure, they are finite in number. The centroids of these components, with respect to the measure ν_0, form the vertices of a nondegenerate convex polygon Q contained in $K \cup K'$. (The nondegeneracy and convexity of Q are established by the same arguments as those in the proofs of Lemmas 1.2.6 and 1.2.8, with Lemma 5.2.6 replacing Lemma 1.2.3.) Furthermore, by Lemma 5.2.6, the polygon Q has the following property: If v is a vertex of Q, and $1 \leq j \leq n$, the line through p_j and v meets a different vertex v' of Q.

Let ϕ be a nonsingular projective transformation taking the line l onto the line at infinity. Identify the set

$$S = \{\phi p_j : 1 \leq j \leq n\}$$

with corresponding directions. The nondegenerate polygon ϕQ is convex (cf. the remarks in Section 0.3). Moreover, it is an S-regular polygon in the sense of Definition 1.2.7. Therefore, by Corollary 1.2.10, there exists a regular polygon P and a $\psi \in GA_n$ such that $\phi Q = \psi P$, or $Q = \phi^{-1} \psi P$. It follows that the set $\{p_j : 1 \leq j \leq n\}$ is projectively equivalent, via $\phi^{-1} \psi$, to a subset of directions of the edges of P. ∎

We now turn to the question of verifying convex bodies by point X-rays; see Definition 5.1.7. By establishing analogues of Lemma 1.2.18 and Theorem 1.2.21, we shall see that verification needs one X-ray less than determination.

Lemma 5.3.10. *Let p_j, $1 \leq j \leq n$, be points on a line l. Suppose that K is a convex body not intersecting l and that Q is a convex polygon inscribed in K whose edges are on lines through the points p_j. Then Q is inscribed in any other convex body K' such that $X_{p_j} K \simeq X_{p_j} K'$, $1 \leq j \leq n$.*

Proof. The proof follows that of Lemma 1.2.18. All that is needed is to replace Lebesgue measure λ_2 by the measure ν_1 having l as base line and to observe that the work done by the Cavalieri principle, Lemma 1.2.2, is now carried out by Lemma 5.2.5. ∎

Theorem 5.3.11. *A convex body can be verified by X-rays at three points.*

Proof. Let l be a line not intersecting K, and let ϕ be a nonsingular projective transformation taking l to the line at infinity. The image ϕK of K is also a convex body (cf. the remarks in Section 0.3). We can therefore choose directions u_1, u_2, and u_3 exactly as in the proof of Theorem 1.2.21, so that ϕK is verified by its parallel X-rays in these directions. Let $p_j = \phi^{-1} u_j$, $1 \leq j \leq 3$.

Suppose that $X_{p_j} K \simeq X_{p_j} K'$, $1 \leq j \leq 3$. Let V be the closed set of all vertices of (possibly degenerate) parallelograms inscribed in ϕK whose edges are parallel to u_1 and u_2. If P is any such parallelogram, then $\phi^{-1} P$ is a quadrilateral inscribed in K whose edges lie on lines through p_1 and p_2. By Lemma 5.3.10, $\phi^{-1}P$ is also inscribed in K', and it follows that P is inscribed in $\phi K'$.

Next, suppose that C is a component of int $(\phi K \triangle \phi K')$, and let $\mathcal{C}$ be the system of components associated to C, as in the proof of Theorem 1.2.21, via the directions u_j, $1 \le j \le 3$. Then $\phi^{-1}C$ is a component of int $(K \triangle K')$. If we define

$$\mathcal{C}' = \{\phi^{-1}C' : C' \in \mathcal{C}\},$$

then $\mathcal{C}'$ is the system of components associated to $\phi^{-1}C$, via the points p_j, $1 \le j \le 3$, as in the proof of Theorem 5.3.9. Moreover, the system $\mathcal{C}'$ is finite, by the same argument we used in that proof. Consequently, the system $\mathcal{C}$ must also be finite.

The previous two paragraphs, together with the incidence-preserving properties of projective transformations, now allow us to follow the proof of Theorem 1.2.21 to conclude that the endpoints of the components in $\mathcal{C}$ form the vertices of a weakly $\{u_1, u_2, u_3\}$-regular polygon. We can then repeat verbatim the last part of the proof of Theorem 1.2.21 to obtain a contradiction to the existence of the component C. ∎

The reader will have noticed that X-rays at points are considerably more awkward to deal with than parallel X-rays. A contribution to the unsolved Problem 5.1 can be made by specializing further the class of sets under scrutiny. To this end, we shall give a necessary and sufficient condition for the X-rays of two (not necessarily convex) polygons to be equal almost everywhere at a point p. We can and shall assume that p is the origin. All polygons are assumed to be nondegenerate.

Before undertaking this task, it is worth pointing out that it is even possible for two convex polygons to have X-rays at o equal almost everywhere, but not equal everywhere. To see this, let Q be any centered convex polygon, and E_j, $j = 1, 2$, two closed half-planes bounded by different lines l_j, $j = 1, 2$, respectively, through o. Then the polygons $P_j = Q \cap E_j$, $j = 1, 2$, have X-rays at o equal for all directions except those parallel to the lines l_j; in the latter directions they differ. We shall use the following lemma.

Lemma 5.3.12. *Let T be the triangle in $\mathbb{E}^2$ with vertices at o, (r, α), and (s, β), $r > 0$, $s > 0$, $0 \le \alpha < \beta \le \pi$, in polar coordinates. For $\alpha \le \theta \le \beta$, the X-ray or directed X-ray of T at o is given by*

$$X_o T(\theta) = D_o T(\theta) = \frac{rs \sin(\beta - \alpha)}{s \sin(\beta - \theta) + r \sin(\theta - \alpha)}.$$

Proof. This follows directly from the equation of the line through (r, α) and (s, β) in polar coordinates. ∎

If P and P' are polygons, we can partition the plane into a finite set of double cones with vertex at o such that neither P nor P' has any of its vertices in the interior of these double cones. Each double cone can be represented in polar coordinates by

$$C(\alpha, \beta) = \{(r, \theta) : \alpha \leq \theta \leq \beta\},$$

where $0 < \beta - \alpha \leq \pi$.

Suppose that $C(\alpha, \beta)$ is such a double cone and that $0 \leq \varphi < \pi$. With the set of all edges of P and P' labeled by e_n, $1 \leq n \leq n_0$, we shall denote by $\{e_n, n \in N(\varphi)\}$ those parallel to φ and meeting the interior of $C(\alpha, \beta)$, and hence both its bounding lines. Suppose that e_n intersects the line $\{\theta = \alpha\}$ at (r_n, α). If $o \in e_n$, define $a_n = 0$. Otherwise, if e_n is an edge of P, define $a_n = +1$ (or -1) if a moving point on the line $\{\theta = \gamma, \alpha < \gamma < \beta\}$ leaves P (or enters P, respectively) as its distance from o increases across e_n; and vice versa for edges e_n of P'.

Theorem 5.3.13. *With notation as before, two polygons P and P' satisfy $X_o P \simeq X_o P'$ if and only if for each appropriate double cone $C(\alpha, \beta)$ and $0 \leq \varphi < \pi$, the family $\{e_n, n \in N(\varphi)\}$ of edges satisfies*

$$\sum \{a_n |r_n| : n \in N(\varphi)\} = 0.$$

Proof. If $C(\alpha, \beta)$ is a suitable double cone associated to general polygons P and P', then $P \cap C(\alpha, \beta)$ and $P' \cap C(\alpha, \beta)$ can each be expressed as an alternating set sum of triangles, modulo parts of their boundaries, of the same form as the triangle T of Lemma 5.3.12. Since $C(\alpha, \beta)$ contains no vertices of P or P' in its interior, the X-rays of $P \cap C(\alpha, \beta)$ and $P' \cap C(\alpha, \beta)$ must actually agree for all θ with $\alpha < \theta < \beta$. From these observations and Lemma 5.3.12, one concludes that $P \cap C(\alpha, \beta)$ and $P' \cap C(\alpha, \beta)$ have X-rays equal almost everywhere at o if and only if

$$\sum \frac{a_n |r_n| |s_n| \sin(\beta - \alpha)}{|s_n| \sin(\beta - \theta) + |r_n| \sin(\theta - \alpha)} = 0, \tag{5.4}$$

for $\alpha < \theta < \beta$, where the sum runs over all edges e_n meeting the interior of $C(\alpha, \beta)$, e_n contains the points (r_n, α) and (s_n, β), and $a_n = \pm 1$ is the appropriate weight. By elementary trigonometry, if e_n is an edge of P or P' parallel to φ and containing (r_n, α) and (s_n, β), then

$$|s_n| \sin(\beta - \varphi) + |r_n| \sin(\varphi - \alpha) = 0.$$

Solving for $|s_n|$ and substituting, we get

$$\frac{a_n |r_n| |s_n| \sin(\beta - \alpha)}{|s_n| \sin(\beta - \theta) + |r_n| \sin(\theta - \alpha)} = a_n |r_n| \frac{\sin(\alpha - \varphi)}{\sin(\theta - \varphi)}.$$

Replacing the terms on the left-hand side of (5.4) by the last expression for all the relevant edges $e_n \in N(\varphi)$ of P and P', we get

$$\sum_{n \in N(\varphi)} a_n |r_n| \frac{\sin(\alpha - \varphi)}{\sin(\theta - \varphi)} = \left(\sum_{n \in N(\varphi)} a_n |r_n| \right) \frac{\sin(\alpha - \varphi)}{\sin(\theta - \varphi)}.$$

It follows that polygons satisfying the conditions of the theorem will have equal X-rays almost everywhere.

The same calculation can be made with s_n instead of r_n. This means that the left-hand side of (5.4) is not changed if we replace the set of edges $\{e_n : n \in N(\varphi)\}$ by a single segment $e'(\varphi)$ joining the points $(\sum \{a_n |r_n| : n \in N(\varphi)\}, \alpha)$ and $(\sum \{a_n |s_n| : n \in N(\varphi)\}, \beta)$. Making this replacement in (5.4) for each direction φ_j parallel to an edge of P or P', and canceling the constant $\sin(\beta - \alpha)$, we obtain the equation

$$\sum_j \frac{p_j q_j}{q_j \sin(\beta - \theta) + p_j \sin(\theta - \alpha)} = 0,$$

for $\alpha < \theta < \beta$. Here the sum runs over all the indices j corresponding to directions φ_j of the relevant edges of P or P', with $p_j = \sum \{a_n |r_n| : n \in N(\varphi_j)\}$ and $q_j = \sum \{a_n |s_n| : n \in N(\varphi_j)\}$. Further, no two of the segments e'_j joining (p_j, α) and (q_j, β) are parallel.

Using a common denominator in the previous equality, we get

$$\sum_j \left(p_j q_j \prod_{k \neq j} (q_k \sin(\beta - \theta) + p_k \sin(\theta - \alpha)) \right) = 0,$$

for $\alpha < \theta < \beta$. Dividing by $\cos\theta$, we see that the left-hand side becomes a polynomial in $\tan\theta$, so it vanishes for all θ.

Setting $\theta = \theta_j$ in this equality, where θ_j is chosen so that

$$q_j \sin(\beta - \theta_j) + p_j \sin(\theta_j - \alpha) = 0,$$

we get

$$p_j q_j \prod_{k \neq j} (q_k \sin(\beta - \theta_j) + p_k \sin(\theta_j - \alpha)) = 0.$$

If $q_k \sin(\beta - \theta_j) + p_k \sin(\theta_j - \alpha) = 0$ for some $k \neq j$, then by elementary trigonometry e'_k is parallel to e'_j. Since this is impossible, we must have $p_j = 0$ (or, equivalently, $q_j = 0$), and consequently

$$\sum \{a_n |r_n| : n \in N(\varphi_j)\} = 0,$$

as required.   ∎

It is now possible to settle the question of determination of convex polygons by point X-rays. The following lemma is useful.

Lemma 5.3.14. *Let P and P' be convex polygons such that $o \notin \text{int } P$ and $X_o P \simeq X_o P'$. For each double cone $C = C(\alpha, \beta)$ as before, either $P \cap C = \pm P' \cap C$, or $P \cap C$ and $P' \cap C$ are intersections of C with parallel strips with equal X-rays at o.*

Proof. Choose a double cone $C = C(\alpha, \beta)$ as in the previous theorem, so that P and P' both meet C but have no vertex in its interior. By convexity, there are two edges e_n, $n = 1, 2$, of P and two edges e_n, $n = 3, 4$, of P' with (r_n, α) and (s_n, β), say, in e_n. We may suppose that $0 \leq r_1 < r_2$ and that $|r_3| \leq |r_4|$. If $P \cap C \neq P' \cap C$, then Theorem 5.3.13 allows two possibilities: Either the four edges e_n consist of two pairs of parallel edges (but not all four are parallel) with e_3 and e_4 on the same side of the origin, or all four edges are parallel. In the first case, it must be that the edges e_2 and e_4 are parallel and $|r_4| = r_2$, and the edges e_1 and e_3 are parallel and $|r_3| = r_1$. Our other assumptions then imply that $r_4 = -r_2$ and $r_3 = -r_1$, so $P \cap C = -P' \cap C$. In the second case, when all four edges are parallel, the intersections of P and P' with C are formed by intersecting C with parallel strips of the same width. ∎

Theorem 5.3.15. *Convex polygons are determined by X-rays at any set of two points.*

Proof. Suppose that P and P' are different convex polygons with X-rays equal almost everywhere at two points, o and p, say. By Theorem 5.3.3(iii), we may assume that one of the points does not belong to the interior of one of the polygons; $o \notin \text{int } P$, say.

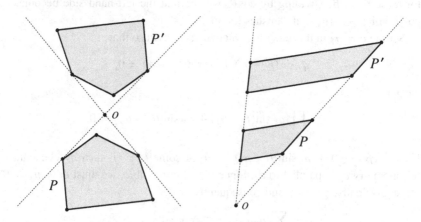

Figure 5.11. Convex polygons with equal X-rays at the origin.

Suppose first that $o \notin P$. We claim that just the fact that the X-rays at o are equal almost everywhere implies that either (i) $P = -P'$ or (ii) both P and P' are intersections of congruent parallel strips with the same double cone with vertex at o; see Figure 5.11. To prove this, choose a double cone $C(\alpha, \beta)$ as in the previous lemma, so that P and P' both meet $C(\alpha, \beta)$ but have no vertex in its interior and $P \cap C(\alpha, \beta) \neq P' \cap C(\alpha, \beta)$. If $P \cap C(\alpha, \beta) = -P' \cap C(\alpha, \beta)$, then since P is convex and $o \notin P$, the same equation holds for any such double cone, and $P = -P'$. If $P \cap C(\alpha, \beta) \neq -P' \cap C(\alpha, \beta)$, then Lemma 5.3.14 implies that $P \cap C(\alpha, \beta)$ and $P' \cap C(\alpha, \beta)$ are intersections of $C(\alpha, \beta)$ with parallel strips of equal width. Reducing α and increasing β, we meet a vertex of P or P', obtaining a maximal double cone $C(\alpha, \beta)$ for which this holds. Using convexity, it can be seen that if $P \neq -P'$, then such a maximal double cone contains P and P', so P and P' are intersections of $C(\alpha, \beta)$ with parallel strips of the same width. This establishes the claim.

In case (i), at least one of the polygons does not contain p in its interior. We can apply the argument of the previous paragraph to conclude that either P' is the reflection of P in p, or both P and P' are intersections of parallel strips with the same double cone with vertex at p. However, these conclusions are incompatible; in fact, it is not even possible for P and P' to have common supporting lines from both p and o, as they must. If (ii) holds, the same proof works, unless $p \in \operatorname{int} P \cap \operatorname{int} P'$, when we replace P and P' by the closures of $P \setminus P'$ and $P' \setminus P$. Since these sets also have X-rays equal almost everywhere at o and p, we can apply the same method as before to reach a contradiction.

The remaining possibility is that both P and P' contain both o and p in their boundaries. Applying Lemma 5.3.14, we conclude that for each double cone C with vertex at o not containing vertices of P or P' in its interior, we have $P \cap C = \pm P' \cap C$. By convexity, this implies that there is a centered polygon Q and closed half-planes H and H' containing o in their boundaries such that $P = Q \cap H$ and $P' = Q \cap H'$. However, the same argument applied to p instead of o implies that p is also the center of Q. Since a bounded set cannot have two centers, this is a contradiction. ∎

In the course of the previous proof, we showed that an X-ray at a single point p exterior to a convex polygon determines it up to a reflection in p, excluding the trivial case of intersections of strips with a double cone with vertex at p. This is not true for an X-ray at an interior point, as the reader will see by examining the polygons depicted in Figure 3.8, where both polygons are positioned so that the point p lies at the center of the indicated hexagon.

5.4. X-rays in the projective plane

The results of the previous section, together with those from Chapter 1, raise natural questions as to what information might be garnered by taking some parallel X-rays and some at points. In this section we briefly indicate some generalizations of earlier theorems to cover problems of this mixed type. The appropriate setting for such generalizations is the projective plane $\mathbb{P}^2$.

As is noted Section 0.1, we can regard $\mathbb{P}^2$ as $\mathbb{E}^2$ with a line (or, strictly speaking, copy of $\mathbb{P}^1$) at infinity added. By a convex body in $\mathbb{P}^2$ we then simply mean a convex body in $\mathbb{E}^2$. Points at infinity can be identified with pairs of antipodal directions in $\mathbb{E}^2$, and an X-ray or directed X-ray at a point at infinity is just the parallel X-ray in (either of) the corresponding directions.

Set up a Cartesian coordinate system in $\mathbb{E}^2$, and let ν_1 be the measure of Definition 5.2.3 with $i = 1$ and the x-axis as base line. Our first task is to investigate the ν_1-measure of sets with equal parallel X-rays. We remind the reader that our convention is that parallel X-rays (in contrast to point X-rays) are called equal if they are equal almost everywhere.

Lemma 5.4.1. *Let E_1 and E_2 be measurable sets contained in $\{(x, y) : y \geq 0\}$ and with equal parallel X-rays in a direction parallel to the x-axis. Then $\nu_1(E_1) = \nu_1(E_2)$.*

Proof. For $j = 1, 2$, denote by $E_{j,y}$ the intersection of E_j with the horizontal line through $(0, y)$. Then

$$\nu_1(E_j) = \int_0^\infty |y|^{-1} \lambda_1(E_{j,y}) \, dy,$$

for $j = 1, 2$. Since the equality of the X-rays implies that $\lambda_1(E_{1,y}) = \lambda_1(E_{2,y})$ for almost all y, we are finished. ∎

For our second lemma, it is convenient to use a not necessarily orthogonal coordinate system. Let (x, w) denote the coordinates of a point with respect to an x-axis and a w-axis at an angle of α, $0 < \alpha \leq \pi/2$. Note that the relation $y = w \sin \alpha$ holds, where y is the usual second coordinate. Suppose that E_1 and E_2 are measurable sets of the form

$$E_j = \{(x, w) : a \leq x \leq b, 0 \leq f_j(x) \leq w \leq g_j(x)\},$$

for $j = 1, 2$, where f_j and g_j are continuous functions on $[a, b]$.

Lemma 5.4.2. *Let E_1 and E_2 be as before, with the same parallel X-rays in a direction parallel to the w-axis. Suppose also that $g_1(x) \leq f_2(x)$ for $a \leq x \leq b$ and that E_1 has positive ν_1-measure. Then $\nu_1(E_1) > \nu_1(E_2)$.*

Proof. By Fubini's theorem,

$$\nu_1(E_j) = \iint_{E_j} |y|^{-1} \, dx \, dy = \int_a^b \int_{f_j(x)}^{g_j(x)} |w|^{-1} \, dw \, dx = \int_a^b \log \frac{g_j(x)}{f_j(x)} \, dx,$$

for $j = 1, 2$. The hypotheses imply that $g_1(x) - f_1(x) = g_2(x) - f_2(x)$, for almost all $x \in [a, b]$. Suppose that we set $h(x) = f_2(x) - f_1(x)$, for $a \leq x \leq b$. Then

$$\log \frac{g_2(x)}{f_2(x)} = \log \frac{h(x) + g_1(x)}{h(x) + f_1(x)} < \log \frac{g_1(x)}{f_1(x)},$$

for almost all $x \in [a, b]$ for which $h(x) > 0$ and $g_1(x) > f_1(x)$, since the function $\log\big((h + c)/(h + d)\big)$ is strictly decreasing in h when $h \geq 0$ and $c > d \geq 0$. Since $\nu_1(E_1) > 0$, $g_1(x) - f_1(x) > 0$, and hence $h(x) > 0$, for a subset of $[a, b]$ of positive λ_1-measure. This immediately gives $\nu_1(E_1) > \nu_1(E_2)$, as required. ∎

Armed with these lemmas, we shall now investigate directed X-rays at points in the projective plane, leaving X-rays for a discussion at the end of this section. The following result generalizes Theorem 5.3.1.

Theorem 5.4.3. *Let K be a convex body in $\mathbb{P}^2$, and let p_1 and p_2 be points in $\mathbb{P}^2$ such that the line l joining them intersects K. Then K is determined by its directed X-rays at p_1 and p_2.*

Proof. If l meets K, then at least one point is finite. If both are finite, however, the conclusion follows from Theorem 5.3.1. We shall therefore assume that l is the x-axis, p_1 is a finite point on l, and p_2 is a point at infinity corresponding to a direction parallel to l.

Suppose that $K' \neq K$ is a convex body with the same directed X-rays as K at p_1 and p_2. Then p_1 does not belong to K or K'. Let C be a component of int $(K \triangle K')$ whose closure meets l, and suppose that $C' = p_1 C$ is defined. Then, by Lemma 5.2.7(i), with $i = 1$ and $k = 2$, we have $\lambda_2(C) < \lambda_2(C')$. But C and C' have equal X-rays in the direction p_2, so by the Cavalieri principle, Lemma 1.2.2, $\lambda_2(C) = \lambda_2(C')$, a contradiction. Next, suppose that C_1 is a component of int $(K \triangle K')$ whose closure does not meet l. We may assume that $p_1^{-1} C_1 = C_2$ is defined. By Lemma 5.2.5, $\nu_1(C_1) = \nu_1(C_2)$. There is a component C_3 of int $(K \triangle K')$, disjoint from C_1 and C_2, having the same X-ray as C_2 in the direction p_2. By Lemma 5.4.1, $\nu_1(C_3) = \nu_1(C_2)$. In this way, we can inductively construct a sequence of disjoint components of int $(K \triangle K')$, all having the same ν_1-measure.

Suppose either that $l \cap \text{int } K \neq \emptyset$ or that l supports K and there is a line $t \neq l$ supporting K at some $q \in K \cap l$. Then the fact, often used in the previous section, that int $(K \triangle K')$ is of finite ν_1-measure provides a contradiction. If l is the only line supporting K at all points of $K \cap l$, we resort to a "chord-chasing" argument similar to that employed in Theorem 5.3.1. ∎

As we mentioned earlier, it is not known if the previous theorem is true if both points are finite and the line joining them does not meet K (see Problem 5.1). Our ignorance here extends to the case when only one point is finite. If neither point is finite, however, Theorem 1.2.5 provides counterexamples.

We can now prove the projective analogue of Theorem 5.3.6.

Theorem 5.4.4. *Convex bodies are determined by directed X-rays at any set of three noncollinear points in the projective plane.*

Proof. Let K and K' be different convex bodies with the same directed X-rays at the noncollinear points p_j, $1 \le j \le 3$. At least one of the points, p_1 say, must be finite. Theorem 5.4.3 allows us to assume that none of the lines joining pairs of the points meets K.

Suppose that p_2 and p_3 are both at infinity. For $j = 2, 3$, let $v_1^{(j)}$ be the measure with the line through p_1 and p_j as base line. Let σ be the measure defined for measurable sets E by

$$\sigma(E) = \lambda_2(E) + v_1^{(2)}(E) + v_1^{(3)}(E).$$

Suppose that C is a component of int $(K \triangle K')$ of maximal σ-measure, and let H be the open half-plane containing C and bounded by the line t through the endpoints of C in bd $K \cap$ bd K'. If $p_1 \in H$, then $p_1 C$ has, by Lemma 5.2.5, the same $v_1^{(j)}$-measure as C for $j = 2, 3$, and larger λ_2-measure, by Lemma 5.2.7(i) with $i = 1$ and $k = 2$. So $\sigma(p_1 C) > \sigma(C)$, a contradiction. If $p_1 \notin H$, then the line t intersects at least one of the two rays s_2, s_3 issuing from p_1 in the directions p_2, p_3, respectively, and bounding an angular region containing K. Suppose that t meets s_2. Let $C' \neq C$ be the component of int $(K \triangle K')$ with the same X-rays as C in the direction p_2. Then C' has the same λ_2-measure as C, by the Cavalieri principle, Lemma 1.2.2, and the same $v_1^{(2)}$-measure, by Lemma 5.4.1. Lemma 5.4.2 tells us, however, that $v_1^{(3)}(C') > v_1^{(3)}(C)$. (Here we are identifying the line through p_1 and p_3 with the x-axis, and the line through p_1 and p_2 with the w-axis, with its direction chosen suitably.) Therefore $\sigma(C') > \sigma(C)$, again a contradiction.

We are left with the case that p_2, say, is also a finite point and p_3 is at infinity. Redefine the measure σ by $\sigma(E) = \lambda_2(E) + v_1(E)$, where v_1 has the line through p_1 and p_2 as base line. Let C be a component of int $(K \triangle K')$ of maximal σ-measure, with t and H defined as before. We argue as in the previous paragraph. If $p_j \in H$ for $j = 1$ or 2, then $p_j C$ has the same v_1-measure as C and larger λ_2-measure. On the other hand, if $p_j \notin H$ for $j = 1$ or 2, then the component $C' \neq C$ of int $(K \triangle K')$, with the same X-ray as C in the direction p_3, has the same λ_2-measure as C, but larger v_1-measure. Either way we have a contradiction. ∎

The same techniques yield the corresponding generalizations of the theorems in the previous section concerning X-rays. For example, we have the following uniqueness theorems.

Theorem 5.4.5. *Convex bodies are determined by X-rays at any set of four points in the projective plane, with no three collinear.*

Theorem 5.4.6. *Convex bodies are determined by X-rays at any set of collinear points in the projective plane not projectively equivalent to a subset of directions of the edges of a regular polygon.*

We omit the proofs of these results.

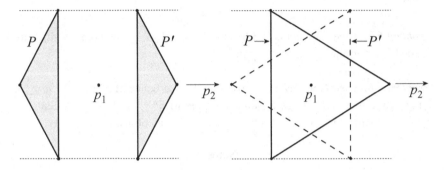

Figure 5.12. Convex bodies with equal X-rays at two points in the projective plane.

It is interesting to note that some of the following open problems can be settled for the projective plane. Consider Problem 5.2, when p_1 is finite and p_2 is at infinity. The examples portrayed in Figure 5.12 answer both parts of this question negatively. It is also easy to answer Problem 5.3 in the projective plane. Let K be a centered regular $(2n + 1)$-gon K, and K' a rotation of K by $\pi/(2n + 1)$ about the origin. Then K and K' have the same X-rays at the interior point o and equal X-rays at $2n + 1$ mutually nonparallel directions parallel to the edges of the convex hull of $K \cup K'$.

Open problems

Problem 5.1. (See Theorem 5.3.15 and Note 5.2.) Is a convex body K determined by two directed X-rays at points p_1 and p_2 if the line through p_1 and p_2 misses K?

Problem 5.2. (See Figure 5.8 and Note 5.2.) Suppose that p_1 and p_2 are points on a line l meeting the interior of a convex body K. Is K determined by its X-rays at p_1 and p_2 if either (i) p_1 and p_2 are not in int K or (ii) $p_1 \in$ int K and $p_2 \notin K$?

Problem 5.3. (See Figure 5.10.) Is a convex body K in $\mathbb{E}^2$ determined by X-rays at p_1, p_2, and p_3 if $p_3 \in$ int K and the line through p_1 and p_2 misses K?

Problem 5.4. Are convex bodies in $\mathbb{E}^2$ determined by X-rays at any set of collinear points not considered in Theorem 5.3.9?

Problem 5.5. (See Note 5.3.) Find a viable method to reconstruct an unknown convex body K in $\mathbb{E}^2$ from its directed X-rays at three noncollinear points, or from its X-rays at four points, with no three collinear.

Problem 5.6. Obtain suitable stability results for point X-rays of planar convex bodies.

Problem 5.7. (See Theorems 5.3.6 and 5.3.8.) How many point X-rays are needed to determine a convex body in $\mathbb{E}^n$?

Problem 5.8. (See Note 5.7.) Suppose that K is a convex body in $\mathbb{E}^n$, $n \geq 3$, lying inside a closed surface S homeomorphic to S^{n-1}. If K appears centrally symmetric when viewed from each point of S, is K an ellipsoid?

Notes

5.1. *Point X-rays and chordal symmetrals.* X-rays and directed X-rays at points can both be found in the literature, but not under separate names. The notion of a chordal symmetral goes back to the study of equichordal bodies, bodies whose chordal symmetrals are balls, and the notorious equichordal problem, the history of which is discussed in Note 6.3. Our notation follows that of Lutwak [537]; here, and in [283], the term "equichordal body" is used instead of chordal symmetral. The term "directed chordal symmetral" is introduced here for the first time. Theorem 5.1.5 was first proved by Longinetti [515, Theorem 2], by a quite different method.

Black et al. [68] undertake a detailed study of the structure of a single directed X-ray at the origin of a planar convex body K not containing the origin. They show, for example, that in addition to the convexity of $D_o K$ (cf. Theorem 5.1.5), the derivatives of the directed X-ray must satisfy certain a priori estimates near the supporting rays and near points of zero curvature. In a continuation of this work, Butcher, Medin, and Solmon [133] investigate geometric and topological properties of the set of convex bodies in the plane with a common directed X-ray.

5.2. *Point X-rays of planar convex bodies.* Hammer [380] first considered the possibility that convex bodies might be determined, or at least verified, by a finite number of X-rays at points; see Note 1.2. The first studies were made, independently, by Falconer [220] and Gardner [256]. Both papers contain versions of Theorem 5.3.1, and also Theorem 5.3.3(i) and (iii), concerning directed X-rays and X-rays at two points. The proofs are different, and different again from the proofs given here, though some notion of "chord chasing" is an essential ingredient to all. (This idea seems to have first been employed by Süss [794]. It was also used by Rogers [695] to prove Corollary 5.3.4 for the case when the two points are interior to the body.) The final form of the author's shorter proof of uniqueness was only written after the preprint of Falconer's paper was circulated; the latter also treats the problem of reconstruction (see Note 5.3).

Falconer returns to the question of determination by X-rays at two points in [218], where he re-proves his uniqueness results by applying the stable manifold theorem of differentiable dynamics. He also shows that differentiability of a convex body can be deduced from that of its X-rays. For example, if $k \in \mathbb{N}$ and the X-rays of a convex body K at the interior points p_1 and p_2 belong to C^k, then K is of class C^k. The same holds if p_1 and p_2 are exterior to K, in the situation of Theorem 5.3.3(i); here, however, differentiability is not assured at the points where the tangent lines to K through p_1 and p_2 meet bd K. Higher-dimensional analogues are also noted in [218].

Falconer suggested to the author that Theorem 5.3.5 should be easy to prove, and the result appeared in [256]. This was only a small step toward the first genuine uniqueness theorem. The idea of replacing Lebesgue measure by the measure ν_1 of Definition 5.2.3, which had seeds in work of Finch, Smith, and Solmon and of Falconer (see [220, Lemma 2]), is due to Volčič. In his important contribution [821], he proved the preliminary Lemmas 5.2.4, 5.2.5, and 5.2.7, for the case $i = 1$, and Theorem 5.3.3 as it is stated here. In the same paper are to be found the uniqueness theorems, Theorems 5.3.6, 5.3.7, and 5.3.8. Volčič also uses Lemma 5.2.6 for $i = 1$, but an explicit proof first appears in work of Gardner [257], where Lemmas 5.2.5–5.2.7 are given in full generality. The latter paper also establishes Theorem 5.3.9.

The author's opinion, that examples similar to those depicted in Figure 5.8 really exist, is based on some interesting computer studies carried out by A. Volčič with the help of G. Michelacci.

Theorem 5.3.11, on verification, was obtained by Volčič [822].

Theorems 5.3.13 and 5.3.15 are proved by Gardner [260]. They give a partial answer to Problem 5.1, but it is unlikely that they significantly contribute to finding its solution, which remains the first priority in this area.

When Volčič's results appeared in [821], they were already formulated as theorems about X-rays in the projective plane. Essentially everything in Section 5.4 can be found in [821], with the exception of Theorem 5.4.6, which is taken from [257]. Many of the results in this chapter extend to spaces of constant curvature. For the hyperbolic plane and sphere, this is carried out for Corollary 5.3.4 by Fejes Tóth and Kemnitz [228], and for Theorem 5.4.6 it is sketched by Kurusa [485]. Dulio and Peri [195], [196] establish a version of Theorem 5.3.3 that holds in spaces of constant curvature. They do the same for Theorem 5.3.8 in [197] by showing that convex bodies in a plane of constant curvature are determined (up to a reflection in the origin, in the case of the sphere) by point X-rays at four points in general position.

5.3. *Reconstruction.* Falconer [220] not only proves the uniqueness theorems given in this chapter for X-rays at two points, but also gives a method for reconstruction from the X-rays in this case. Naturally, this is more involved. Consider, for example, Theorem 5.3.3(iii), where the points p_1 and p_2 belong to the interior of the unknown convex body K, and suppose that the line through the points is the x-axis and meets the boundary of K at points q_1 and q_2, where q_1, p_1, p_2, and q_2 lie on the x-axis in that order. It is first shown in [220] that by considering certain integrals of the X-rays of K at p_1 and p_2, one can find two equations, one nonlinear, from which q_1 and q_2 can be found. Next, maps P_1 and P_2 are defined, as follows. Suppose that w is any point. For $j = 1, 2$, let $u_j(w)$ be a direction parallel to the line through p_j and w, and let $P_j(w)$ be the point, if it exists, on this line such that the line segment $[w, P_j(w)]$ contains p_j and has length $X_{p_j}K\left(u_j(w)\right)$. By using an argument reminiscent of the proof of Lemma 6.3.9 in the next chapter, it is shown in [220] that there is a neighborhood U of q_2 such that if $w \in U$, the sequence $\{(P_2 P_1)^n(w)\}$ converges to a point $F(w)$ in the intersection of U with the x-axis. Furthermore, it is proved that F is a continuous map on the intersection of U with any line parallel to the x-axis. From this it is easy to conclude that the X-rays of K at p_1 and p_2 determine bd K inside U, and hence that the whole of bd K can be determined. We quote from the introduction of [220]: "Production of a practical reconstruction algorithm and the error analysis, whilst feasible, would be complicated since this would involve the solution of non-linear simultaneous equations and equations defined by the limit of an iterative procedure." In fact, an attempt to implement Falconer's method has been made by

Dartmann [178], who easily locates the points q_1 and q_2, but has to resort to guessing bd K inside U in order to reconstruct K.

In a rather exhaustive study, Lam and Solmon [490] provide an algorithm for the reconstruction of a convex polygon from one directed X-ray at the origin. They include a Matlab program and some examples of the implemented algorithm. The latter is based on a result, discovered independently by the author, that says that a "wedge" of the polygon (i.e., its intersection with a double cone with vertex at the origin containing no vertices of the polygon in its interior, as in Theorem 5.3.13) can be reconstructed from four values of the directed X-ray at angles in this double cone.

Michelacci [610] gives a method for reconstructing several points on the boundary of a strictly convex set from two parallel X-rays and one X-ray at a point.

As was outlined in Note 1.5, algorithms for reconstructing convex bodies from possibly noisy parallel X-rays have been implemented by electrical engineers, and some of these will easily adapt to point X-rays. None of this work solves Problem 5.5, in which the number of X-rays is limited. However, it is anticipated that the method of Gardner and Kiderlen [273] (see Note 1.5) should adapt to provide a solution to this problem.

5.4. *Well-posedness.* There are two papers devoted to the well-posedness problem associated with Theorem 5.3.6. Given three noncollinear points p_j, $1 \leq j \leq 3$, one can consider the map taking a convex body $K \in \mathcal{K}_0^2$ to the triple $(D_{p_1}K, D_{p_2}K, D_{p_3}K)$, which belongs to $(\mathcal{K}_0^2)^3$ by virtue of Theorem 5.1.5. The injectivity of this map is provided by Theorem 5.3.6, of course, and its continuity was established by Mägerl and Volčič [558]. One then knows that the problem of reconstructing K from the data $(D_{p_1}K, D_{p_2}K, D_{p_3}K)$ is well posed if it can be proved that the inverse of the map is also continuous. In [558], this was only shown under a certain boundedness assumption, but this defect is removed by Bianchi and Volčič [63]. The latter paper also deals with the well-posedness problem associated with Theorem 5.4.4.

5.5. *Point X-rays in higher dimensions.* Few results seem to be available for point X-rays in $\mathbb{E}^n$, $n \geq 3$, and none that do not follow trivially from those in this chapter. The latter include all those concerning X-rays at two points, such as Theorems 5.3.1 and 5.3.3 and Corollary 5.3.4. Suppose, for example, that K is a convex body in $\mathbb{E}^n$, $n \geq 3$, and that p_1 and p_2 are points. Theorem 5.3.1 or 5.3.3 can be applied, under the conditions stated in them, to determine all the 2-dimensional sections of K containing the line through p_1 and p_2, and so K itself is determined. Similarly, suppose that the hypotheses of Corollary 5.3.4 hold for a convex body K in $\mathbb{E}^n$. Then Corollary 5.3.4 says that each 2-dimensional section of K containing the line through p_1 and p_2 is centrally symmetric about the midpoint of $[p_1, p_2]$, and it follows that K is also.

5.6. *Discrete point X-rays.* Let E be a finite set in $\mathbb{E}^n$ and let $p \in \mathbb{E}^n$. The *discrete point X-ray of E at p* is the function giving the number of points in E on each line containing p; this corresponds to replacing λ_1 by λ_0 in Definition 5.1.2. A systematic study of discrete point X-rays of lattice sets (i.e., finite subsets of $\mathbb{Z}^n$) was initiated by Dulio, Gardner, and Peri [193]. They first establish the corresponding version of Lemma 2.3.2, proving that for any finite set P in $\mathbb{Z}^2$, there are different lattice sets with the same discrete point X-rays at the points in P. The proof is considerably more involved than that of Lemma 2.3.2, utilizing the existence of arbitrarily long arithmetic progressions of relatively prime numbers.

The remaining results in [193] focus on the determination of convex lattice sets in $\mathbb{Z}^2$ from their discrete point X-rays at points in a set P in $\mathbb{Z}^2$. It is easy to see that such sets are not generally determined if P is an arbitrary (even infinite) subset of $\mathbb{Z}^2$ contained in a line, if the sets are allowed to meet the line. Henceforth, then, in this note it is assumed that the convex lattice sets are restricted so that they do not meet any line joining two points of P. With this understanding, a version of Theorem 5.3.3 concerning determination by X-rays at two points is obtained, and, moreover, it is shown by examples that the assumptions are necessary, so that unlike the continuous case (cf. Problems 5.1 and 5.2) no open problems remain. More number theory is used to show that convex lattice sets in $\mathbb{Z}^2$ are not determined by their discrete point X-rays at any three collinear lattice points. Projective transformations allow results analogous to those of Gardner and Gritzmann [267] (see Note 2.2) to be obtained. Specifically, convex lattice sets in $\mathbb{Z}^2$ are determined by discrete point X-rays at four lattice points in order on a line whose cross ratio is not equal to 4/3, 3/2, 2, 3, or 4; discrete point X-rays at six collinear lattice points are generally not enough, but discrete point X-rays at any seven collinear lattice points determine convex lattice sets in $\mathbb{Z}^2$. Finally, classical projective geometry is employed in [193] to show that the discrete version of Theorem 5.3.8 is false: There is a set P of four points in general position in $\mathbb{Z}^2$ such that convex lattice sets are not determined by their discrete point X-rays at the points in P. It is not known at present whether discrete point X-rays at some three noncollinear lattice points, or any seven lattice points, suffice for determining convex lattice sets in $\mathbb{Z}^2$.

It is convenient to mention here that Dulio [192] has introduced the notion of a point X-ray in a graph, though the concept is not directly related to the discrete point X-ray as defined above.

5.7. *Point projections.* Suppose that K is a convex body in $\mathbb{E}^n$. If $p \notin K$, call the support of the directed X-ray $D_p K$ the *point projection* of K at p. One can ask which sets P of points have the property that K is determined by the λ_{n-1}-measure of the point projections of K for $p \in P$. In the plane, this just means that the angle between the tangent lines to K through p is known for each $p \in P$. Green [331] shows that if this angle is constant for all p belonging to a circle containing K in its interior, then K must be a disk if and only if the angle is an irrational multiple of π or a reduced rational multiple of π in which the numerator is even. Nitsche [650] proves that K is a disk if this angle is constant for points on each of two concentric circles containing K in their interiors. In work of Kincses and Kurusa in [433], [434], [435], [486], and [487], conditions are found under which planar convex polygons or bodies are determined by point projections from points on one or more curves.

In higher dimensions, little seems to be known about inverse problems in which the data concern the λ_{n-1}-measure of point projections from sets of points. Suppose, however, that a convex body K is contained inside a closed surface S in $\mathbb{E}^n$, $n \geq 3$, and that for each $p \in S$ the point projection of K at p is the intersection of S^{n-1} and a spherical convex cone with vertex p; in other words, K looks spherical from each point of S. Matsuura [583] proves that K is a ball, at least when S is the surface of a convex body. Bianchi and Gruber [60] find (their result is actually more general) that if K looks ellipsoidal (in the obvious sense) from each point of S, then K is an ellipsoid. They conjecture that the answer to Problem 5.8 is positive, and note that this would have the interesting corollary that the only n-dimensional convex billiard tables with caustics are ellipsoidal, thereby solving an open problem discussed by Gruber [363] (see also [449, Problem 1.8, p. 76]). J. Höbinger asked if a convex body must be an ellipsoid if it lies between two parallel hyperplanes and its projection from each point of one hyperplane onto the other is centrally symmetric. Burton and Larman [121] give examples that show this is not true, but provide a positive answer under a stronger hypothesis. Arocha, Montejano, and Morales [13] find a short proof of latter.

5.8. *Wilhelm Süss (1895–1958) and the Japanese school.* (Sources: [285], [482], [722].) Wilhelm Süss was born in Frankfurt, Germany. His studies in Freiburg and Göttingen were interrupted by service in the first world war, and he obtained his doctorate in Frankfurt in 1920. (While a soldier, he used to carry Hilbert's *Grundlagen der Geometrie (Foundations of Geometry)* with him.) After a year as assistant to Bieberbach in Berlin, he took a position in Kagoshima, Japan, where he supervised

German language and literature. Here he came into contact with Japanese geometers, and found enough time for mathematical research to make his Habilitation a formality when he returned to Germany in 1928. Süss spent six years in Greifswald, and then became professor at the University of Freiburg, later serving as rector. He was president of the German Mathematical Society from 1938 to 1945.

Figure 5.13. Wilhelm Süss.

In addition to these accomplishments, Süss was the founder of the Mathematical Institute at Oberwolfach in 1944, which he directed until his death in 1958. Some astute political maneuvering, and compromise, were necessary during World War II (see [556] for commentary), and Süss and his wife endured post-war years when invited speakers were reluctant to attend the poorly funded and inadequately heated institute. Persistence was rewarded, and today Oberwolfach, in its beautiful Black Forest setting, is perhaps the single most famous – and agreeable – venue in the world, offering generously subsidized weekly conferences in most areas of mathematics.

Apparently it was Soji Nakajima (Matsumura), a graduate of Tohoku Imperial University, and later professor at the Taipei University of Taiwan, who drew Süss into the Tohoku group of geometers while a colleague of his at Kagoshima. Two others of this group are worthy of special mention; like Süss, they were responsible for several key ideas in the next two chapters.

Matsusaburo Fujiwara (1881–1946) was born in Tsu. His studies were rewarded with a silver watch from the Japanese Emperor Meiji on graduation from Tokyo Imperial University, where he became a lecturer in 1905. In 1908, Fujiwara traveled to Europe, where he studied in Göttingen, Berlin, and Paris. On his return in 1911, he became professor at the Tohoku Imperial University at Sendai, founded the same year. Here he stayed until his retirement in 1945. His house was destroyed during an air raid on Sendai, so during his final year he lived with his son's family. During his career, Fujiwara wrote more than a hundred papers on geometry, analysis, number theory, and applied mathematics, as well as several textbooks. In 1925 he was elected to the Imperial Academy of Japan, at whose request he studied the history of old Japanese mathematics in his later years. He lectured on his findings in the presence of the emperor.

Tadahiko Kubota (1885–1952) was born in Tokyo, where he finished his studies at the Tokyo Imperial University in 1908. After teaching at the First Higher School of Tokyo, he was appointed to the Tohoku Imperial University at Sendai in 1911. The following year, the Japanese government sent Kubota to study in England, France, and Germany. On his return in 1915 he became full professor at the Tohoku Imperial University, where he served until his retirement in 1946. He was then elected to the Japan Academy of Science. In 1945 Kubota, like Fujiwara, saw his house destroyed in an air raid on Sendai, and some time after moved to Tokyo, where from 1948 on he directed the Institute for Mathematical Statistics. Kubota's bibliography lists nearly 200 items, mainly in geometry and analysis.

6

Chord functions and equichordal problems

Section 5.3 was devoted to point X-rays of planar convex bodies. This chapter takes a wider perspective in two important ways. First, the objects of interest will sometimes be star bodies. The class of star bodies contains the class of convex bodies, and the larger class is not only natural, but, as we shall see in Chapter 8, essential in certain circumstances. Second, we shall consider data of a more general type, the i-chord functions of a star body, where i is a real number. Such a function deals with ith powers of distances from the origin to the boundary of the body, and, when $i = 1$, reduces to the ordinary X-ray at the origin.

Again, there is a very definite purpose for this generalization. Many of the results in Sections 6.1 and 6.2 will yield, via Theorem 7.2.3 in the next chapter, corresponding information on the determination of bodies by their ith section functions. For example, glance ahead at Corollary 7.2.11, which implies that there are nonspherical convex bodies in $\mathbb{E}^3$ with all of their sections by planes through the origin having the same area. The connection is simple; to calculate the area of such a section, one integrates the square of the radial function, so 2-chord functions are implicated.

By allowing i to take other values, problems of the equichordal type can also be treated in a unified manner. The action here centers around the famous equichordal problem. One form of this asks whether a convex body can contain two equichordal points, points having the same property as the center of a disk, in that all the chords passing through them have the same length. The problem dates back to 1916, and well-known variations of the problem correspond to the values $i = 0$ and $i = -1$. Easy to state, the equichordal problem resisted many assaults and has only recently capitulated. (C. A. Rogers's first line of advice to those contemplating working on the problem – "Don't!" – was often quoted.) The latest results are too long and complicated to present here, but Theorem 6.3.10 is a significant ingredient, and Note 6.3 offers some farther information.

Some of the first part of the chapter is intimately connected to Chapter 5, with no extra demands on the reader regarding measure theory. However, several rather involved definitions must be absorbed before any progress can be made. Every attempt is made to render these more palatable, since, as we indicated earlier, the level of generality is dictated by genuine concerns.

6.1. i-chord functions and i-chordal symmetrals

We begin with the key definition for this chapter, formulated in terms of the radial function, to which there is a brief introduction in Section 0.7 (see also Figure 0.3). Also present in the definition is the concept of a star set, defined in Section 0.7. A star set is a star body in its linear hull; every star body is a body star-shaped at o, but the converse is not true. For a body L to qualify as a star body, it must also have a radial function ρ_L continuous on its support S_L. This helps eliminate certain difficulties explained after Theorem 6.1.3, but still guarantees that every convex body, regardless of its position with respect to the origin, is a star body. Similarly, the assumption that $L - p$ is a star body is stronger than saying that L is a body star-shaped at the point p.

Definition 6.1.1. Let $i \in \mathbb{R}$, and suppose that L is a star set in $\mathbb{E}^n$. If $i \leq 0$, we assume here and throughout that $o \in \operatorname{relint} L$ or $o \notin L$. The i-*chord function* $\rho_{i,L}$ of L at o is defined for $u \in S^{n-1}$ as follows. If u is not in the domain of ρ_L, we define $\rho_{i,L} = 0$. Otherwise, if $i \neq 0$, we let

$$\rho_{i,L}(u) = \begin{cases} \rho_L(u)^i + \rho_L(-u)^i & \text{if } o \in L, \\ \left| |\rho_L(u)|^i - |\rho_L(-u)|^i \right| & \text{if } o \notin L. \end{cases}$$

For $i = 0$, we define the 0-*chord function of L at o* for $u \in S^{n-1}$ by

$$\rho_{0,L}(u) = \begin{cases} \rho_L(u)\rho_L(-u) & \text{if } o \in L, \\ \exp\left| \log|\rho_L(u)/\rho_L(-u)| \right| & \text{if } o \notin L. \end{cases}$$

Now suppose that p is any point and that $L - p$ is a star set. Then the i-*chord function of L at p* is $\rho_{i,L-p}$.

Definition 6.1.1 may seem a little intimidating, but it is actually quite simple. Let p be a point, and consider the distances from p to the boundary of L along a line through p parallel to u. If $p \in L$, the i-chord function of L at p, when defined, gives the sum of the ith powers of these distances or the product of these distances, according as $i \neq 0$ or $i = 0$, respectively. If, on the other hand, $p \notin L$, then it provides the difference (the greater less the smaller) of the ith powers of these distances or the quotient (the greater over the smaller) of these distances, according as $i \neq 0$ or $i = 0$, respectively. The extra assumption that $p \in \operatorname{relint} L$ or $p \notin L$ (in other words, L does not contain p in its relative boundary) is inserted to avoid singularities.

Observe that when $i = 1$, the i-chord function of L at p is just the X-ray of L at p. In fact, the i-chord function is closely related to the X-ray of order i (cf. Definition 5.2.1). When singularities are avoided – for example, when L is a star set not containing the origin, or when $i > 0$ – Lemma 5.2.2 shows that the X-ray of order i is the i-chord function divided by i, if $i \neq 0$, and the log of it, when $i = 0$. This also justifies the definition of the 0-chord function, when $o \notin L$; if $o \in L$, the definition becomes natural when it is observed that

$$\lim_{i \to 0} \left(\frac{1}{2} \rho_{i,L}(u) \right)^{1/i} = \sqrt{\rho_{0,L}(u)},$$

where the quantity in the limit on the left is the ith mean (cf. (B.2)) of $\rho_L(u)$ and $\rho_L(-u)$, while that on the right is the zeroth or geometric mean of $\rho_L(u)$ and $\rho_L(-u)$.

The support of $\rho_{i,L}$ is $S_L \cup S_{-L}$. (This is clear from the definition of $\rho_{i,L}$ if $o \in L$, and when $o \notin L$ it follows from the observation that if $u \in S_L$, then there is a sequence $\{u_m\}$ of points in S^{n-1}, converging to u, such that $|\rho_L(u_m)| \neq |\rho_L(-u_m)|$, $m \in \mathbb{N}$.) Since ρ_L is a Borel function on S^{n-1}, we conclude from this that $\rho_{i,L}$ is a bounded even Borel function on S^{n-1}.

To each of the tomographic functions discussed in this book, we have associated an appropriate symmetral. For parallel X-rays, this was the Steiner symmetral; for the width function, the central symmetral; for the brightness function, the Blaschke body; and for the X-ray at a point, the chordal symmetral at that point. In each case, the symmetrized set has the same corresponding tomographic function as the original set. The following concept of symmetral is suitable for the i-chord function.

Definition 6.1.2. Let $i \in \mathbb{R}$, and let L be a star body in $\mathbb{E}^n$. Suppose either that $i > 0$ or that $i \leq 0$ and $o \in \text{int } L$. The i-*chordal symmetral* $\tilde{\nabla}_i L$ of L is the centered set defined by $\tilde{\nabla}_L$

$$\rho_{i, \tilde{\nabla}_i L}(u) = \rho_{i,L}(u),$$

for all $u \in S^{n-1}$. If L is an n-dimensional star set in $\mathbb{E}^m$, $m \geq n$, then $\tilde{\nabla}_i L$ is defined as before, where the linear hull of L is identified with $\mathbb{E}^n$. When $i = n-1$, we write $\tilde{\nabla}_{n-1} L = \tilde{\nabla} L$, and refer to $\tilde{\nabla} L$ as the *dual Blaschke body* of L.

If $i = 1$, then $\tilde{\nabla}_1 L = \tilde{\Delta} L$ is just the chordal symmetral of L. For $i \leq 0$, we assume that $o \in \text{int } L$ in order that $\tilde{\nabla}_i L$ be bounded. When $o \in \text{int } L$, the definition says that

$$\rho_{\tilde{\nabla}_i L}(u) = \left(\frac{\rho_L(u)^i + \rho_L(-u)^i}{2} \right)^{1/i},$$

if $i \neq 0$, and

$$\rho_{\tilde{\nabla}_0 L}(u) = \sqrt{\rho_L(u)\rho_L(-u)},$$

for all $u \in S^{n-1}$.

Theorem 6.1.3. *If L is a star set (or star body) in $\mathbb{E}^n$ such that S_L is centered, then $\tilde{\nabla}_i L$ is also a star set (or star body, respectively).*

Proof. The sets L and $\tilde{\nabla}_i L$ have the same linear hull, so we may assume that this is $\mathbb{E}^n$, and therefore L is a star body. Since S_L is centered, the support of $\rho_{i,L}$ is S_L, and this also implies that the restriction of $\rho_{i,L}$ to S_L is continuous. We deduce that $S_{\tilde{\nabla}_i L} = S_L$, and the restriction of $\rho_{\tilde{\nabla}_i L}$ to $S_{\tilde{\nabla}_i L}$ is continuous. Moreover, $S_{\tilde{\nabla}_i L}$ is compact and regular in S^{n-1}, since this is true of S_L. It is a routine matter to show that the compactness and regularity of $\tilde{\nabla}_i L$ now follow from the corresponding properties of $S_{\tilde{\nabla}_i L}$. ∎

The previous theorem warrants some discussion. Observe that it implies that $\tilde{\nabla}_i L$ is a star body whenever L is a star body and $o \in \operatorname{int} L$ or $o \notin L$. Simple examples show that this convenient property is not shared by bodies merely star-shaped at the interior or exterior point o, whose chordal symmetrals, for example, need not even be bodies. On the other hand, as we saw after Definition 5.1.3, the chordal symmetral of a star (even convex) body L need not be a body if $o \in \operatorname{bd} L$, so the same is true for the i-chordal symmetral. This difficulty is eliminated by the extra assumption that S_L is centered. Notice that if $o \notin L$ and $i > 0$, $\tilde{\nabla}_i L$ is itself a set containing o, but not in its interior, whose radial function has a centered support.

If $o \in \operatorname{int} L$, and $i \neq 0$,

$$\rho_{\tilde{\nabla}_i L}(u)^i = \frac{1}{2}\left(\rho_L(u)^i + \rho_{-L}(u)^i\right),$$

for all $u \in S^{n-1}$. When $i = n - 1$, it becomes clear that the $(n-1)$th powers of radial functions in the definition of the dual Blaschke body correspond to the surface area measures in Definition 3.3.8 of the Blaschke body. Because of this, the dual Blaschke body is aptly named. As was explained in Note 3.4, the ith area measures for integers i with $1 < i < n - 1$ behave differently from those with $i = 1$ or $i = n - 1$, and there is no "ith Blaschke body" for these values of i. In complete contrast to this, there are no extra difficulties with ith powers of radial functions for $1 < i < n - 1$.

Note that $\tilde{\nabla}_i L = L$ if and only if L is centered. Figure 6.1 shows some of the i-chordal symmetrals of an equilateral triangle K whose centroid is at o. By Jensen's inequality (B.3) for means, the i-chordal symmetrals increase in size as i increases. For $i > -1$, they are rather similar in general shape to the chordal symmetral depicted for the same triangle in Figure 5.2; in particular, they are not convex. As i decreases to $-\infty$, the i-chordal symmetrals approach the centered hexagon inscribed in K, and as the pictures for $i = -1$ and $i = -2$ in Figure 6.1 suggest, they are convex for $i \leq -1$. These are instances of the next theorem.

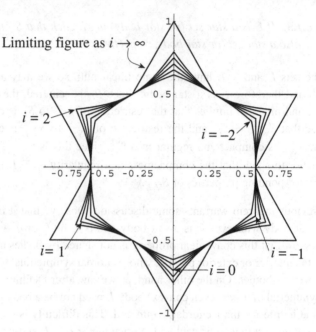

Figure 6.1. i-chordal symmetrals of an equilateral triangle.

Theorem 6.1.4. *If $i \leq -1$ and K is a convex body in $\mathbb{E}^n$ with $o \in \operatorname{int} K$, then the i-chordal symmetral $\tilde{\nabla}_i K$ is convex.*

Proof. Let $u, v \in S^{n-1}$. Using first Lemma 5.1.4 and then Minkowski's inequality (B.4) with $p = -i \geq 1$, we obtain

$$\rho_{\tilde{\nabla}_i K} (u + v)^{-1}$$

$$= \left(\frac{1}{2} \rho_K (u + v)^i + \frac{1}{2} \rho_{-K} (u + v)^i \right)^{-1/i}$$

$$\leq \left(\frac{1}{2} \left(\rho_K (u)^{-1} + \rho_K (v)^{-1} \right)^{-i} + \frac{1}{2} \left(\rho_{-K} (u)^{-1} + \rho_{-K} (v)^{-1} \right)^{-i} \right)^{-1/i}$$

$$\leq \left(\frac{1}{2} \rho_K (u)^i + \frac{1}{2} \rho_{-K} (u)^i \right)^{-1/i} + \left(\frac{1}{2} \rho_K (v)^i + \frac{1}{2} \rho_{-K} (v)^i \right)^{-1/i}$$

$$= \rho_{\tilde{\nabla}_i K} (u)^{-1} + \rho_{\tilde{\nabla}_i K} (v)^{-1}.$$

By Lemma 5.1.4, this yields the convexity of $\tilde{\nabla}_i K$. ∎

It is interesting to note that when K is a convex body with $o \in \text{int } K$, $\tilde{\nabla}_{-1}K$ is just the polar body of the central symmetral of the polar body of K! (Compare Figures 3.3 and 6.1.) This is because (0.36) gives

$$h_{(\tilde{\nabla}_{-1}K)^*} = \rho_{\tilde{\nabla}_{-1}K}^{-1} = \frac{1}{2}\rho_K^{-1} + \frac{1}{2}\rho_{-K}^{-1} = \frac{1}{2}h_{K^*} + \frac{1}{2}h_{-K^*} = h_{\Delta(K^*)}.$$

When L is a star body with $o \notin L$, $\tilde{\nabla}_i L$ is (in the plane, at least) figure-eight shaped, as Figure 5.3 shows for L a disk and $i = 1$. As i decreases to zero, $\tilde{\nabla}_i L$ converges to the singleton set $\{o\}$, and as i increases to ∞, $\tilde{\nabla}_i L$ converges to the set M with radial function

$$\rho_M = \max\{\rho_L, 0\}.$$

Moreover, each "half" of $\tilde{\nabla}_i L$ is convex; to prove this, just follow the proof for $i = 1$ given in Theorem 5.1.5, using Minkowski's inequality (B.4) for $p = -i < 0$ instead of for $p = -1$.

There is a volume inequality for $\tilde{\nabla}_i L$; see Theorem 7.2.2.

6.2. Chord functions of star sets

In this section i-chord functions are used to determine star sets. We shall work in the plane, though it should be clear that some of the results extend to $\mathbb{E}^n$.

The first issue to be addressed is the contentious one of equality versus equality almost everywhere for i-chord functions. In Chapter 5, we noted that the upper half of the unit disk and a centered disk of radius $1/2$ have X-rays at the origin equal almost everywhere, but not everywhere. It was also remarked that this phenomenon cannot occur for convex bodies not containing the origin in their boundaries. The following lemma shows that this remains true for star bodies.

Lemma 6.2.1. *Let $i \in \mathbb{R}$, and for $j = 1, 2$, let L_j be a star body in $\mathbb{E}^n$ such that S_{L_j} is centered. If $\rho_{i,L_1}(u) = \rho_{i,L_2}(u)$ for λ_{n-1}-almost all $u \in S^{n-1}$, then $\rho_{i,L_1} = \rho_{i,L_2}$.*

Proof. For $j = 1, 2$, S_{L_j} is centered, so the support of ρ_{i,L_j} is S_{L_j}, and the latter is compact and regular in S^{n-1}. The set

$$S = \{u \in S_{L_1} \cup S_{L_2} : \rho_{L_1}(u) = \rho_{L_2}(u)\}$$

is the complement in $S_{L_1} \cup S_{L_2}$ of a set of λ_{n-1}-measure zero. From the regularity of S_{L_1} and S_{L_2} it follows that S is dense in $S_{L_1} \cup S_{L_2}$, and compactness implies that $S_{L_1} = S_{L_2}$. Furthermore $\rho_{i,L_1} = \rho_{i,L_2}$ on $S_{L_1} = S_{L_2}$, since ρ_{i,L_j} is continuous on its support, for $j = 1, 2$. The lemma follows. ∎

This means that if the i-chord functions of two star bodies at a point p agree almost everywhere, and p is either contained in the interior of the bodies or exterior to them, then the i-chord functions at p agree everywhere. When p is exterior to both bodies, equality of the i-chord functions at p also implies that the bodies have common supporting lines through p.

Our investigations proceed in much the same fashion as for the special case $i = 1$; the connection with X-rays of order i permits the use of the lemmas proved in Section 5.2. The next result is a generalization of parts (i) and (iii) of Theorem 5.3.3; the generalization of part (ii) of that theorem involves some technical difficulties we wish to avoid.

Theorem 6.2.2. *Let $i \in \mathbb{R}$, and suppose that K, K' are convex bodies and that p_1, p_2 are distinct points in $\mathbb{E}^2$ such that K and K' have i-chord functions equal almost everywhere at p_j, $j = 1, 2$. Suppose further that*

(i) *the line l through p_1 and p_2 meets int K, p_1 and p_2 do not belong to int K, and K and K' either both meet $[p_1, p_2]$ or are both disjoint from $[p_1, p_2]$, or*

(ii) *p_1, p_2 belong to int K.*

Then $K = K'$, if $i > 0$. Further, $K = K'$ if $i \leq 0$ and $v_i(K \triangle K')$ is finite, where l is the base line of the measure v_i of Definition 5.2.3.

Proof. For $i \neq 2$, the proof is a modification of that of the relevant parts of Theorem 5.3.3. For Theorem 5.3.3(i), we appealed to Theorem 5.3.1, involving directed X-rays. The analogous notion of a directed i-chord function, at a point p not in the interior of a body, can be obtained from the i-chord function by setting it equal to zero at $u \in S^1$ if the ray issuing from p in the direction u does not meet the body in a point other than p itself. With this understanding, the result corresponding to Theorem 5.3.1 can be obtained simply by replacing the measure v_1 by v_i. (Note that Lemma 5.2.5 and Lemma 5.2.7(i) apply to all i, and Lemma 5.2.7(ii) to all $i \neq 2$; moreover, our assumptions make Lemma 5.2.4 redundant for $i \leq 0$.) The arguments of Theorem 5.3.3(i) and (iii) then transfer in the same way to yield the proof for $i \neq 2$.

When $i = 2$, Lemma 5.2.7(ii) is unavailable. This was only used in the first part of the proof of Theorem 5.3.1, but we must resort to other methods to replace it. We have to show that no component C of int $(K \triangle K')$ has its closure meeting l in case (i), when the points p_1 and p_2 are exterior to and lie on the same side of K and K'.

Suppose, initially, that $l \cap C \neq \emptyset$, and also that bd K and bd K' each meet l at points at distances r_1 and s_1, and r_2 and s_2, respectively, from p_1, with $r_j < s_j$, $j = 1, 2$. Then $s_1^2 - r_1^2 = s_2^2 - r_2^2$. If p_1 and p_2 are a distance b apart, with p_2 farther from $K \cap l$, we also have

$$(s_1 + b)^2 - (r_1 + b)^2 = (s_2 + b)^2 - (r_2 + b)^2.$$

Now for $j = 1, 2$,

$$(s_j + b)^2 - (r_j + b)^2 = \int_{r_j^2}^{s_j^2} (1 + bt^{-1/2})\, dt.$$

For $j = 1, 2$, the interval of integration is of the same length, and the integrand decreases as t increases. Therefore $r_1 = r_2$ and $s_1 = s_2$, contradicting the assumption on C.

Therefore $l \cap C = \emptyset$. There is a component C' of int $(K \triangle K')$, disjoint from C, such that C and C' have equal 2-chord functions at p_1 and p_2. In case (i) we know that the ν_1-measure of C and C' is finite, by Lemma 5.2.4 and convexity, so Lemma 5.2.6 tells us that the centroids c, c' of C and C', with respect to the measure ν_1, lie on the same line $l_j \neq l$ through p_j, for $j = 1, 2$. But then $c = c' = l_1 \cap l_2$, which is impossible since C and C' are disjoint. ∎

We shall see in Theorem 6.2.10 (see also Figure 6.3 and Remark 6.3.6) that knowledge of the i-chord functions of a convex body at two points on a line meeting its interior does not, in general, determine the body uniquely.

Our next goal is to relax the convexity condition. For the applications we have in mind, we only need to consider i-chord functions at interior points.

Theorem 6.2.3. *Let $i \in \mathbb{R}$. Suppose that L_1 and L_2 are bodies in $\mathbb{E}^2$ with common interior points p_1 and p_2 such that $L_j - p_k$ is a star body, for $j = 1, 2$ and $k = 1, 2$. Suppose in addition that L_1 and L_2 have equal i-chord functions at these points. Then $L_1 = L_2$, if $i > 1$. Further, $L_1 = L_2$ if $i \leq 1$ and $\nu_i(L_1 \triangle L_2)$ is finite, where ν_i is the measure of Definition 5.2.3 with the line l through p_1 and p_2 as base line.*

Proof. Our assumptions on L_1 and L_2 imply that each is star-shaped at p_1 and p_2 and that the boundary of each meets l in exactly two points. The proof of Theorem 6.2.2(ii), just that of Theorem 5.3.3(iii), still works. In that proof, convexity is only used to ensure that the ν_1-measure of the symmetric difference of the bodies is finite. Here, ν_i is finite on all bounded sets if $i > 1$, and when $i \leq 1$ our assumption guarantees that $\nu_i(L_1 \triangle L_2)$ is finite. ∎

The assumption in Theorem 6.2.2 that $\nu_i(K \triangle K')$ is finite for $i \leq 0$ (and the similar hypothesis in Theorem 6.2.3) is in general necessary; see Remark 6.3.6. Note that Theorem 6.2.3 requires this for $i \leq 1$, since it is possible that the ν_1-measure, for example, of $L_1 \triangle L_2$ is infinite, if the boundary of L_1 or L_2 is tangent to the line through p_1 and p_2. It may seem tempting to ignore at least the nonpositive values of i, but we shall see in the next section that these can be of natural interest. In Theorem 6.2.14, we obtain differentiability conditions on L_1 and L_2 ensuring that $\nu_i(L_1 \triangle L_2)$ is finite.

Corollary 6.2.4. *If $i \in \mathbb{R}$ is positive, and the i-chord functions of a convex body K in $\mathbb{E}^2$ at distinct points p_1, p_2 are equal, then K is centrally symmetric about the midpoint of the segment $[p_1, p_2]$. If $i > 1$, this also holds for a body L such that $L - p_j$ is a star body for $j = 1, 2$.*

Proof. The proof of the first statement is completely analogous to the proof of Corollary 5.3.4 when Theorem 6.2.2 is used in place of parts (i) and (iii) of Theorem 5.3.3. For the second statement, we use Theorem 6.2.3 instead. ∎

The previous corollary has application in Theorem 7.2.17. Some of the other uniqueness results of Chapter 5 also extend to i-chord functions. We shall just state a partial generalization of Theorem 5.3.9.

Theorem 6.2.5. *Let $i \in \mathbb{R}$. Let P be a finite set of points on a line l in $\mathbb{E}^2$, not projectively equivalent to a subset of directions of the edges of a regular polygon, and suppose that K and K' are convex bodies in $\mathbb{E}^2$ not meeting l and with equal i-chord functions at the points in P. Then $K = K'$.*

Proof. This is exactly the same as the last part of the proof of Theorem 5.3.9, once we replace the measures ν_1 and ν_0 in that proof by ν_i and ν_{i-1}, respectively. ∎

The next two theorems generalize Theorem 5.3.13 and Theorem 5.3.15 from X-rays to i-chord functions; we shall apply the results in Chapter 7. The notation is exactly as in those theorems, and all polygons are assumed to be nondegenerate.

Theorem 6.2.6. *Suppose that $i \in \mathbb{R}$ is positive. Two polygons P and P' have X-rays of order i at o that are equal almost everywhere if and only if for each appropriate double cone $C(\alpha, \beta)$ and $0 \le \varphi < \pi$, the family $\{e_n, n \in N(\varphi)\}$ of edges satisfies*

$$\sum \{a_n |r_n|^i : n \in N(\varphi)\} = 0.$$

Proof. Suppose that $i > 0$. By Lemma 5.2.2, the X-ray of order i of a *convex* polygon Q, when multiplied by i, is precisely the i-chord function of Q. A general polygon P is a union of triangles with disjoint interiors; the X-ray of order i of Q is merely the sum of the X-rays of order i of these triangles, and hence, when multiplied by i, the sum of their i-chord functions. By Lemma 5.3.12, the i-chord function of the triangle T with vertices at o, (r, α), and (s, β) $(r > 0, s > 0, 0 \le \alpha < \beta \le \pi)$ is given, for $\alpha \le \theta \le \beta$, by

$$\left(\frac{rs \sin(\beta - \alpha)}{s \sin(\beta - \theta) + r \sin(\theta - \alpha)} \right)^i .$$

As in the proof of Theorem 5.3.13, it follows that $P \cap C(\alpha, \beta)$ and $P' \cap C(\alpha, \beta)$ have X-rays of order i at o that are equal almost everywhere if and only if

$$\sum a_n \left(\frac{|r_n||s_n| \sin(\beta - \alpha)}{|s_n| \sin(\beta - \theta) + |r_n| \sin(\theta - \alpha)} \right)^i = 0,$$

for $\alpha < \theta < \beta$, where the sum runs over all edges e_n meeting the interior of $C(\alpha, \beta)$, e_n contains the points (r_n, α) and (s_n, β), and $a_n = \pm 1$ is the appropriate weight. Using the condition $i > 0$, we can now follow the rest of the proof of Theorem 5.3.13 almost verbatim, when ith powers are taken of the appropriate expressions in that proof. The resulting condition is that stated previously.   ∎

Theorem 6.2.7. *Let $i \in \mathbb{R}$ be positive. Two convex polygons are identical if their i-chord functions agree almost everywhere at each of two distinct points.*

Proof. The proof of Theorem 5.3.15 can be modified to apply to i-chord functions. One must first replace the phrase "parallel strips with equal X-rays" with "parallel strips with equal i-chord functions," and substitute the previous theorem for Theorem 5.3.13, to obtain the corresponding version of Lemma 5.3.14. The proof is completed by making a similar replacement and the substitution of Theorem 6.2.2(ii) for Theorem 5.3.3(iii) in the proof of Theorem 5.3.15.   ∎

It seems that i-chord functions for $i \leq 0$ behave somewhat differently. In particular, the last result is not true for $i = -1$; see Theorems 6.2.9 and 6.2.10 and Figures 6.2 and 6.3. To prove this, we shall employ an interesting duality between -1-chord functions and parallel X-rays, described in the first part of the next theorem. As usual, we regard the projective plane $\mathbb{P}^2$ as $\mathbb{E}^2$ with a line at infinity adjoined.

Theorem 6.2.8. *Let l be a line not at infinity in $\mathbb{P}^2$. Suppose that ϕ is a nonsingular projective transformation taking l to the line at infinity, and let K and K' be convex bodies in $\mathbb{E}^2$ not meeting l.*

(i) If p is a finite point on l, then K and K' have equal -1-chord functions at p if and only if ϕK and $\phi K'$ are convex bodies with equal parallel X-rays in a direction corresponding to ϕp.

(ii) If p is a finite point with $p \notin l$, then K and K' have equal -1-chord functions at p if and only if ϕK and $\phi K'$ have equal -1-chord functions at the finite point ϕp.

Proof. We can assume that l is the x-axis, so ϕ is of the form

$$\phi(x, y) = \left(\frac{a_1 x + b_1 y + c_1}{y}, \frac{a_2 x + b_2 y + c_2}{y} \right).$$

(Compare (0.2).) Then ϕ is nonsingular if and only if $a_1 c_2 \neq a_2 c_1$. Since ϕ is permissible for K and K', both ϕK and $\phi K'$ are convex bodies (see Section 0.3).

To prove (i), we may suppose that p is the origin. A small calculation shows that ϕ maps lines through p onto lines with slope c_2/c_1 (vertical lines if $c_1 = 0$). A line m through p with slope t meets the boundary of K in points $q_j = (x_j, tx_j)$, $j = 1, 2$, where x_1 and x_2 must have the same sign. Then

$$(\|q_1 - p\|^{-1} - \|q_2 - p\|^{-1})^2 = \left(\frac{1}{1+t^2} \right) \left(\frac{1}{x_1} - \frac{1}{x_2} \right)^2.$$

Further, we find that

$$\|\phi q_1 - \phi q_2\|^2 = \left(\frac{c_1^2 + c_2^2}{t^2}\right)\left(\frac{1}{x_1} - \frac{1}{x_2}\right)^2.$$

Similar expressions can be derived when m is the y-axis. It follows that the value of the -1-chord function of K along m is a constant (depending only on ϕ and the slope of m) times the value of the parallel X-ray of the body ϕK along the line ϕm. This fact suffices to prove (i).

For the proof of (ii), we may suppose that $p = (0, d), d \neq 0$. A line m through p with slope t meets the boundary of K in points $q_j = (x_j, tx_j + d), j = 1, 2$. After some computation we find that

$$(\|q_1 - p\|^{-1} \pm \|q_2 - p\|^{-1})^2 = \left(\frac{1}{1 + t^2}\right)\left(\frac{1}{|x_1|} \pm \frac{1}{|x_2|}\right)^2$$

and

$$(\|\phi q_1 - \phi p\|^{-1} \pm \|\phi q_2 - \phi p\|^{-1})^2 = A\left(\left|\frac{tx_1 + d}{x_1}\right| \pm \left|\frac{tx_2 + d}{x_2}\right|\right)^2,$$

where $A = d^2/((a_1 d - c_1 t)^2 + (a_2 d - c_2 t)^2)$. Again, similar expressions apply when m is the y-axis. Suppose that $p \notin K$. Then $\phi p \notin \phi K$; moreover, x_1 and x_2 have the same sign, and $tx_1 + d$ and $tx_2 + d$ also have the same sign. Using these facts, and selecting the minus sign in the previous expressions, we see that the value of the -1-chord function of K in a direction parallel to m is a constant (depending only on ϕ and the slope of m) times the value of the -1-chord function of ϕK in a direction parallel to ϕm. If $p \in K$, we note that $\phi p \in \phi K$, that x_1 and x_2 have opposite signs, and that $tx_1 + d$ and $tx_2 + d$ have the same sign. Selecting the plus sign in the previous expressions, we arrive at the same conclusion. Part (ii) of the theorem follows.   ∎

Theorem 6.2.9. *Suppose that P is a finite set of points on a line l in $\mathbb{E}^2$, projectively equivalent to a subset of directions of the edges of a regular polygon. Then there are different convex polygons disjoint from l and with equal -1-chord functions at the points in P.*

Proof. Let $p_j, 1 \leq j \leq k$, be a finite set of points, projectively equivalent to a subset of directions of the edges of a regular polygon and lying on a line l. Then there is a nonsingular projective transformation ϕ, taking l to the line at infinity, such that $\phi p_j, 1 \leq j \leq k$, correspond to a subset of directions of the edges of a regular polygon. As in Theorem 1.2.5, there are different regular polygons Q and Q' with equal parallel X-rays in these directions. By translating these polygons, if necessary, we may assume that ϕ^{-1} is permissible for Q and Q'.

Then Theorem 6.2.8(i) shows that the sets $\phi^{-1}Q$ and $\phi^{-1}Q'$ are convex polygons (cf. Section 0.3) with the same -1-chord functions at each of the points p_j, $1 \leq j \leq k$.   ∎

Examples such as those just constructed are shown in Figure 6.2.

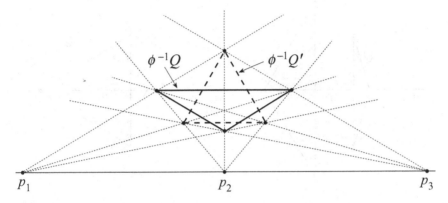

Figure 6.2. Different polygons with equal -1-chord functions.

Theorem 6.2.10. *There are distinct convex polygons Q and Q' with equal -1-chord functions at two points p_1 and p_2 such that the line through p_1 and p_2 meets the interiors of Q and Q'.*

Proof. There are distinct convex polygons P and P' contained in the open half-plane $\{(x, y) : x > 0\}$ with the same parallel X-rays in the direction of the x-axis and the same -1-chord functions at a point $q \neq o$ on the x-axis (see Figure 5.12, for example, where in that figure p_1 corresponds to the point we are now calling q, and the line through p_1 parallel to the direction p_2 is the x-axis). In Theorem 6.2.8, let l be the y-axis; then the projective transformation ϕ defined by

$$\phi(x, y) = \left(\frac{1}{x}, \frac{y}{x}\right)$$

is a suitable choice. Define $p_1 = \phi^{-1}q$. Then p_1 is on the x-axis, and by Theorem 6.2.8(ii), the convex polygons $Q = \phi^{-1}P$ and $Q' = \phi^{-1}P'$ have equal -1-chord functions at p_1. Using Theorem 6.2.8(i), and setting $p_2 = o$, we note that ϕp_2 corresponds to a direction parallel to the x-axis, so Q and Q' also have equal -1-chord functions at p_2.   ∎

Polygons constructed as in Theorem 6.2.10 are depicted in Figure 6.3. A similar example exists for $i = 0$, where the points p_1 and p_2 are the centers of similitude for two different disks, whose centers lie on the line through p_1 and p_2. It is interesting to note that the existence of corresponding examples for $i = 1$ has not yet been established; see Problem 5.2.

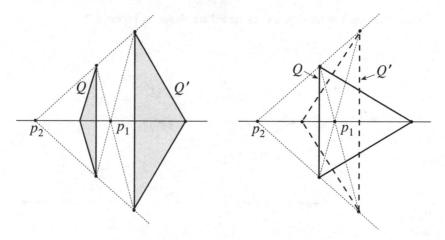

Figure 6.3. More polygons with equal -1-chord functions.

In Theorems 6.2.2 and 6.2.3 we were forced to assume that for some values of i the ν_i-measure of the symmetric difference of two bodies is finite. We now undertake to find suitable differentiability properties that obviate the need for the irritating extra condition.

Let L be a body in $\mathbb{E}^2$ with two distinct interior points p_1 and p_2 such that $L - p_j$ is a star body, $j = 1, 2$. Then L is star-shaped at p_1 and p_2. Moreover, if p_1 and p_2 lie on a line l, then the boundary of L meets l at points q_1 and q_2, with $q_1, p_1, p_2,$ and q_2 in that order on l, say. Let the boundary of L have equation $r = r(\theta)$ in polar coordinates centered at p_1, and equation $s = s(\varphi)$ in polar coordinates centered at p_2, both with polar axes lying on l and similarly oriented, so that

$$q_1 = \left(r(\pi), \pi\right) = \left(s(\pi), \pi\right)$$

and

$$q_2 = \left(r(0), 0\right) = \left(s(0), 0\right).$$

In the next two lemmas we collect some useful information about derivatives of r and s and refer the reader to Figure 6.4.

Lemma 6.2.11. *Let L be as before. Suppose that $p = (r, \theta) = (s, \varphi), 0 \le \varphi \le \pi$, and let $\theta = \theta(\varphi)$ be defined by this equation. If r and s are differentiable, then*
(i) $d\theta/d\varphi = s/r$, *when* $\varphi = 0$ *and* π, *and*
(ii) $\mathrm{arccot}\left(-r'/r\right) - \theta = \mathrm{arccot}\left(-s'/s\right) - \varphi.$

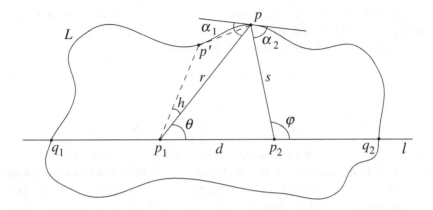

Figure 6.4. Illustration for Lemma 6.2.11.

Proof. Let d be the distance between p_1 and p_2. Then

$$\theta = \arctan\left(\frac{s\,\sin\varphi}{d + s\,\cos\varphi}\right).$$

Differentiating with respect to φ, and setting $\varphi = 0$ (or $\varphi = \pi$), we get $s/(s+d)$ (or $s/(s-d)$, respectively). Both the latter expressions equal s/r, at $\varphi = 0$ or π, respectively, proving (i).

Let α_j be the angle between the tangent line to bd L at p and the line segment $[p, p_j]$, $j = 1, 2$, as in Figure 6.4. By considering the triangle formed by p_1, p, and a point $p' = (r, \theta+h)$, and letting h tend to zero, we find that $\cot\alpha_1 = -r'/r$; similarly, we obtain $\cot(\pi - \alpha_2) = -s'/s$. Now $\alpha_1 - \theta = \pi - \alpha_2 - \varphi$, and (ii) follows. ∎

Lemma 6.2.12. *Let L be as before, let $i \in \mathbb{R}$ be negative, and let $k \in \mathbb{N}$ satisfy $1 \le k < -i + 1$. If r and s are at least k-times differentiable, then for $j = 1, 2$, the i-chord function of L determines the kth derivative of r and s at q_j.*

Proof. The i-chord function of L at p_1 is $r(\theta)^i + r(\theta + \pi)^i$. Taking the kth derivative, we get

$$ir(\theta)^{i-1}r^{(k)}(\theta) + ir(\theta + \pi)^{i-1}r^{(k)}(\theta + \pi) + f(\theta),$$

where f is an expression involving r and its first $(k-1)$ derivatives at θ and $(\theta + \pi)$. After deriving a similar expression for the kth derivative of the i-chord function of L at p_2, and substituting $\theta = 0$ and $\varphi = 0$, we obtain

$$r(0)^{i-1}r^{(k)}(0) + r(\pi)^{i-1}r^{(k)}(\pi) = a \qquad (6.1)$$

and

$$s(0)^{i-1}s^{(k)}(0) + s(\pi)^{i-1}s^{(k)}(\pi) = b. \qquad (6.2)$$

Here a (or b) depends only on r and its first $(k-1)$ derivatives (or on s and its first $(k-1)$ derivatives, respectively), evaluated at 0 and π.

We claim that two further equations hold, namely,

$$s(0)^k r^{(k)}(0) - r(0)^k s^{(k)}(0) = a' \tag{6.3}$$

and

$$s(\pi)^k r^{(k)}(\pi) - r(\pi)^k s^{(k)}(\pi) = b', \tag{6.4}$$

where a' and b' are constants depending only on the values of r and s and their first $(k-1)$ derivatives at 0 and π. When $k = 1$, these equations hold with $a' = b' = 0$, as can be seen by putting θ and φ both equal to 0 and both equal to π in Lemma 6.2.11(ii). Suppose that $k > 1$. Differentiating the equation in Lemma 6.2.11(ii) $(k-1)$ times with respect to φ, we get an equation of the form

$$U\left(\frac{d\theta}{d\varphi}\right)^{k-1} r^{(k)} + A = V s^{(k)} + C. \tag{6.5}$$

Here, A (or C) depends only on r and θ (or s, respectively) and their first $(k-1)$ derivatives, and

$$U = \frac{1}{r\left(1 + (r'/r)^2\right)} \quad \text{and} \quad V = \frac{1}{s\left(1 + (s'/s)^2\right)}.$$

Now suppose that $\theta = 0$ or π. Then Lemma 6.2.11(ii) yields $r'/r = s'/s$, and we also have

$$\left(\frac{d\theta}{d\varphi}\right)^{k-1} = \left(\frac{s}{r}\right)^{k-1},$$

by Lemma 6.2.11(i). Substituting these expressions into (6.5) and simplifying, we obtain (6.3) and (6.4), when $\theta = 0$ and π, respectively.

The determinant of the system of equations (6.1)–(6.4) has absolute value

$$\left| r(0)^{i+k-1} s(\pi)^{i+k-1} - r(\pi)^{i+k-1} s(0)^{i+k-1} \right|.$$

Since $i + k - 1 \neq 0$, this vanishes if and only if

$$\frac{r(\pi)s(0)}{r(0)s(\pi)} = 1.$$

This is impossible, since the expression on the left, the cross ratio $\langle p_1, p_2, q_1, q_2 \rangle$, is different from one. Therefore the system (6.1)–(6.4) determines the kth derivatives of r and s at the points q_j, $j = 1, 2$. When $k = 1$, the right-hand sides of (6.1)–(6.4) depend only on the i-chord functions of L at p_1 and p_2, so r' and s' can be found when $\theta = 0$ or π. Once the derivatives of r and s of orders up to $(k-1)$ have been determined at $\theta = 0$ and π, the system (6.1)–(6.4) then furnishes the kth derivatives. The proof of the lemma is now completed by induction on k. ■

Lemma 6.2.13. *Let $i \in \mathbb{R}$. Suppose that L_1 and L_2 are bodies in $\mathbb{E}^2$ with common interior points p_1 and p_2 such that $L_j - p_k$ is a star body, for $j = 1, 2$ and $k = 1, 2$. Suppose also that L_1 and L_2 have equal i-chord functions at these points. Then the boundaries of L_1 and L_2 meet the line l through p_1 and p_2 at common points q_1 and q_2.*

Proof. For $j = 1, 2$, the assumption that $L_j - p_1$ is a star body implies that the boundary of L_j meets l in exactly two points. The boundaries of L_1 and L_2 must intersect; let $x_1 \in \operatorname{bd} L_1 \cap \operatorname{bd} L_2$. If $x_1 \in l$, then we are finished. If not, we can inductively construct a sequence $\{x_n\}$ of points as follows. For each $n \in \mathbb{N}$, let $y_n \neq x_n$ be the point in $\operatorname{bd} L_1 \cap \operatorname{bd} L_2$ on the line through x_n and p_2, and let x_{n+1} be the point in $\operatorname{bd} L_1 \cap \operatorname{bd} L_2$ on the line through y_n and p_1. Then $\{x_n\}$ converges to a point $q_1 \in l \cap \operatorname{bd} L_1 \cap \operatorname{bd} L_2$, and it follows that there must be a second such point q_2. ∎

When $i > 0$, the convexity of L_1 and L_2 is enough to ensure that the hypotheses of the previous lemma imply that $L_1 = L_2$, as was proved in Theorem 6.2.2(ii). The next theorem provides conditions suitable for $i \leq 0$.

Theorem 6.2.14. *Let $i \in \mathbb{R}$, $i \leq 0$, and let $m = [i]$. Suppose that L_1 and L_2 are bodies in $\mathbb{E}^2$ with common interior points p_1 and p_2 such that $L_j - p_k$ is a star body, for $j = 1, 2$ and $k = 1, 2$. Suppose also that L_1 and L_2 have equal i-chord functions at these points. Assume that the boundaries of L_1 and L_2 belong to $C^{1-m+\varepsilon}(S^1)$, for $j = 1, 2$ and some ε with $0 < \varepsilon < 1$, and that they have equal $(1 - m)$th derivatives, in polar coordinates centered at p_1, say, at their common boundary point q_1 on the line l. Then $L_1 = L_2$.*

Proof. The point q_1 in the statement of the theorem is one of the points whose existence is guaranteed by Lemma 6.2.13.

We may assume that l is the x-axis and p_1 is the origin. Suppose that the equation of $\operatorname{bd} L_j$ is given in polar coordinates by $r = r_j$, $j = 1, 2$. Our assumptions imply that there is a θ_0 with $0 < \theta_0 < \pi/2$ such that for $|\theta| \leq \theta_0$, r_1 and r_2 are positive $(1 - m)$-times differentiable functions and

$$|r_j^{(1-m)}(\theta) - r_j^{(1-m)}(0)| \leq c_j |\theta|^\varepsilon,$$

for suitable constants c_1 and c_2 (cf. (0.39)). Let $A = \{(r, \theta) : r > 0, |\theta| \leq \theta_0\} \cap \operatorname{int}(L_1 \triangle L_2)$. It will suffice to prove that $\nu_i(A)$ is finite, where ν_i is the measure of Definition 5.2.3 with base line l. (This is because Lemma 5.2.5 can then be used to show that $A' = \{(r, \theta) : r > 0, \pi - \theta_0 \leq \theta \leq \pi + \theta_0\} \cap \operatorname{int}(L_1 \triangle L_2)$, and hence $L_1 \triangle L_2$, is of finite ν_i-measure, and then Theorem 6.2.3 can be applied.) Since $m \leq i$, it is clear from Definition 5.2.3 that it will suffice to show that $\nu_m(A)$

is finite. We have

$$v_m(A) = \int_{-\theta_0}^{\theta_0} \left| \int_{r_1(\theta)}^{r_2(\theta)} r^{m-1} \sin^{m-2}\theta \, dr \right| d\theta$$

$$= \int_{-\theta_0}^{\theta_0} |t_1(\theta) - t_2(\theta)| \sin^{m-2}\theta \, d\theta,$$

where we have set $t_j = r_j^m/m$ when $m \neq 0$ and $t_j = \log r_j$ when $m = 0$, $j = 1, 2$.

Since r_j is nonzero for $|\theta| \leq \theta_0$, it follows that for $j = 1, 2, t_j$ is also $(1-m)$-times differentiable and satisfies

$$|t_j^{(1-m)}(\theta) - t_j^{(1-m)}(0)| \leq c_j' |\theta|^\varepsilon,$$

for $|\theta| \leq \theta_0$ and suitable constants c_1' and c_2'.

By Taylor's theorem, for $j = 1, 2$,

$$t_j(\theta) = \sum_{k=0}^{-m} t_j^{(k)}(0)\theta^k/k! + t_j^{(1-m)}(\theta_j)\theta^{1-m}/(1-m)!,$$

for $|\theta| \leq \theta_0$ and some θ_j strictly between 0 and θ. Since $r_1(0) = r_2(0)$, we have $t_1(0) = t_2(0)$. The relation $m = [i]$ implies that $-m < -i + 1$, so when $i < 0$, Lemma 6.2.12 and the equality of the i-chord functions yield $r_1^{(k)}(0) = r_2^{(k)}(0)$, for $1 \leq k \leq -m$. Our assumptions allow us to conclude that this equation, and hence also the equation $t_1^{(k)}(0) = t_2^{(k)}(0)$, holds for $0 \leq k \leq 1 - m$. Together with the previous paragraph, this gives

$$|t_1(\theta) - t_2(\theta)|$$

$$= \left| t_1^{(1-m)}(\theta_1) - t_2^{(1-m)}(\theta_2) \right| |\theta|^{1-m}/(1-m)!$$

$$= \left| \left(t_1^{(1-m)}(\theta_1) - t_1^{(1-m)}(0) \right) - \left(t_2^{(1-m)}(\theta_2) - t_2^{(1-m)}(0) \right) \right| |\theta|^{1-m}/(1-m)!$$

$$\leq (c_1'|\theta_1|^\varepsilon + c_2'|\theta_2|^\varepsilon)|\theta|^{1-m}/(1-m)! \leq c|\theta|^{1-m+\varepsilon},$$

for $|\theta| \leq \theta_0$. This establishes the finiteness of $v_m(A)$, since the previous integrand is $O(|\theta|^{\varepsilon-1})$ for $|\theta| \leq \theta_0$. ∎

The assumptions we make in the last theorem cannot be weakened too much; see Remarks 6.3.6 and 6.3.14.

The results presented so far in this section are necessary because it is generally impossible to determine a body by its i-chord function at a single point. Instead of collecting data at more than one point, we might choose to measure the i-chord function at one point, but for different values of i. We now turn to some theorems of this type. Our first aim is to show that the assumption that two star sets have equal i-chord functions for two different values of i is enough to ensure that they are equal for all i. We need the following simple lemma.

Lemma 6.2.15. *Suppose that $p \in \mathbb{R}$, $p \neq 0, 1$. If x_j and y_j, $j = 1, 2$, are nonnegative reals satisfying*

$$x_1 + y_1 = x_2 + y_2$$

and

$$x_1^p + y_1^p = x_2^p + y_2^p,$$

then $\{x_1, y_1\} = \{x_2, y_2\}$.

Proof. We may suppose that $x_1 \leq x_2$ and that $y_2 \leq y_1$. Using the first equation, we see that x_2 and y_2 lie between x_1 and y_1 (or vice versa). We shall therefore assume that

$$x_1 \leq x_2 \leq y_2 \leq y_1$$

and prove that $x_1 = x_2$. The other possibilities can be dealt with similarly.

Suppose that $x_1 < x_2$, so $y_2 < y_1$. By the mean value theorem,

$$x_2^p - x_1^p = pa^{p-1}(x_2 - x_1)$$

and

$$y_1^p - y_2^p = pb^{p-1}(y_1 - y_2),$$

for some $a \in (x_1, x_2)$ and $b \in (y_2, y_1)$. The hypotheses of the theorem now imply that $a^{p-1} = b^{p-1}$ and hence that $a = b$. This is impossible since a and b belong to disjoint intervals. ∎

Theorem 6.2.16. *Let i and j be distinct real numbers, and suppose that L_1, L_2 are star sets in $\mathbb{E}^n$. If $i \leq 0$ or $j \leq 0$, assume that L_1 and L_2 do not contain the origin in their relative boundaries. Then $\rho_{i,L_1} = \rho_{i,L_2}$ and $\rho_{j,L_1} = \rho_{j,L_2}$ if and only if*

$$L_1 \cap l = \pm L_2 \cap l,$$

for each line l containing o, and hence if and only if $\rho_{t,L_1} = \rho_{t,L_2}$ for all $t \in \mathbb{R}$.

Proof. The condition that $L_1 \cap l = \pm L_2 \cap l$ for each line l containing o is equivalent to saying that ρ_{L_1} and ρ_{L_2} have the same domain D and

$$\{\rho_{L_1}(u), \rho_{L_1}(-u)\} = \{\rho_{L_2}(u), \rho_{L_2}(-u)\}, \tag{6.6}$$

for all $u \in D$. It follows immediately from Definition 6.1.1 that $\rho_{t,L_1} = \rho_{t,L_2}$ for all $t \in \mathbb{R}$, and in particular for $t = i$ and $t = j$.

Let us prove the converse. The equation $\rho_{i,L_1} = \rho_{i,L_2}$ implies that these functions have the same domain, so $S_{L_1} \cup S_{-L_1} = S_{L_2} \cup S_{-L_2}$, implying that the linear hulls of L_1 and L_2 are the same. We can therefore assume that this common linear hull is $\mathbb{E}^n$, and hence that both L_1 and L_2 are star bodies.

We claim that $o \in L_1$ if and only if $o \in L_2$. If $i \leq 0$ or $j \leq 0$, then $o \in \operatorname{int} L_1$, and our assumptions then imply that $o \in \operatorname{int} L_2$ also. Suppose that i and j are positive and that $o \in L_1$ but $o \notin L_2$. Let $u \in S^{n-1}$ be such that $\rho_{i,L_2}(u) \neq 0$. Then $\rho_{i,L_1}(u) \neq 0$, and we can assume that $\rho_{L_2}(u) > -\rho_{L_2}(-u) > 0$. Consequently

$$\rho_{L_1}(u)^i + \rho_{L_1}(-u)^i = \rho_{L_2}(u)^i - |\rho_{L_2}(-u)|^i$$

and

$$\rho_{L_1}(u)^j + \rho_{L_1}(-u)^j = \rho_{L_2}(u)^j - |\rho_{L_2}(-u)|^j.$$

Let $x_1 = \rho_{L_1}(u)^i$, $y_1 = \rho_{L_1}(-u)^i$, $x_2 = \rho_{L_2}(u)^i$, and $y_2 = |\rho_{L_2}(-u)|^i$. With $p = j/i$, we have $p > 0$, $p \neq 1$,

$$x_2 = x_1 + y_1 + y_2,$$

and

$$x_2 = (x_1^p + y_1^p + y_2^p)^{1/p}.$$

Since $p \neq 1$, y_2 is nonzero, and x_1 or y_1 is nonzero, these equations contradict Jensen's inequality (B.6) for sums, proving the claim.

Let D_{L_m} denote the domain of ρ_{L_m}, $m = 1, 2$. If L_1 and L_2 contain o, then $D_{L_1} = D_{L_2} = S^{n-1}$. If neither L_1 nor L_2 contains o, then for $m = 1, 2$ we have $S_{L_m} \cup S_{-L_m} = S_{L_m} = D_{L_m}$, and hence $D_{L_1} = D_{L_2}$ again. Let $D = D_{L_1} = D_{L_2}$.

If i and j are nonzero, and both L_1 and L_2 contain o, then

$$\rho_{L_1}(u)^i + \rho_{L_1}(-u)^i = \rho_{L_2}(u)^i + \rho_{L_2}(-u)^i$$

and

$$\rho_{L_1}(u)^j + \rho_{L_1}(-u)^j = \rho_{L_2}(u)^j + \rho_{L_2}(-u)^j,$$

for each $u \in D$. Set $x_1 = \rho_{L_1}(u)^i$, $y_1 = \rho_{L_1}(-u)^i$, $x_2 = \rho_{L_2}(u)^i$, and $y_2 = \rho_{L_2}(-u)^i$. With $p = j/i$, we can apply Lemma 6.2.15 to obtain (6.6) for all $u \in D$. When $j = 0$, say, the same substitutions lead to the equations

$$x_1 + y_1 = x_2 + y_2$$

and (taking ith roots)

$$x_1 y_1 = x_2 y_2,$$

from which the same conclusion follows.

Now suppose that i and j are nonzero and that neither L_1 nor L_2 contains the origin. Let u be such that $\rho_{i,L_1}(u) \neq 0$; then $\rho_{i,L_2}(u) \neq 0$ also. Suppose initially that $\rho_{L_1}(u) > -\rho_{L_1}(-u) > 0$ and that $\rho_{L_2}(u) > -\rho_{L_2}(-u) > 0$. Then

$$\rho_{L_1}(u)^i - |\rho_{L_1}(-u)|^i = \rho_{L_2}(u)^i - |\rho_{L_2}(-u)|^i$$

and

$$\rho_{L_1}(u)^j - |\rho_{L_1}(-u)|^j = \rho_{L_2}(u)^j - |\rho_{L_2}(-u)|^j,$$

for each $u \in D$. Set $x_1 = \rho_{L_1}(u)^i$, $y_1 = |\rho_{L_2}(-u)|^i$, $x_2 = \rho_{L_2}(u)^i$, and $y_2 = |\rho_{L_1}(-u)|^i$. With $p = j/i$, we apply Lemma 6.2.15 to obtain

$$\{\rho_{L_1}(u), -\rho_{L_2}(-u)\} = \{\rho_{L_2}(u), -\rho_{L_1}(-u)\}.$$

Since $\rho_{L_1}(u) \neq -\rho_{L_1}(-u)$, we have $\rho_{L_1}(u) = \rho_{L_2}(u)$. By the continuity of the radial functions on their common domain D (which in this case equals their supports), the same equality holds for all u in the same component of D. Now suppose that $\rho_{L_1}(u) > 0$ and that $\rho_{L_2}(u) < 0$, so $\rho_{L_1}(u) > -\rho_{L_1}(-u) > 0$ and $\rho_{L_2}(-u) > -\rho_{L_2}(u) > 0$. The same considerations now lead to $\rho_{L_1}(u) = \rho_{L_2}(-u)$, for all u in the same component of D. Therefore (6.6) holds for all $u \in D$.

If $j = 0$, say, the other assumptions and the same substitutions lead to the equations

$$x_1 + y_1 = x_2 + y_2$$

and (taking ith roots)

$$x_1/y_2 = x_2/y_1.$$

One can check that the desired conclusion follows as before.   ∎

Corollary 6.2.17. *Let i and j be distinct real numbers, and suppose that L_1, L_2 are star sets in $\mathbb{E}^n$ with equal i-chord functions and equal j-chord functions. If L_2 is centered, then $L_1 = L_2$.*

Proof. Since $L_2 \cap l$ is centered for each line l containing o, Theorem 6.2.16 implies that L_1 is centered and hence that $\rho_{L_1}(u) = \rho_{L_2}(u)$ for each $u \in S^{n-1}$.   ∎

The next two theorems exploit a construction from Chapter 3.

Theorem 6.2.18. *For $n \geq 2$, there are noncongruent convex polytopes P_1 and P_2 in $\mathbb{E}^n$ containing the origin in their interiors and with equal i-chord functions for all $i \in \mathbb{R}$.*

Proof. Let P_1 and P_2 be the noncongruent polytopes of Theorem 3.3.17. These were constructed so that

$$P_1 \cap l = \pm P_2 \cap l,$$

for all lines l containing o. It follows that

$$\{\rho_{P_1}(u), \rho_{P_1}(-u)\} = \{\rho_{P_2}(u), \rho_{P_2}(-u)\},$$

for all $u \in S^{n-1}$, which by Definition 6.1.1 implies that P_1 and P_2 have equal i-chord functions for each $i \in \mathbb{R}$. ∎

Theorem 6.2.19. *For $n \geq 2$, there are noncongruent convex bodies of revolution K_1 and K_2, of class C_+^∞ in $\mathbb{E}^n$, containing the origin in their interiors and with equal i-chord functions for all $i \in \mathbb{R}$.*

Proof. The bodies K_1 and K_2 of Theorem 3.3.18 can be seen to satisfy

$$K_1 \cap l = \pm K_2 \cap l,$$

for all lines l containing o; as in the previous theorem, this suffices. ∎

The last two theorems indicate that in the absence of central symmetry about the origin, strong additional conditions must be imposed in order to obtain uniqueness (up to reflection in the origin) from the equality of i-chord functions for two different values of i. The next two results are of this type.

Theorem 6.2.20. *Let i and j be distinct real numbers, and suppose that L_1, L_2 are star sets in $\mathbb{E}^n$ with equal i-chord functions and equal j-chord functions. If L_2 does not contain the origin and is connected, then $L_1 = \pm L_2$.*

Proof. By Theorem 6.2.16, neither L_1 nor L_2 contains o, and for each u in the common domain D of ρ_{L_1} and ρ_{L_2}, we have

$$\{\rho_{L_1}(u), \rho_{L_1}(-u)\} = \{\rho_{L_2}(u), \rho_{L_2}(-u)\}.$$

By the connectedness of L_2, it is clear that D has two components, U and $-U$, say. If $\rho_{L_1}(v) = \rho_{L_2}(v)$ and $\rho_{L_1}(w) = \rho_{L_2}(-w)$ for some v and w in U, then by the continuity of the radial functions on their supports, there must be some $u \in U$ with $\rho_{L_2}(u) = \rho_{L_2}(-u)$. This can only hold if $o \in L_2$, a contradiction. So either $\rho_{L_1}(u) = \rho_{L_2}(u)$ for each $u \in U$, or $\rho_{L_1}(-u) = \rho_{L_2}(u)$ for each $u \in U$, and the same applies to $-U$. If $\rho_{L_1}(u) = \rho_{L_2}(u)$ for each $u \in U$ and $\rho_{L_1}(-u) = \rho_{L_2}(u)$ for each $u \in -U$, then $\rho_{L_2}(u) = \rho_{L_2}(-u)$ for each $u \in U$, a contradiction as before. It follows that $L_1 = \pm L_2$. ∎

Note that the previous theorem applies when L_2 is a convex body not containing the origin.

The following example is instructive. Let L_1 be the union of three disjoint disks in $\mathbb{E}^2$, each of radius less than $1/2$, whose centers lie in the unit circle and are equally spaced around it. Let L_2 be the set obtained from L_1 by replacing one of the disks with its reflection in the origin. The star bodies L_1 and L_2 satisfy all the other assumptions of Theorem 6.2.20 (for any i and j), but L_2 is not connected, and $L_1 \neq \pm L_2$.

Theorem 6.2.21. *Let i and j be distinct real numbers, and suppose that L_1, L_2 are star sets in $\mathbb{E}^n$ with equal i-chord functions and equal j-chord functions. If $o \in \operatorname{relint} L_2$ and ρ_{L_m}, restricted to its support, is real analytic, for $m = 1, 2$, then $L_1 = \pm L_2$.*

Proof. As in the proof of Theorem 6.2.16, we may assume that L_1 and L_2 are star bodies. Since $o \in \operatorname{int} L_2$, $o \in \operatorname{int} L_1$ also. By Theorem 6.2.16, for each $u \in S^{n-1}$, we have

$$\{\rho_{L_1}(u), \rho_{L_1}(-u)\} = \{\rho_{L_2}(u), \rho_{L_2}(-u)\}.$$

Suppose that $L_1 \neq L_2$. There exists a $u_0 \in S^{n-1}$ with $\rho_{L_1}(u_0) \neq \rho_{L_2}(u_0)$. By the continuity of the radial functions on their supports, there is a neighborhood U of u_0 in S^{n-1} such that $\rho_{L_1}(u) \neq \rho_{L_2}(u)$ and hence $\rho_{L_1}(u) = \rho_{L_2}(-u)$, for all $u \in U$. Two real-analytic functions agreeing on a connected open set must coincide, so $\rho_{L_1}(u) = \rho_{L_2}(-u)$ for all $u \in S^{n-1}$. It follows that $L_1 = -L_2$. ∎

Let

$$L_1 = \{(r, \theta) : 0 \leq r \leq 1,\ m\pi/6 \leq \theta \leq (m+1)\pi/6,\ m = 0, 4, 8\}$$

and

$$L_2 = \{(r, \theta) : 0 \leq r \leq 1,\ m\pi/6 \leq \theta \leq (m+1)\pi/6,\ m = 0, 2, 4\},$$

so that L_1 and L_2 are unions of three sectors of the unit disk in $\mathbb{E}^2$, with L_2 obtained from L_1 by reflecting one of the sectors in the origin. The star bodies L_1 and L_2 satisfy all the assumptions of Theorem 6.2.21 (for any i and j), except that $o \notin \operatorname{relint} L_2$, and $L_1 \neq \pm L_2$.

Theorem 6.2.22. *Let i and j be distinct real numbers, and suppose that L_1, L_2 are star sets in $\mathbb{E}^n$. Suppose further that L_2 is centered and $o \in \operatorname{relint} L_2$, and that there exist positive constants a and b such that*

$$\rho_{i,L_1}(u) = a\rho_{i,L_2}(u)$$

and

$$\rho_{j,L_1}(u) = b\rho_{j,L_2}(u),$$

for all $u \in S^{n-1}$. Then L_1 is a dilatate of L_2.

Proof. As in the proof of Theorem 6.2.16, we may assume that L_1 and L_2 are star bodies. Then, since $o \in \operatorname{int} L_2$, we have $o \in \operatorname{int} L_1$. Suppose, initially, that both i and j are nonzero. Using the fact that L_2 is centered, we see that

$$\rho_{L_1}(u)^i + \rho_{L_1}(-u)^i = 2a\,\rho_{L_2}(u)^i$$

and

$$\rho_{L_1}(u)^j + \rho_{L_1}(-u)^j = 2b\,\rho_{L_2}(u)^j,$$

for all $u \in S^{n-1}$. By continuity, there must be a $u_0 \in S^{n-1}$ such that $\rho_{L_1}(u_0) = \rho_{L_1}(-u_0)$. Substituting this u_0 into the last two displayed equations, we see that $b = a^{j/i}$. Let $x = \rho_{L_1}(u)^i/\rho_{L_2}(u)^i$ and $y = \rho_{L_1}(-u)^i/\rho_{L_2}(u)^i$. Then, with $p = j/i$, we obtain

$$\frac{x+y}{2} = a = \left(\frac{x^p + y^p}{2}\right)^{1/p}.$$

Since $p \neq 1$, Jensen's inequality (B.3) for means implies that $x = y$. Therefore $\rho_{L_1}(u) = \rho_{L_1}(-u)$, and the conclusion of the theorem follows from the previous equations.

If $j = 0$, say, the second displayed equation in the previous paragraph must be replaced by

$$\rho_{L_1}(u)\rho_{L_1}(-u) = b\,\rho_{L_2}(u)^2.$$

Setting $u = u_0$, we find that $b = a^{2/i}$. The same substitutions as before yield the equation

$$\frac{x+y}{2} = a = \sqrt{xy},$$

and Jensen's inequality (B.3) for means (here reducing to the arithmetic–geometric mean inequality) applies as before to complete the proof. ∎

The last theorem does not hold for arbitrary centered sets L_2. To see this, consider the star bodies L_1 and L_2 in $\mathbb{E}^2$ defined by

$$L_1 = \{(r, \theta) : 1 \leq r \leq 3,\ 0 \leq \theta \leq \pi/4\}$$

and

$$L_2 = \{(r, \theta) : -1 \leq r \leq 1,\ 0 \leq \theta \leq \pi/4\}.$$

The hypotheses of Theorem 6.2.22 apply with $i = 1$, $j = 2$, $a = 1$, and $b = 4$.

6.3. Equichordal problems

Definition 6.3.1. Let $i \in \mathbb{R}$, and let p be a point in the interior of a body L in $\mathbb{E}^n$. Suppose that $L - p$ is a star body and that there is an $a > 0$ such that the i-chord function of L at p has the constant value a. Then p is called an i-*equichordal point* of L (with constant a). If $i = 1$, we simply refer to p as an *equichordal point* of L. If $i = 0$ (or -1), we also call p an *equiproduct point* (or an *equireciprocal point*, respectively) of L.

For $i = 1$, we can rephrase this definition by saying that p is an equichordal point of L if every chord of L through p has the same length. When $o \in \operatorname{int} L$, o is an i-equichordal point of L if and only if the i-chordal symmetral $\tilde{\nabla}_i L$ is an n-dimensional ball, in view of Definition 6.1.2.

For any $i \in \mathbb{R}$, the origin is an i-equichordal point of a centered ball. There are many other examples, however. It is easy to construct nonspherical star bodies with the origin as an equichordal point, as follows. In polar coordinates, let $r = r(\theta), 0 \le \theta \le \pi, 0 < r(\theta) < 2$, define a curve in the upper half-plane joining the points $(1, 0)$ and $(1, \pi)$. If we let $r(\theta) = 2 - r(\theta - \pi)$, for $\pi < \theta < 2\pi$, then the star body whose boundary is given by $r = r(\theta)$ has the required property. The next theorem improves upon this observation by providing nonspherical convex examples.

Theorem 6.3.2. *For each $i \in \mathbb{R}$ and $n \ge 2$, there is a nonspherical convex body in $\mathbb{E}^n$ with the origin as an i-equichordal point.*

Proof. Let $i \in \mathbb{R}$. Suppose that K' is a nonspherical compact convex set in the $\{x_1, x_n\}$-plane in $\mathbb{E}^n$ with o as an i-equichordal point in that plane and symmetric with respect to the x_n-axis. Let K be the convex body obtained by rotating K' about the x_n-axis; the intersection of K with a hyperplane H orthogonal to this axis is an $(n-1)$-dimensional ball whose diameter is the same as that of the intersection of K' and H. If u is any direction, the vectors u and $-u$ belong to the same 2-dimensional subspace containing the x_n-axis. It follows immediately that K also has o as an i-equichordal point.

Therefore we only need to prove the existence of nonspherical convex bodies in $\mathbb{E}^2$ that are symmetric with respect to the y-axis and have o as an i-equichordal point. To this end, suppose that $f(\theta)$ is a continuous function, not identically zero, that is (i) odd, (ii) of period 2π, and (iii) such that $f(\theta - \pi/2)$ is even. It then follows that

$$f\left(\theta - \frac{\pi}{2}\right) = f\left(-\theta - \frac{\pi}{2}\right) = -f\left(\theta + \frac{\pi}{2}\right),$$

for $0 \le \theta \le 2\pi$, where we have used properties (iii) and (i) of f. Consequently, f also satisfies (iv) $f(\theta) = -f(\theta + \pi)$, for $0 \le \theta \le 2\pi$.

Let $c > 0$ be such that $f(\theta) + c > 0$ for all θ, and

$$r(\theta) = \begin{cases} \left(f(\theta) + c\right)^{1/i} & \text{if } i \ne 0, \\ \exp\left(f(\theta)/c\right) & \text{if } i = 0, \end{cases}$$

for all θ. We then see that for $i \neq 0$,

$$r(\theta)^i + r(\theta + \pi)^i = \left(f(\theta) + c\right) + \left(f(\theta + \pi) + c\right) = 2c,$$

and for $i = 0$,

$$r(\theta)r(\theta + \pi) = \exp\left(\frac{1}{c}f(\theta) + \frac{1}{c}f(\theta + \pi)\right) = 1.$$

Furthermore,

$$f(\theta) = -f(\theta + \pi) = -f(\theta - \pi) = f(\pi - \theta),$$

where we have used properties (iv), (ii), and (i) of f, and therefore

$$r(\theta) = r(\pi - \theta),$$

for all θ. From these properties of $r(\theta)$ we see that if K is a star body whose radial function is $r(\theta)$, then K is symmetric about the y-axis and has o as an i-equichordal point. Furthermore, one can check that K is not a disk.

It remains to show that for suitable f and c, the star body K just defined is actually convex. Suppose that $f \in C^2(\mathbb{R})$. A straightforward calculation shows that for $i \neq 0$,

$$r(\theta) - r''(\theta) =$$

$$\left(f(\theta) + c\right)^{-2+1/i} \left(\left(f(\theta) + c\right)^2 - \frac{1}{i}\left(\frac{1}{i} - 1\right)\left(f'(\theta)\right)^2 - \frac{1}{i}\left(f(\theta) + c\right)f''(\theta)\right),$$

for all θ. When $i = 0$, we obtain

$$r(\theta) - r''(\theta) = \left(1 - \left(\frac{f'(\theta)}{c}\right)^2 - \frac{f''(\theta)}{c}\right)\exp\left(f(\theta)/c\right).$$

In view of the formula (0.41) for curvature in polar coordinates, it suffices to show that $r - r'' \geq 0$, and this is true for sufficiently large c. ∎

The function $f(\theta) = a\sin\theta$, for suitable $a > 0$, provides the simplest example of a function with the required properties. Then, if $i = 1$ or $i = 2$, for example, one may take $a = 1/2$, $c = 1$ or $a = 1$, $c = 3/2$, respectively. The resulting convex bodies for $n = 2$ are illustrated in Figure 6.5. For $i = 0$ one can choose $a = 1$ and $c = 2$.

We now know that for any $i \in \mathbb{R}$ there are many bodies containing one i-equichordal point. The motivating question for the rest of this section is the following one, whose answer is only known for certain values of i (see Problem 6.4).

Question 6.3.3. *Let $i \in \mathbb{R}$. Does there exist a body with two i-equichordal points?*

The answer is known to be negative when $i = 1$; see Note 6.3. Before discussing some results that point to this fact, we shall examine two special values, $i = -1$ and 0, for which this question has a positive answer.

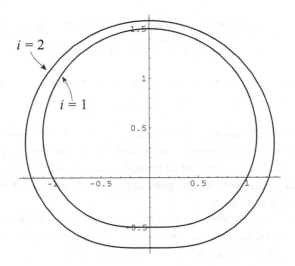

Figure 6.5. Convex bodies with the origin as an i-equichordal point.

Theorem 6.3.4. *An ellipse contains two equireciprocal points, one at each focus.*

Proof. We may assume that the ellipse has eccentricity $e < 1$, one focus at o, and the other at the point $(0, -2c)$. Its equation

$$\frac{e^2(y+c)^2}{c^2} + \frac{e^2 x^2}{c^2(1-e^2)} = 1$$

may be transformed into the polar equation

$$\frac{1}{r} = \frac{e}{c(1-e^2)}(1 + e \sin \theta).$$

(Note that this is of the form $r = \left(f(\theta) + c'\right)^{-1}$, as in the case $i = -1$ of Theorem 6.3.2, where f is a constant multiple of $\sin \theta$.) This polar coordinate equation shows that o is an equireciprocal point, and by symmetry the other focus must be also. ∎

Theorem 6.3.5. *Let p be a point at a distance $d < 1$ from the center of a disk of radius 1. Then p is an equiproduct point, with constant $(1 - d^2)$.*

Proof. Suppose, without loss of generality, that the disk has equation

$$(x - d)^2 + y^2 = 1,$$

where $d < 1$, and $p = o$. Transforming to polar coordinates, we obtain

$$r = \sqrt{1 - d^2 \sin^2 \theta} + d \cos \theta.$$

The boundary of the disk meets the line $\tan\theta = t$ through o at points where

$$r = \left(\sqrt{1+t^2(1-d^2)} \pm d\right)/\sqrt{1+t^2}.$$

Multiplying these two values of r, we get $(1-d^2)$, showing that o is an equiproduct point of the disk, with this constant.   ∎

The trick used in Theorem 6.3.2, of rotating a 2-dimensional body in $\mathbb{E}^n$ around a line about which it is symmetric, can be applied to generate other examples. In this way we see, for example, that an n-dimensional ellipsoid in $\mathbb{E}^n$ contains two equireciprocal points, and that any point in the interior of an n-dimensional ball in $\mathbb{E}^n$ is an equiproduct point.

Theorem 6.3.5 shows that without restriction on i, there is no limit to the number of i-equichordal points, even with the same constant, that a body may contain. The case $i = 0$ may be exceptional, however, as we shall soon see.

Remark 6.3.6. The previous theorem pertains to Theorem 6.2.2 (and similarly to Theorem 6.2.3), which involves an assumption that $v_i(K \bigtriangleup K')$ is finite for $i \leq 0$. In general, this is unavoidable. To see this, consider two congruent and intersecting disks. Any point p on the line segment joining the two points where their boundaries intersect is equidistant from the two centers. From Theorem 6.3.5, we conclude that the two disks have equal (and, in fact, constant) 0-chord functions at p. This also shows that the assumption in Theorem 6.2.14 that the boundaries have equal $(1-i)$th derivatives at a common boundary point is, in general, necessary.

Theorem 6.3.7. *Suppose that $i \in \mathbb{R}$, $i > 1$, that p_1 and p_2 are distinct points in $\mathbb{E}^2$, and that a_1 and a_2 are constants. There is at most one body L in $\mathbb{E}^2$ for which p_j is an i-equichordal point with constant a_j, $j = 1, 2$, and L must be symmetric about the line through p_1 and p_2. Moreover, if $a_1 = a_2$, then L is also centrally symmetric, with center at the midpoint of the segment $[p_1, p_2]$.*

Proof. If L, L' are two bodies for which p_j is an i-equichordal point with constant a_j, $j = 1, 2$, then L and L' have the same i-chord functions at p_1 and p_2. Therefore $L = L'$, by Theorem 6.2.3. Applying this argument to L and its reflection in the line through p_1 and p_2, we see that L must be symmetric about this line.

If $a_1 = a_2$, the i-chord functions of L at p_1 and p_2 are equal. It follows from Corollary 6.2.4 that L is symmetric about the midpoint of $[p_1, p_2]$.   ∎

Corollary 6.3.8. *For $i \in \mathbb{R}$, $i > 1$, there is no body in $\mathbb{E}^n$ with more than two i-equichordal points with the same constant.*

Proof. Suppose that $i > 1$ and that L is a body in $\mathbb{E}^n$ with three i-equichordal points p_j, $1 \leq j \leq 3$, all with the same constant. If J is a 2-dimensional plane containing the three points, then in J each p_j is also an i-equichordal point of $L \cap J$, with the same constant. By the previous theorem, $L \cap J$ is centrally symmetric with center at the midpoint of $[p_j, p_k]$, whenever $j \neq k$. This is impossible, since a bounded set can only have one center. ∎

So far we have no information relevant to Question 6.3.3 concerning the case $i = 1$. For convex bodies, an easy modification remedies this, since Theorem 5.3.3(iii) and Corollary 5.3.4 can then be used to obtain corresponding versions of the previous theorem and corollary. In fact, the results do hold for $i = 1$ without the convexity restriction, and we begin to investigate this now.

Suppose that L is a body with two equichordal points, p_1 and p_2. We assume that the line through the points is the x-axis, with $p_1 = (-c, 0)$ and $p_2 = (c, 0)$, $c > 0$, and that bd L meets the x-axis at the points q_1 and q_2, with q_1, p_1, p_2, and q_2 in that order on the axis. Obviously p_1 and p_2 have the same constant a, and we shall take $a = 2$, so $c < 1$.

The first task is to study the properties of a certain mapping. If w is any point, and $j = 1, 2$, let $P_j(w)$ be the point, if it exists, on the line through p_j and w such that the line segment $[w, P_j(w)]$ contains p_j and has length 2. Then P_j is defined in a neighborhood of bd L. See Figure 6.6.

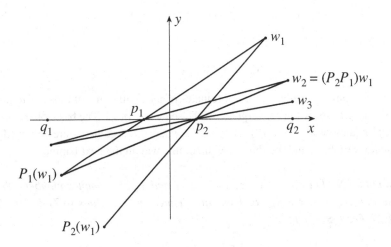

Figure 6.6. The equichordal mappings.

We claim that $q_1 = (-1, 0)$ and $q_2 = (1, 0)$. To see this, consider a point w at the intersection of bd L and the y-axis. The sequences $\{(P_1 P_2)^n w\}$ and $\{(P_2 P_1)^n w\}$ converge to q_1 and q_2, respectively. However, $(P_2 P_1)^n w$ is the reflection of $(P_1 P_2)^n w$ in the y-axis, so $q_1 = -q_2$. This proves the claim, since q_1 and q_2 are a distance 2 apart.

Suppose that $w_1 = (x_1, y_1)$ is a point in the first quadrant. For w_1 sufficiently close to $(1, 0)$, the point $P_1(w_1)$ lies in the third quadrant, and $w_2 = (P_2P_1)w_1$ lies in the first quadrant. We can obtain formulas for the coordinates of w_2 in terms of those of w_1 as follows. If $P_1(w_1) = (-a_1, -b_1)$, then

$$(x_1 + a_1)^2 + (y_1 + b_1)^2 = 4$$

and

$$(x_1 + c)b_1 = y_1(a_1 - c).$$

By substituting from the second equation into the first, we obtain $a_1 = f(x_1, y_1)$ and $b_1 = g(x_1, y_1)$, where

$$f(x, y) = c + (x + c)\left(2/\sqrt{(x + c)^2 + y^2} - 1\right) \tag{6.7}$$

and

$$g(x, y) = y\left(2/\sqrt{(x + c)^2 + y^2} - 1\right). \tag{6.8}$$

Furthermore, if $w_2 = (x_2, y_2)$, then $x_2 = f(a_1, b_1)$ and $y_2 = g(a_1, b_1)$, since the geometry is the same. Let $w_{m+1} = (P_2P_1)w_m$ and $P_1(w_m) = (-a_m, -b_m)$ for each $m \in \mathbb{N}$. Then the equations

$$w_{m+1} = (x_{m+1}, y_{m+1}) = \big(f(a_m, b_m), g(a_m, b_m)\big) \tag{6.9}$$

and

$$(a_m, b_m) = \big(f(x_m, y_m), g(x_m, y_m)\big) \tag{6.10}$$

give a recursive formula for the coordinates of w_{m+1} in terms of those of w_m.

We have to study the convergence of the sequence $\{w_m\}$. The basic observation is that equations (6.7)–(6.10) make sense when the arguments are regarded as *complex* variables, and we shall view them this way in the next lemma.

Lemma 6.3.9. *Let* $w_1 = (x_1, y_1)$, *where* x_1 *and* y_1 *are complex numbers. Suppose that* a_m, b_m, x_m, y_m, *and* w_m *are defined by equations (6.7)–(6.10), for* $m \in \mathbb{N}$. *For* $\delta > 0$, *let*

$$U(\delta) = \{(x, y) \in \mathbb{C}^2 : |x - 1| < \delta^2, |y| < \delta\}.$$

Then there is a $\delta_0 > 0$ *such that if* $w_1 \in U(\delta_0)$, *the sequence* $\{y_n\}$ *converges to zero, while the sequence* $\{x_n\}$ *converges to a complex number* $F_c(w_1)$, *where* F_c *is an analytic function of* x_1 *and of* y_1.

Proof. We begin by noting that for (x, y) in a sufficiently small neighborhood of $(1, 0)$, equations (6.7) and (6.8) show that both f and g are analytic functions of the complex variable x or y, when the other variable is fixed. The Taylor series of f in powers of y yields

$$f(x, y) = 2 - x + O(|y|^2).$$

Expanding g in a Taylor series in powers of $(x - 1)$, we have

$$g(x, y) = g(1, y) + O(|x - 1|) = y\left(2/\sqrt{(1 + c)^2 + y^2} - 1\right) + O(|x - 1|)$$
$$= y\left((1 - c)/(1 + c) + O(|y|^2)\right) + O(|x - 1|).$$

Let $\delta > 0$. For $(x, y) \in U(\delta)$, the previous equation shows that

$$g(x, y) = y\left((1 - c)/(1 + c) + O(\delta^2)\right).$$

Suppose that $(x_1, y_1) \in U(\delta)$. Then we have $b_1 = O(|y_1|)$, and

$$x_2 = 2 - a_1 + O(|b_1|^2) = 2 - \left(2 - x_1 + O(|y_1|^2)\right) + O(|b_1|^2) = x_1 + O(|y_1|^2).$$

Also,

$$y_2 = b_1 \left(\frac{1 - c}{1 + c} + O(\delta^2)\right) = y_1 \left(\left(\frac{1 - c}{1 + c}\right)^2 + O(\delta^2)\right).$$

Choose d with

$$\left(\frac{1 - c}{1 + c}\right)^2 < d < 1.$$

By the last two displayed equations, there is a $\delta_1 > 0$ and a C such that $|x_2 - x_1| < C|y_1|^2$ and $|y_2| < d|y_1|$ whenever $(x_1, y_1) \in U(\delta_1)$.

Define $\delta_0 = \delta_1/\sqrt{1 + C/(1 - d^2)}$. We claim that if $(x_1, y_1) \in U(\delta_0)$, then $(x_m, y_m) \in U(\delta_1)$ for all $m \in \mathbb{N}$. This is certainly true for $m = 1$, since $\delta_0 < \delta_1$. If it is true for all $n \leq m$, then the preceding arguments show that $|x_{n+1} - x_n| < C|y_n|^2$ and $|y_{n+1}| < d|y_n|$, for $n \leq m$. Now repeated use of the last inequality gives

$$|y_{m+1}| < d^m|y_1| < d^m\delta_0 < \delta_1.$$

Also,

$$|x_{m+1} - 1| \leq |x_1 - 1| + \sum_{n=1}^{m} |x_{n+1} - x_n| < \delta_0^2 + \sum_{n=1}^{m} C|y_n|^2$$
$$< \delta_0^2 + C \sum_{n=0}^{m-1} d^{2n}\delta_0^2 < \delta_0^2\left(1 + C/(1 - d^2)\right) = \delta_1^2.$$

Therefore $(x_{m+1}, y_{m+1}) \in U(\delta_1)$, and we have proved our claim, by induction.

It follows that the inequalities $|x_{n+1} - x_n| < C|y_n|^2$ and $|y_{n+1}| < d|y_n|$ hold for all $n \in \mathbb{N}$. From these, we see that $y_m \to 0$ as $m \to \infty$ and that

$|x_{n+1} - x_n| < C d^{2n} \delta_0^2$ for each n, whenever $w_1 \in U = U(\delta_0)$. Now

$$x_m = x_1 + \sum_{n=1}^{m-1} (x_{n+1} - x_n)$$

expresses x_m, via equations (6.7)–(6.10), as an analytic function of x_1 and of y_1. The previous inequality, together with Weierstrass's M-test, shows that x_m converges uniformly to a complex number $F_c(w_1)$. The uniform convergence also implies that F_c is an analytic function of x_1 and of y_1.   ∎

Theorem 6.3.10. *Suppose that p_1 and p_2 are points in $\mathbb{E}^2$ and that a is a constant. There is at most one body L in $\mathbb{E}^2$ for which p_j is an equichordal point with constant a, $j = 1, 2$. The boundary of L is real analytic, and L is symmetric about the line through p_1 and p_2 and about the midpoint of the segment $[p_1, p_2]$.*

Proof. With Lemma 6.3.9 in hand, we now revert to real variables, and let $U = U(\delta_0)$ be the corresponding neighborhood in $\mathbb{E}^2$ of $(1, 0)$, and F_c the corresponding function, from that lemma. Suppose initially that a body L exists containing two equichordal points p_1 and p_2 with constant $a = 2$, as before. Then the maps P_1 and P_2 are well defined on bd L. If $w \in$ bd L, the maps P_1 and P_2, applied alternately a finite number of times, will take w to a point in bd $L \cap U$. From equations (6.7)–(6.10), we see that to prove bd L is real analytic, it will suffice to show that bd $L \cap U$ is real analytic.

If $w_1 = (x, 0) \in U$, then of course $F_c(w_1) = x$, so $\partial F_c / \partial x = 1 \neq 0$ there. Therefore bd $L \cap U$, the solution set of the equation $F_c(x, y) = 1$ in U, is indeed real analytic. Furthermore, the function F_c, and hence bd $L \cap U$ and bd L itself, is uniquely determined by our choice of p_1, p_2, and a. Next, we note that $F_c(x, y) = F_c(x, -y)$, so L is symmetric about the x-axis. In addition, we have $F_c(x, y) = 1$ if and only if $F_c(-x, -y) = 1$, so L is also centrally symmetric, with center at o.

Suppose now that p_1 and p_2 are arbitrary points in the plane, that a is any number greater than the distance between p_1 and p_2, and that L is a body with p_1 and p_2 as equichordal points with constant a. Defining c, the *eccentricity*, to be the ratio of this distance to a, we can apply the previous method to obtain a corresponding function F_c. The fact that this function is uniquely defined shows that there will be at most one corresponding body L, and the arguments used earlier show that it must have the properties stated in the theorem.   ∎

Corollary 6.3.11. *There is no body in $\mathbb{E}^n$ with more than two equichordal points.*

Proof. The corollary follows from the last theorem, just as Corollary 6.3.8 was deduced from Theorem 6.3.7.   ∎

Remark 6.3.12. It is known that there is no body with two equichordal points, though the proof is well outside the scope of this book. In view of this, it may seem strange that we have bothered to prove Theorem 6.3.10 at all; but, as we indicate in Note 6.3, the result is actually an important contribution toward a proof of nonexistence.

We now return to the values $i = 0$ and -1, and ask whether disks and ellipses are the only convex bodies with two equiproduct or equireciprocal points, respectively.

Theorem 6.3.13. *Let L be a body in $\mathbb{E}^2$ with two equiproduct points, p_1 and p_2, and suppose that* bd L *meets the line l through p_1 and p_2 at q_1 and q_2. If L is convex, or, more generally, if L has a unique left or right tangent at q_1 and q_2, then L is a disk.*

Proof. Let us suppose that q_1, p_1, p_2, and q_2 lie on l in that order and that L has a unique right tangent at q_1. For $j = 1, 2$, the constant a_j for p_j is determined by the distances from p_j to q_1 and q_2, so if D is a disk whose boundary contains the points q_1 and q_2, then p_j is an equiproduct point for D, also with constant a_j.

For any point $x \in$ bd L, and $j = 1, 2$, define $P_j(x)$ to be the point in bd L, distinct from x, on the line through x and p_j. Let x_1 be a point other than q_1 or q_2 in bd L. For each $n \in \mathbb{N}$, define $x_n = (P_1 P_2)^{n-1}(x_1)$. Then the sequence $\{x_n\}$ converges to q_1, and we may assume, without loss of generality, that it does so on the same side of q_1 as the right tangent to L at q_1.

Let D be the disk containing q_1, q_2, and x_1 in its boundary. Then, for each n, the point x_n belongs both to bd L and to bd D, and so D has the same right tangent as L at q_1. Applying the same argument to any other point $x' \in$ bd L, we see that $x' \in$ bd D', where D' is a disk containing q_1 and q_2 in its boundary and with the same right tangent as L, and therefore as D, at q_1. Therefore $D' = D$, and $x' \in$ bd D. This proves that $L = D$, as required. ∎

Remark 6.3.14. A substantial amount is known about bodies with two equireciprocal points, which is summarized but not proved here. We have seen, in Theorem 6.3.4, that an ellipse is such a body. Suppose that L has two equireciprocal points, p_1 and p_2, with constants a_1 and a_2, and that bd L meets the line l through p_1 and p_2 at q_1 and q_2. The following facts are known. There are unique tangents to L at q_1 and q_2, and these are perpendicular to l. If L is convex, then $a_1 = a_2$. The latter also holds if bd L has finite curvature at q_1 or q_2, and in this case L must actually be an ellipse, with foci at p_1 and p_2. Some such extra assumptions are necessary, since L can be convex and of class C^1, but nonelliptical.

It is interesting to compare these results with Theorem 6.2.14. Let us set $i = -1$ in that theorem and suppose that L_1 is an ellipse with foci p_1 and p_2, so the -1-chord function of L_1 is equal to the same constant at each of these points. The assumption that L_2 is any other body with -1-chord functions equal to those of L_1 at p_1 and p_2 will imply that $L_2 = L_1$, and therefore that L_2 is an ellipse, providing the other hypotheses of Theorem 6.2.14 are satisfied. These are rather stronger than those of the previous paragraph, but the nonelliptical example mentioned there shows that they cannot be weakened too much.

Open problems

Problem 6.1. (See Problem 5.1.) If $i \in \mathbb{R}$ is positive, is a convex body determined by its i-chord functions at two points?

Problem 6.2. (See Note 6.2.) Characterize those pairs of functions that are the i-chord functions at two interior points of some convex body.

Problem 6.3. Improve Theorem 6.2.14 by weakening its assumptions, or find examples placing limits on how this can be done.

Problem 6.4. Let $i \in \mathbb{R}$, $i \neq -1$, 0, or 1. Does there exist a body with two i-equichordal points?

Notes

6.1. *Chord functions, i-chordal symmetrals, and ith radial sums.* The Definition 6.1.2 of the i-chordal symmetral and Theorem 6.1.3 are taken from the article [283] of Gardner and Volčič, where the term "ith dual Blaschke body" is used instead. As for the chordal symmetral, the notation is that of Lutwak [537]. Firey [235] introduced these sets, for $i \leq -1$, and proved Theorem 6.1.4. In [538], Lutwak also considers the case $i = -1$ (which he terms the "equireciprocal body") and uses the $(n+1)$-chordal symmetral in connection with centroid bodies (see Note 9.1). Lutwak always works with convex bodies or star bodies containing the origin in their interiors.

Gardner, Vedel Jensen, and Volčič [280] define for $i > 0$ the i-chordal symmetral of a bounded Borel set as an ingredient of their extension of the dual Brunn–Minkowski theory mentioned in Section A.7.

The dual form of Problem 3.5 is answered by Gardner and Volčič [283] as follows. If i and j are distinct real numbers, and L is a star body in $\mathbb{E}^n$, containing the origin in its interior, such that $\tilde{\nabla}_i L$ is homothetic to $\tilde{\nabla}_j L$, then L is centered. The proof is a simple application of Jensen's inequality (B.3) for means.

Firey [235] introduced the following concepts for $i \leq -1$ and convex bodies. Suppose that $i \in \mathbb{R}$ is nonzero, that L_1 and L_2 are star bodies in $\mathbb{E}^n$ containing the origin in their interiors, and that $a_j \geq 0$, $j = 1, 2$. Then there is a unique star body L such that

$$\rho_L^i = a_1 \rho_{L_1}^i + a_2 \rho_{L_2}^i.$$

We can rewrite this equation as

$$L = (a_1 \cdot_i L_1) \tilde{+}_i (a_2 \cdot_i L_2),$$

where $\tilde{+}_i$ and $\cdot_i$ are called ith radial addition and multiplication. When $a_1 = a_2 = 1$, L is called the *ith radial sum* of L_1 and L_2. Note that by (0.30) we have $\tilde{+}_1 = \tilde{+}$.

Also, $\cdot_i$ is related to ordinary multiplication by $r \cdot_i L = r^{1/i} L$. In this notation, we can then write

$$\tilde{\nabla}_i L = \left(\frac{1}{2} \cdot_i L \right) \tilde{+}_i \left(\frac{1}{2} \cdot_i (-L) \right).$$

6.2. *Chord functions of star sets.* Special instances of the i-chord functions of Definition 6.1.1 were used decades ago by Kubota [481] and Süss [794] for the same purpose that we employ them in this and the next chapter. In more recent times, they were investigated by Larman and Tamvakis [496], where Corollary 6.2.4 (which clearly holds in $\mathbb{E}^n$) is established for $i \in \mathbb{N}$.

Falconer [218] was the first to define i-chord functions, at least for $i > 0$, as we do here. He proved Theorem 6.2.2 for $i \geq 1$, by applying the stable manifold theorem of differentiable dynamics. Falconer also studies i-chord functions in [220] (see Note 7.4). Our approach to Theorem 6.2.2 follows the more accessible techniques of [257]; there, only integer values of i are considered, and this allows the corresponding generalization of part (ii) of Theorem 5.3.3, at the cost of an extra "chord-chasing" argument for $i = 2$. (Probably this generalization can be proved for all real i.)

Most of the other contents of Section 6.1 leading up to Theorem 6.2.14 are proved by Gardner [257], though Theorem 6.2.8 is only hinted at in the last paragraph of Section 7 of that paper. The exceptions are the Lemmas 6.2.11 and 6.2.12 of Zuccheri [868]. These allow our formulation of Theorem 6.2.14, which is not strictly stronger, but is perhaps more appealing, than the version presented in [257].

Soranzo [779] extends Definition 6.1.1 of the i-chord function to $i = \pm\infty$, when K is a convex body, by $\rho_{+\infty, K}(u) = \max\{|\rho_K(u)|, |\rho_K(-u)|\}$ and $\rho_{-\infty, K}(u) = \min\{|\rho_K(u)|, |\rho_K(-u)|\}$, and finds some results concerning the determination of convex bodies by such functions.

The material concerning inverse problems where i-chord functions are specified for more than one value of i, from Lemma 6.2.15 to the end of Section 6.1, is developed by Gardner and Volčič [283]. The general framework adopted in this chapter, including the definitions in Section 6.1 and Lemma 6.2.1, is also set out in [283]. Soranzo and Volčič [780] study the question of when convex bodies are determined by their i-chord function at one point and j-chord function at another point, where $i \neq j$.

Problem 6.2, communicated by K. J. Falconer and one of many possible problems of the same type, is in the same spirit as Problem 2.13.

Groemer [355], [356] obtains stability versions of several of the uniqueness results in this chapter.

Dulio and Peri [196] extend the notion of an i-chord function to spaces of constant curvature and establish a corresponding version of Theorem 6.2.2.

6.3. *Equichordal problems.* Equichordal and equiproduct points were first studied many years ago. Fujiwara [251] and Yanagihara [846] noted that there are noncircular planar convex bodies containing one equichordal point, or equiproduct point, respectively. Kelly [424] constructed a whole family of such examples with one equichordal point. (Kelly's result is generalized by Martinez-Maure [571].) Equireciprocal points entered the stage rather later, in Klee's article [446]. The natural generalization, the i-equichordal points for arbitrary i, are first defined explicitly in [257], but are implicit in many earlier papers. Similarly, Theorem 6.3.2 does not seem to have appeared before its proof by Gardner and Volčič [283]. Note, however, that convex bodies with the origin as an equireciprocal point are just polar duals of convex bodies of constant width, so the case $i = -1$ of Theorem 6.3.2 follows from Theorem 3.2.5.

Question 6.3.3 appeared in [258] and in the book [173, Problem A1] of Croft, Falconer, and Guy. The positive answer for $i = 0$, via Theorem 6.3.5, known to Yanagihara [846] and others, is essentially the *intersecting chord theorem*, Proposition 35 in Book III of Euclid! (See [384, p. 71], for example.) Yanagihara [847] also established Theorem 6.3.13 for convex bodies of class C^1. (The convexity condition was dropped by Kelly [423], and Theorem 6.3.13 as stated here is proved by Zuccheri [869].) These papers provide some results on equiproduct points in higher dimensions, and the latter paper defines and studies exterior equiproduct points.

Theorem 6.3.4, providing the positive answer to Question 6.3.3 for $i = -1$, was noted by Falconer [219]. His remains the only significant paper on equireciprocal points. All the results mentioned in Remark 6.3.14 are given there; the arguments make use of the stable manifold theorem.

The special case $i = 1$ of Question 6.3.3 is the famous *equichordal problem*, for a long time the oldest unsolved planar problem in this branch of geometry. It was posed independently by Fujiwara [251] and Blaschke, Rothe, and Weitzenböck [74]. The problem is stated in many collections of open problems; see, for example, the book [449, Problem 2] of Klee and Wagon for an excellent report on the history of the problem up to 1991. More references are listed in [449] and by Heil and Martini [386].

Until recently, Theorem 6.3.10, due to Wirsing [842], was the dominant work on the problem. (For convex bodies, the symmetry part of Theorem 6.3.10 was proved earlier by Süss [794], and the uniqueness part by Dirac [188].) However, the equichordal problem has now be settled, as we noted in Remark 6.3.12, and we wish to give a very brief indication of this development here. (It is worth noting that although the answer to the case $i = 1$ of Question 6.3.3 is negative, Petty and Crotty [666] showed that the corresponding question in certain Minkowski spaces has a positive answer; see also the book of Thompson [803].)

Let $p_1 = (-c, 0)$, $p_2 = (c, 0)$, and $a = 2$, as we had before Lemma 6.3.9. If a body L exists containing p_1 and p_2 as equichordal points with constant 2, the point $(c, 1)$ must belong to its boundary, since L is symmetric about the x-axis by Theorem 6.3.10. This means that c, the eccentricity defined in the proof of Theorem 6.3.10, cannot be too large. In fact, $c < \sqrt{3}/2$, since otherwise the distance from $(c, 1)$ to the point p_1 would be larger than 2. Using more elaborate arguments, Ehrhart [209] proved that for convex bodies, $c < 0.5$; this was significantly improved by Michelacci [607] and by Michelacci and Volčič [611], who show without the convexity assumption that $c < 0.235$.

Defining $h(c) = F_c(c, 1)$, with F_c as in Lemma 6.3.9, we must have $h(c) = 1$, if L is to exist. Suppose that $0 < c < \sqrt{3}/2$. Wirsing notes that with the methods of Theorem 6.3.10 one can easily show that $h(c)$ exists; in fact, the real sequence $\{x_n\}$ is monotonic and converges to $h(c)$ with $c < h(c) < 2 - c$. The previous estimates then show that $h(c) = 1$ is not an identity, as $h(.5) \neq 1$, for example. (Wirsing refers to earlier computations in [188] to draw the same conclusion.) Wirsing also remarks that his methods show that $h(c)$ is an analytic function for $0 < c < \sqrt{3}/2$, regardless of whether or not L exists. Putting all these facts together, we deduce that there can only be countably many values of c for which L exists. This could be improved to finitely many, if h were analytic at 0. However, Wirsing [842] shows that this is not true; in fact, he proves that $h(c) = 1 + O(c^n)$, for all $n \in \mathbb{N}$, as c tends to zero. The idea is to construct curves that approximate the boundary of L, if it exists, sufficiently closely.

The same idea is pushed a great deal further by Schäfke and Volkmer in their long paper [723] (and also, independently, by Rogers [697]). By a complicated asymptotic analysis, Schäfke and Volkmer find that there exists an ω with $1.359 \le \omega \le 1.361$ such that

$$ h(c) = 1 + \omega e^{-\pi^2/4c} \left(1 + \frac{\pi^2}{12} c + O(c^2) \right), $$

as $c \to 0$. We conclude from this result that $h(c) > 1$ for sufficiently small c, implying that L cannot exist when c is too small (and therefore that there are actually only finitely many values of c for which L can exist). In a note added in proof to [723], the authors state that they will compute error bounds implying that $c > 0.03$ and, by a different method, that $h(c) > 1$ for $0.03 \le c \le 0.24$. This, together with the estimate $c < 0.235$ mentioned earlier, would mean that L cannot exist. However, these computations have not been carried out at the present time.

Finally, in 1997, Rychlik [713] solved the equichordal problem completely. Like Schäfke and Volkmer, and Wirsing before them, Rychlik changes to complex variables to carry out his proof. Although the latter is less quantitative than that in [723], use is made of invariant manifold theory and some fairly heavy machinery from complex function theory. H. Volkmer, in his *Mathematical Reviews* report, summarizes the proof as follows:

"The paper under review proves the nonexistence of equichordal curves by applying methods from the theory of dynamical systems and Riemann surfaces. It is shown that an equichordal curve, if it existed, would be a heteroclinic connection of a map T naturally associated with the equichordal problem. A heteroclinic connection of T is an invariant curve connecting two fixed points of T. The map is then complexified. The main idea of the proof is to consider a Riemann surface associated with the heteroclinic connection. This surface would have to be compact, which would force an equichordal curve to be algebraic. It is then quite straightforward to show that T does not admit an invariant algebraic curve. As one would expect, the paper is long (72 pages) and involved. The reading is easy at the beginning but becomes more difficult towards the end of the paper.

The paper represents an extraordinary effort of the author that, without doubt, will find high appreciation by everyone in the mathematical world."

Figure 6.7. Wilhelm Blaschke.

6.4. *Wilhelm Blaschke (1885–1962)*. (Sources: [686], [760].) Wilhelm Johann Eugen Blaschke was born in Graz, Austria. His father, who taught mathematics and descriptive geometry at a high school, introduced Blaschke to the work of Steiner (cf. Note 1.7). Blaschke studied civil engineering and mathematics at Graz and Vienna, where in 1908 he was awarded his doctorate under W. Wirtinger. Despite a handicap due to a bout of infantile paralysis, Blaschke traveled very widely, studying at Brünn, Pisa and Göttingen (under Klein and Hilbert) and taking positions in Bonn, Greifswald, Prague, Leipzig (where he met Herglotz), Königsberg, and Tübingen. Finally, in 1919, he accepted the challenge of building a new department in Hamburg, staying until his retirement in 1953. Blaschke's interest in travel continued, however, and he was also a visiting professor at Johns Hopkins University in 1931, the University of Chigago in 1932, and the University of Istanbul in 1953–5. In fact, he is the author of *Reden und Reisen eines Geometers (Discourses and Travels of a Geometer)*, one of the few autobiographical books by a mathematician, published in 1957.

During the second world war, he lost his home and possessions in a bombing raid on Hamburg. Shortly after the war ended, Blaschke was dismissed from his position in Hamburg, but reinstated about a year later, an episode about which an extensive commentary can be found in [556]. He married twice and had two children.

Blaschke was one of the great mathematicians (and perhaps the foremost geometer) of his time. His presence in Hamburg attracted others of similar stature, such as E. Artin, H. Hasse, and E. Hecke (of the Funk–Hecke theorem), and he founded the journal *Abhandlungen aus dem mathematischen Seminar der Universität Hamburg*. Blaschke was the author of over 200 papers and more than a dozen highly influential books touching every area of geometry. In his first book, the celebrated *Kreis und Kugel* (*Circle and Sphere*) [71], originally published in 1916, Blaschke took up a theme initiated by Steiner. Unlike Steiner, however, Blaschke employed tools from analysis and algebra with extraordinary power. In [71] can be found, among many other things, the most satisfying proof (via Blaschke's selection theorem, Theorem 0.4.1) of the existence of the extremum in the isoperimetric inequality. Perhaps his most famous book is the three-volume *Vorlesungen über Differentialgeometrie* (*Lectures on Differential Geometry*); others treated topological differential geometry, projective geometry, integral geometry, kinematics, and mechanics. With his work on integral geometry, Blaschke founded a school which built a new subject from a few sporadic prior results, and which, fostered by students of his such as Hadwiger (cf. Note 9.12) and Santaló, still flourishes today. Comprehensive overviews of Blaschke's work are given by Leichtweiss [503] and Strubecker [792].

Most of Blaschke's books have appeared in Russian; amazingly, *none* have been translated into English.

7

Sections, section functions, and point X-rays

The various themes explored in earlier chapters have both synthesis and counterpoint in this one. Just as Chapter 3 examined inverse problems where the given data involve orthogonal projections, so this chapter deals with information about sections through one or more points.

Much of the first section mirrors Section 3.1. Though star bodies would be the natural objects to consider, we have had to retreat to convexity for lack of more general results. For example, Theorem 7.1.9 implies that two convex bodies in $\mathbb{E}^n$, $n \geq 3$, must be homothetic if their intersections with any hyperplane through the origin are also homothetic. In the realm of convex bodies, a useful duality is provided by polar bodies (see Section 0.8). Via (0.38), this sometimes enables one to convert a theorem concerning projections to one about sections through the origin, or vice versa. We apply this technique in Theorem 7.1.11, which shows that "homothetic" cannot be replaced by "similar" in the foregoing statement. Unfortunately polar duality is of limited relevance in geometric tomography (cf. the last paragraph of Section 0.8), and we can only apply it on a couple of occasions. This means that separate proofs usually have to be constructed, even when there is a direct analogy with a result in Chapter 3. Moreover, the "mirror" is imperfect; for example, Theorem 7.1.9 actually draws the stronger conclusion that the bodies must either be dilatates or homothetic ellipsoids. Similarly, Theorem 7.1.10, sometimes called the false center theorem, states that if the intersection of a convex body with any hyperplane through the origin is centrally symmetric, the body must either be centered or an ellipsoid.

Section 7.2 introduces section functions. These include X-rays at the origin and i-chord functions, as well as functions giving the i-dimensional volume of i-dimensional sections through the origin. The latter generalize the ordinary X-ray at the origin and are also dual to ith projection functions. We can therefore call them i-dimensional X-rays at the origin, or ith section functions, and we use both terms. Consequently, Section 7.2 is at once an extension of the material in Chapter 5 on point X-rays of convex bodies, and a complement to Section 3.3, where we studied determination of convex bodies by projection functions. After the pivotal Theorem 7.2.3 provides a link between ith section functions and i-chord functions, results generally follow with just a few lines of proof. (Behind Theorem 7.2.3, however, is Theorem C.2.4, the injectivity property of the spherical Radon transform; recall that the injectivity property of the cosine transform played an important role in Section 3.3.) For example, we obtain the generalized Funk section theorem, Theorem 7.2.6, stating that different centered star bodies have different ith section functions (compare Aleksandrov's projection theorem, Theorem 3.3.6). The i-chordal symmetrals and bodies of constant i-section show that this result cannot be extended to arbitrary star bodies. Corollary 7.2.16 answers the dual form of the recently solved problem of the existence of a nonspherical convex body of constant width and constant brightness (cf. Note 3.6). The interesting Theorem 7.2.17, which says, roughly, that only centered balls have mutually congruent sections through the origin, is a meeting place for several techniques developed in the last two chapters.

If Theorem C.2.4 is taken for granted, we only ask the reader to make acquaintance with the language of dual mixed volumes, since no other new methods are required.

In the last section, we return to point X-rays in order to show that some of the uniqueness theorems proved for planar convex bodies in Chapter 5 are not true for measurable sets.

7.1. Homothetic and similar sections

It is quite obvious that an arbitrary set is determined by complete knowledge of all its sections through a point. We therefore begin with the result for sections analogous to Theorem 3.1.3. The remarks before Theorem 3.1.1 apply to sections, too; however, we are faced with the additional difficulty that many natural properties are not preserved by taking sections. For example, if L_1 is a translate of L_2, and S is a subspace, it is not generally true that $L_1 \cap S$ is a translate of $L_2 \cap S$.

Theorem 7.1.1. *Suppose that K_j, $j = 1, 2$, are compact convex sets in $\mathbb{E}^n$, containing the origin in their relative interiors. Let $2 \le k \le n - 1$. If $K_1 \cap S$ is homothetic to (or a translate of) $K_2 \cap S$ for each $S \in \mathcal{G}(n, k)$, then K_1 is homothetic to K_2 (or a translate of K_2, respectively).*

Proof. Since K_1 and K_2 must have the same linear hull, we can assume that this is $\mathbb{E}^n$, and hence that K_1 and K_2 are convex bodies.

Although equivalence up to homothety is preserved by projections, it is not preserved by sections. Despite this, it suffices to prove the theorem for $k = 2$. To see this, suppose that the result is true for sections by 2-dimensional subspaces and that the assumptions of the theorem hold for some $k > 2$. Let T be an $(n-k+2)$-dimensional subspace and S any 2-dimensional subspace with $S \subset T$. Denote by U the k-dimensional subspace containing S and $T^\perp$, so $K_1 \cap U$ is homothetic to (or a translate of) $K_2 \cap U$. The projections of these sets on T have the same properties. So $(K_1|T) \cap S = (K_1 \cap U)|T$ is homothetic to (or a translate of, respectively) $(K_2|T) \cap S = (K_2 \cap U)|T$. This holds for each $S \in \mathcal{G}(n, 2)$ with $S \subset T$, so we can use the theorem with $k = 2$ and with n replaced by $n-k+2$ to conclude that $K_1|T$ is homothetic to (or a translate of, respectively) $K_2|T$. Since this holds for each $T \in \mathcal{G}(n, n-k+2)$, the corresponding theorem for projections (Theorem 3.1.3) implies that K_1 is homothetic to (or a translate of, respectively) K_2.

We therefore assume that $K_1 \cap S$ is homothetic to $K_2 \cap S$ for all $S \in \mathcal{G}(n, 2)$. Let r_0 be the largest number such that $r_0 K_2 \subset K_1$. Then there is a common supporting hyperplane H to both K_1 and $r_0 K_2$. If $C_1 = K_1 \cap H$ and $C_2 = r_0 K_2 \cap H$, we have $C_2 \subset C_1$.

Suppose that there is a line t in H such that $t \cap C_1$ is a single point, but $t \cap C_2$ is a line segment. Let S be the 2-dimensional subspace containing t. Then $K_1 \cap S$ and $r_0 K_2 \cap S$ are not homothetic, as they must be. So both C_1 and C_2 are points, or both are line segments, or each exposed point of C_2 belongs to the boundary of C_1. In the latter case we have $C_1 = C_2$. In every case we can choose a point p in both boundaries (in H) of C_1 and C_2 such that, with the possible exception of one line t_0, all lines $t \subset H$ through p satisfy the condition that either

(i) $t \cap C_1 = t \cap C_2 = \{p\}$ or

(ii) $t \cap C_j$, $j = 1, 2$, are line segments with one endpoint at p that both extend on the same side of p.

Let t_1 be the line through o and p, and let q_1, q_2 be the points other than p where t_1 meets the boundaries of K_1 and $r_0 K_2$, respectively. Then for some $r > 0$ we have

$$q_2 - p = r(q_1 - p). \tag{7.1}$$

Suppose that S is a 2-dimensional subspace containing t_1 and meeting H in a line $t \neq t_0$. Then t supports $K_1 \cap S$ and $r_0 K_2 \cap S$, and these sets lie on the same side of t in S. Also, $K_1 \cap S$ and $r_0 K_2 \cap S$ meet t either in the single point p or in a line segment with p as common endpoint. This means that the homothety taking $K_1 \cap S$ to $r_0 K_2 \cap S$ must arise from translation by $-p$ and a dilatation, which from (7.1) is by a factor r. Therefore

$$(r_0 K_2 \cap S) - p = r((K_1 \cap S) - p).$$

This holds for all $S \in \mathcal{G}(n, 2)$ containing t_1, with at most one exception; but, by continuity, there can actually be no exceptions. It follows that

$$r_0 K_2 - p = r(K_1 - p), \tag{7.2}$$

so K_1 is homothetic to K_2.

Now suppose that $K_1 \cap S$ is a translate of $K_2 \cap S$, for all $S \in \mathcal{G}(n, 2)$. Following the previous argument, we see that the dilatation factor between $K_1 \cap S$ and $r_0 K_2 \cap S$ is r, for all S containing t_1. But $K_1 \cap S$ is a translate of $K_2 \cap S$, so $r = r_0$, and K_1 is a translate of K_2, by (7.2). ∎

The assumption that sections of K_1 are homothets of those of K_2 does not imply that K_1 is a dilatate of K_2, since K_1 and K_2 might be any two balls containing the origin in their interiors. This example is almost generic, however, as the much stronger conclusion of Theorem 7.1.9 demonstrates. It seems to be unknown whether an extension to star bodies is possible; see Problem 7.1.

Remark 7.1.2. Theorem 7.1.1 remains true for arbitrary compact convex sets K_1 and K_2; see Note 7.1.

Corollary 7.1.3. *Let K be a convex body in $\mathbb{E}^n$. Suppose that $2 \le k \le n - 1$ and that $K \cap S$ is centrally symmetric for each $S \in \mathcal{G}(n, k)$. Then K is centrally symmetric.*

Proof. Let $K_1 = K$ and $K_2 = -K$. For each $S \in \mathcal{G}(n, k)$, $K_2 \cap S = -K_1 \cap S$. Since $K_1 \cap S$ is centrally symmetric, $K_1 \cap S$ is a translate of $K_2 \cap S$, by Lemma 3.1.4. By Theorem 7.1.1 and Remark 7.1.2, K_1 is a translate of K_2, and by Lemma 3.1.4 again, K is centrally symmetric. ∎

It is important to realize that the sections $K \cap S$ in Corollary 7.1.3 are not assumed to be centered. If they are, then of course K must be centered. If not, however, K need not be centered; for an example, let K be any ball containing the origin in its interior, but not at its center. Again, this example is almost generic, since it turns out that if K is not centered, it must be an ellipsoid (see Theorem 7.1.10).

Corollary 7.1.4. *Let K be a convex body in $\mathbb{E}^n$. Suppose that $2 \le k \le n - 1$ and that $K \cap S$ is a ball for each $S \in \mathcal{G}(n, k)$. Then K is a ball.*

Proof. Let $K_1 = K$ and $K_2 = B$. Then Theorem 7.1.1 and Remark 7.1.2 imply that K is homothetic to B, so it must be a ball. ∎

The proof of the following theorem employs polar duality. For an introduction to polar bodies and basic facts about them, see Section 0.8.

Theorem 7.1.5. *Let K be a compact convex set in $\mathbb{E}^n$ containing the origin in its relative interior. Let $2 \le k \le n - 1$, and suppose that every section $K \cap S$, $S \in \mathcal{G}(n, k)$, is an ellipsoid. Then K is an ellipsoid.*

Proof. It suffices to consider the case $k = n - 1$. If $\dim K < n$, then there is an $S \in \mathcal{G}(n, n - 1)$ containing K, so K is an ellipsoid. Suppose that K is a convex body. Since $K^{**} = K$, (0.38) says that $(K^*|S)^*$ is an ellipsoid for all $S \in \mathcal{G}(n, n - 1)$. The polar body of an ellipsoid is again an ellipsoid (see Section 0.8), so $K^*|S$ is also an ellipsoid for all $S \in \mathcal{G}(n, n - 1)$. By Theorem 3.1.7, K^* is an ellipsoid, and this means that K must also be an ellipsoid. ∎

Remark 7.1.6. Theorem 7.1.5 remains true for an arbitrary compact convex set K; see Note 7.1.

The starting point for the next development is the observation that the converse to Corollary 7.1.3 is not true in general. For example, sections of a cube through an interior point are not always centrally symmetric. It turns out that Corollary 7.1.3, as well as Theorem 7.1.1, can be considerably strengthened.

Theorem 7.1.7. *Let K be a convex body in $\mathbb{E}^n$ and $2 \le k \le n - 1$. Suppose that there is an $x \ne o$ such that for each $S \in \mathcal{G}(n, k)$, $(K + x) \cap S$ is homothetic to $K \cap S$. Then K is an ellipsoid.*

We omit the proof of this difficult theorem, contenting ourselves with the following lemma.

Lemma 7.1.8. *It suffices to prove Theorem 7.1.7 for $n = 3$.*

Proof. Suppose that Theorem 7.1.7 is true for $n = 3$. We first consider the case when $k = 2$ and $n > 3$. Let $S \in \mathcal{G}(n, 2)$. Choose $T \in \mathcal{G}(n, 3)$ such that $S \subset T$ and $x \in T$. Note that $(K + x) \cap T = (K \cap T) + x$. If S' is any 2-dimensional subspace contained in T, then by assumption $((K \cap T) + x) \cap S' = (K + x) \cap S'$ is homothetic to $(K \cap T) \cap S' = K \cap S'$. Therefore $K \cap T$ is an ellipsoid, by the case $n = 3$. This implies that $K \cap S = (K \cap T) \cap S$ is an ellipse, and since S was arbitrary, this holds for all $S \in \mathcal{G}(n, 2)$. Theorem 7.1.5 and Remark 7.1.6 now imply that K is an ellipsoid.

Assume now that $k > 2$. Let T be any $(n - k + 2)$-dimensional subspace such that the projection y of x on T is not the origin. Observe that the set of such subspaces is dense in $\mathcal{G}(n, n - k + 2)$. Consider a 2-dimensional subspace S with $S \subset T$. Denote by U the k-dimensional subspace containing S and $T^\perp$, so by assumption $(K + x) \cap U$ is homothetic to $K \cap U$. The projections of these sets on T are also homothetic, so $((K|T) + y) \cap S = ((K + y) \cap U)|T$ is homothetic to $(K|T) \cap S = (K \cap U)|T$, and this is true for each $S \in \mathcal{G}(n, 2)$ with $S \subset T$.

By the theorem for $k = 2$ and n replaced by $n - k + 2$, $K|T$ is an ellipsoid. Since this holds for a dense set of $T \in \mathcal{G}(n, n - k + 2)$, it must be true for all $T \in \mathcal{G}(n, n - k + 2)$, by continuity. It now follows from Theorem 3.1.7 that K is an ellipsoid. ∎

Theorem 7.1.9. *Let K_1 and K_2 be convex bodies in $\mathbb{E}^n$ and $2 \leq k \leq n - 1$. If $K_1 \cap S$ is homothetic to $K_2 \cap S$ for each $S \in \mathcal{G}(n, k)$, then either $K_2 = r K_1$ for some $r > 0$ or K_1 and K_2 are homothetic ellipsoids.*

Proof. By Theorem 7.1.1 and Remark 7.1.2, K_1 and K_2 are homothetic, so $K_2 = r K_1 + x$ for some $r > 0$ and $x \in \mathbb{E}^n$. If $x \neq o$, then we note that by assumption $K_1 \cap S$ is homothetic to $(r K_1 + x) \cap S$, for each $S \in \mathcal{G}(n, k)$. Consequently, $r K_1 \cap S$ is also homothetic to $(r K_1 + x) \cap S$, for each $S \in \mathcal{G}(n, k)$. So the hypotheses of Theorem 7.1.7 are satisfied, with $K = r K_1$, and it follows that K_1 is an ellipsoid. ∎

Theorem 7.1.10 (false center theorem). *Let K be a convex body in $\mathbb{E}^n$ and $2 \leq k \leq n - 1$. If $K \cap S$ is centrally symmetric for each $S \in \mathcal{G}(n, k)$, then K is either centered or an ellipsoid.*

Proof. By Corollary 7.1.3, K has a center, c say. If $c \neq o$, we note, using Lemma 3.1.4, that

$$(K - 2c) \cap S = (-K) \cap S = -(K \cap S),$$

for any $S \in \mathcal{G}(n, k)$. But $K \cap S$ also has a center, c' say, so $-(K \cap S) = (K \cap S) - 2c'$. It follows that $(K - 2c) \cap S$ is a translate of $K \cap S$, for each $S \in \mathcal{G}(n, k)$, and then K is an ellipsoid, by Theorem 7.1.7. ∎

We now use a pair of convex bodies constructed in Chapter 3, and polar duality, to show that homothety cannot be replaced by similarity in Theorem 7.1.1.

Theorem 7.1.11. *There are centered, coaxial convex bodies of revolution K_1 and K_2 in $\mathbb{E}^n$, $n \geq 3$, such that for each $S \in \mathcal{G}(n, 2)$, $K_1 \cap S$ is similar to $K_2 \cap S$, yet K_1 is not affinely equivalent to K_2.*

Proof. Let K_1 and K_2 be the polar bodies of the centered convex bodies in Theorem 3.1.8. For centered bodies, affine equivalence is the same as linear equivalence (see Section 0.2). In view of (0.38), all that is needed now is the observation that (0.37) implies that two centered convex bodies are similar (or linearly equivalent) if and only if their polar bodies are similar (or linearly equivalent, respectively). ∎

We turn next to convex bodies whose sections through an interior point are of constant width. As we shall see, each such body must itself be of constant width. However, unlike the case of projections (Theorem 3.2.6), more is true; the only bodies with this property are balls (see Remark 7.1.16 – we shall not prove this stronger result), and therefore not every section of a nonspherical convex body of constant width is of constant width. Despite this, it suffices to consider sections by $(n-1)$-dimensional subspaces, as we noted (for projections) in the remarks before Theorem 3.1.1, since the general result then follows by induction. We shall need the following definition.

Definition 7.1.12. A chord c of a convex body K is called a *normal* if it is perpendicular, at one of its endpoints, to a supporting hyperplane to K; and a *double normal* if it is perpendicular at both endpoints to supporting hyperplanes to K.

Every diameter is a double normal, and every double normal is a normal.

Lemma 7.1.13. *Suppose that K is a convex body of constant width. Then every normal of K is a diameter, and hence a double normal. Any two diameters of K lying in the same 2-dimensional plane must intersect. In addition, K is strictly convex.*

Proof. Suppose that K has constant width b. Let c be a normal of K perpendicular at its endpoint p_1 to a supporting hyperplane H_1 to K. Let H_2 be the supporting hyperplane to K parallel to H_1, and suppose that H_2 meets K at p_2. If $c \neq [p_1, p_2]$, then the length of $[p_1, p_2]$ is greater than the distance b between H_1 and H_2, the width of K in the direction parallel to c. But then the width of K in the direction parallel to $[p_1, p_2]$ is greater than b, a contradiction. Therefore $c = [p_1, p_2]$ is a double normal. Since K has constant width, its diameter must be b, so c is a diameter.

If there are two nonintersecting diameters of K in the same 2-dimensional plane, then their convex hull is of diameter greater than that of K, which is impossible.

Now suppose that K contains a segment in its boundary, and let u be a direction orthogonal to this segment. Two normals of K parallel to u containing distinct points of the segment must both be diameters, and this is impossible, by the previous paragraph. ∎

Lemma 7.1.14. *Let K be a convex body in $\mathbb{E}^n$, $n \geq 3$, containing the origin in its interior. Suppose that $K \cap S$ is of constant width (in S) for each $S \in \mathcal{G}(n, n-1)$. Then there is a diameter c of K containing the origin. Furthermore, if $S \in \mathcal{G}(n, n-1)$ contains c, then each normal of $K \cap S$ is a diameter of K.*

Proof. To prove the first statement, suppose that c_0 is a diameter of K, of length b. If $o \in c_0$, we are finished. If $o \notin c_0$, let $S \in \mathcal{G}(n, n-1)$ contain c_0; the diameter of $K \cap S$ must also be b. Let x be a point in $K \cap S$ where $\|x\|$ is maximum, and let c be the chord of K containing x and o. Then there is a supporting $(n-2)$-dimensional plane to $K \cap S$ at x orthogonal to c. Therefore c is a normal of $K \cap S$, and so a diameter of it, by Lemma 7.1.13. It follows that the length of c is b, so c is a diameter of K.

Suppose that c is a diameter of K containing o and contained in $S \in \mathcal{G}(n, n-1)$. Any normal of $K \cap S$ is, by Lemma 7.1.13, a diameter of $K \cap S$. But $K \cap S$ contains c, and so its diameter is the same as that of K. ∎

Corollary 7.1.15. *Let $2 \le k \le n-1$, and let K be a convex body in $\mathbb{E}^n$ containing the origin in its interior. Suppose that $K \cap S$ is of constant width (in S) for each $S \in \mathcal{G}(n, k)$. Then K is of constant width.*

Proof. As we noted before, it suffices to consider the case $k = n - 1$. By Lemma 7.1.14, there is a diameter c of K, of length b, say, containing o. Suppose that u is any direction, and let $S \in \mathcal{G}(n, n-1)$ contain u and c. Suppose that $y \in \operatorname{bd} K \cap S$ is such that the supporting hyperplane to K at y is orthogonal to u, and let c' be a normal of $K \cap S$ at y. Then c' is also a diameter of K, by Lemma 7.1.14. Consequently, the width of K in the direction u, which equals the length of c', is b, and so K is of constant width. ∎

Remark 7.1.16. The following much stronger theorem can be proved. Let $2 \le k \le n - 1$, and let K be a convex body in $\mathbb{E}^n$ such that $K \cap S$ is of constant width (in S) for each $S \in \mathcal{G}(n, k)$. Then K is a ball.

Theorem 7.1.1 implies that if $2 \le k \le n - 1$ and the k-dimensional sections of a convex body through an interior point are each known up to a translation, then the body is also determined, up to a translation. This is not true for $k = 1$, since any convex body K has the same 1-dimensional sections through the origin, up to a translation, as $-K$. Moreover, determination is not even possible up to congruence. To see this, note that any nonspherical convex body with the origin as an equichordal point (cf. Theorem 6.3.2) has the same 1-dimensional sections through the origin, up to a translation, as a centered ball. Such bodies cannot be centered, and we shall soon see that, for the class of centered bodies, a strong uniqueness theorem, the generalized Funk section theorem, Theorem 7.2.6, is available. In order to obtain the most powerful results of this type, it is necessary to introduce the higher-dimensional section functions.

7.2. Section functions and point X-rays

Knowing the intersection of a star set L with a 1-dimensional subspace up to a translation is equivalent to knowing the length of this intersection. If we measure

the lengths of intersections with all 1-dimensional subspaces, we have precisely
the X-ray of L at the origin (cf. Definition 5.1.2). Just as we introduced projection functions in Section 3.3 as generalizations of the width function, so we now
introduce generalizations of the X-ray at the origin appropriate for the study of
higher-dimensional sections of a set. There are at least two possibilities here.

If $1 \leq k \leq n - 1$, E is a bounded λ_n-measurable set in $\mathbb{E}^n$, and p is a point,
we can define the *k-dimensional X-ray of E at p* to be the function giving the λ_k-
measure of $E \cap (S + p)$, for each $S \in \mathcal{G}(n, k)$ for which this exists. For star sets,
however, a more general notion employs the theory of dual mixed volumes. As the
name suggests, this is a dual form of the classical theory of mixed volumes, and it
provides appropriate tools for dealing with metrical problems involving sections;
a summary can be found in Section A.7. If $1 \leq k \leq n - 1$, L is a star set in $\mathbb{E}^n$,
and $i \in \mathbb{R}$ is nonzero, the dual volume $\tilde{V}_{i,k}(L \cap S)$ is given for $S \in \mathcal{G}(n, k)$ by

$$\tilde{V}_{i,k}(L \cap S) = \frac{1}{2k} \int_{S^{n-1} \cap S} \rho_{i,L}(u)\, du.$$

The function $\tilde{V}_{i,k}(L \cap \cdot)$ is called a *section function*. When $i = k$, we call it the *i*th
section function, by analogy with the *i*th projection function defined in Chapter 3.
However, the polar coordinate formula for volume implies that

$$\tilde{V}_{i,i}(L \cap S) = \lambda_i(L \cap S),$$

for each $S \in \mathcal{G}(n, i)$, so the *i*th section function is nothing other than the *i*-
dimensional X-ray of L at the origin. More generally, we may use the term *i*th
section function to mean *i*-dimensional X-ray at the origin for any bounded λ_n-
measurable set. For $i \neq 0$, section functions also include the *i*-chord function.
Indeed, the latter is just the special case $k = 1$. Specializing further by taking
$i = k = 1$, we retrieve the ordinary X-ray at the origin. Other basic functions
are also included; for example, $\tilde{V}_{1,k}(L \cap S)$ gives a constant multiple of the mean
length (measured in S) of sections by lines through the origin.

With this new array of tomographic functions in hand, it is natural to inquire
about the corresponding symmetrals. We shall prove in Corollary 7.2.4 that these
are just the *i*-chordal symmetrals encountered in Chapter 6. To acclimatize to the
new notation of dual mixed volumes, let us prove the volume inequality dual to
Theorems 3.2.3 and 3.3.9, established for the central symmetral and Blaschke
body. We need a lemma.

Lemma 7.2.1. *Let $i \in \mathbb{R}$ be nonzero, and let L and M be star bodies in $\mathbb{E}^n$
containing the origin in their interiors. If M is centered, then*

$$\tilde{V}_i(M, \tilde{\nabla}_i L) = \tilde{V}_i(M, L).$$

Proof. Using (A.56), Definition 6.1.2 of the i-chordal symmetral $\tilde{\nabla}_i L$, and the fact that ρ_M is even, for each $u \in S^{n-1}$ we have

$$
\begin{aligned}
\tilde{V}_i(M, \tilde{\nabla}_i L) &= \frac{1}{n} \int_{S^{n-1}} \rho_M(u)^{n-i} \rho_{\tilde{\nabla}_i L}(u)^i \, du \\
&= \frac{1}{2n} \int_{S^{n-1}} \rho_M(u)^{n-i} \rho_L(u)^i \, du + \frac{1}{2n} \int_{S^{n-1}} \rho_M(u)^{n-i} \rho_{-L}(u)^i \, du \\
&= \frac{1}{2n} \int_{S^{n-1}} \rho_M(u)^{n-i} \rho_L(u)^i \, du + \frac{1}{2n} \int_{S^{n-1}} \rho_M(-u)^{n-i} \rho_L(u)^i \, du \\
&= \frac{1}{n} \int_{S^{n-1}} \rho_M(u)^{n-i} \rho_L(u)^i \, du \\
&= \tilde{V}_i(M, L).
\end{aligned}
$$
■

Theorem 7.2.2. *Let $i \in \mathbb{R}$ be nonzero, and let L be a star body in $\mathbb{E}^n$ with $o \in \operatorname{int} L$. If $i \leq n$, then*

$$
\lambda_n(\tilde{\nabla}_i L) \leq \lambda_n(L),
$$

whereas the reverse inequality holds when $i > n$; equality holds if and only if $i = n$ or $i \neq n$ and L is centered.

Proof. In Lemma 7.2.1, let $M = \tilde{\nabla}_i L$. Suppose that $0 < i \leq n$. Then using (B.29), we obtain

$$
\lambda_n(\tilde{\nabla}_i L)^n = \tilde{V}_i(\tilde{\nabla}_i L, \tilde{\nabla}_i L)^n = \tilde{V}_i(\tilde{\nabla}_i L, L)^n \leq \lambda_n(\tilde{\nabla}_i L)^{n-i} \lambda_n(L)^i,
$$

yielding the required inequality. When $i < 0$ or $i > n$, the previous displayed inequality is reversed; for $i < 0$, this again provides the inequality in the statement of the theorem, whereas for $i > n$, it reverses it. Equality holds in (B.29) when $i = n$, and when $i \neq n$, if and only if L is a dilatate of $\tilde{\nabla}_i L$, and hence if and only if L is centered. ■

The object now is to study the determination of a star set by its section functions. For readers not familiar with dual mixed volumes, it may be helpful initially to interpret the results for the case $i = k$, for which $\tilde{V}_{i,k} = \tilde{V}_{i,i} = \lambda_i$.

The next important theorem provides the main link between this chapter and the previous one; it implies that under mild restrictions, two star sets have the same ith section functions (in other words, i-dimensional X-rays at the origin) if and only if their i-chord functions are equal. This will permit us to reap the benefits of work done on i-chord functions in Chapter 6.

Theorem 7.2.3. *Let $i \in \mathbb{R}$ be nonzero, let $1 \leq k \leq n - 1$, and for $j = 1, 2$, let L_j be a star set in $\mathbb{E}^n$ of dimension at least $k + 1$ and such that S_{L_j} is centered. Then*

$$
\tilde{V}_{i,k}(L_1 \cap S) = \tilde{V}_{i,k}(L_2 \cap S),
$$

for all $S \in \mathcal{G}(n, k)$, if and only if

$$\rho_{i,L_1} = \rho_{i,L_2}.$$

Proof. Suppose that the first equality holds. Let the linear hull of L_1 be S_0. If the linear hull of L_2 is not S_0, then there is an $S \in \mathcal{G}(n, k)$ such that $\dim(L_1 \cap S) = k$ and $\dim(L_2 \cap S) < k$, or vice versa. Since $\tilde{V}_{i,k}(L) > 0$ if and only if L is a star set of dimension k (cf. Section A.7), this is a contradiction. Therefore the linear hull of L_2 is also S_0, and we may assume that $S_0 = \mathbb{E}^n$, so that both L_1 and L_2 are star bodies.

Let $T \in \mathcal{G}(n, k + 1)$. If $S \in \mathcal{G}(n, k)$ and $S \subset T$, then by (A.59),

$$\tilde{V}_{i,k}(L_j \cap S) = \frac{1}{2k} \int_{S^{n-1} \cap S} \rho_{i,L_j}(u)\, du,$$

for $j = 1, 2$. Let us put

$$f = \rho_{i,L_1} - \rho_{i,L_2}.$$

Then f is a bounded even Borel function on S^{n-1}, and

$$\int_{S^{n-1} \cap S} f(u)\, du = 0,$$

for all $S \in \mathcal{G}(n, k)$ with $S \subset T$. Identifying T with $\mathbb{E}^{k+1}$, and using Theorem C.2.4, we see that $f(u)$ is zero for λ_k-almost all u in $S^{n-1} \cap T$.

We claim that $f(u)$ is zero for λ_{n-1}-almost all u in S^{n-1}. Note that this is already established when $k = n - 1$. Suppose that $k < n - 1$. Using the fact that λ_{n-1} is the unique Borel-regular, rotation-invariant measure in S^{n-1} such that S^{n-1} has measure $\omega_n = n\kappa_n$ (see Section 0.5), we see that for any bounded Borel function g on S^{n-1},

$$\int_{S^{n-1} \cap u^\perp} g(v)\, dv = \frac{(n-1)\kappa_{n-1}}{(k+1)\kappa_{k+1}} \int_{\mathcal{G}(n-1, k+1)} \int_{S^{n-1} \cap T} g(v)\, dv\, dT,$$

for each $u \in S^{n-1}$, where the outer integral is over the set of all $T \in \mathcal{G}(n, k+1)$ with $T \subset u^\perp$, which is identified with $\mathcal{G}(n-1, k+1)$. (This is because both sides of the equation can be used to define such measures in $S^{n-1} \cap u^\perp$, which we can identify with S^{n-2}.) When $g = f$, the right-hand integral is zero, so

$$\int_{S^{n-1} \cap u^\perp} f(v)\, dv = 0.$$

By Theorem C.2.4 again, this proves the claim. We can now apply Lemma 6.2.1 to deduce that f is identically zero. This gives the second equality in the statement of the theorem.

The converse is trivial, so the proof is complete. ∎

Notice that in contrast to its counterpart Theorem 3.3.2, the first equality of the previous theorem is not restricted to the case $k = n - 1$; indeed, the second equality is independent of k, and for $i > 0$ may be expressed by saying $\tilde{\nabla}_i L_1 = \tilde{\nabla}_i L_2$.

The restriction to sets whose radial functions have centered supports is necessary. To see this, let E be the closed half of the unit ball B above the xy-plane in $\mathbb{E}^3$, and denote by M_1 and M_2 the intersection of E with the sets $\{(x, y, z) : x^2 \geq y^2 + z^2, x \geq 0\}$ and $\{(x, y, z) : y^2 \geq x^2 + z^2, y \geq 0\}$, respectively. Let

$$L_1 = (-M_1 \cup -M_2) \cup E \setminus (\operatorname{int} M_1 \cup \operatorname{int} M_2).$$

If S is a plane through the origin, then the area of $L_1 \cap S$ is $\pi/2$. Let $L_2 = c B$, where $c = 1/\sqrt{2}$. Then the first equality of Theorem 7.2.3 is satisfied, when $i = k = 2$. On the other hand, we have $\rho_{2,L_1}(u) = 1$ for all $u \in S^2$ except for those in a certain nonempty set A of zero λ_2-measure. In fact, if A_x (or A_y) is the intersection of S^2 with the half-plane $\{(x, y, z) : x = 1/\sqrt{2}, z \geq 0\}$ (or $\{(x, y, z) : y = 1/\sqrt{2}, z \geq 0\}$, respectively), and $A = A_x \cup A_y \cup -A_x \cup -A_y$, then $\rho_{2,L_1}(u) = 2$ for all $u \in A$.

The following corollary shows that for $i \neq 0$ the i-chordal symmetral is the appropriate symmetral for the corresponding section function.

Corollary 7.2.4. *Let $i \in \mathbb{R}$ be nonzero, and let L be a star body in $\mathbb{E}^n$ such that S_L is centered. Then $\tilde{\nabla}_i L$ is the unique centered star body such that for some (or for any) k with $1 \leq k \leq n - 1$, we have*

$$\tilde{V}_{i,k}(\tilde{\nabla}_i L \cap S) = \tilde{V}_{i,k}(L \cap S),$$

for all $S \in \mathcal{G}(n, k)$.

Proof. In Theorem 7.2.3, set $L_1 = \tilde{\nabla}_i L$ and $L_2 = L$. It then follows from the definition of $\tilde{\nabla}_i L$ that it has the stated property.

Suppose that L' is a centered star body with

$$\tilde{V}_{i,k}(L' \cap S) = \tilde{V}_{i,k}(L \cap S),$$

for all $S \in \mathcal{G}(n, k)$. By Theorem 7.2.3, we have $\rho_{i,L'} = \rho_{i,L}$. Therefore $\rho_{i,L'} = \rho_{i,\tilde{\nabla}_i L}$. Since L' and $\tilde{\nabla}_i L$ are centered, this implies that $\rho_{L'} = \rho_{\tilde{\nabla}_i L}$ and hence that $L' = \tilde{\nabla}_i L$. ∎

Remark 7.2.5. In Chapter 8, we only require a theory applying to bodies containing the origin. In this case, there is an alternative to the assumption that the support S_L of a star body L is centered. If L is a star body with a continuous radial function, then it follows immediately from Definition 6.1.2 that $\tilde{\nabla}_i L$ is a star body, for $i > 0$ (when $i \leq 0$ it is always assumed that $o \in \operatorname{int} L$). Lemma 6.2.1

holds for star bodies with continuous radial functions, and from this one obtains Theorem 7.2.3 and Corollary 7.2.4 for such bodies. Note that Lemma 7.2.1 and Theorem 7.2.2 also remain true for star bodies with continuous radial functions.

A dual form of Aleksandrov's projection theorem, Theorem 3.3.6, is another immediate consequence of Theorem 7.2.3.

Theorem 7.2.6 (generalized Funk section theorem). *Let $i \in \mathbb{R}$ be nonzero, let $1 \leq k \leq n - 1$, and suppose that L_1, L_2 are centered star sets, of dimension at least $k + 1$, in $\mathbb{E}^n$. If $\tilde{V}_{i,k}(L_1 \cap S) = \tilde{V}_{i,k}(L_2 \cap S)$ for all $S \in \mathcal{G}(n, k)$, then $L_1 = L_2$.*

Proof. Theorem 7.2.3 implies that $\rho_{i,L_1} = \rho_{i,L_2}$, giving $\rho_{L_1} = \rho_{L_2}$, since L_1 and L_2 are centered. ∎

If L_1 and L_2 are noncongruent centered 2-dimensional star sets contained in the same plane in $\mathbb{E}^3$ and of equal area, then $\tilde{V}_{2,2}(L_1 \cap S) = \tilde{V}_{2,2}(L_2 \cap S)$ for all $S \in \mathcal{G}(3, 2)$. Such examples show that the restriction on dimension is necessary in Theorems 7.2.3 and 7.2.6.

Corollary 7.2.7. *Let $1 \leq k \leq n - 1$, and suppose that L is a centered star body in $\mathbb{E}^n$. Then L is determined, among all centered star bodies in $\mathbb{E}^n$, by the λ_k-measures of all its sections by k-dimensional subspaces.*

We now introduce the star bodies of constant section.

Definition 7.2.8. If L is a star body in $\mathbb{E}^n$ such that for some i with $1 \leq i \leq n-1$, $\lambda_i(L \cap S)$ has the same value for each $S \in \mathcal{G}(n, i)$, we say L is of *constant i-section*. If $i = n - 1$, we simply say L has *constant section*.

Theorem 7.2.9. *The only centered star bodies of constant i-section are n-dimensional balls.*

Proof. This follows from the generalized Funk section theorem, Theorem 7.2.6, on taking $L_1 = L$ and L_2 to be a centered ball of the appropriate radius. ∎

Recall the definition (Definition 6.3.1) of an i-equichordal point.

Theorem 7.2.10. *A star body L in $\mathbb{E}^n$ with $o \in$ int L is of constant i-section if and only if it has the origin as an i-equichordal point.*

Proof. In Theorem 7.2.3, take $L_1 = L$ and $L_2 = rB$ for suitable $r > 0$. ∎

Corollary 7.2.11. *For each $n \geq 2$ and integer i with $1 \leq i \leq n - 1$, there is a nonspherical convex body in $\mathbb{E}^n$ containing the origin in its interior and of constant i-section.*

Proof. Theorem 7.2.10 shows that the nonspherical convex bodies in $\mathbb{E}^n$ with o as an i-equichordal point, constructed in Theorem 6.3.2, are also of constant i-section.   ∎

If $i = 2$, for example, one may take $f(\theta) = \sin\theta$ and $c = 3/2$ in Theorem 6.3.2. The resulting convex body in $\mathbb{E}^3$ of constant section is illustrated in Figure 7.1; it is obtained by revolving the curve corresponding to $i = 2$ in Figure 6.5 around the vertical axis.

Figure 7.1. A nonspherical convex body of constant section.

Having established the generalized Funk theorem and the preceding results concerning bodies of constant i-section, we have made substantial progress toward the dual forms of results in Section 3.3. We now continue to harvest the rewards of effort expended on i-chord functions in the previous chapter.

Theorem 7.2.12. *Let $1 \le k \le n - 1$, and let i and j be distinct nonzero real numbers. Suppose that L_1, L_2 are star sets in $\mathbb{E}^n$, of dimension at least $k + 1$, such that S_{L_1} is centered and*

$$\tilde{V}_{i,k}(L_1 \cap S) = \tilde{V}_{i,k}(L_2 \cap S)$$

and

$$\tilde{V}_{j,k}(L_1 \cap S) = \tilde{V}_{j,k}(L_2 \cap S),$$

for all $S \in \mathcal{G}(n, k)$.

(i) *If L_2 is centered, then $L_1 = L_2$.*

(ii) *If L_2 does not contain the origin and is connected, then $L_1 = \pm L_2$.*

(iii) *If $o \in \operatorname{relint} L_2$ and ρ_{L_m}, restricted to its support, is real analytic, for $m = 1, 2$, then $L_1 = \pm L_2$.*

Proof. The three statements in the theorem follow directly from Theorem 7.2.3 and Corollary 6.2.17, Theorem 6.2.20, and Theorem 6.2.21, respectively. ∎

The next two examples follow straight from Theorem 7.2.3 and Theorems 6.2.18 and 6.2.19, and put limitations on how much the previous theorem might be improved.

Theorem 7.2.13. *For $n \geq 2$, there are noncongruent convex polytopes P_1 and P_2 in $\mathbb{E}^n$ containing the origin in their interiors and such that for any nonzero $i \in \mathbb{R}$ and k with $1 \leq k \leq n-1$,*

$$\tilde{V}_{i,k}(P_1 \cap S) = \tilde{V}_{i,k}(P_2 \cap S),$$

for all $S \in \mathcal{G}(n, k)$.

Theorem 7.2.14. *For $n \geq 2$, there are noncongruent convex bodies of revolution K_1 and K_2, of class C_+^∞ in $\mathbb{E}^n$, containing the origin in their interiors and such that for any nonzero $i \in \mathbb{R}$ and k with $1 \leq k \leq n-1$,*

$$\tilde{V}_{i,k}(K_1 \cap S) = \tilde{V}_{i,k}(K_2 \cap S),$$

for all $S \in \mathcal{G}(n, k)$.

The next theorem answers the dual version of Problem 3.9 and yields a corollary disposing of the dual version of the old problem on constant width and constant brightness (cf. Note 3.6).

Theorem 7.2.15. *Let i and j be distinct nonzero real numbers, and let k_i, k_j be integers with $1 \leq k_i, k_j \leq n-1$. Suppose that L_1 and L_2 are star bodies in $\mathbb{E}^n$ containing the origin in their interiors, that L_2 is centered, and that there exist positive constants a and b such that*

$$\tilde{V}_{i,k_i}(L_1 \cap S) = a\tilde{V}_{i,k_i}(L_2 \cap S),$$

for all $S \in \mathcal{G}(n, k_i)$ and

$$\tilde{V}_{j,k_j}(L_1 \cap S) = b\tilde{V}_{j,k_j}(L_2 \cap S),$$

for all $S \in \mathcal{G}(n, k_j)$. Then L_1 is a dilatate of L_2.

Proof. Note that in view of (A.59), Theorem 7.2.3 remains true if the right-hand sides of the two equations in its statement are multiplied by the same constant. The proof follows directly from this observation and Theorem 6.2.22. ∎

Corollary 7.2.16. *Let* $1 \leq i \neq j \leq n - 1$, *and suppose that* L *is a star body in* $\mathbb{E}^n$ *with* $o \in \text{int } L$. *If* L *is of constant* i-*section and constant* j-*section, then* L *is a centered ball.*

Proof. Let $L_1 = L$ and $L_2 = B$, $k_i = i$, and $k_j = j$ in Theorem 7.2.15. ∎

We move to a theorem contributing to an unsolved problem (see Problem 7.4 and Note 7.2) and to be compared with the results listed in Note 3.2 concerning the dual Problem 3.3.

Theorem 7.2.17. *Suppose that* L *is a star body in* $\mathbb{E}^n$ *with* $o \in \text{int } L$ *and that* $2 \leq i \leq n - 1$. *If all the sections of* L *by* i-*dimensional subspaces are congruent, then* L *is a centered ball.*

Proof. Since L is of constant i-section, o is an i-equichordal point of L, by Theorem 7.2.10. If S, $S' \in \mathcal{G}(n, i)$, then there is a rigid motion ϕ such that $\phi(L \cap S) = L \cap S'$.

Suppose that $\phi o = p$; of course, $p \in L$. We assert that there is no loss of generality in assuming that $p = o$, for suppose that $p \neq o$; then

$$\rho_L(u)^i + \rho_L(-u)^i = c,$$

for all $u \in S^{n-1} \cap S'$, because o is an i-equichordal point of L. However, this equation also holds for all $u \in S^{n-1} \cap S$, implying that

$$\rho_{L-p}(u)^i + \rho_{L-p}(-u)^i = c,$$

for all $u \in S^{n-1} \cap S'$. Let T be a 2-dimensional subspace containing p and contained in S'. Then the two previous equations show that $L \cap T$ has equal i-chord functions at o and p. By Corollary 6.2.4, $L \cap T$ is centrally symmetric, with center $p/2$. Therefore $L \cap S'$ is also symmetric about $p/2$. Let ϕ_1 denote reflection in the point $p/2$, and $\phi_2 = \phi_1\phi$. Then ϕ_2 is a rigid motion such that $\phi_2(L \cap S) = L \cap S'$ and $\phi_2 o = o$, proving our assertion.

We now know that if S, $S' \in \mathcal{G}(n, i)$, then there is a rigid motion ϕ, preserving the origin, such that $\phi(L \cap S) = L \cap S'$. Consequently, for each j with $1 \leq j \leq i$, $\tilde{V}_{j,i}(L \cap S)$ is constant for all $S \in \mathcal{G}(n, i)$. The result now follows from Theorem 7.2.15 with $L_1 = L$ and $L_2 = B$. ∎

In the last part of this section, we shall use the techniques we have developed to generalize some of the results of Chapter 5. Though we continue to state our results in terms of dual volumes, we again stress the significance of the special case $i = k$; for example, the next theorem, a generalization of parts (i) and (iii) of Theorem 5.3.3, then gives some conditions under which a convex body in $\mathbb{E}^n$ is determined by its k-dimensional X-rays at two points.

Theorem 7.2.18. *Let* $i \in \mathbb{R}$ *be positive, let* $1 \le k \le n - 1$, *and suppose that* K *is a convex body in* $\mathbb{E}^n$. *Suppose also that* p_1 *and* p_2 *are distinct points such that one of the following conditions holds:*

(i) *The line* l *through* p_1 *and* p_2 *meets* int K, p_1 *and* p_2 *do not belong to* int K, *and it is specified whether or not* K *meets* $[p_1, p_2]$;

(ii) p_1, p_2 *belong to* int K.

Then K *is determined by the dual volumes* $\tilde{V}_{i,k}\big((K - p_j) \cap S\big)$, $S \in \mathcal{G}(n, k)$, $j = 1, 2$, *of its sections through the two points.*

Proof. Suppose that $K' \ne K$ is another convex body with

$$\tilde{V}_{i,k}\big((K - p_j) \cap S\big) = \tilde{V}_{i,k}\big((K' - p_j) \cap S\big),$$

for all $S \in \mathcal{G}(n, k)$ and $j = 1, 2$. The proof of Theorem 7.2.3 shows that K and K' have i-chord functions at p_j equal for λ_{n-1}-almost all $u \in S^{n-1}$, $j = 1, 2$. (The assumptions in Theorem 7.2.3 on the supports of the radial functions are not needed for this weaker conclusion.) By Fubini's theorem, there is a 2-dimensional plane T containing p_1 and p_2 such that the planar convex bodies $K \cap T$ and $K' \cap T$ are different, but have i-chord functions at p_j equal for λ_1-almost all $u \in S^{n-1} \cap T$, $j = 1, 2$. This contradicts Theorem 6.2.2. ∎

Corollary 7.2.19. *Let* $i \in \mathbb{R}$ *be positive, let* $1 \le k \le n - 1$, *and suppose that* K *is a convex body in* $\mathbb{E}^n$. *Suppose that* p_1 *and* p_2 *are distinct points and that*

$$\tilde{V}_{i,k}\big((K - p_1) \cap S\big) = \tilde{V}_{i,k}\big((K - p_2) \cap S\big),$$

for all $S \in \mathcal{G}(n, k)$. *Then* K *is centrally symmetric about the midpoint of the segment* $[p_1, p_2]$.

Proof. The hypotheses imply that parallel k-dimensional planes through p_1 and p_2 meet K in sections with equal dual volume $\tilde{V}_{i,k}$. The argument in the proof of Corollary 5.3.4 can now be applied, where the previous theorem is used in place of Theorem 5.3.3. ∎

Theorem 7.2.20. *Let* $i \in \mathbb{R}$ *be positive, let* $1 \le k \le n - 1$, *and suppose that* P_1 *and* P_2 *are* n-*dimensional convex polytopes in* $\mathbb{E}^n$. *Assume that there are distinct points* p_1 *and* p_2 *such that*

$$\tilde{V}_{i,k}\big((P_1 - p_j) \cap S\big) = \tilde{V}_{i,k}\big((P_2 - p_j) \cap S\big),$$

for all $S \in \mathcal{G}(n, k)$ *and* $j = 1, 2$. *Then* $P_1 = P_2$.

Proof. Suppose that $P_1 \ne P_2$. The proof of Theorem 7.2.3 shows that P_1 and P_2 have i-chord functions at p_j equal for λ_{n-1}-almost all $u \in S^{n-1}$, $j = 1, 2$. By Fubini's theorem, there is a 2-dimensional plane T containing p_1 and p_2 such that convex polygons $P_1 \cap T$ and $P_2 \cap T$ are different, but have i-chord functions at p_j equal for λ_1-almost all $u \in S^{n-1} \cap T$, $j = 1, 2$. This contradicts Theorem 6.2.7.

∎

Note that the restriction to positive i in the last few results cannot be completely removed, since Theorem 6.2.10 provides counterexamples for $i = -1$.

7.3. Point X-rays of measurable sets

As in Chapter 2, we find that outside the natural class of bodies, some of the uniqueness results fail. For example, let L_1 be the unit disk in $\mathbb{E}^2$ and let L_2 be the annulus bounded by the centered circles of radius 1 and 2. These bodies show that the generalized Funk section theorem, Theorem 7.2.6, is not true when $i = k = 1$, $n = 2$, and L_1 and L_2 are measurable sets. The purpose here is not to attempt to extend the whole theory to measurable sets, though this might be a worthwhile endeavor, but rather to return to point X-rays and present a couple of instances where uniqueness fails for measurable sets. The principal result is the following theorem.

Theorem 7.3.1. *Let P be a finite set of points in $\mathbb{E}^n$. There are two disjoint bounded nonnull λ_n-measurable sets with the same X-rays at points in P.*

Put another way, measurable sets are not determined (cf. Definition 5.1.6) by X-rays at any finite set of points. We shall not present the proof of this theorem, owing to its length, but the main ideas are discussed in Note 7.6.

Instead, we shall study in detail an interesting result producing examples in which the sets have fractal boundaries (see Figure 7.2, where the white and black sets have equal X-rays at the points p_j, $1 \le j \le 3$).

We need a little notation. Suppose that E is a bounded Borel set and that $P = \{p_1, \ldots, p_m\}$ is a finite set of points in $\mathbb{E}^2$. Suppose also that

$$(E - p_j) \cap (E - p_k) = \emptyset \tag{7.3}$$

and that

$$\frac{1}{2}(E + p_j) \subset E, \tag{7.4}$$

for $1 \le j \ne k \le m$. Let

$$A = E \setminus \bigcup_{j=1}^{m} \frac{1}{2}(E + p_j), \tag{7.5}$$

and assume that $\lambda_2(A) > 0$. (We refer the reader to the special cases considered after the next theorem.)

Figure 7.2. Measurable sets with equal point X-rays.

Theorem 7.3.2. *Let the bounded Borel set E and finite set P of points in $\mathbb{E}^2$ have the properties (7.3)–(7.5). Then there are two disjoint Borel subsets E_1 and E_2 of E with positive λ_2-measure and the same X-rays at each of the points in P.*

Proof. Define the Borel set A by (7.5). For $1 \le j \le m$, let ϕ_j be the homothety defined for $x \in \mathbb{E}^2$ by $\phi_j x = (x + p_j)/2$. Then (7.3)–(7.5) imply that for $j \ne k$ the sets $\phi_j E$, $\phi_k E$, and A are disjoint. By (7.4) and (7.5), for $1 \le j \le m$ and $n \in \mathbb{N}$ we have $\phi_j^n A \subset \phi_j^n E \setminus \phi_j^{n+1} E$, and so if $n_1 \ne n_2$, the sets $\phi_j^{n_1} A$ and $\phi_j^{n_2} A$ are disjoint. From these observations, one can check that all the sets of the form

$$\phi_{j_k}^{n_k} \cdots \phi_{j_1}^{n_1} A,$$

where $1 \le j_i \le m$, $j_i \ne j_{i+1}$, and $n_i \in \mathbb{N}$, are disjoint.
    Define

$$E_1 = A \cup \bigcup \{\phi_{j_k}^{n_k} \cdots \phi_{j_1}^{n_1} A : k \text{ even}, 1 \le j_i \le m, \ j_i \ne j_{i+1}, \ n_i \in \mathbb{N}\}$$

and

$$E_2 = \bigcup \{\phi_{j_k}^{n_k} \cdots \phi_{j_1}^{n_1} A : k \text{ odd}, 1 \le j_i \le m, \ j_i \ne j_{i+1}, \ n_i \in \mathbb{N}\}.$$

The sets E_1 and E_2 are Borel, being countable unions of sets all homothetic to the Borel set A. From what we have already proved, we also know that E_1 and E_2 are disjoint subsets of E.

Let $1 \le j \le m$, and suppose that t is a line through p_j. Then

$$\lambda_1(E_1 \cap t) = \lambda_1(A \cap t) + \sum \{\lambda_1(t \cap \phi_j^{n_k} \phi_{j_{k-1}}^{n_{k-1}} \cdots \phi_{j_1}^{n_1} A) : k \ge 2 \text{ even}, \ j_{k-1} \ne j\}$$
$$+ \sum \{\lambda_1(t \cap \phi_{j_k}^{n_k} \cdots \phi_{j_1}^{n_1} A) : k \ge 2 \text{ even}, \ j_k \ne j\},$$

where in such sums it will always be assumed that $1 \le j_i \le m$, $j_i \ne j_{i+1}$, and $n_i \in \mathbb{N}$. Now, by (0.8), we have

$$\sum_{n=1}^{\infty} \lambda_1(t \cap \phi_j^n A) = \sum_{n=1}^{\infty} 2^{-n} \lambda_1(A \cap t) = \lambda_1(A \cap t).$$

Using this, and amalgamating the first two terms in the previous expression for $\lambda_1(E_1 \cap t)$, we obtain

$$\lambda_1(E_1 \cap t) = \sum \{\lambda_1(t \cap \phi_j^{n_k} \phi_{j_{k-1}}^{n_{k-1}} \cdots \phi_{j_0}^{n_0} A) : k \ge 0 \text{ even}, \ j_{k-1} \ne j\}$$
$$+ \sum \{\lambda_1(t \cap \phi_{j_k}^{n_k} \cdots \phi_{j_0}^{n_0} A) : k \ge 0 \text{ even}, \ j_k \ne j\}$$
$$= \lambda_1(E_2 \cap t).$$

This shows that E_1 and E_2 have the same X-rays at each of the points in P. ∎

Let E be an open triangle with vertices $p_1 = o$, $p_2 = (1, 0)$, and $p_3 = (1/2, \sqrt{3}/2)$. Figure 7.2 then results, in which the common boundary of the sets E_1 and E_2 is the fractal known as the Sierpiński triangle.

The conditions (7.3)–(7.5) placed on E and P before Theorem 7.3.2 impose a hidden restriction on the cardinality of P, since by (0.8) and the definition of A,

$$\lambda_2(A) = \lambda_2(E)(1 - m2^{-2}).$$

Since $A \subset E$, this implies that $m \le 3$. Theorem 7.3.2 holds in $\mathbb{E}^n$, with the same proof, and in this case the restriction $m \le 2^n - 1$ applies.

The last theorem shows that Theorem 5.3.6, on the determination of convex bodies by three directed X-rays, does not extend to measurable sets. Of course, the stronger Theorem 7.3.1 also shows that Theorem 5.3.8 fails for measurable sets.

Open problems

Problem 7.1. (See Note 7.1.) Does Theorem 7.1.1 hold for star bodies?

Problem 7.2. (See Theorems 7.1.1 and 7.1.11.) Suppose that $2 < k \le n - 1$ and that L_1, L_2 are star bodies in $\mathbb{E}^n$ with $L_1 \cap S$ similar to $L_2 \cap S$, for all $S \in \mathcal{G}(n, k)$. Is L_1 homothetic to $\pm L_2$?

Problem 7.3. (See Theorems 7.1.1 and 7.1.11 and Note 7.1.) Suppose that $2 \leq k \leq n-1$ and that L_1 and L_2 are star bodies in $\mathbb{E}^n$ such that $L_1 \cap S$ is congruent to $L_2 \cap S$, for all $S \in \mathcal{G}(n,k)$. Is $L_1 = \pm L_2$?

Problem 7.4. (See Theorem 7.2.17 and Note 7.2.) Suppose that L is a star body in $\mathbb{E}^n$ and that $2 \leq i \leq n-1$. If all the sections $L \cap S$, $S \in \mathcal{G}(n,i)$, are affinely equivalent, is L an ellipsoid?

Problem 7.5. (See Note 7.4.) Let $1 \leq i \leq n-1$. Are most convex bodies containing the origin in their interior determined (among all such bodies) by their ith section function?

Problem 7.6. Let L_1 and L_2 be centered convex bodies in $\mathbb{E}^3$ whose sections by any plane through the origin have equal perimeters. Is $L_1 = L_2$? If the answer is positive, is the natural generalization to star bodies in $\mathbb{E}^n$ true?

Notes

7.1. *Homothetic and similar sections.* Theorem 7.1.1 was proved by Rogers [693]. Rogers established his result for the case $k = 2$, and noted that a similar proof would deal with other values of k. We use Rogers's proof for $k = 2$, but give the clever reduction to this case found by Burton and Mani [122, Lemma 2]. A direct proof of the extension of Rogers's theorem noted in Remark 7.1.2 was given by Burton [117]. The special case Corollary 7.1.4 of Rogers's theorem was known for $n = 3$ by Süss; see [794].

We deduced Theorem 7.1.5 from Theorem 3.1.7 by polar duality, but Busemann [127, Theorem 16.12] provides a direct proof. Grinberg [334, Theorem 3] indicates an interesting way of obtaining Theorem 7.1.5 from Corollary 7.1.3 and the parallelogram law of Banach space theory. Burton [117] extends this theorem as in Remark 7.1.6, referring also to several auxiliary results in [127].

As we noted in the text, Corollary 7.1.3 is far from the best possible. Rogers was fully aware of this fact and conjectured Theorem 7.1.10, at least for sections through an interior point; this became known as the *false center problem*. The latter was solved by Aitchison, Petty, and Rogers [1]. The paper is quite long and technical, though its authors note that a considerably shorter proof would be possible for strictly convex bodies. According to Larman [493], J. Höbinger carried out the extension of the Aitchison–Petty–Rogers result to sections through an exterior point, but only in the case of a smooth convex body. Larman's paper, also long and involved, removes this restriction, thereby demonstrating Theorem 7.1.10, the false center theorem, in its full generality. Morales and Montejano [635] find a shorter proof of this result.

The most powerful result of the type under discussion is undoubtedly Theorem 7.1.7, a conjecture of P. M. Gruber confirmed by Burton and Mani [122]. For the present, at least, this would seem to be the end of the story. We have omitted the proof of the 3-dimensional case of Burton and Mani's theorem, which occupies nearly twenty pages; but we hope to have consoled the reader by reproducing their ingenious reduction of the general case to this one, and the contortions needed to wring from their main result both Larman's theorem and Theorem 7.1.9 (a generalization of [117, Theorem 3]).

Theorem 7.1.11 was obtained by polar duality from the corresponding result for projections, Theorem 3.1.8, and like that example is due to Petty and McKinney [667].

The result of Remark 7.1.16 on sections of constant width is due to Montejano [630]. (Wegner [828] dealt with the case when $n = 3$ and the body K is smooth, when generalizing the much earlier theorem of Süss [795], who assumed additionally that the sections are taken through an interior point.) The proof is too long to give in this book. This is partly due to the removal of the smoothness assumption, but it also indicates that complications arise when one transfers a concept more naturally related to projections to a setting involving sections. The result we do prove, Corollary 7.1.15, is employed in [630] as a lemma toward the more general theorem. Basic observations about normals and double normals, and Lemma 7.1.13, can be found in the book [83] of Bonnesen and Fenchel (see especially Sections 33 and 63).

The author can show that Problem 7.1 has a positive answer for the translation case. It is quite likely that Problem 7.3 is amenable to the methods used by Golubyatnikov to obtain partial answers to the analogous Problem 3.2 for projections; see Note 3.1.

7.2. *Bodies with congruent or affinely equivalent sections.* Problem 7.4, like its counterpart Problem 3.3, remains open, but many special cases are known to be true. Süss [794] proved Theorem 7.2.17, concerning congruent sections, for convex bodies in $\mathbb{E}^3$. The result of Mani [567] on complete turnings (see Note 3.2) gives Theorem 7.2.17 for n odd and $i = n - 1$, and Burton [119] applies Theorem 7.1.10, the false center theorem, to prove Theorem 7.2.17 when $i = 3$ and $n \geq 4$. Theorem 7.2.17 as stated is due to Schneider [734]. (Of course, Corollary 7.1.4 is a very special case.) For convex bodies, at least, still more powerful results have been obtained by Montejano [631]. Using his theorem on complete turnings mentioned in Note 3.2, he shows that if all the sections $K \cap S$, $S \in \mathcal{G}(n, i)$, of a convex body K in $\mathbb{E}^n$ with $o \in K$ are affinely equivalent, then K is either an ellipsoid or centered; moreover, K must be a ball if all these sections are equivalent under volume-preserving affine transformations, or if all are similar. Schneider notes in his paper that Gromov [359] provides a positive answer to Problem 7.4, when L is convex and $o \in L$, if either $i \leq n - 2$ or n is odd and $i = n - 1$. The simplest unsettled case is therefore $n = 4$ and $i = 3$.

7.3. *Sets of constant section.* There are several references in the literature to bodies of constant section. The connection, Theorem 7.2.10, between these and $(n - 1)$-equichordal points goes back at least to Kubota [481], for $n = 3$. (See also the references in Note 7.4, as well as the paper [537] of Lutwak for the case $i = n - 1$.) The author is not aware of any explicit construction, as in Corollary 7.2.11, of nonspherical convex bodies of constant i-section before that of Gardner and Volčič [283]. However, Hadwiger [373, p. 196] essentially dealt with the case $i = 2$ and $n = 3$.

For convex bodies in $\mathbb{E}^3$ and with $i = 1$ and 2, Corollary 7.2.16 was proved by Süss [794]; see [283] for the general result.

7.4. *Determination by section functions.* Theorem 7.2.2, giving a volume inequality for the i-chordal symmetral, was deduced by Firey [235], for $i \leq -1$ and L convex, from a dual Brunn–Minkowski inequality (a version of (B.30) appropriate for $\tilde{+}_i$ instead of $\tilde{+}$). For $i = 1$ and $i = n - 1$, Theorem 7.2.2 was proved by Lutwak [537, eqs. (4.5), (6.5)]. A version of Theorem 7.2.2 applying to any bounded Borel set is established by Gardner, Vedel Jensen, and Volčič [280] in the course of their extension of the dual Brunn–Minkowski theory mentioned in Section A.7.

With the exception of Theorem 7.2.17, the other results in Section 7.2 were obtained by Gardner and Volčič [283]. The driving force is Theorem 7.2.3, which has many antecedents in the literature. We have mentioned the work of Kubota [481] and Lutwak [537] on bodies of constant section. Larman and Tamvakis [496] proved Theorem 7.2.3 for convex bodies and $i = k = n - 1$, by a long, direct proof. Shortly afterward, Falconer employed the injectivity of the spherical Radon transform (Theorem C.2.4), as we do, to give a much shorter proof of Larman and Tamvakis's result; in [220] and again in [218], he treated convex bodies when $i = k$. Nevertheless, it must be pointed out that this only moves the difficult part of the proof into Appendix C, as happened with the analogous result for projections, Theorem 3.3.2, requiring the injectivity of the cosine transform. The earlier works [729] and [731] of Schneider on geometrical applications of the spherical Radon and cosine transforms should not be overlooked.

Lutwak [537] proves Corollary 7.2.4 for $i = n-1$ in connection with intersection bodies (compare Theorem 8.1.2).

We call Theorem 7.2.6 the generalized Funk section theorem because it was proved by Funk [252] for convex bodies when $i = k = 2$ and $n = 3$. Motivated by the study of Fermi surfaces (see Note 8.12), Lifshitz and Pogorelov [507] proved Theorem 7.2.6 for $i = k = n - 1$. (This was rediscovered by Petty [661] and again by Falconer [217], for convex bodies, and the special case Theorem 7.2.9 appeared in Helgason's book [389, Corollary 4.16].)

Gardner, Soranzo, and Volčič [279] observe that for $1 \leq i \leq n - 1$, no star body in $\mathbb{E}^n$ with a continuous radial function is determined, up to reflection in the origin, by its ith section function. They show that the set of all such star bodies that are determined in this sense by their ith sections functions for all i, $1 \leq i \leq n - 1$, is nowhere dense. They also raise Problem 7.5 and obtain some partial results.

In [335], Grinberg and Quinto show that there is an $\varepsilon > 0$ such that a centered ball in $\mathbb{E}^3$ is not determined by the values of its section function corresponding to sections by planes whose angle with the vertical axis is less than ε. They also obtain conditions on a pair of centered star bodies so that they must be equal if their section functions agree on such a set. Compare the results of Schneider and Weil [744] for projections discussed in Note 3.5.

Soranzo and Volčič [780] study the question of when convex bodies are determined by their ith section function at one point and jth section function at another point, where $i \neq j$.

If $u \in S^{n-1}$ and $q = 0$, the derivative $A_K^{(q)}(0, u)$ of the parallel section function of a convex body K containing the origin in its interior is just the value of its section function at u. Koldobsky and Shane [469] use spherical harmonics to find conditions under which K is determined by derivatives of order q at zero of parallel section functions taken with respect to one or two interior points. Their results include a partial generalization of the case $i = k = n - 1$ of Theorem 7.2.18(ii).

Groemer [349], [354, Section 6.3], [356, Section 5.6] considers the stability of volume inequalities connected with bodies of constant 1-section. In [355] and [356, Section 5.6], he also obtains some stability results for integral transforms and applies these to derive stability theorems related to section functions. (Earlier work of Campi [136] also dealt with stability for the spherical Radon transform.)

Dulio and Peri [194] study star-shaped sets in the sphere S^{n-1}, introduce spherical dual volumes, and prove a spherical version of the generalized Funk section theorem (Theorem 7.2.6), as well as a spherical dual Kubota integral recursion (see Theorem A.7.2). In [196], these authors further extend these definitions and results to a Riemannian manifold setting, obtaining versions of Theorems 7.2.3 and 7.2.6 and Corollary 7.2.19 that hold in spaces of constant curvature.

7.5. *Determination by half-volumes.* Let $u \in S^{n-1}$ be a direction, and suppose that L is a star body in $\mathbb{E}^n$ with $o \in \text{int } L$. Denote by C_u the open hemisphere

$$C_u = \{v \in S^{n-1} : |u \cdot v| > 0\},$$

and let J_u be the closed half-space containing C_u and with boundary $u^\perp$. We call the quantity

$$\lambda_n(L \cap J_u) = \frac{1}{n} \int_{C_u} \rho_L(v)^n \, dv$$

a *half-volume*. Groemer [355], [356, Theorem 5.6.9] proves the following theorem (along with a stability version), to be compared with Theorem 7.2.3. Let L_1 and L_2 be star bodies in $\mathbb{E}^n$, containing the origin in their interiors. Then

$$\lambda_n(L_1 \cap J_u) = \lambda_n(L_2 \cap J_u),$$

for all $u \in S^{n-1}$, if and only if

$$\rho_{L_1}(u)^n - \rho_{L_1}(-u)^n = \rho_{L_2}(u)^n - \rho_{L_2}(-u)^n,$$

for all $u \in S^{n-1}$. The proof uses the *hemispherical transform* H defined by

$$Hf(u) = \int_{C_u} f(u) \, du,$$

for $f \in C(S^{n-1})$, in place of the spherical Radon transform. This transform is injective on *odd* functions, rather than even functions. In fact, more generally, it can be proved that if μ is an odd measure in S^{n-1} such that $\mu(C_u) = 0$ for all $u \in S^{n-1}$, then $\mu = 0$; see the book of Groemer [356, Proposition 3.4.11]. This was originally demonstrated by Funk [253] for $n = 3$, and by Schneider [729], [731], who found a common approach to this and Theorems C.2.1 and C.2.4. (The latter two papers also address the question of whether the injectivity still holds when the transform H is modified so that integration is over caps of S^{n-1} of fixed radius r, $0 < r < 1$. Ungar [809] initiated this topic, which invites us to consider intersections by sets other than planes; see the fascinating article [855] of Zalcman.)

It follows readily that if L is a star body with $o \in \text{int } L$ whose half-volumes are all constant, then L must be centered; see [731, p. 62] for this result and its history, which goes back to Funk [253]. The survey [386, Section 4.1] of Heil and Martini has references to related characterizations of centered bodies.

Campi [137] and Groemer [356, Theorem 5.6.9] investigate the stability and well-posedness of the problem of reconstructing a star body from its half-volumes. By proving stability results for yet another integral transform on $C(S^{n-1})$, Groemer [358] proves that two (not necessarily centered) star bodies containing the origin in their interiors must be equal if the volumes of their intersections with each half-hyperplane containing the origin in its boundary are always equal.

The reader interested in results of this type should also consult the works of Rubin [701], [702], [703], where the related integral transforms are analyzed.

Let L be a star body with $o \in \text{int } L$, and let $2 \leq k \leq n - 1$. Goodey and Weil [323] note that Groemer's result just described immediately implies that L is determined by the volumes of its intersections with half-spaces of dimension k. They introduce the *average half-section function* $\bar{s}_{L,k}(u)$, $u \in S^{n-1}$, giving the average of the volumes of the intersections of L with k-dimensional half-spaces H containing u with u orthogonal to the boundary of H. They prove that $\bar{s}_{L,k}$ determines L when $k = 2$ and $n = 3$ or 4, and when $n \geq 5$ and either $k \leq (n+2)/2$ or $k > (2n+1)/3$, obtaining some stability results as well. Goodey and Weil also show that $\bar{s}_{L,k}(u)$ does not determine L if $k = (2n+1)/3$ is an integer and provide an analysis for some other pairs (n, k) that implicates algebraic number theory, as in their investigation of determination by directed projection functions (see Note 3.7).

In the language of geometric probing (cf. Note 1.6), a *half-space probe* returns the volume of the intersection of an object with a half-space. Skiena [768] proves that $7n+7$ half-space probes suffice, and $2n$ are necessary, to successively determine a convex n-gon in $\mathbb{E}^2$, among all convex polygons; for verification, corresponding bounds of $n + 1$ and $[2n/3] + 1$ are found.

7.6. *Point X-rays of measurable sets.* Theorem 7.3.1 is due to Kemperman [426]. (The result was announced by Gutman et al. [368, Theorem 3], but their proof covers only the case of two points in $\mathbb{E}^2$.) The basic goal is to construct, for each bounded λ_n-measurable set D in $\mathbb{E}^n$, a bounded measurable function F on D, not almost everywhere zero, but with an X-ray (i.e., line transform, cf. (C.6)) zero at each point in P. If this is accomplished, then we can assume that $-1 \le F \le 1$. Consider the set $M(F^+)$ of measurable functions g on D with $0 \le g \le 1$ and with the same X-rays as the positive part F^+ of F at points in P. This is a compact convex subset of L^∞ with the weak* topology, and contains F^+ and F^-, so it has at least two distinct extreme points. These are provided by the Krein–Milman theorem, as in Note 2.7, and one similarly concludes that these extreme points are characteristic functions of distinct measurable subsets E and E' of D. The X-rays of these sets are equal at each point in P.

To achieve the basic goal just mentioned, consider first the case when all the points in P lie in a hyperplane H. We can assume that $H = \{x = (x_1, \ldots, x_n) : x_n = 0\}$. A projective transformation ϕ will take H to the hyperplane at infinity (cf. Section 0.2); in fact, we can take

$$\phi(x_1, \ldots, x_n) = \left(\frac{x_1}{x_n}, \frac{x_2}{x_n}, \ldots, \frac{x_{n-1}}{x_n}, \frac{1}{x_n}\right).$$

We can assume that D is compact and that $D \cap H = \emptyset$. Then ϕD is a compact set in $\mathbb{E}^n$, so we can choose a bounded measurable function h on ϕD, not almost everywhere zero, with zero parallel X-rays in the directions ϕp for $p \in P$. (This h is chosen in the same way as the h in Note 2.7.) Now define

$$F(x) = h(\phi x)/x_n^2.$$

A straightforward computation shows that the integral of F along a line l through a point p in H is a constant multiple, depending only on l, of the integral of h along the line ϕl. (This is the same phenomenon that we saw in Theorem 6.2.8(i); one can see this by identifying h with the characteristic function of the convex body K in that theorem.) Since for $p \in P$, the X-ray of h from the direction ϕp is zero, it follows that the X-ray of F at p is also zero. This completes the proof in this case, because it is easily seen that F cannot be almost everywhere zero. The general case is basically dealt with by the clever trick of regarding $\mathbb{E}^n$ as a hyperplane in $\mathbb{E}^{n+1}$, but we can only refer the reader to [426] for the details.

Theorem 7.3.2 and the consequent beautiful Figure 7.2, together with another figure based on a square instead of a triangle, are due to Brehm [96]. Brehm also proves Theorem 7.3.1 when no three of the points in P are collinear; the proof, though still geometrical in flavor, is more complicated than that of Theorem 7.3.2, and does not seem to produce the same appealing fractal-like pictures.

Perhaps the only other result known about point X-rays of nonconvex sets is due to Brehm [98]. Consider the points $p_1 = o$ and $p_2 = (1, 0)$ in the plane. Brehm constructs two different sets of ten disjoint disks, the union of each having the same X-rays at p_1 and p_2. This is done as follows. Suppose that $0 < \theta_i < \pi/2$, $1 \le i, j \le m$, and that q_{ij} is the point at the intersection of the line through p_1, at angle θ_i with the positive x-axis, and the line through p_2, at angle $\pi - \theta_j$ with the positive x-axis. Let $r_{ij} = \|p_1 - q_{ij}\|$ and $s_{ij} = r_{ji} = \|p_2 - q_{ij}\|$. If we let $x_i = \cot \theta_i$, then

$$r_{ij} \sin \theta_i = s_{ij} \sin \theta_j = \frac{1}{x_i + x_j}.$$

Suppose that $\varepsilon > 0$, that Q is some subset of the points q_{ij}, and that at each point q_{ij} in Q we place a disk centered at q_{ij} of radius equal to ε times the y-coordinate of q_{ij}. If ε is sufficiently small, the disks are disjoint. If t is a line through p_1 or p_2, one can check that the λ_1-measure of the intersection of t with the union of these disks is a constant multiple (depending only on t) of $\sum\{1/(x_i + x_j) : q_{ij} \in Q\}$. Then, to obtain different finite sets of disks with equal X-rays at p_1 and p_2, it suffices to find distinct positive reals x_i, $1 \le i \le m$, and a symmetric $m \times m$ matrix $A = (a_{ij})$ with the following properties: A is not identically zero, a_{ij} is -1, 0, or 1 for each i and j, and

$$\sum_{i=1}^{m} \frac{a_{ij}}{x_i + x_j} = 0.$$

Brehm calls this a *harmonic system*; the corresponding points q_{ij} form a sort of "point-switching component" for p_1 and p_2 (cf. Definition 2.3.1). Brehm has found several harmonic systems, computationally, by means of a recursive procedure.

It is a simple matter to check that one can use homothetic copies of a fixed triangle instead of disks in this construction of Brehm. An argument very similar to that of Theorem 2.3.4 then yields two distinct polygons that are star-shaped at o with equal X-rays at each of two points. Brehm also finds harmonic systems in which $x_i \in \mathbb{N}$, $1 \le i \le m$; using them, he finds integer lattice points p_1 and p_2 such that there are two distinct polygons, each a finite union of integer lattice squares in the same open half-plane bounded by the line through p_1 and p_2, with equal X-rays at p_1 and p_2. An example is shown in Figure 7.3, in which the two polygons appear in different shades of grey.

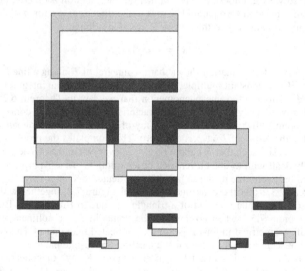

$p_1 \bullet$ $\bullet\, p_2$

Figure 7.3. Polygons with equal point X-rays.

7.7. *Sections of convex polytopes.* The papers [444] and [445] of Klee, with results on projections summarized in Note 3.10, contain corresponding results for sections through a given point. See also Grünbaum's book [367, Section 5.1]. Results exist concerning the expected number of j-dimensional faces of a random section of a regular convex polytope by a k-dimensional subspace. See the paper of Lonke [520] and the references given there.

In Note 3.10, we remarked that when additional a priori information is available, a finite set of projections can determine a convex polytope. By polar duality, this transfers to sections, called *cross-section probes* by Skiena [770] in his survey on geometric probing. There are also methods, such as that of Boissonnat [77], that are designed to reconstruct a polyhedron from a finite set of planar sections.

7.8. *Critical sections.* The problem of finding the sections of a convex body with minimal or maximal volume – *critical sections* – has attracted much attention. Let C^n denote the centered unit cube in $\mathbb{E}^n$. In their entertaining article [155], Chakerian and Logothetti attribute to Laplace a formula for the volume of a section of C^n orthogonal to a main diagonal, and to G. Pólya a general formula for the volume of sections; in the case of central sections, the formulas involve oscillatory integrals of the form

$$\int_0^\infty \left(\frac{\sin t}{t}\right)^n dt.$$

It is now known that

$$1 \leq \lambda_k(C^n \cap S) \leq (\sqrt{2})^{n-k},$$

for all $S \in \mathcal{G}(n,k)$, $1 \leq k \leq n-1$. The lower bound, conjectured by A. Good, follows from work of Hadwiger [374] when $k = n-1$ and Vaaler [810] for general k. For $k = n-1$, an upper bound of 5 was obtained by Hensley [391], but the best possible bound of $\sqrt{2}$ is due to Ball [19]. These results for $k = n-1$ are also presented in Koldobsky's book [465, Chapter 7]; the proof there for the upper bound is based on that of Nazarov and Podkorytov [649]. Ball establishes the upper bound for general k in [22], along with the estimate

$$\lambda_k(C^n \cap S) \leq \left(\frac{n}{k}\right)^{k/2},$$

for all $S \in \mathcal{G}(n,k)$, $1 \leq k \leq n-1$, which is the best possible when k divides n. Despite this, there are values of k and n for which the best bound is not known. Neither lower nor upper bound is easily obtained, and both have applications; the lower bound is useful in number theory and the upper bound has consequences for the Busemann–Petty problem, as explained in Note 8.9. There is also a connection to the slicing problem, Problem 8.3, explained in some detail in Note 9.8. Gluskin [293] builds on Ball's work to find an estimate for the maximal Lebesgue measure of central sections of certain Cartesian products of measurable sets.

Suppose that $0 < p \leq \infty$ and that B_p^n is the unit ball in l_p^n, defined as in Note 3.11. Let $S \in \mathcal{G}(n,k)$, $1 \leq k \leq n-1$, and let $h(p) = \lambda_k(B_p^n \cap S)/\lambda_k(B_p^k)$. Meyer and Pajor [604] prove that $h(p)$ increases for $p \geq 1$ (Barthe [37] shows that this is true for $p > 0$). Since $h(2) = 1$, this gives

$$\lambda_k(B_p^n \cap S) \geq \lambda_k(B_p^k),$$

for $p \geq 2$, so the minimal k-dimensional central section of B_p^n is B_p^k for $p \geq 2$. (The case $p = \infty$ is just the lower bound in the previous paragraph.) Of course, for $p \leq 2$, the inequality is reversed.

For $p = 1$, when the unit ball is a centered cross-polytope, Meyer and Pajor [604] show that

$$\frac{\sqrt{n}}{4^{n-1}}\binom{2n-2}{n-1}\frac{2^{n-1}}{(n-1)!} \leq \lambda_{n-1}(B_1^n \cap u^\perp) \leq \frac{2^{n-1}}{(n-1)!},$$

for all $u \in S^{n-1}$. For each inequality they also characterize those u for which equality holds, and conjecture that the situation is the same for all $0 < p < 2$. Koldobsky [455] proves this conjecture and presents it in his book [465, Chapter 7], where several additional references on this topic can be found. Only the problem of finding the maximal sections of B_p^n for $2 < p < \infty$ remains open.

Suppose that K is a regular simplex with edge length $\sqrt{2}$ and centroid at the origin. Webb [826] shows that

$$\lambda_{n-1}(K \cap u^{\perp}) \leq \frac{\sqrt{n+1}}{\Gamma(n)\sqrt{2}},$$

for all $u \in S^{n-1}$, and this upper bound is attained if and only if $u^{\perp}$ contains $n-1$ of the vertices of K.

Filliman [232] takes a quite different approach to finding critical sections of a convex polytope. He employs polar duality to find a formula, akin to that of Lawrence [498] mentioned in Note 2.1, giving (under certain conditions) the kth section function (k-dimensional X-ray at o) of a polytope P in $\mathbb{E}^n$ with $o \in \text{int } P$. He uses this to locate geometrically all minimal and maximal sections of a regular simplex, and notes that for general polytopes there is an important difference between the two: Minimal sections should occur where the section function is differentiable, and maximal sections where it is not differentiable.

Computational aspects are discussed by Gritzmann and Klee [341, Section 8.1].

7.9. *Almost-spherical or almost-ellipsoidal sections.* Dvoretzky [201] proved the following theorem. Given $\varepsilon > 0$ and $k \in \mathbb{N}$, there is an N such that for each $n > N$ and centered convex body K in $\mathbb{E}^n$, there is an $S \in \mathcal{G}(n, k)$ such that

$$B \cap S \subset K \cap S \subset (1 + \varepsilon)B \cap S.$$

In other words, each centered convex body of sufficiently high dimension has an almost-spherical k-dimensional central section. Dvoretzky's theorem is of cardinal importance in the local theory of Banach spaces, often appearing in the following form:

Given $\varepsilon > 0$ and $k \in \mathbb{N}$, there is an N such that every Banach space X of dimension $n > N$ contains a k-dimensional subspace S such that $d(S, l_2^k) \leq (1+\varepsilon)$.

Here, d is the Banach–Mazur distance (cf. Note 4.10). Milman [620] proved that the theorem holds for $k \geq f(\varepsilon)\log N$, where $f(\varepsilon) > 0$ depends only on ε, and it is easy to see that the factor $\log N$ cannot be improved upon. (Take K to be a centered cube, or, equivalently, $X = l_\infty^n$.) Dvoretzky [202] noted that by polar duality, each centered convex body of sufficiently high dimension also has an almost-spherical k-dimensional projection; Straus [790] showed that there is actually a subspace for which both the section and projection are almost spherical, and Larman and Mani [494] proved that there are large sets of subspaces for which sections and projections are almost ellipsoidal.

Dvoretzky's theorem has stimulated a good deal of research. It is difficult to imagine a better introduction to this highly technical subject than the elegant and insightful account of Ball [27]. For further information, see the surveys of Giannopoulos and Milman [287], Lindenstrauss [510], and Lindenstrauss and Milman [511] and the books of Milman and Schechtman [622] and Pisier [673]. Naturally, several of the results have purely geometrical interpretations. For example, each centered cross-polytope of sufficiently high dimension has a k-dimensional central section almost affinely equivalent to the unit ball of l_p^k, $1 < p \leq 2$. See the article [511, p. 1173] for references. There are also some results concerning the Banach–Mazur distance between sections or projections; see the article of Rudelson [709] and the references given there.

The paper of Groemer [347], with results on projections summarized in Note 3.12, also addresses sections through a given point.

Related to this topic are some results of Makeev. In [564], for example, he proves that if K is a convex body in $\mathbb{E}^3$ with $o \in \text{int}\, K$, then there is a 2-dimensional subspace S with a centered affinely regular hexagon inscribed in $K \cap S$. (It is known that a centered affinely regular hexagon can be inscribed in every planar centered convex body.) He also shows that if K is a convex body in $\mathbb{E}^4$ with $o \in \text{int}\, K$, then there is a 2-dimensional subspace S with a centered affinely regular octagon inscribed in $K \cap S$.

7.10. *A characterization of star-shaped sets.* A set is *acyclic* if it is homologically equiv-
alent to a point, meaning that it is connected and all its homology groups are trivial. Let C be a compact set in $\mathbb{E}^n$ and $1 \le k \le n-1$. Kosinski [472] shows that if $C \cap S$ is nonempty and acyclic for each $S \in \mathcal{G}(n, k)$ meeting C, then the line segment $[o, x]$ belongs to C for each $x \in C$.

A planar set is acyclic if and only if it is simply connected, so if $k = 2$, "acyclic" can be replaced by "simply connected" in Kosinski's theorem. The case $n = 3$ of this result was discovered earlier by Valentine, and the proof is given in his book [811, p. 92].

7.11. *Sections by other sets of planes.* Up to now, we have mainly discussed inverse prob-
lems concerning sections through one point, and have observed a duality between these and inverse problems about projections. Occasionally, sections through a finite number of points have also been considered. Here we browse some work on sections by other sets of planes, for example, the set of *all* planes of a fixed dimension (a complete survey would be an overwhelming task).

The existence of the radial map (0.3) shows that the relative boundary of every k-dimensional compact convex set is homeomorphic to S^{k-1}, and so all k-
dimensional sections of a convex body in $\mathbb{E}^n$ have this property. In Problem 68 of *The Scottish Book* [585], S. M. Ulam asked if convex bodies are character-
ized by this property when $k = n - 1$. Several results of this type are available. For example, suppose that C is a compact set in $\mathbb{E}^n$ and that $1 \le k \le n - 1$. Aumann [16] proved that if $C \cap F$ is acyclic for each k-dimensional plane F meet-
ing C, then C is convex. (Observe that this is generalized by Kosinski's theorem in Note 7.10; in fact, Kosinski [473] re-proves Aumann's theorem.) As in Kosinski's theorem, "acyclic" can be replaced by "simply connected" when $k = 2$. Montejano and Shchepin [634] extend Aumann's theorem to weakly closed subsets of a locally convex linear space and note that Kosinksi's theorem does not generalize in this way. Fáry [222] proves a weaker version of Aumann's theorem, and more references are given in the survey [568] of Mani-Levitska.

An alternative answer to Ulam's question was provided by Montejano [629], who proved that if M is a closed, connected $(n - 1)$-manifold topologically embedded in $\mathbb{E}^n$ such that $H \setminus M$ has exactly two components for each hyperplane H meeting M in more than one point, then M is the boundary of a convex body.

A *topological n-sphere* is a set homeomorphic to S^n. In [629], Montejano gives characterizations of topological n-spheres in $\mathbb{E}^{n+1}$ in terms of properties of horizontal sections. Here he uses techniques of Daverman [181]. This line of work appar-
ently began with the following question of Bing [66]: If each horizontal section of a topological 2-sphere S in $\mathbb{E}^3$ is either a point or a simple closed curve, must S be tame, that is, is there a homeomorphism of $\mathbb{E}^3$ to itself that takes S to S^2? This was later answered affirmatively; the interested reader may consult [181], work of Pax [658], and the references given there.

In [724], Schapira finds a general integral transform inversion formula that im-
plies, in particular, that a suitably smooth body in $\mathbb{E}^3$ is determined by the number of connected components minus the number of holes of its sections by all planes.

Stein [786] shows that if E is an open set in $\mathbb{E}^n$ and $1 \leq k \leq n-1$, then E is convex if there is a continuous function f from $\mathcal{G}(n,k)$ to $\mathbb{E}^n$ such that for each $S \in \mathcal{G}(n,k)$, $E \cap \left(S + f(S) \right)$ is nonempty and convex. Montejano [632] provides analogous characterizations of balls and sets of constant width, and Bianchi and Gruber [60] similarly treat ellipsoids.

Other characterizations of ellipsoids are contained in Petty's survey [664] and the paper by Alonso and Martin [6]. For example, suppose that K is a convex body in $\mathbb{E}^n$, $n \geq 3$, and that $2 \leq k \leq n-1$. The false center theorem, Theorem 7.1.10, implies that if every k-dimensional section of K is centrally symmetric, then K is an ellipsoid. However, the hypothesis can be restricted to k-dimensional sections of K of diameter less than some $a > 0$, as Burton [118] demonstrates. Burton [116] also proves that a convex body K in $\mathbb{E}^n$, $n \geq 3$, has the property that for each $u \in S^{n-1}$, there is an $a_u > 0$ such that

$$K \cap \{x : u \cdot x = h_K(u) - t\}$$

is centrally symmetric for all t with $0 < t < a_u$, if and only if K is the Minkowski sum of a zonotope and an ellipsoid. It follows that the same property characterizes ellipsoids among strictly convex bodies.

Inspired by an old result of S. Olovjanischnikov, Barker and Larman [34] make two interesting conjectures. The first is that if K and K_1 are convex bodies in $\mathbb{E}^n$, $n \geq 3$ with $K \subset \operatorname{int} K_1$, and $K_1 \cap H$ is centrally symmetric whenever H is a hyperplane supporting K, then K_1 is an ellipsoid. The second is that if K, K_1, and K_2 are convex bodies in $\mathbb{E}^n$ with $K \subset \operatorname{int} K_1 \cap \operatorname{int} K_2$, and $\lambda_{n-1}(K_1 \cap H) = \lambda_{n-1}(K_2 \cap H)$ whenever H is a hyperplane supporting K, then $K_1 = K_2$. Some partial results concerning the second conjecture are given.

It is certain that other characterizations in terms of sections by planes are scattered here and there in the literature. For example, Molzon, Shiffman, and Siborny [627] prove that an analytic subvariety of $\mathbb{C}^n$ is algebraic if its sections by many (in a certain technical sense) hyperplanes are algebraic.

Given a polytope, one can ask for characterizations of sections of largest volume among all sections by planes of a given dimension (as opposed to those, considered in Note 7.8, through one fixed point). For a simplex in $\mathbb{E}^n$, it is known that a maximal $(n-1)$-dimensional section must be a facet, when $n \leq 4$. For $n \leq 3$, references are given by Eggleston, Grünbaum, and Klee [207, Section 7], who also show that this is true for arbitrary n and sections having at most $n+1$ vertices; and Philip [672] settled the case $n = 4$. On the other hand, for $n \geq 5$, there is an n-dimensional simplex K with an $(n-1)$-dimensional section of larger volume than any facet of K; Walkup [825] found a 5-dimensional example and Philip [672] established the general statement by induction. Compare also a negative result of Makai and Martini [559]. Filliman [232, p. 79] conjectured that maximal sections of a simplex are identical to its minimal projections.

Dalla and Larman [174] construct a convex body K in $\mathbb{E}^3$ such that almost all 2-dimensional sections of K are polygons, but the set of extreme points of K has Hausdorff dimension one; in particular, K does not have a countable set of facets.

If C is a planar compact convex set, let $q(C) = \pi \lambda_2(C)/V_1(C)^2$, and note that $q(C) \leq 1$, with equality if and only if C is a disk, by the isoperimetric inequality (B.14). According to Chakerian [153], the following question of H. Steinhaus is still unanswered. If K is a convex body in $\mathbb{E}^3$, is it true that if $q(K \cap H) \geq c$ for each plane H, then

$$\frac{9\pi \lambda_3(K)^2}{2V_2(K)^3} \geq c\,?$$

(The analogous question for projections is Problem 4.3.) Chakerian proves that the same assumptions imply that

$$\frac{3\pi V_1(K)\lambda_3(K)}{4V_2(K)^2} \geq \frac{3\pi^2}{32}c$$

and conjectures that the constant $3\pi^2/32$ can be replaced by 1.

Frantz [249] recalls the nonmeasurable subset of the plane constructed by Sierpiński [766], which meets every line in at most two points. He proves that a planar set must be measurable, however, if each of its linear sections is a union of at most two disjoint open intervals. An old open problem, revived by Croft, Falconer, and Guy [173, Problem G9] (see also the article of Mauldin [586]), asks whether there is a planar Borel set meeting each line in exactly two points.

H. T. Croft asked whether for each planar measurable set E of positive area there is a $t_0 > 0$ such that for all t with $0 < t < t_0$, there is a line l with $\lambda_1(E \cap l) = t$. This is not known, but interestingly, Falconer [215] has shown that this is true in $\mathbb{E}^3$ with lines replaced by planes and λ_1 by λ_2; see also [173, Problem G7].

7.12. *Integral geometry.* Integral geometry is an extensive subject. A standard reference is Santaló's book [721], but the excellent text [746] of Schneider and Weil is perhaps more compatible with our standpoint. The survey [747] of Schneider and Wieacker and Chapter 4 of Schneider's book [737] also provide substantial introductions, while Palamodov [654] focuses on integral transforms. Integral geometry has intimate connections with geometric tomography. For example, Kubota's integral recursion (A.46) belongs to a family of integral-geometric formulas and in fact can be deduced from the so-called *principal kinematic formula* (see [737, p. 229, eq. (4.5.3)] or [746, Korollar 3.2.4]) giving the measure of the set of rigid motions for which a moving convex body meets a fixed convex body, in terms of the intrinsic volumes of the two bodies. Zhang [863] finds kinematic formulas involving dual quermassintegrals.

Though we have not needed them in this book, other formulas can also be deduced and deserve mention. Perhaps the best known, *Crofton's intersection formulas*, are as follows. Let K be a convex body in $\mathbb{E}^n$ and $0 \leq i \leq k \leq n$. Then

$$V_{n+i-k}(K) = \frac{\binom{n}{k-i}\kappa_i\kappa_n}{\binom{k}{i}\kappa_{n+i-k}\kappa_k} \int V_i(K \cap F)\,dF.$$

Here, integration is over the set of all k-dimensional planes in $\mathbb{E}^n$, with respect to the canonical measure, normalized so that the measure of the set of such planes meeting B is κ_{n-k}; see [737, p. 235, eq. (4.5.8)] or [746, Korollar 3.3.2].

Another example, this time from *translative integral geometry*, states that for a convex body K in $\mathbb{E}^n$, $1 \leq i \leq n-1$, and $S \in \mathcal{G}(n,k)$,

$$V(K, n+i-k; B^k, k-i) = \frac{\kappa_{k-i}}{\binom{n}{k-i}} \int_{S^\perp} V_i\left(K \cap (S+x)\right) dx.$$

See [737, p. 255, eq. (4.5.35)] and [746, Abschnitt 3.1].

For other integral-geometric formulas related to these, see the sources already cited. It is a significant fact that such formulas are not restricted only to convex bodies. Extensions are possible to special classes of sets, such as the *convex ring*, comprising finite unions of convex bodies, and *sets of positive reach*, with the property that for some $r > 0$ and for each x whose distance from the set is less than r, there is a unique point in the set nearest to x.

The latter sets, introduced by H. Federer, take us to the brink of geometric measure theory. For an example of the sort of generality possible, consider the following version of Crofton's intersection formulas, corresponding to the case $n = 2, k = 1$, and $i = 0$. If E is a subset of $\mathbb{E}^2$, and l is a line, define $n_E(l) = |E \cap l|$. Then, for a *rectifiable* set E (a subset of a countable union of rectifiable arcs of finite total length),

$$\lambda_1(E) = \frac{1}{2} \int_0^{2\pi} \int_0^{\infty} n_E(\theta, t) \, d\theta \, dt,$$

where $n_E(\theta, t)$ corresponds to the line l making an angle θ with a fixed axis and of distance t from the origin. H. Steinhaus asked if n_E determines E, up to a set of λ_1-measure zero; see [173, Problem G8] for comments and references. Fast [223] presents a positive answer to Steinhaus's question. Richardson's independent solution in [690] and [691] applies only to rectifiable curves, but also provides a method for reconstructing a closed rectifiable curve E from n_E.

For such versions of Crofton's intersection formulas in higher dimensions, see, for example, the book [637, Theorem 3.16, p. 32] of Morgan.

Suppose K is a convex body in $\mathbb{E}^n$ and $p \geq 0$. The quantity

$$I_p(K) = \frac{\omega_n}{2} \int \lambda_1(K \cap G)^p \, dG,$$

where integration is over the set of all lines G in $\mathbb{E}^n$, with respect to the canonical measure, is called a *chord-power integral*. The Crofton intersection formulas with $i = k = 1$ yield

$$I_1(K) = \frac{\omega_n}{2} \lambda_n(K).$$

The *Crofton–Hadwiger formula*

$$I_{n+1}(K) = \frac{n(n+1)}{2} \lambda_n(K)^2$$

can be found in [721, p. 237, eq. (14.24)] and [746, p. 176], for example, along with several related formulas. Pohl [674] is led to consider chord-power integrals in attempting to find the probability that two closed strands of DNA are linked. Mallows and Clark [565] answer a question of Blaschke by finding two noncongruent convex planar polygons with chord-power integrals equal for all p; see also the work of Waksman [824]. For other material pertaining to chord-power integrals, consult the works of Cabo [134], Goodey and Weil [319], Hansen and Reitzner [381], Nagel [641], Schneider and Wieacker [747, Section 7], and Zhang [856], [859], [863]. See also Note 9.3 for a discussion of the related covariogram problem.

Burton [120] finds integral-geometric formulas relating the Hausdorff dimension of the set of extreme points of a convex body to the integral, over all sections, of the function giving the number of extreme points of a section of the body. See also [737, p. 69].

7.13. *Mean section bodies.* Suppose that K is a convex body in $\mathbb{E}^n$, that $1 \leq k \leq n - 1$, and that $F_1, \ldots, F_m$ are k-dimensional planes. If we define

$$K_m = \frac{1}{m} \big((K \cap F_1) + \cdots + (K \cap F_m) \big),$$

then K_m is a compact convex set, known up to a translation if the sections $K \cap F_j$, $1 \leq j \leq m$, are also known up to a translation. Goodey and Weil [319] consider the problem of determining a convex body K in $\mathbb{E}^n$ from random sections by planes $F_1, \ldots, F_m$, by estimating it from K_m, as $m \to \infty$. This leads them to define, for

$1 \leq k \leq n - 1$, the *kth mean section body* $M_k K$ of K, whose support function is obtained by integrating the support functions of the sections $K \cap F$ over all k-dimensional sections by planes F, and to consider when K is determined by $M_k K$. They show that K is never determined by $M_1 K$, since the latter is a translate of $(\kappa_{n-1}\lambda_n(K)/n\kappa_n)B$. On the other hand, if K is centrally symmetric, they prove that

$$M_2 K = \frac{\pi}{2n\kappa_n}\Pi_1\Pi_{n-1}K,$$

so such K are determined by $M_2 K$ in view of Aleksandrov's projection theorem, Theorem 3.3.6. Goodey [308] proves that a centrally symmetric convex body K in $\mathbb{E}^n$ is also determined by $M_k K$ when $3 \leq k \leq n - 1$, and shows that most centrally symmetric convex bodies are so determined among all convex bodies. Another interpretation of $M_k K$ is found by Weil [837] (see also the sequel by Goodey and Weil [322]).

Motivated by a question in stereology, Weil [838] introduces the *kth Blaschke section body* $B_k K$ of a convex body K in $\mathbb{E}^n$. If $\leq k \leq n - 1$, this is the convex body whose surface area measure is obtained by integrating the surface area measures of the sections $K \cap F$ over all k-dimensional planes F. He also defines mean section bodies and Blaschke section bodies relative to a subspace S; here, the integration is restricted to sections by translates of S. (When K is a polytope, the mean section body of K relative to S is just the *fiber polytope* first studied by Billera and Sturmfels [65] for its useful combinatorial properties.) Goodey, Kiderlen, and Weil [311] continue this investigation, proving that for $2 \leq k \leq n - 1$, a convex body K in $\mathbb{E}^n$ is determined by its kth Blaschke section body $B_k K$ if it is determined by its kth mean projection body $P_k K$. See Note 3.9 and work of Kiderlen [429], [431] related to these topics.

7.14. *Stereology*. The term "stereology" was coined in 1961 at an informal meeting organized by H. Elias on the Feldberg, Germany, to address mutual interests of metallurgists and biologists. The following definition of stereology is due to Elias and reproduced in the preface of Underwood's book [808]: Stereology, *sensu stricto*, deals with a body of methods for the exploration of three-dimensional space, when only two-dimensional sections through solid bodies or their projections on a surface are available. A similar definition, offered by Weibel [829, p. 2], allows for k-dimensional sections or projections in n-dimensional space.

For geometers, the most valuable introductions to stereology are the works of Weil [836] and Schneider and Weil [746, Kapitel 5]. Weil's survey makes it clear that other test sets can be used besides planes for intersections and projections, and indicates that although stereology draws on integral geometry and geometric probability, it has also stimulated research into stochastic geometry. The probabilistic or statistical nature of stereology sets it apart from geometric tomography. However, there is some overlap. For example, stereology makes heavy use of kinematic formulas, such as [737, Theorem 4.5.2], and these are intimately related to formulas of central importance in geometric tomography, as we noted earlier. Also, tables such as [808, Tables 4.1 and 4.2] of volume, surface area, and so on, of various bodies of revolution and regular polyhedra, indicate the common attention given to such quantities. The long list of references provided in [836] and [746, Kapitel 5] should satisfy those wishing to pursue this topic.

7.15. *Local stereology*. In the late 1980s, a new branch of stereology called *local stereology* was pioneered by Vedel Jensen and Gundersen (see [406], for example) and has already achieved significant medical results in neuroscience and cancer grading. Local stereology, surveyed in Vedel Jensen's book [815], is a collection of stereological designs based on sections through a fixed reference point. As such, it relates especially with the part of geometric tomography that concerns intersections with subspaces, and in particular, with the dual Brunn–Minkowski theory.

Local stereology was developed for the study of biological tissue in cases where the tissue is transparent and physical sections can be replaced by optical sections. The procedure in the laboratory is typically as follows. The tissue sample of interest (for example, kidney, brain, or skin) is cut into a small number of blocks. Each block is subsequently cut isotropically into slabs of thickness 50–100 μm. A subset of the slabs is selected for microscopic analysis. When such a slab is transparent, it is possible to focus down through the slab and thereby generate optical sections which can be displayed on a video screen. By moving the focal plane up and down in the slab, a whole continuum of optical sections is generated. The general aim of local stereology is to estimate, from measurements in these optical sections, various quantitative properties of a cell population such as average volume or surface area, without making any special assumptions about shape. In fact many structures are far from being even star-shaped (see Figure 7.15), and this was a major motivation for Gardner, Vedel Jensen, and Volčič [280] to develop the extension of the dual Brunn–Minkowski theory mentioned in Section A.7.

For an idea of the connection between local stereology and the dual Brunn–Minkowski theory, consider the problem of estimating the volume of a bounded Borel set C in $\mathbb{E}^n$ from measurements in random k-dimensional sections through a fixed point, which can be taken to be the origin. The random subspaces are assumed to be isotropic, that is, their common probability distribution is the Haar measure in $\mathcal{G}(n, k)$.

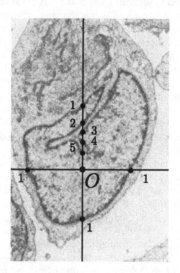

Figure 7.4. Outline from an epithelial cell nucleus in a rat kidney glomerulus.

The so-called Horvitz–Thompson procedure involves calculating the probability that an isotropic subspace meets an arbitrary volume element of C. This leads, for example, to the following local volume estimator of $\lambda_n(C)$, based on an isotropic subspace $S \in \mathcal{G}(n, k)$ (see [815, (4.12)] with $p = k$ and $r = 0$):

$$\widehat{V}_{n,k}(C \cap S) = \frac{\omega_n}{\omega_k} \int_{C \cap S} \|x\|^{n-k} \, dx.$$

By (A.63), we have

$$\widehat{V}_{n,k}(C \cap S) = \frac{\kappa_n}{\kappa_k} \widetilde{V}_{n,k}(C \cap S),$$

so the local volume estimator is proportional to the corresponding dual volume. The local volume estimators are unbiased, that is, the mean value of $\widehat{V}_{n,k}(C \cap S)$ with respect to the distribution of S is equal to $\lambda_n(C)$. This follows directly from an extension of the dual Kubota integral recursion, Theorem A.7.2, presented in [280]. Sets encountered in biology, such as that in Figure 7.15, meet each line in a finite union of line segments. If C is such a set, and $n = 3$ and $k = 1$, for example, the previous formulas give integral expressions that reduce, up to a constant, to alternating sums of powers of distances from the fixed reference point along a line through the point (see Figure 7.15 and [280, Section 7]). By averaging these quantities over a random sample of lines containing the fixed reference point, the volume of C can be estimated.

7.16. *Paul Funk (1886–1969).* (Source: [399].) Paul Funk was born in Vienna. He studied mathematics and physics at Tübingen, Vienna, and Göttingen, receiving his doctorate in 1911 for a thesis written under the direction of Hilbert. (The degree was officially recognized by the University of Vienna in 1913.) After a year teaching at a high school in Salzburg, Funk took a position at the German Technical University in Prague, where he became full professor in 1928. He married and spent happy years with his wife and two sons, before being pensioned due to the occupation of Czechoslovakia in 1939. During the war, Funk suffered a period of internment at the Theresienstadt concentration camp. Afterward, from 1945 on, Funk worked at the Technical University in Vienna. His research interests were complemented by an enthusiasm for the history of mathematics; he is credited, for example,

Figure 7.5. Paul Funk.

with a revival of interest in the work of Bolzano. Funk was a member of the Austrian Academy of Sciences, and in 1967 his work was recognized with a national award.

Funk's later years were made difficult by the loss of his wife, and by eyesight that became so weak that he could barely read. With generous help from colleagues, he wrote several books, on difference equations, the Laplace transform, and the calculus of variations, always with heavy emphasis on applications and including historical references. His name is associated with the powerful Funk–Hecke theorem on spherical harmonics, from which one can deduce the injectivity properties of the cosine and spherical Radon transform, as in Theorems C.2.1 and C.2.4, as well as the hemispherical transform H (cf. Note 7.5); see the works of Groemer [356, Section 3.4], Schneider [731], [737, p. 431], and Falconer [217]. Funk's approximately 40 papers include articles on Finsler geometry and various areas of applied mathematics and physics, such as elasticity and electronmicroscopy.

8

Intersection bodies and volume inequalities

The first goal of this chapter is to introduce the concept of the intersection body of a star body. Such bodies serve for sections as projection bodies do for projections. The idea of an intersection body is a relatively new one, but the topic has been the focus of intense study in recent years. To facilitate the development, we restrict the discussion to star bodies with continuous radial functions, though some of the theory extends to arbitrary star bodies and indeed to bounded Borel sets (see Notes 8.1 and 8.9).

The radial function of the intersection body of a star body gives the volumes of its intersections with hyperplanes through the origin. The intersection body of a convex body need not be convex, but Busemann's theorem, Theorem 8.1.10, implies that the intersection body of a centered convex body is convex. Examples of intersection bodies of star bodies are centered ellipsoids or any sufficiently smooth centered convex body in $\mathbb{E}^n$, $n \leq 4$ (see Theorem 8.1.17). On the other hand, we see in Theorem 8.1.18 that in four or more dimensions, a cylinder is not the intersection body of a star body.

In the second section, intersection bodies are applied to the Busemann–Petty problem. The analogue of Shephard's problem for projections, this asks whether a centered convex body with central sections of larger volume than another must also have larger volume. Theorem 8.2.5 shows, by explicit computation, that in $\mathbb{E}^7$ a counterexample is provided by a suitable cylinder and ball. If the body whose sections have smaller volume is the intersection body of a star body, however, the answer is positive, by Theorem 8.2.8. In fact, Corollary 8.2.11 affords a direct link to intersection bodies. This connection yields Theorem 8.2.12, a complete solution to the Busemann–Petty problem; the answer is positive only in four or fewer dimensions. What made the problem difficult is the convexity requirement, supporting our philosophy that star bodies are more natural for questions involving sections through the origin.

The final section introduces an even newer concept, the cross-section body. For planar bodies, this is identical to the projection body, whereas for centered bodies, the cross-section body is just the intersection body. Not much is known about this mysterious and captivating hybrid.

Since the material is quite complicated, we have had to omit some details here and there. Frequent references are made to properties of the spherical Radon transform, including some rather elaborate inversion formulas collected together in Appendix C.

8.1. Intersection bodies of star bodies

This section expounds the basic properties of intersection bodies of star bodies. These are defined as follows.

Definition 8.1.1. Let L be a star body in $\mathbb{E}^n$, $n \geq 2$, with $\rho_L \in C(S^{n-1})$. We call the star body IL such that

$$\rho_{IL}(u) = \lambda_{n-1}(L \cap u^\perp) = \frac{1}{n-1} \int_{S^{n-1} \cap u^\perp} \rho_L(v)^{n-1} \, dv$$

$$= R\left(\frac{1}{n-1} \rho_L^{n-1}\right)(u)$$

the *intersection body* of L.

In other words, the intersection body IL intersects the line l_u through o parallel to u in a centered line segment of length equal to twice the $(n-1)$-dimensional volume of the intersection of L with $u^\perp$.

Notice the form of the definition involving the spherical Radon transform R defined by (C.11). This is an interpretation we will find extremely useful throughout the chapter.

The reader should be aware that we restrict the previous definition to star bodies L with continuous radial functions. If $o \in \text{int} L$, then L has a continuous radial function, and the latter implies that $o \in L$. If $o \in \text{bd} L$, however, L may or may not have a continuous radial function. We do not wish to dwell on this matter here, but it will be discussed again in Remark 8.1.19. For now, let us stress that throughout the chapter, the phrase *intersection body of a star body* means *intersection body of a star body with continuous radial function*.

Using the fact that ρ_L is continuous, it is easy to prove from Definition 8.1.1 that ρ_{IL} is also continuous, so IL is a centered star body, and $I(-L) = IL$. Several examples are offered in this section, but it will be convenient to exhibit them as we develop some properties of the map I taking a star body to its intersection body. We already know one property; it follows from the generalized Funk section theorem, Theorem 7.2.6, with $i = k = n - 1$ that if L_1 and L_2 are centered star bodies in $\mathbb{E}^n$, then

$$IL_1 = IL_2 \Rightarrow L_1 = L_2;$$

that is, I is injective when restricted to centered star bodies.

We now establish dual forms of a couple of facts about the Minkowski map Π noted at the beginning of Chapter 4. The first indicates the relationship between the dual Blaschke body and the intersection body.

Theorem 8.1.2. *Let L be a star body in $\mathbb{E}^n$, $n \geq 2$, with $\rho_L \in C(S^{n-1})$. Then $\tilde{\nabla}L$ is the unique centered star body such that*

$$I(\tilde{\nabla}L) = IL.$$

Proof. By Corollary 7.2.4 with $i = k = n - 1$ and Remark 7.2.5,

$$\rho_{I(\tilde{\nabla}L)}(u) = \lambda_{n-1}(\tilde{\nabla}L \cap u^{\perp}) = \lambda_{n-1}(L \cap u^{\perp}) = \rho_{IL}(u),$$

for each $u \in S^{n-1}$.   ∎

The next theorem characterizes the equality of intersection bodies in terms of that of dual Blaschke bodies or dual mixed volumes.

Theorem 8.1.3. *Let L_1 and L_2 be star bodies in $\mathbb{E}^n$, $n \geq 2$, with continuous radial functions. The following statements are equivalent:*

$$IL_1 = IL_2; \tag{8.1}$$

$$\tilde{\nabla}L_1 = \tilde{\nabla}L_2; \tag{8.2}$$

$$\tilde{V}(L_1, n - 1; L) = \tilde{V}(L_2, n - 1; L), \tag{8.3}$$

for each centered star body L with $\rho_L \in C(S^{n-1})$.

Proof. The fact that (8.1)⇔(8.2) follows from Theorem 7.2.3 with $i = k = n - 1$ and Remark 7.2.5, in view of Definition 6.1.2 of the dual Blaschke body.

Suppose that (8.1) holds. In proving (8.3), Theorem 8.1.2, Lemma 7.2.1 (with $i = n - 1$), and Remark 7.2.5 show that we may assume that L_j is centered, for $j = 1, 2$, since otherwise we can replace L_j by $\tilde{\nabla}L_j$. But then (8.1) implies that $L_1 = L_2$, by the generalized Funk section theorem, Theorem 7.2.6, so (8.3) is automatically true.

Assume (8.3). Let $f \in C_e(S^{n-1})$ be nonnegative. Define a centered star body L by

$$\rho_L(u) = \frac{1}{n-1} \int_{S^{n-1} \cap u^{\perp}} f(v) \, dv = \frac{1}{n-1} Rf(u),$$

for all $u \in S^{n-1}$. Then for $j = 1, 2$,

$$
\tilde{V}(L_j, n - 1; L) = \frac{1}{n} \int_{S^{n-1}} \rho_{L_j}(u)^{n-1} \rho_L(u) \, du
$$

$$
= \frac{1}{n(n-1)} \int_{S^{n-1}} \rho_{L_j}(u)^{n-1} Rf(u) \, du
$$

$$
= \frac{1}{n(n-1)} \int_{S^{n-1}} R(\rho_{L_j}^{n-1})(u) f(u) \, du
$$

$$
= \frac{1}{n} \int_{S^{n-1}} \lambda_{n-1}(L_j \cap u^{\perp}) f(u) \, du,
$$

where we have used the definition (A.51) of a dual mixed volume and Theorem C.2.6, stating that R is self-adjoint. Therefore

$$
\int_{S^{n-1}} \left(\lambda_{n-1}(L_1 \cap u^{\perp}) - \lambda_{n-1}(L_2 \cap u^{\perp}) \right) f(u) \, du = 0,
$$

for all nonnegative $f \in C_e(S^{n-1})$. But this last equation must hold for all $f \in C_e(S^{n-1})$, because any such function is the difference of its positive part f^+ and negative part f^-, which are nonnegative functions in $C_e(S^{n-1})$. If we now set

$$
f(u) = \lambda_{n-1}(L_1 \cap u^{\perp}) - \lambda_{n-1}(L_2 \cap u^{\perp}),
$$

for all $u \in S^{n-1}$, we see that

$$
\rho_{IL_1}(u) - \rho_{IL_2}(u) = \lambda_{n-1}(L_1 \cap u^{\perp}) - \lambda_{n-1}(L_2 \cap u^{\perp}) = 0,
$$

for all $u \in S^{n-1}$. This proves (8.1).   ∎

It is time to look at some examples. We begin with intersection bodies of planar star bodies.

Theorem 8.1.4. *Let L be a star body in $\mathbb{E}^2$ with $o \in \operatorname{int} L$. Then IL is a rotation by $\pi/2$ about the origin of the magnified chordal symmetral $2\tilde{\Delta}L$. Consequently, every centered star body in $\mathbb{E}^2$ containing the origin in its interior is the intersection body of a star body.*

Proof. Let L be any star body in $\mathbb{E}^2$ with $o \in \operatorname{int} L$, let $u \in S^1$, and let v be orthogonal to u. Then

$$
\rho_{IL}(u) = \lambda_1(L \cap u^{\perp}) = \lambda_1(\tilde{\Delta}L \cap u^{\perp}) = \rho_{2\tilde{\Delta}L}(v).
$$

The middle equation follows straight from the definition of the chordal symmetral $\tilde{\Delta}L$ (cf. Definition 5.1.3 and the remarks thereafter). This proves the first statement of the theorem. If L is centered, then $\tilde{\Delta}L = L$. It follows that $L = IL_0$, where L_0 is $\frac{1}{2}L$ rotated by $\pi/2$ about the origin.   ∎

The previous theorem is false in higher dimensions; see Figure 8.3 and Theorem 8.1.18.

As a first example in $\mathbb{E}^n$, let us note that the intersection body of the unit ball B in $\mathbb{E}^n$ is $\kappa_{n-1}B$ and that every centered n-dimensional ball is the intersection body of a star body (another centered ball). To extend these observations to centered ellipsoids, we need to study the behavior of intersection bodies under linear maps.

Lemma 8.1.5. *Let $\phi \in GA_n$, and let E be a λ_{n-1}-measurable subset of a hyperplane H in $\mathbb{E}^n$ orthogonal to $u \in S^{n-1}$. Then*

$$\lambda_{n-1}(\phi E) = \|\phi^{-t}u\| \,|\det\phi|\lambda_{n-1}(E).$$

Proof. It suffices to consider $\phi \in GL_n$. Since ϕ preserves the ratio of λ_{n-1}-measures of sets in parallel hyperplanes (cf. Section 0.5), it follows that this is true for pairs of subsets of H, so $\lambda_{n-1}(\phi E) = a\lambda_{n-1}(E)$, where a is independent of E. Let E be an $(n-1)$-dimensional cube in H. If $u \in S^{n-1}$ is orthogonal to H, then ϕ takes the n-dimensional cube $E + [o, u]$ to the parallelepiped $\phi(E + [o, u])$, so by (0.7),

$$\lambda_n\big(\phi(E + [o, u])\big) = |\det\phi|\lambda_n(E + [o, u]) = |\det\phi|\lambda_{n-1}(E).$$

Since $\phi(E + [o, u]) = \phi E + \phi[o, u]$, the height of this parallelepiped orthogonal to ϕH is $\phi u \cdot \phi^{-t}u/\|\phi^{-t}u\|$. By (0.1), the latter is $1/\|\phi^{-t}u\|$, so

$$\lambda_n\big(\phi(E + [o, u])\big) = \lambda_{n-1}(\phi E)/\|\phi^{-t}u\|.$$

The lemma follows from the two displayed equations. ∎

Theorem 8.1.6. *The intersection bodies of linearly equivalent star bodies are linearly equivalent. In fact, if $\phi \in GL_n$, then*

$$I(\phi L) = |\det\phi|\phi^{-t}(IL).$$

Proof. Let $u \in S^{n-1}$. Then, by (0.1), $x \cdot u = 0$ if and only if $\phi^{-1}x \cdot \phi^t u = 0$. Therefore if w is the unit vector in the direction of $\phi^t u$, we have $\phi^{-1}u^\perp = w^\perp$. Also, if F is a λ_{n-1}-measurable subset of $u^\perp$, then by Lemma 8.1.5,

$$\lambda_{n-1}(\phi F) = \|\phi^{-t}u\| \,|\det\phi|\lambda_{n-1}(F).$$

Using these facts, (0.29), and the formula (0.34) for the change in a radial function under a linear transformation, we obtain

$$\begin{aligned}
\rho_{I(\phi L)}(u) &= \lambda_{n-1}(\phi L \cap u^\perp)\\
&= \lambda_{n-1}\big(\phi(L \cap \phi^{-1}u^\perp)\big)\\
&= \lambda_{n-1}\big(\phi(L \cap w^\perp)\big)\\
&= \|\phi^{-t}w\| \,|\det\phi|\lambda_{n-1}(L \cap w^\perp)\\
&= |\det\phi|\rho_{IL}(w)/\|\phi^t u\|
\end{aligned}$$

$$= |\det \phi| \rho_{IL}(\phi^t u)$$

$$= |\det \phi| \rho_{\phi^{-t}(IL)}(u).$$

This completes the proof. ∎

Corollary 8.1.7. *The intersection body of a centered n-dimensional ellipsoid is a centered ellipsoid. In fact,*

$$I E = \frac{\kappa_{n-1}\lambda_n(E)}{\kappa_n} E^*,$$

for each such ellipsoid E in $\mathbb{E}^n$. Conversely, every centered n-dimensional ellipsoid is the intersection body of another centered ellipsoid.

Proof. In view of the last lemma, this is exactly the same as the proof of the corresponding result for projection bodies (Corollary 4.1.6), when Π is replaced by I. ∎

It is natural to ask whether there is a characterization of intersection bodies of star bodies analogous to that of projection bodies provided by Corollary 4.1.12. Such a characterization has been found, for a slightly more general notion of intersection body; see Note 8.1.

We now encounter a dramatic difference between projection bodies and intersection bodies. Whereas the former are always convex, the intersection body of even a convex body is not generally convex.

Theorem 8.1.8. *Let K be a convex body in $\mathbb{E}^n$. Then there is a translate K' of K, with $o \in \mathrm{int}\, K'$, such that $I K'$ is not convex.*

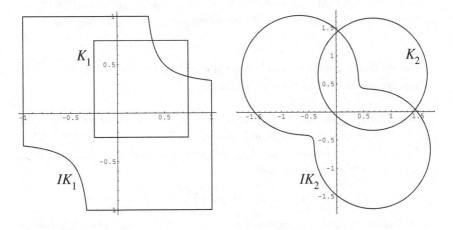

Figure 8.1. Nonconvex intersection bodies.

Proof. There is an exposed point $x \in \operatorname{bd} K$; see Section 0.3. This means that there is a hyperplane H supporting K at the single point x. Let H_1 be a hyperplane through x such that $a_1 = \lambda_{n-1}(K \cap H_1) \neq 0$. Choose a hyperplane $H_2 \neq H_1$ containing the $(n-2)$-dimensional plane $H_1 \cap H$ and with $a_2 = \lambda_{n-1}(K \cap H_2) \neq 0$. Then the unit vectors v, v_1, and v_2 orthogonal to H, H_1, and H_2, respectively, lie in the same 2-dimensional subspace. Choose a sequence $\{x_m\}$ of points in $\operatorname{int} K$ converging to x. For any $u \in S^{n-1}$, the sets $(K - x_m) \cap u^\perp$ converge to $(K - x) \cap u^\perp$ in the Hausdorff metric. Let $\varepsilon > 0$. For $m \geq N(\varepsilon)$, we have

$$\rho_{I(K-x_m)}(\pm v) = \lambda_{n-1}\big((K - x_m) \cap v^\perp\big) < \varepsilon$$

and

$$\rho_{I(K-x_m)}(\pm v_j) = \lambda_{n-1}\big((K - x_m) \cap v_j^\perp\big) \geq a_j - \varepsilon,$$

for $j = 1, 2$; this is because λ_{n-1} is continuous with respect to the Hausdorff metric in $u^\perp$ (cf. Section 0.4). By taking ε sufficiently small, we can therefore ensure that $I(K - x_{N(\varepsilon)})$ is not convex. ∎

Theorem 8.1.8 shows that the intersection body of a ball need not be convex, as in Figure 8.1.

Theorem 8.1.9. *There is a convex body (or centered convex body) K in $\mathbb{E}^2$ such that every translate (or nonzero translate, respectively) of K containing the origin in its interior has a nonconvex intersection body.*

Proof. An equilateral triangle whose centroid is at o has this property. By an easy exercise in polar coordinates, one sees that for any translate containing o in its interior, the boundary of the intersection body consists of hyperbolic arcs, concave toward o. Similarly, one can show that the centered square has the required property (see Figure 8.1; in this case, the boundary may also contain line segments). ∎

The following important result bears a corollary standing in contrast to the examples provided by the last two theorems.

Theorem 8.1.10 (Busemann's theorem). *Let K be a convex body with $o \in \operatorname{int} K$, and let S be an $(n-2)$-dimensional subspace in $\mathbb{E}^n$. For each $u \in S^{n-1} \cap S^\perp$, let S_u be the $(n-1)$-dimensional half-subspace containing S and u and with S as boundary. For each such u, define*

$$r(u) = \lambda_{n-1}(K \cap S_u).$$

Then the curve with polar equation $r = r(u)$ in $S^\perp$ is the boundary of a convex body in $S^\perp$.

Proof. It is easy to see that $r = r(u)$ is continuous on $S^{n-1} \cap S^\perp$. Let $u_j \in S^{n-1}$, $j = 1, 2$, be nonparallel vectors in $S^{n-1} \cap S^\perp$, and suppose that u_3 is the unit vector in the direction $u_1 + u_2$. Let $r_j \in \mathbb{R}$, $j = 1, 2$, be positive but otherwise arbitrary real numbers. If $r_3 u_3 = (1 - t) r_1 u_1 + t r_2 u_2$ is the point where the line through o parallel to u_3 meets the line segment $[r_1 u_1, r_2 u_2]$, then, as in the proof of Lemma 5.1.4, we have

$$t = \frac{r_1}{r_1 + r_2} \quad \text{and} \quad \frac{r_3}{\|u_1 + u_2\|} = (r_1^{-1} + r_2^{-1})^{-1}. \tag{8.4}$$

Lemma 5.1.4 also implies that to show that the body in $S^\perp$ bounded by the curve $r = r(u)$ is convex, it will suffice to prove that

$$\frac{r(u_3)}{\|u_1 + u_2\|} \geq \left(r(u_1)^{-1} + r(u_2)^{-1} \right)^{-1}.$$

(Of course, this also follows from (8.4) on setting $r_j = r(u_j)$, $j = 1, 2$.) See Figure 8.2.

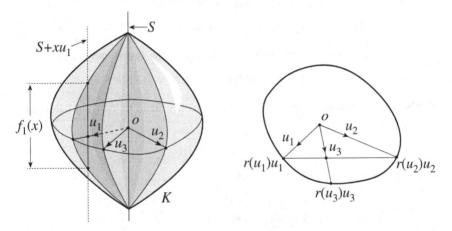

Figure 8.2. Busemann's theorem.

For each nonnegative $x \in \mathbb{R}$, let $f_j(x) = \lambda_{n-2}\big(K \cap (S + x u_j)\big)$, for $1 \leq j \leq 3$. Then

$$r(u_j) = \int_0^\infty f_j(x)\, dx,$$

for $1 \leq j \leq 3$. Now for each s with $0 \leq s \leq 1$ and $j = 1, 2$, define $r_j = r_j(s)$ by

$$s = \frac{1}{r(u_j)} \int_0^{r_j} f_j(x)\, dx.$$

It follows that

$$\frac{ds}{dr_j} = \frac{1}{r(u_j)} f_j(r_j), \tag{8.5}$$

for $j = 1, 2$. Defining $r_3 = r_3(s)$ by equation (8.4), and differentiating that equation with respect to s, we also obtain

$$\frac{\|u_1 + u_2\|}{r_3^2} \frac{dr_3}{ds} = \frac{1}{r_1^2} \frac{dr_1}{ds} + \frac{1}{r_2^2} \frac{dr_2}{ds}. \qquad (8.6)$$

Now

$$\frac{r(u_3)}{\|u_1 + u_2\|} = \int_0^\infty \frac{f_3(x)}{\|u_1 + u_2\|} dx$$

$$\geq \int_0^1 \frac{f_3(r_3(s))}{\|u_1 + u_2\|} \frac{dr_3}{ds} ds$$

$$= \int_0^1 f_3(r_3) \frac{r_3^2}{\|u_1 + u_2\|^2} \left(\frac{1}{r_1^2} \frac{dr_1}{ds} + \frac{1}{r_2^2} \frac{dr_2}{ds} \right) ds,$$

by (8.6).

We denote the last integrand by I and estimate it as follows. Note that the sets $K \cap (S + r_j u_j)$, $1 \leq j \leq 3$, all lie in the same hyperplane, and by convexity,

$$(1 - t)(K \cap (S + r_1 u_1)) + t(K \cap (S + r_2 u_2)) \subset K \cap (S + r_3 u_3),$$

where t is as in (8.4). The Brunn–Minkowski inequality (B.10) now yields

$$f_3(r_3)^{1/(n-2)} \geq (1 - t) f_1(r_1)^{1/(n-2)} + t f_2(r_2)^{1/(n-2)}.$$

Applying the arithmetic–geometric mean inequality (B.1) to the right-hand side, we see that

$$f_3(r_3) \geq f_1(r_1)^{(1-t)} f_2(r_2)^t.$$

Using this and equations (8.4) and (8.5), we obtain

$$I = f_3(r_3) \frac{r_3^2}{\|u_1 + u_2\|^2} \left(\frac{1}{r_1^2} \frac{dr_1}{ds} + \frac{1}{r_2^2} \frac{dr_2}{ds} \right)$$

$$\geq f_1(r_1)^{(1-t)} f_2(r_2)^t \frac{r_1^2 r_2^2}{(r_1 + r_2)^2} \left(\frac{1}{r_1^2} \frac{r(u_1)}{f_1(r_1)} + \frac{1}{r_2^2} \frac{r(u_2)}{f_2(r_2)} \right)$$

$$= f_1(r_1)^{w_1/(w_1 + w_2)} f_2(r_2)^{w_2/(w_1 + w_2)} \frac{1}{(w_1 + w_2)^2} \left(w_1^2 \frac{r(u_1)}{f_1(r_1)} + w_2^2 \frac{r(u_2)}{f_2(r_2)} \right),$$

where $w_j = 1/r_j$, $j = 1, 2$.

We now apply the arithmetic–geometric inequality (B.1) twice, first with $a_j = w_j r(u_j)/f_j(r_j)$, $j = 1, 2$, and then with $a_j = 1/(w_j r(u_j))$, to get

$$I \geq \frac{1}{(w_1 + w_2)} (w_1 r(u_1))^{w_1/(w_1 + w_2)} (w_2 r(u_2))^{w_2/(w_1 + w_2)}$$

$$\geq (r(u_1)^{-1} + r(u_2)^{-1})^{-1}.$$

This estimate for I yields the required inequality at once.   ∎

Corollary 8.1.11. *Let* K *be a centered convex body in* $\mathbb{E}^n$. *Then* IK *is also a centered convex body.*

Proof. Let S be any $(n-2)$-dimensional subspace. Let $u \in S^{n-1} \cap S^{\perp}$, and let $v \in S^{n-1} \cap S^{\perp}$ be orthogonal to u. Then

$$\rho_{IK}(u) = \lambda_{n-1}(K \cap u^{\perp}) = 2\lambda_{n-1}(K \cap S_v) = 2r(v),$$

in the notation of Busemann's theorem, Theorem 8.1.10. By that theorem, $IK \cap S^{\perp}$ is convex. A set whose intersection with any 2-dimensional subspace is convex is itself convex, so IK is convex. ∎

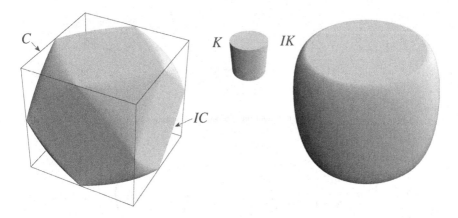

Figure 8.3. Convex intersection bodies.

Some examples of convex intersection bodies are illustrated in Figure 8.3. On the left C is the centered unit cube, dilatated by a factor of $1/2$; on the right, a centered cylinder K of radius 1 and height 2 is shown with its (translated) intersection body.

There is another major difference between projection bodies and intersection bodies. While there are examples of arbitrarily smooth centered convex bodies in $\mathbb{E}^3$ that are not projection bodies (cf. Remark 4.1.15), we shall see in Theorem 8.1.17 that every sufficiently smooth centered convex body in $\mathbb{E}^3$ is the intersection body of a star body. In order to prove this, we shall first study bodies of revolution, beginning with an explicit formula for the volume of central sections of a centered body of revolution. The latter uses a general formula of Theorem C.2.9 for the spherical Radon transform. The reader may find Figure 8.4 helpful for the particular case required here.

Lemma 8.1.12. *Let* L *be a centered star body of revolution about the* x_n-*axis in* $\mathbb{E}^n$, *so its radial function* ρ_L *may be considered as a function of the angle* φ *from*

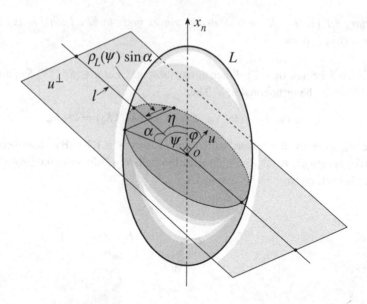

Figure 8.4. Section of a body of revolution.

the x_n-axis. Then

$$\lambda_{n-1}(L \cap u^{\perp}) = \frac{2\omega_{n-2}}{(n-1)\sin\varphi} \int_{\frac{\pi}{2}-\varphi}^{\frac{\pi}{2}} \rho_L(\psi)^{n-1} \left(1 - \frac{\cos^2\psi}{\sin^2\varphi}\right)^{(n-4)/2} \sin\psi \, d\psi,$$

for each $u \in S^{n-1}$ whose angle from the x_n-axis is φ, where $0 < \varphi \le \pi/2$. For $\varphi = 0$, we have

$$\lambda_{n-1}(L \cap u^{\perp}) = \kappa_{n-1}\rho_L\left(\frac{\pi}{2}\right)^{n-1}.$$

Proof. Since $\lambda_{n-1}(L \cap u^{\perp}) = Rg(u)$, where $g = \rho_L^{n-1}/(n-1)$, the first equation in the statement of the lemma follows directly from Theorem C.2.9. The expression for $\varphi = 0$ follows from (0.8) and the fact that the corresponding section of L is an $(n-1)$-dimensional ball of radius $\rho_L(\pi/2)$. ∎

Suppose that L is a star body of revolution in $\mathbb{E}^n$, $n \ge 3$. We say that L is *axis-convex* if each line parallel to its axis that meets it does so in a (possibly degenerate) line segment. Note that a convex body of revolution is axis-convex.

Theorem 8.1.13. *Let L be a centered axis-convex body of revolution in $\mathbb{E}^n$, $n \le 4$, such that $\rho_L \in C^{n-2}(S^{n-1})$. Then L is the intersection body of a star body.*

Proof. We have to exhibit a star body M such that $L = IM$. Suppose that the axis of L is the x_n-axis. We shall prove that a suitable M, also a body of revolution about the x_n-axis, exists, so the radial functions of both L and M may be regarded as functions of the angle from the x_n-axis. Applying the last lemma and its notation with L replaced by M, we see that we must find an M satisfying

$$
\rho_L(\varphi) = \lambda_{n-1}(M \cap u^\perp)
$$

$$
= \frac{2\omega_{n-2}}{(n-1)\sin\varphi} \int_{\frac{\pi}{2}-\varphi}^{\frac{\pi}{2}} \rho_M(\psi)^{n-1} \left(1 - \frac{\cos^2\psi}{\sin^2\varphi}\right)^{(n-4)/2} \sin\psi \, d\psi,
$$

for $0 < \varphi \le \pi/2$, and $\rho_L(0) = \kappa_{n-1}\rho_M(\pi/2)^{n-1}$. We substitute $x = \sin\varphi$ and $t = \cos\psi$ in the integral to obtain

$$
\rho_L(\sin^{-1}x) = \frac{2\omega_{n-2}}{(n-1)x^{n-3}} \int_0^x \rho_M(\cos^{-1}t)^{n-1}(x^2 - t^2)^{(n-4)/2} \, dt, \qquad (8.7)
$$

for $0 < x \le 1$, and $\rho_L(\sin^{-1}0) = \kappa_{n-1}\rho_M(\cos^{-1}0)^{n-1}$.

There is an inversion formula (C.17) for the integral equation (8.7), but we shall actually produce a nonnegative continuous solution.

Suppose that $n = 4$. Rather than apply the inversion formula directly, it is easier to note that (8.7) becomes

$$
\rho_L(\sin^{-1}x) = \frac{4\pi}{3x} \int_0^x \rho_M(\cos^{-1}t)^3 \, dt,
$$

for $0 < x \le 1$, and $\rho_L(\sin^{-1}0) = 4\pi\rho_M(\cos^{-1}0)^3/3$. Multiplying by x and differentiating with respect to x, we obtain

$$
\rho_M(\cos^{-1}x)^3 = \frac{3}{4\pi}\frac{d}{dx}\left(x\rho_L(\sin^{-1}x)\right) = \frac{3}{4\pi}\left(\frac{x\rho_L'(\sin^{-1}x)}{\sqrt{1-x^2}} + \rho_L(\sin^{-1}x)\right),
$$

for $0 < x < 1$, and $\rho_M(\cos^{-1}0)^3 = 3\rho_L(\sin^{-1}0)/4\pi$. We define $\rho_M(\cos^{-1}x)^3$ at $x = 1$ to be the limit as $x \to 1-$ of the right-hand side of the last displayed equation. Using L'Hôpital's rule and our assumptions about L, this gives

$$
\rho_M(\cos^{-1}1)^3 = \rho_M(0)^3 = \frac{3}{4\pi}\left(\rho_L\left(\frac{\pi}{2}\right) - \rho_L''\left(\frac{\pi}{2}\right)\right).
$$

Then ρ_M is continuous. Moreover, the fact that L is axis-convex and centered means that $x\rho_L(\sin^{-1}x)$ increases with x, since this quantity is the distance from the x_n-axis to a point on bd L whose angle at o with this axis is $\sin^{-1}x$. Consequently, ρ_M is nonnegative, and the proof for $n = 4$ is complete.

Now assume that $n = 3$. L'Hôpital's rule and our assumptions about L show that $\rho_L(\sin^{-1}x)$ is continuously differentiable for $0 \le x \le 1$. For $n = 3$, equation (8.7) reduces to the Abel integral equation (C.18) with $F(x) = \rho_L(\sin^{-1}x)$ and $G(t) = \rho_M(\cos^{-1}t)^2/2$, so by the remarks in Section C.2 after Corollary C.2.11, a continuous solution exists of the form

$$
\rho_M(\cos^{-1}t)^2 = \frac{1}{\pi}\frac{d}{dt}\int_0^t \frac{x\rho_L(\sin^{-1}x)}{\sqrt{t^2 - x^2}} \, dx,
$$

for $0 < t \leq 1$, and $\rho_M(\cos^{-1} 0)^2 = \rho_L(\sin^{-1} 0)/\pi$. We have to show that ρ_M is nonnegative, and this will follow if the integral increases with t. Under the substitution $s = x/t$, the integral becomes

$$\int_0^1 \frac{st\,\rho_L\left(\sin^{-1}(st)\right)}{\sqrt{1 - s^2}}\, ds.$$

Again, the axis-convexity and symmetry of L mean that $x\rho_L(\sin^{-1} x)$ increases with x, and the result for $n = 3$ follows. ∎

Some restriction such as axis-convexity is needed in the last theorem, for the double bounded cone L_1 in Figure 8.7 (which can be "fattened" slightly so that it contains the origin in its interior) is not the intersection body of any star body; this will follow from Theorems 8.2.4 and 8.2.8.

The next main goal is to prove Theorem 8.1.17. This will require several lemmas. The first of these sees the reappearance of the X-ray $X_u K$ of a convex body K in a direction u. Another ingredient is a function A_K, just a convenient notation for the Radon transform $\widetilde{1_K}$ of the characteristic function of K (cf. (C.2)). Note that for fixed u, $A_K(t, u)$ is the $(n-1)$-dimensional X-ray of K parallel to $u^\perp$. It is sometimes called the *parallel section function* of K.

Lemma 8.1.14. *Let K be a convex body in $\mathbb{E}^n$. For each $t \in \mathbb{R}$ and $u \in S^{n-1}$, let $A_K(t, u) = \lambda_{n-1}(K \cap (u^\perp + tu))$ and let R denote the spherical Radon transform. Then*

$$RA_K(t, \cdot)(u) = \omega_{n-2} \int_{S^{n-1} \cap u^\perp} \int_t^\infty X_u K(rv)\left(r^2 - t^2\right)^{(n-4)/2} r\, dr\, dv. \quad (8.8)$$

Proof. Let $u \in S^{n-1}$ and recall that l_u denotes the line through o and u. By the definition of R (see (C.11)) and A_K, Fubini's theorem, and Definition 2.1.1 of the parallel X-ray $X_u K$, we have

$$
\begin{aligned}
RA_K(t, \cdot)(u) &= \int_{S^{n-1} \cap u^\perp} A_K(t, v)\, dv \\
&= \int_{S^{n-1} \cap u^\perp} \int_{K \cap (v^\perp + tv)} 1\, dx\, dv \\
&= \int_{S^{n-1} \cap u^\perp} \int_{u^\perp \cap (v^\perp + tv)} \int_{K \cap (l_u + y)} 1\, dz\, dy\, dv \\
&= \int_{S^{n-1} \cap u^\perp} \int_{u^\perp \cap (v^\perp + tv)} X_u K(y)\, dy\, dv \\
&= \int_{S^{n-1} \cap u^\perp} \int_{S^{n-1} \cap u^\perp \cap v^\perp} \int_0^\infty X_u K(tv + sw)s^{n-3}\, ds\, dw\, dv,
\end{aligned}
$$

where in the previous expression we have changed to polar coordinates.

We now change the coordinates (v, w) to (θ, η), where $\theta \in S^{n-1} \cap u^\perp$ and $\eta \in S^{n-1} \cap u^\perp \cap \theta^\perp$ are the unit vectors in the directions $tv + sw$ and $-sv + tw$, respectively. Then $dw\,dv = d\eta\,d\theta$, so we obtain

$$
\begin{aligned}
RA_K(t,\cdot)(u) &= \int_{S^{n-1}\cap u^{\perp}} \int_{S^{n-1}\cap u^{\perp}\cap\theta^{\perp}} \int_0^{\infty} X_u K\left(\sqrt{s^2+t^2}\,\theta\right) s^{n-3}\,ds\,d\eta\,d\theta \\
&= \omega_{n-2} \int_{S^{n-1}\cap u^{\perp}} \int_0^{\infty} X_u K\left(\sqrt{s^2+t^2}\,\theta\right) s^{n-3}\,ds\,d\theta \\
&= \omega_{n-2} \int_{S^{n-1}\cap u^{\perp}} \int_t^{\infty} X_u K(rv)\left(r^2-t^2\right)^{(n-4)/2} r\,dr\,dv,
\end{aligned}
$$

the final expression achieved by substituting v for θ and $r^2 = s^2+t^2$. ∎

The following lemma provides some smoothness properties of A_K needed in the sequel.

Lemma 8.1.15. *Let K be a centered convex body of class C^k, $1 \le k \le \infty$, and let $rB \subset K$, $r > 0$. If $1 \le i \le k$ and $u \in S^{n-1}$, then*

$$
A_K^{(i)}(t,u) = \frac{d^i}{dt^i} A_K(t,u)
$$

is continuous on $(-r,r) \times S^{n-1}$.

Proof. We first claim that the *extended radial function* $\rho_K(x,u)$ of K, defined by

$$
\rho_K(x,u) = \max\{c : x + cu \in K\},
$$

where $x \in \operatorname{int} K$ and $u \in S^{n-1}$, is in $C^k\left(\operatorname{int} K \times S^{n-1}\right)$. To see this, suppose that $x \in \operatorname{int} K$ and $y \in \operatorname{bd} K$, and let

$$
u = f(x,y) = \frac{y-x}{\|y-x\|} \in S^{n-1}.
$$

Since $\operatorname{bd} K$ is of class C^k, $f \in C^k\left(\operatorname{int} K \times \operatorname{bd} K\right)$. For fixed x, $f(x,\cdot)$ is a C^k diffeomorphism from $\operatorname{bd} K$ to S^{n-1}. By the implicit function theorem, $y = h(x,u)$ is in $C^k\left(\operatorname{int} K \times S^{n-1}\right)$. Therefore

$$
\rho_K(x,u) = \|y-x\| = \|h(x,u)-x\|
$$

is in $C^k\left(\operatorname{int} K \times S^{n-1}\right)$, proving the claim.

Let $-r < t < r$. Then

$$
A_K(t,u) = \lambda_{n-1}(K \cap (u^{\perp}+tu)) = \frac{1}{n-1}\int_{S^{n-1}\cap u^{\perp}} \rho_K(tu,v)^{n-1}\,dv.
$$

The result now follows immediately from the claim proved in the previous paragraph. ∎

Next comes a crucial lemma relating the radial function ρ_K to the spherical Radon transform of expressions involving A_K. For a generalization, see Note 8.1.

Lemma 8.1.16. *Let K be a centered convex body in $\mathbb{E}^n$ of class C^2, and suppose that there is an even Borel function g on S^{n-1} such that $\rho_K = Rg$. For λ_{n-1}-almost all $u \in S^{n-1}$, we have*

$$g(u) = -\frac{1}{4\pi^2} \int_0^\infty \frac{A_K(t, u) - A_K(0, u)}{t^2}\, dt, \qquad \text{if } n = 3 \qquad (8.9)$$

and

$$g(u) = -\frac{1}{16\pi^2} A_K''(0, u), \qquad \text{if } n = 4. \qquad (8.10)$$

Proof. We begin with the easier case when $n = 4$. Setting $n = 4$ in (8.8) and differentiating with respect to t, we obtain

$$\frac{d}{dt} R A_K(t, \cdot)(u) = -\omega_2 \int_{S^3 \cap u^\perp} X_u K(tv) t\, dv.$$

Differentiating again with respect to t, setting $t = 0$, and noting that $X_u K(o) = 2\rho_K(u)$, we find that

$$\frac{d^2}{dt^2} R A_K(t, \cdot)(u)\Big|_{t=0} = -\omega_2 \int_{S^3 \cap u^\perp} X_u K(o)\, dv$$
$$= -\omega_2 \omega_3 X_u K(o) = -16\pi^2 \rho_K(u).$$

By Lemma 8.1.15 with $i = 2$, we may differentiate under the integral sign in the definition of the spherical Radon transform to obtain

$$Rg(u) = \rho_K(u) = -\frac{1}{16\pi^2} \frac{d^2}{dt^2} R A_K(t, \cdot)(u)\Big|_{t=0} = -\frac{1}{16\pi^2} R A_K''(0, \cdot)(u).$$

Now (8.10) follows from the injectivity of the spherical Radon transform on even functions (see Theorem C.2.4).

Suppose that $n = 3$. The smoothness assumption on K implies that for fixed u and v, $X_u K(rv)$ is a continuously differentiable function of r except on the boundary of its support, and that it vanishes there for λ_{n-1}-almost all $u \in S^{n-1}$. Using these facts and (8.8) with $n = 3$, and integrating by parts with respect to r, for λ_{n-1}-almost all $u \in S^{n-1}$ we have

$$R A_K(t, \cdot)(u) = 2 \int_{S^2 \cap u^\perp} \int_t^\infty X_u K(rv) \left(r^2 - t^2\right)^{-1/2} r\, dr\, dv.$$
$$= -2 \int_{S^2 \cap u^\perp} \int_t^\infty X_u K'(rv) \left(r^2 - t^2\right)^{1/2} dr\, dv,$$

where the prime denotes differentiation with respect to r. When $t = 0$, we get

$$R A_K(0, \cdot)(u) = -2 \int_{S^2 \cap u^\perp} \int_0^\infty X_u K'(rv) r\, dr\, dv.$$

By substituting $s = r/t$, we obtain

$$RA_K(t, \cdot)(u) = -2 \int_{S^2 \cap u^\perp} \int_1^\infty X_u K'(stv) \left(s^2 - 1\right)^{1/2} t^2 \, ds \, dv$$

and

$$RA_K(0, \cdot)(u) = -2 \int_{S^2 \cap u^\perp} \int_0^\infty X_u K'(stv) st^2 \, ds \, dv,$$

where the prime denotes differentiation with respect to st. Now using Fubini's theorem to interchange integrals, we find that

$$R\left(\int_0^\infty \frac{A_K(t, \cdot) - A_K(0, \cdot)}{t^2} \, dt\right)(u) = \int_0^\infty \frac{RA_K(t, u) - RA_K(0, u)}{t^2} \, dt$$

$$= 2 \int_{S^2 \cap u^\perp} \left(\int_1^\infty \left(\int_0^\infty X_u K'(stv) \, dt\right)\left(s - \sqrt{s^2 - 1}\right) ds + \right.$$

$$\left. + \int_0^1 \left(\int_0^\infty X_u K'(stv) \, dt\right) s \, ds\right) dv$$

$$= 2 \int_{S^2 \cap u^\perp} \left(-\int_1^\infty \frac{X_u K(o)}{s} \left(s - \sqrt{s^2 - 1}\right) ds - \int_0^1 \frac{X_u K(o)}{s} s \, ds\right) dv$$

$$= -4\pi X_u K(o) \left(\int_1^\infty \frac{s - \sqrt{s^2 - 1}}{s} \, ds + 1\right)$$

$$= -4\pi \, (2\rho_K(u)) \, (\pi/2 - 1 + 1) = -4\pi^2 \rho_K(u).$$

(The final integral can be evaluated by substituting $z = 1/s$ and then using trigonometric substitution.) This formula and the injectivity of the spherical Radon transform on even functions yield (8.9). ∎

Theorem 8.1.17. *Let $n \leq 4$ and let K be a centered convex body in $\mathbb{E}^n$ of class C^∞. Then K is the intersection body of a star body.*

Proof. Since K is of class C^∞, we have $\rho_K \in C_e^\infty(S^{n-1})$, so Theorem C.2.5 guarantees the existence of a $g \in C_e^\infty(S^{n-1})$ such that $\rho_K = Rg$. It suffices to show that $g \geq 0$, for by Definition 8.1.1, K is then the intersection body of the star body M defined by $\rho_M^{n-1}/(n-1) = g$.

Let $u \in S^{n-1}$. Equations (8.9) and (8.10) give formulas for $g(u)$ when $n = 3$ and 4, respectively. By the Brunn–Minkowski inequality (B.10), $A_K(t, u)^{1/(n-1)}$ is a concave function of t on its support. Therefore the integrand in (8.9) is not positive for any t and so $g(u) \geq 0$ when $n = 3$. Similarly, the concavity of $A_K(t, u)^{1/(n-1)}$ at $t = 0$ implies that $A_K''(0, u) \leq 0$ and by (8.10), $g(u) \geq 0$ when $n = 4$. ∎

The next theorem provides examples of centered convex bodies that are *not* intersection bodies of any star bodies.

Theorem 8.1.18. *A cylinder in $\mathbb{E}^n$ is the intersection body of a star body if and only if it is centered and $n \leq 3$.*

Proof. Let K be a cylinder in $\mathbb{E}^n$. In view of Theorem 8.1.6 and the fact that every intersection body is centered, we may assume that K is the Cartesian product of the $(n-1)$-dimensional centered unit ball in the plane $x_n = 0$ and the line segment $[-1, 1]$ in the x_n-axis.

Suppose that K is the intersection body of a star body. Then, by Theorem 8.1.2, there is a unique centered star body M such that $K = IM$. By Definition 8.1.1, this is equivalent to $f = Rg$, where $f = \rho_K$ and $g = \rho_M^{n-1}/(n-1)$ is nonnegative, even, and continuous. Because R commutes with rotations, by Lemma C.2.7, and has the injectivity property of Theorem C.2.4, M is also a body of revolution about the x_n-axis. We may therefore regard the radial functions ρ_K and ρ_M as functions of the vertical angle φ from the x_n-axis. Now $\rho_K(0) = 1$, so $\rho_M(\pi/2) = \kappa_{n-1}^{-1/(n-1)}$. Furthermore, $\rho_K(\varphi) = \sec\varphi$ for $0 \leq \varphi \leq \pi/4$, and by (C.17), these values of ρ_K determine $\rho_M(\varphi)$ for $\pi/4 \leq \varphi \leq \pi/2$. But then we must have $\rho_M(\varphi) = \kappa_{n-1}^{-1/(n-1)} \csc\varphi$ for $\pi/4 \leq \varphi \leq \pi/2$, since for these values of φ, M coincides with a centered cylinder with its axis along the x_n-axis and of radius $\kappa_{n-1}^{-1/(n-1)}$, and the corresponding central sections of M then have the correct λ_{n-1}-measure.

Geometrically, this means that M must contain the body M' obtained by rotating about the x_n-axis the triangle in the (x_1, x_n)-plane with vertices at o and $(\kappa_{n-1}^{-1/(n-1)}, \pm\kappa_{n-1}^{-1/(n-1)})$. Furthermore, $M \neq M'$, since ρ_M is continuous. Let us compute $\lambda_{n-1}(M' \cap e_1^\perp)$, where e_1 is the unit vector in the direction of the positive x_1-axis. This is an $(n-1)$-dimensional cylinder with two $(n-1)$-dimensional bounded cones removed, so we find, using (0.14) and (0.8), that

$$\lambda_{n-1}(M' \cap e_1^\perp) = 2\left(\kappa_{n-1}^{-1/(n-1)}\right)^{n-1}\kappa_{n-2} - \frac{2}{n-1}\left(\kappa_{n-1}^{-1/(n-1)}\right)^{n-1}\kappa_{n-2}$$

$$= \frac{2(n-2)\kappa_{n-2}}{(n-1)\kappa_{n-1}} = a_n,$$

say. Now

$$\lambda_{n-1}(M' \cap e_1^\perp) < \lambda_{n-1}(M \cap e_1^\perp) = \rho_K(\pi/2) = 1,$$

so $a_n < 1$. Using only the formula (0.9) for κ_n and the fact that

$$\Gamma\left(1 + \frac{n}{2}\right) = \frac{n}{2}\Gamma\left(1 + \frac{n-2}{2}\right),$$

one can easily prove that $a_{n+2} > a_n$ for all $n \in \mathbb{N}$. But (0.9) also shows that $a_4 = 1$ and $a_5 = 4/\pi > 1$, so for $n \geq 4$, K is not the intersection body of a star body.

Let $n = 3$. By direct computation (the details of which we omit) using the inversion formula (C.15) for $n = 3$, it can be shown that $K = IM$, where M is the centered star body of revolution about the x_3-axis whose radial function is

$$\rho_M(\varphi) = \begin{cases} \dfrac{1}{\sqrt{\pi}}\left(\dfrac{1}{1-\cos\varphi} + \dfrac{\sin^2\varphi\sec\varphi}{\sqrt{\cos 2\varphi - \cos^2\varphi}}\right)^{1/2} & \text{if } 0 < \varphi \le \dfrac{\pi}{4}, \\[4mm] \csc\varphi/\sqrt{\pi} & \text{if } \dfrac{\pi}{4} \le \varphi \le \dfrac{\pi}{2}, \end{cases}$$

and $\rho_M(0) = 1/\sqrt{2\pi}$. Therefore M is a centered cylinder of radius $1/\sqrt{\pi}$ and height $2/\sqrt{\pi}$ whose flat top and bottom are replaced by certain surfaces that are concave toward its center – a "dented tin can" with the points $(0, 0, \pm 1/\sqrt{2\pi})$ in its boundary. This is illustrated in Figure 8.5. ∎

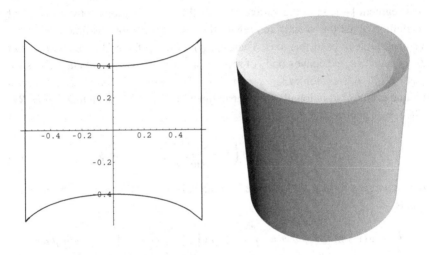

Figure 8.5. A star body whose intersection body is a cylinder.

Remark 8.1.19. The case $n = 4$ of the previous theorem deserves special attention. We computed that $a_4 = 1$, and the geometry behind this is clear. The volume of a 3-dimensional circular cylinder of equal height and radius, less that of a bounded cone of the same height and base, equals the volume contained in a hemisphere of the same radius as the cylinder, as Archimedes knew well. Let K and M' be as in the previous proof. When $n = 4$, M' is the star body of revolution about the x_4-axis obtained by rotating the triangle in the (x_1, x_4)-plane with vertices at o and $(\kappa_3^{-1/3}, \pm\kappa_3^{-1/3})$. Therefore $\rho_{M'}(\phi) = 0$ for $0 \le \phi < \pi/4$ and $\rho_{M'}(\phi) = \kappa_3^{-1/3}\csc\phi$ for $\pi/4 \le \phi \le \pi/2$, where ϕ is the angle from the x_4-axis. Using Lemma 8.1.12 with L replaced by M', and noting that this lemma is valid even though M' does not have a continuous radial function, it is easy to calculate that in fact

$$\rho_K(u) = \lambda_{n-1}(M' \cap u^\perp),$$

for all $u \in S^{n-1}$. We cannot say that $K = IM'$, however, since although $o \in M'$, $\rho_{M'}$ is not continuous.

Theorem 8.1.20. *For $n \geq 5$, there is a centered convex body in $\mathbb{E}^n$ of class C_+^∞ that is not the intersection body of a star body.*

Proof. Our aim is to show that a suitable approximation to the centered cylinder K of Theorem 8.1.18 satisfies these requirements. Recall that K is the Cartesian product of the $(n-1)$-dimensional centered unit ball in the plane $x_n = 0$ and the line segment $[-1, 1]$ in the x_n-axis. We shall consider various convex bodies of revolution about the x_n-axis and regard their radial functions as functions of the vertical angle φ from this axis. Then $\rho_K(\varphi) = \csc \varphi$ for $\pi/4 \leq \varphi \leq \pi/2$. Let K' be any centered convex body of revolution about the x_n-axis of class C^∞ such that $\rho_{K'}(\varphi) = \csc \varphi$ for $\pi/3 \leq \varphi \leq \pi/2$. By Theorems C.2.5 and C.2.8, there is a unique even C^∞ rotationally symmetric function g on S^{n-1} such that $\rho_{K'} = Rg$. Then, by Theorem C.2.9, we have

$$1 = 2\omega_{n-2} \int_{\frac{\pi}{2}-\varphi}^{\frac{\pi}{2}} g(\psi) \left(1 - \frac{\cos^2 \psi}{\sin^2 \varphi}\right)^{(n-4)/2} \sin \psi \, d\psi, \qquad (8.11)$$

for each $\pi/3 < \varphi \leq \pi/2$. From the limiting values $\varphi = \pi/2$ and $\varphi = \pi/3$, (8.11) yields

$$\int_0^{\pi/2} g(\psi) \sin^{n-3} \psi \, d\psi = \int_{\pi/6}^{\pi/2} g(\psi) \left(1 - \frac{4}{3}\cos^2 \psi\right)^{(n-4)/2} \sin \psi \, d\psi.$$

When $n \geq 5$, this can be rearranged in the form

$$\int_0^{\pi/2} g(\psi) h(\psi) \, d\psi = 0,$$

where $h(\psi) > 0$ for $0 < \psi < \pi/2$. If g is nonnegative, it must be identically zero, which is impossible. Therefore there is a ψ_0 with $0 < \psi_0 < \pi/2$ such that $g(\psi_0) < 0$, proving that K' is not the intersection body of a star body.

For $\varepsilon > 0$, let $\rho_{K_\varepsilon} = \rho_{K'} - \varepsilon\rho_{K'}^2$. For sufficiently small ε, ρ_{K_ε} is the C^∞ radial function of a centered body of revolution about the x_n-axis. We claim that for sufficiently small ε, K_ε is a convex body of class C_+^∞. Let $\rho = \rho_{K_\varepsilon}$ and regard ρ as a function of φ. A simple calculation shows that

$$2\rho'^2 - \rho\rho'' + \rho^2 = (1 - 3\varepsilon\rho_{K'})\left(2(\rho_{K'}')^2 - \rho_{K'}\rho_{K'}'' + \rho_{K'}^2\right) + \varepsilon\rho_{K'}^3 + O(\varepsilon^2).$$

By (0.41), there is an $\varepsilon_0 > 0$ such that the right-hand side is positive for $\varepsilon \leq \varepsilon_0$, and this implies, by (0.41) again, that for $\varepsilon \leq \varepsilon_0$, K_ε is of class C_+^∞, proving the claim. Let $\varepsilon \leq \varepsilon_0$, and suppose that g_1 and h are the unique even C^∞ rotationally symmetric functions on S^{n-1} such that $\rho_{K'}^2 = Rg_1$ and $\rho_{K_\varepsilon} = Rh$. Since $\rho_{K_\varepsilon} = \rho_{K'} - \varepsilon\rho_{K'}^2$, we have $Rh = Rg - \varepsilon Rg_1 = R(g - \varepsilon g_1)$, which yields $h = g - \varepsilon g_1$ by Theorem A.7.2. For sufficiently small ε, $h(\psi_0) = g(\psi_0) - \varepsilon g_1(\psi_0) < 0$. For such an ε, the body K_ε is not the intersection body of a star body. ∎

The clear relation between a star body M and its intersection body $L = IM$ is certainly very appealing from a geometrical point of view. However, the requirement of an intersection body L of a star body that $\rho_L = Rg$ for some nonnegative *continuous function* g on S^{n-1} creates some complications, as we have seen above. From an analytical point of view, the following definition is more natural. A star body L in $\mathbb{E}^n$ with $\rho_L \in C(S^{n-1})$ is called an *intersection body* if $\rho_L = R\mu$, where μ is a finite even Borel measure in S^{n-1}. (To make sense of this, one has to extend the domain of the spherical Radon transform R to the class of signed finite Borel measures in S^{n-1}.)

An intersection body of a star body is also an intersection body, but the converse is not true. For example, a cylinder in $\mathbb{E}^4$ is not the intersection body of a star body, by Theorem 8.1.18. However, Remark 8.1.19 shows that it is an intersection body, since $\rho_K = Rg$, where $g = \rho_{M'}^{n-1}/(n-1)$ is a finite nonnegative even Borel function on S^{n-1}. (Another example is provided by the intersection body of the doubly infinite "horn" in $\mathbb{E}^3$ formed by rotating the curve $|y| = -1 + 1/\sqrt{x}$ about the y-axis.)

In fact a cylinder in $\mathbb{E}^n$ is an intersection body if and only if it is centered and $n \leq 4$ (see Note 8.1). The next remark collects some other fundamental properties of intersection bodies and intersection bodies of star bodies. The two classes are quite different in some respects; compare (i) and (viii), for example. References and further results can be found in Note 8.1.

Remark 8.1.21. (i) A convex polytope in $\mathbb{E}^n$, $n \geq 3$, is not the intersection body of a star body.

(ii) Theorem 8.1.17 holds under the weaker assumption that K is of class C^2. On the other hand, (i) shows that some smoothness assumption is needed in this theorem.

(iii) An intersection body L in $\mathbb{E}^n$ with $\rho_L \in C^{n-1}(S^{n-1})$ is the intersection body of a star body.

(iv) A star body in $\mathbb{E}^n$ with a continuous radial function is an intersection body if and only if it is a limit (with respect to the radial metric) of finite radial sums of n-dimensional ellipsoids.

(v) The case of a centered cylinder in $\mathbb{E}^4$ shows that the class of intersection bodies of star bodies is not closed (with respect to the radial metric) in the class of star bodies with continuous radial functions. In contrast to this, it follows from (iv) that the larger class of intersection bodies is closed in this sense.

(vi) If $n \leq 4$, each centered convex body in $\mathbb{E}^n$ is an intersection body. Indeed, since each centered convex body in $\mathbb{E}^n$ can be approximated arbitrarily closely in the radial metric by a centered convex body of class C^∞, the result follows from Theorem 8.1.17 and (v).

(vii) It follows from (iv) that if L is an intersection body in $\mathbb{E}^n$ and $1 \leq k \leq n - 1$, then $L \cap S$ is also an intersection body (in S, identified with $\mathbb{E}^k$) for each $S \in \mathcal{G}(n, k)$; in short, a central section of an intersection body is also an intersection body.

(viii) Every polar projection body is an intersection body. In particular, a centered cross-polytope in $\mathbb{E}^n$, $n \geq 2$, is an intersection body. The converse is not true, since a centered cube in $\mathbb{E}^3$ is an intersection body by (vi), but its polar body is a regular octahedron, which is not a projection body (see Remark 4.1.19).

(ix) Suppose that B_p^n is the unit ball in l_p^n (see Note 3.11). Then B_p^n is not an intersection body if $2 < p \leq \infty$ and $n \geq 5$. In particular, a cube in $\mathbb{E}^n$ is not an intersection body if $n \geq 5$.

8.2. Larger bodies with smaller sections

The motivation for this section is the following innocent-sounding question.

Question 8.2.1. *Let $1 \leq i \leq n - 1$, and let L_1 and L_2 be star bodies in $\mathbb{E}^n$ with continuous radial functions. If $\lambda_i(L_1 \cap S) \leq \lambda_i(L_2 \cap S)$ for all $S \in \mathcal{G}(n, i)$, is it true that $\lambda_n(L_1) \leq \lambda_n(L_2)$?*

The following theorem demonstrates that the answer is in general negative, even if L_1 is convex and L_2 is a centered ball.

Theorem 8.2.2. *Let $1 \leq i \leq n - 1$, and suppose that L_1 is any star body in $\mathbb{E}^n$ that is not centered and has $o \in \operatorname{int} L_1$. There is a centered star body L_2 with $o \in \operatorname{int} L_2$ such that $\lambda_i(L_1 \cap S) < \lambda_i(L_2 \cap S)$ for all $S \in \mathcal{G}(n, i)$, but $\lambda_n(L_1) > \lambda_n(L_2)$.*

Proof. By Corollary 7.2.4 with $k = i$, the i-chordal symmetral $\tilde{\nabla}_i L_1$ of L_1 satisfies $\lambda_i(L_1 \cap S) = \lambda_i(\tilde{\nabla}_i L_1 \cap S)$ for all $S \in \mathcal{G}(n, i)$. Moreover, by Theorem 7.2.2, we also have $\lambda_n(L_1) > \lambda_n(\tilde{\nabla}_i L_1)$, since equality could only hold if L_1 were centered. Now we can choose an $r > 1$ so that the star body we seek is $L_2 = r \tilde{\nabla}_i L_1$. ∎

We saw in Corollary 7.2.11 that there exists a nonspherical convex body L_1 with $o \in \operatorname{int} L_1$ and of the same constant i-section as the unit ball. As L_1 is not centered, the proof of the previous theorem shows that we can take this L_1 and L_2 a suitable dilatation of $\tilde{\nabla}_i L_1 = B$. See Figure 8.6, where the boundary of L_1 has polar equation $r = 1 + (\sin \theta)/2$.

We show next that when $i = 1$, it is enough to assume that L_1 is centered.

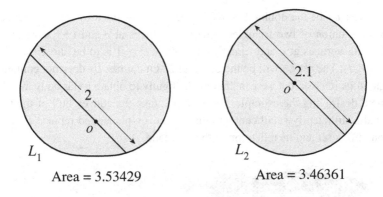

Area = 3.53429   Area = 3.46361

Figure 8.6. A larger convex body with smaller sections.

Theorem 8.2.3. *Suppose that L_1 and L_2 are star bodies in $\mathbb{E}^n$, with continuous radial functions, such that*

$$\lambda_1(L_1 \cap S) \leq \lambda_1(L_2 \cap S),$$

for all $S \in \mathcal{G}(n, 1)$. If L_1 is centered, then

$$\lambda_n(L_1) \leq \lambda_n(L_2),$$

with equality if and only if $L_1 = L_2$.

Proof. By Definition 5.1.3, we have $\lambda_1(\tilde{\Delta} L_j \cap S) = \lambda_1(L_j \cap S)$, for each $S \in \mathcal{G}(n, 1)$ and $j = 1, 2$, and therefore

$$\lambda_1(\tilde{\Delta} L_1 \cap S) \leq \lambda_1(\tilde{\Delta} L_2 \cap S),$$

for all $S \in \mathcal{G}(n, 1)$. Since L_1, $\tilde{\Delta} L_1$, and $\tilde{\Delta} L_2$ are centered, this implies that $L_1 = \tilde{\Delta} L_1 \subset \tilde{\Delta} L_2$. Therefore

$$\lambda_n(L_1) \leq \lambda_n(\tilde{\Delta} L_2) \leq \lambda_n(L_2),$$

by Theorem 7.2.2 with $i = 1$ and Remark 7.2.5.

   If L_1 and L_2 have equal volumes, then equality holds in Theorem 7.2.2, meaning that L_2 is centered and $L_2 = \tilde{\Delta} L_2$. It follows that $L_1 \subset L_2$, so $L_1 = L_2$. ∎

   The following example shows that the previous result generally fails for larger values of i, even when L_2 is a centered ball.

Theorem 8.2.4. *There is a centered star body of revolution L_1 with $o \in \text{int}\, L_1$ and a centered ball L_2 in $\mathbb{E}^3$ such that $\lambda_2(L_1 \cap u^\perp) < \lambda_2(L_2 \cap u^\perp)$ for all $u \in S^2$, but $\lambda_3(L_1) > \lambda_3(L_2)$.*

Proof. Let L_1 be the double bounded cone in $\mathbb{E}^3$ obtained by rotating about the z-axis the union of two triangles, one with vertices at o and $(\pm c, 0, 1)$ and the other with vertices at o and $(\pm c, 0, -1)$, where $c \geq 1$ is to be chosen later (see Figure 8.7). The set L_2 will be the unit ball B. Of course, L_1 does not contain the origin in its interior, but we can "fatten" it slightly to obtain a star body that does, without destroying the example; for example, take the convex hull of the top of L_1 and a sufficiently small centered disk in the xy-plane, and replace L_1 by the union of this set and its reflection in the xy-plane.

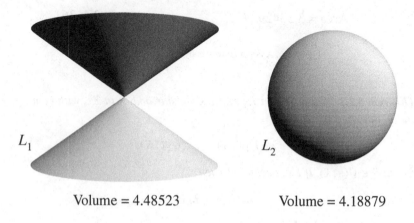

L_1 L_2

Volume = 4.48523   Volume = 4.18879

Figure 8.7. A larger centered star body with smaller sections.

We need to find the central section of L_1 with the largest area. To do this, note that only sections meeting the circular top and bottom of L_1 have nonzero area. Each such section of L_1 is the union of two triangles. If x is the distance from the base of one of these triangles to the z-axis, then $0 \leq x \leq c$ and each triangle has base of length $2(c^2 - x^2)^{1/2}$ and height $(1 + x^2)^{1/2}$. Differentiating, we see that the section of maximum area occurs when $x^2 = (c^2 - 1)/2$, and the corresponding area is $c^2 + 1$. We therefore require that $c^2 + 1 < \pi$ and $2\pi c^2/3 > 4\pi/3$. Any choice of c with

$$\sqrt{2} < c < \sqrt{\pi - 1}$$

will suffice to complete the proof.   ■

In the previous example, the body L_1 is not convex. For a long time it was thought that the central symmetry and convexity of L_1 and L_2 would ensure a positive answer to Question 8.2.1. The next example shows that this is not generally so, even when L_2 is a ball, at least if the dimension of the space is sufficiently large.

Theorem 8.2.5. *Let $n = 7$. There is a centered cylinder K_1 and a centered ball K_2 in $\mathbb{E}^n$ such that $\lambda_{n-1}(K_1 \cap u^\perp) < \lambda_{n-1}(K_2 \cap u^\perp)$ for all $u \in S^{n-1}$, but $\lambda_n(K_1) > \lambda_n(K_2)$.*

Proof. We shall at first work with an arbitrary n, $n \geq 3$, and only later take $n = 7$.

The body K_1 will be a right cylinder over an $(n-1)$-dimensional ball; specifically,

$$K_1 = \left\{ x = (x_1, \ldots, x_n) : \sum_{i=1}^{n-1} x_i^2 \leq a^2, \ |x_n| \leq b \right\},$$

where a and b are constants to be chosen. Our first aim is to obtain an upper bound for the $(n-1)$-dimensional volumes of the central sections of K_1 by hyperplanes, and we shall do this without appealing to the formula of Lemma 8.1.12. We shall need some notation. We denote by e_i the unit vector in the ith coordinate direction. Let f be defined by

$$f(s) = \frac{(b^2 + a^2 s^2)^{1/2}}{bs} \int_0^s (1 - t^2)^{(n-2)/2} \, dt,$$

for $0 < s \leq 1$, and $f(0) = 1$. Further, let $M = \sup_{0 \leq s \leq 1} f(s)$.

The function f is related to the volume κ_n of the unit ball as follows. We have

$$f(1) = \frac{(a^2 + b^2)^{1/2}}{b} \int_0^1 (1 - t^2)^{(n-2)/2} \, dt = \frac{(a^2 + b^2)^{1/2}}{2b} I_{n-1},$$

say, where a little calculus gives

$$I_n = \int_{-1}^1 (1 - t^2)^{(n-1)/2} \, dt = \int_{-\frac{\pi}{2}}^{\frac{\pi}{2}} \cos^n \theta \, d\theta = \frac{\kappa_n}{\kappa_{n-1}}.$$

We claim that

$$\lambda_{n-1}(K_1 \cap u^\perp) \leq 2a^{n-2} b \kappa_{n-2} M, \tag{8.12}$$

for all $u \in S^{n-1}$. To prove this claim, we must consider two cases.

The first is where the section does not meet the top or bottom of the cylinder K_1 (see Figure 8.8(i)). Here, the nth coordinate u_n of the unit vector u satisfies $|u_n| > a/(a^2 + b^2)^{1/2}$. The projection of the section $K_1 \cap u^\perp$ on the plane $x_n = 0$ equals the section of K_1 by that plane, an $(n-1)$-dimensional ball of radius a. Consequently,

$$\begin{aligned}
\lambda_{n-1}(K_1 \cap u^\perp) &= a^{n-1} \kappa_{n-1} / |u \cdot e_n| \\
&< a^{n-2} \kappa_{n-1}(a^2 + b^2)^{1/2} \\
&= a^{n-2} \kappa_{n-2} I_{n-1}(a^2 + b^2)^{1/2} \\
&= 2a^{n-2} b \kappa_{n-2} f(1) \leq 2a^{n-2} b \kappa_{n-2} M,
\end{aligned}$$

as required.

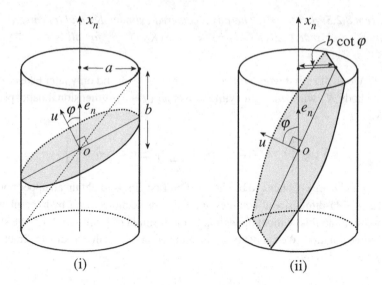

Figure 8.8. The two cases of Theorem 8.2.5.

The second case is where the section meets the top (and therefore, by symmetry, also the bottom) of K_1 (see Figure 8.8(ii)). This occurs when $|u_n| \le a/(a^2 + b^2)^{1/2}$. Let the projection of the section $K_1 \cap u^\perp$ on the plane $x_n = 0$ be E. Then E is the part of a centered $(n-1)$-dimensional ball of radius a between parallel hyperplanes H_1 and H_2, each a distance $b \cot \varphi$ from o, where φ is the angle between u and e_n. The section of E by a hyperplane parallel to H_1 and at a distance z from o is an $(n-2)$-dimensional ball of radius $(a^2 - z^2)^{1/2}$. Therefore

$$\lambda_{n-1}(K_1 \cap u^\perp) = \lambda_{n-1}(E)/|u \cdot e_n|$$

$$= \frac{1}{|u \cdot e_n|} \int_{-b \cot \varphi}^{b \cot \varphi} (a^2 - z^2)^{(n-2)/2} \kappa_{n-2} \, dz.$$

Substituting $t = z/a$ and $s = b \cot \varphi/a$, and noting that $|u \cdot e_n| = \cos \varphi = as/(b^2 + a^2 s^2)^{1/2}$, we obtain

$$\lambda_{n-1}(K_1 \cap u^\perp) = a^{n-2} \kappa_{n-2} \frac{(b^2 + a^2 s^2)^{1/2}}{s} \int_{-s}^{s} (1 - t^2)^{(n-2)/2} \, dt$$

$$= 2a^{n-2} b \kappa_{n-2} f(s) \le 2a^{n-2} b \kappa_{n-2} M,$$

completing the proof of (8.12).

The idea now is to find values of a and b for which K_1 satisfies two requirements: $\lambda_n(K_1) = \kappa_n$ and $\lambda_{n-1}(K_1 \cap u^\perp) < \kappa_{n-1}$ for each $u \in S^{n-1}$. The example is then obtained by taking this K_1 and $K_2 = rB$ for a suitable $r < 1$. It is here we need to assume that $n = 7$. Let us define $c = a/b$. For the first requirement we need $2a^6 b\kappa_6 = \kappa_7$, implying both

$$a = \left(\frac{c\kappa_7}{2\kappa_6}\right)^{1/7}$$

and

$$2a^5 b\kappa_5 = \kappa_5\kappa_7/a\kappa_6.$$

In view of (8.12), the second requirement is then satisfied if

$$\frac{\kappa_5\kappa_7}{\kappa_6^2}\left(\frac{2\kappa_6}{c\kappa_7}\right)^{1/7} M < 1.$$

Now

$$M = \sup_{0 \le s \le 1} \frac{(1+c^2s^2)^{1/2}}{s} \int_0^s (1-t^2)^{5/2}\, dt$$

$$= \sup_{0 \le x \le 1} (1+c^2x)^{1/2} \int_0^1 (1-xt^2)^{5/2}\, dt,$$

where we have replaced t by st in the integral and then set $x = s^2$. By expanding the integrand as a binomial series, we see that

$$\int_0^1 \left(1 - xt^2\right)^{5/2}\, dt \le 1 - \frac{5}{6}x + \frac{3}{8}x^2.$$

Let $g(x) = (1+c^2x)^{1/2}(1 - 5x/6 + 3x^2/8)$. By differentiating, one can check that for $c^2 = 3056/1689$, g attains its maximum value when $x = 1/20$. For this value of c^2 we have

$$\left(\frac{\kappa_5\kappa_7}{\kappa_6^2}\right)\left(\frac{2\kappa_6}{c\kappa_7}\right)^{1/7} M \le \left(\frac{\kappa_5\kappa_7}{\kappa_6^2}\right)\left(\frac{2\kappa_6}{\kappa_7}\right)^{1/7}\left(\frac{1}{c^2}\right)^{1/14} g\left(\frac{1}{20}\right)$$

$$= \left(\frac{512}{175\pi}\right)\left(\frac{35}{16}\right)^{1/7}\left(\frac{1689}{3056}\right)^{1/14}\left(\frac{9209}{8445}\right)^{1/2}\left(\frac{9209}{9600}\right)$$

$$= 0.999998\ldots < 1.$$

This completes the proof.   ∎

Remark 8.2.6. (i) Theorem 8.2.5 is true for all $n \ge 7$; for any such n, suitable constants a and b can be found so that the cylinder K_1 and the ball K_2 just defined have the same properties in $\mathbb{E}^n$. The proof requires some straightforward but quite tedious computations concerning the values of κ_n, $I_n = \kappa_n/\kappa_{n-1}$, and other numbers related to these, as n tends to infinity.

(ii) It can be shown that for each $n \ge 7$ there is a centered convex body K_1 arbitrarily close to B such that this K_1 and $K_2 = B$ yield a negative answer to Question 8.2.1 with $i = n - 1$. See Note 8.9.

The next lemma will be useful in obtaining conditions under which Question 8.2.1 has a positive answer.

Lemma 8.2.7. *Let* $1 \leq i \leq n-1$, *and let* L *be the intersection body of some star body in* $\mathbb{E}^n$. *Then*

$$\tilde{V}(L_1, i; B, n-i-1; L) \leq \tilde{V}(L_2, i; B, n-i-1; L)$$

for all star bodies L_1, L_2 *with continuous radial functions such that*

$$\tilde{V}_{i,n-1}(L_1 \cap u^\perp) \leq \tilde{V}_{i,n-1}(L_2 \cap u^\perp),$$

for all $u \in S^{n-1}$.

Proof. By the definition (A.59) of $\tilde{V}_{i,n-1}$, our assumptions imply that

$$\int_{S^{n-1} \cap u^\perp} \rho_{L_1}(v)^i \, dv \leq \int_{S^{n-1} \cap u^\perp} \rho_{L_2}(v)^i \, dv,$$

for all $u \in S^{n-1}$. Using the definition (C.11) of the spherical Radon transform R, we can rewrite this as $R\rho_{L_1}^i \leq R\rho_{L_2}^i$. Let $L = IL_0$, for some star body L_0, so that $\rho_L = R\left(\rho_{L_0}^{n-1}/(n-1)\right)$. Then for $j = 1, 2$,

$$\tilde{V}(L_j, i; B, n-i-1; L) = \frac{1}{n} \int_{S^{n-1}} \rho_{L_j}(u)^i \rho_L(u) \, du$$

$$= \frac{1}{n(n-1)} \int_{S^{n-1}} \rho_{L_j}(u)^i (R\rho_{L_0}^{n-1})(u) \, du$$

$$= \frac{1}{n(n-1)} \int_{S^{n-1}} (R\rho_{L_j}^i)(u) \rho_{L_0}(u)^{n-1} \, du,$$

where we have used (A.51) and the self-adjointness of R (Theorem C.2.6). This, together with the previous inequality, completes the proof. ∎

The next important theorem is an analogue of the case $i = k = n-1$ of Theorem 4.2.6. It can be generalized; see Note 8.9.

Theorem 8.2.8. *Let* L_1 *and* L_2 *be star bodies in* $\mathbb{E}^n$ *with continuous radial functions. Suppose that*

$$\lambda_{n-1}(L_1 \cap u^\perp) \leq \lambda_{n-1}(L_2 \cap u^\perp),$$

for all $u \in S^{n-1}$. *If* L_1 *is the intersection body of a star body, then*

$$\lambda_n(L_1) \leq \lambda_n(L_2),$$

with equality if and only if $L_1 = L_2$.

Proof. By Lemma 8.2.7 with $i = n-1$, we obtain

$$\tilde{V}(L_1, n-1; L) \leq \tilde{V}(L_2, n-1; L),$$

whenever L is the intersection body of a star body. Setting $L = L_1$, and using (A.58) and the dual Minkowski inequality (B.25), we infer that

$$\lambda_n(L_1)^n \leq \tilde{V}(L_2, n-1; L_1)^n \leq \lambda_n(L_2)^{n-1}\lambda_n(L_1),$$

yielding the desired inequality.

If the volumes of L_1 and L_2 are equal, then we have equality in the dual Minkowski inequality. This occurs if and only if L_1 is a dilatate of L_2, and this then implies that $L_1 = L_2$. ∎

The previous theorem gives a positive answer to Question 8.2.1 when $i = n - 1$ and L_1 is the intersection body of a star body. From it, we can deduce the following result, indicating that the use of cylinders in Theorem 8.2.5 will not succeed in three or four dimensions.

Theorem 8.2.9. *Let $n \leq 4$, and let L_1 and L_2 be star bodies in $\mathbb{E}^n$ with continuous radial functions. Suppose that*

$$\lambda_{n-1}(L_1 \cap u^{\perp}) \leq \lambda_{n-1}(L_2 \cap u^{\perp}),$$

for all $u \in S^{n-1}$. If L_1 is a centered axis-convex body of revolution, then

$$\lambda_n(L_1) \leq \lambda_n(L_2),$$

with equality if and only if $L_1 = L_2$.

Proof. Assume that the hypotheses of the theorem hold, but that $\lambda_n(L_1) > \lambda_n(L_2)$. By approximating, we can find a centered axis-convex star body of revolution L_1', satisfying the hypotheses of Theorem 8.1.13, such that $L_1' \subset L_1$ and $\lambda_n(L_1') > \lambda_n(L_2)$. (To see this, note that L_1 is formed by rotating a copy of a centered planar body M_1 about the x_n-axis. Since M_1 is axis-convex and symmetric with respect to the y-axis in $\mathbb{E}^2$, we can find a centered planar body $M_1' \subset M_1$, arbitrarily close to M_1, that is axis-convex and symmetric with respect to the y-axis and for which $\rho_{M_1'} \in C^2(S^1)$. If L_1' is the body obtained by rotating a copy of M_1' about the x_n-axis, then $\rho_{L_1'} \in C^2(S^{n-1})$, so the curvature of L_1' exists on the hyperplane through o orthogonal to the axis of L_1'.) Since

$$\lambda_{n-1}(L_1' \cap u^{\perp}) \leq \lambda_{n-1}(L_2 \cap u^{\perp}),$$

for all $u \in S^{n-1}$, and L_1' is the intersection body of a star body by Theorem 8.1.13, this contradicts Theorem 8.2.8. The equality condition follows from that of Theorem 8.2.8. ∎

Theorem 8.2.10. *Let L_2 be a centered star body in $\mathbb{E}^n$ with $o \in \text{int } L_2$ and $\rho_{L_2} \in C_e^{\infty}(S^{n-1})$ that is not the intersection body of a star body. Then there is a centered star body L_1 such that $\lambda_{n-1}(L_1 \cap u^{\perp}) \leq \lambda_{n-1}(L_2 \cap u^{\perp})$ for all $u \in S^{n-1}$, but $\lambda_n(L_1) > \lambda_n(L_2)$. Furthermore, if L_2 has positive Gauss curvature everywhere, then an L_1 also exists with this additional property; in particular, L_1 is convex.*

Proof. By Theorem C.2.5, there is a $g \in C_e(S^{n-1})$ such that $\rho_{L_2} = Rg$. We must have $g(u) < 0$ for some $u \in S^{n-1}$, for otherwise L_2 would be the intersection body of the star body M defined by setting $\rho_M^{n-1}/(n-1) = g$. Since g is continuous, it is negative on an open subset of S^{n-1}, and using this fact, one can find a nonconstant $h \in C_e^{\infty}(S^{n-1})$ such that $g(u) < 0 \Rightarrow h(u) \geq 0$ and $g(u) \geq 0 \Rightarrow h(u) = 0$. By Theorem C.2.5 again, there is a $G \in C_e(S^{n-1})$ with $h = RG$.

Define a star body L_1 by

$$\frac{1}{n-1}\rho_{L_1}^{n-1} = \frac{1}{n-1}\rho_{L_2}^{n-1} - tG, \tag{8.13}$$

where $t > 0$ is chosen so that the right-hand side is always positive. Then

$$\rho_{IL_1} = R\left(\frac{1}{n-1}\rho_{L_1}^{n-1}\right)$$

$$= R\left(\frac{1}{n-1}\rho_{L_2}^{n-1} - tG\right) = \rho_{IL_2} - th.$$

This shows that $\rho_{IL_1}(u) \leq \rho_{IL_2}(u)$ when $g(u) < 0$, while $\rho_{IL_1}(u) = \rho_{IL_2}(u)$ when $g(u) \geq 0$. Therefore $IL_1 \subset IL_2$, implying that

$$\lambda_{n-1}(L_1 \cap u^{\perp}) \leq \lambda_{n-1}(L_2 \cap u^{\perp}),$$

for all $u \in S^{n-1}$.

Now using (A.58), (A.51), and the self-adjointness of R (Theorem C.2.6), we obtain

$$\lambda_n(L_2) - \tilde{V}(L_1, n-1; L_2) = \frac{1}{n}\int_{S^{n-1}}\left(\rho_{L_2}(u)^{n-1} - \rho_{L_1}(u)^{n-1}\right)\rho_{L_2}(u)\,du$$

$$= \frac{1}{n}\int_{S^{n-1}}\left(\rho_{L_2}(u)^{n-1} - \rho_{L_1}(u)^{n-1}\right)Rg(u)\,du$$

$$= \frac{1}{n}\int_{S^{n-1}}R\left(\rho_{L_2}^{n-1} - \rho_{L_1}^{n-1}\right)(u)g(u)\,du$$

$$= \frac{n-1}{n}\int_{S^{n-1}}\left(\rho_{IL_2}(u) - \rho_{IL_1}(u)\right)g(u)\,du < 0.$$

Therefore, by the dual Minkowski inequality (B.25),

$$\lambda_n(L_2) < \tilde{V}(L_1, n-1; L_2) \leq \lambda_n(L_1)^{(n-1)/n}\lambda_n(L_2)^{1/n},$$

yielding $\lambda_n(L_1) > \lambda_n(L_2)$. This establishes the first statement of the theorem.

The second statement requires material we cannot treat in detail here. The Gauss curvature of a hypersurface in $\mathbb{E}^n$ is defined in [452, Chapter 7, Section 5], for example. From (8.13) it follows that ρ_{L_1} and any of its derivatives converge uniformly to ρ_{L_2} and its corresponding derivatives as t tends to zero. In [651, eq. (2.5)] one can find a formula for the Gauss curvature of a star body L with $o \in \text{int}\, L$, in terms of ρ_L. The numerator in this formula is a determinant whose entries involve only ρ_L and its first and second derivatives. Consequently, since the Gauss curvature of L_2 is positive everywhere, we can find a $t > 0$ small enough to ensure that that of L_1 is also positive everywhere. It then follows from [452, Theorem 5.6, p. 41] that L is (strictly) convex.   ∎

Corollary 8.2.11. *For centered convex bodies in $\mathbb{E}^n$, and $i = n - 1$, Question 8.2.1 has a positive answer if and only if every centered convex body K in $\mathbb{E}^n$ of class C_+^∞ is the intersection body of a star body.*

Proof. Suppose that K_1 and K_2 are centered convex bodies in $\mathbb{E}^n$ for which Question 8.2.1 has a negative answer when $i = n - 1$; in particular, $\lambda_n(K_1) > \lambda_n(K_2)$. By the last paragraph in Section 0.9, we can find a centered convex body $K_1' \subset K_1$ of class C_+^∞ such that $\lambda_n(K_1') > \lambda_n(K_2)$. Theorem 8.2.8 implies that K_1' cannot be the intersection body of a star body. The converse follows directly from the previous theorem.   ∎

Theorem 8.2.12. *For centered convex bodies in $\mathbb{E}^n$, and $i = n-1$, Question 8.2.1 has a positive answer if and only if $n \leq 4$.*

Proof. The positive answer for $n \leq 4$ follows from the previous corollary and Theorem 8.1.17. A negative answer for $n = 7$ is already available, thanks to Theorem 8.2.5 (and by Remark 8.2.6 this can be extended to $n \geq 7$). This approach does not work for $n = 5$ and 6, however. By Corollary 8.2.11, a negative answer for all $n \geq 5$ follows instead from the existence of a centered convex body in $\mathbb{E}^n$ of class C_+^∞ that is not the intersection of a star body, proved in Theorem 8.1.20.   ∎

Theorem 8.2.13. *Suppose that K is a centered convex body and that L is a star body in $\mathbb{E}^n$ with $\rho_L \in C(S^{n-1})$. If*

$$\lambda_{n-1}(K \cap u^\perp) \leq \lambda_{n-1}(L \cap u^\perp),$$

for all $u \in S^{n-1}$, then

$$\lambda_n(K) \leq \sqrt{n}\lambda_n(L).$$

Proof. Let E be the John ellipsoid of K, so that

$$E \subset K \subset \sqrt{n}E,$$

by Theorem 4.2.12. Note that E is centered. Since centered ellipsoids are intersection bodies by Corollary 8.1.7, we can apply Lemma 8.2.7, with $i = n - 1$ and $L_1, L_2,$ and L replaced by $K, L,$ and E, respectively. Using these facts, (A.58),

(A.55), and the dual Minkowski inequality (B.25), we obtain

$$\lambda_n(L)^{(n-1)/n}\lambda_n(E)^{1/n} \geq \tilde{V}(L, n-1; E)$$

$$\geq \tilde{V}(K, n-1; E)$$

$$= \tilde{V}(K, n-1; E)^{1/n}\tilde{V}(K, n-1; E)^{(n-1)/n}$$

$$\geq \lambda_n(E)^{1/n}\left(\frac{\lambda_n(K)}{\sqrt{n}}\right)^{(n-1)/n},$$

yielding the required inequality.   ∎

See Problem 8.3, as well as Notes 8.9 and 9.8, for discussions concerning improvements to the preceding theorem.

8.3. Cross-section bodies

In this section, we shall study a fascinating relative of the concept of an intersection body.

Definition 8.3.1. Let K be a convex body in $\mathbb{E}^n$, $n \geq 2$. Define the *inner section function* m_K of K by

$$m_K(u) = \max_{t \in \mathbb{R}} \lambda_{n-1}\big(K \cap (u^{\perp} + tu)\big),$$

for $u \in S^{n-1}$. The *cross-section body* of K is the centered star body CK defined by

$$\rho_{CK}(u) = m_K(u),$$

for $u \in S^{n-1}$.

The cross-section body CK has the property that it intersects the line l_u through o parallel to u in a centered line segment of length equal to twice the maximal $(n-1)$-dimensional volume of the intersection of K with a hyperplane orthogonal to u. If K is centered, then since K is also convex, the Brunn–Minkowski inequality (B.10) (cf. Figure B.1) implies that this maximum is attained when the hyperplane passes through the origin and hence that $CK = IK$. Even when K is not centered, CK can be linked to IK via the next theorem.

Theorem 8.3.2. *If K is a convex body in $\mathbb{E}^n$, then $CK = \mathrm{cl}\, \bigcup_{x \in \mathrm{int}\, K} I(K - x)$. In particular, if $o \in \mathrm{int}\, K$,*

$$IK \subset CK.$$

Proof. Let $x \in \mathrm{int}\, K$ and $u \in S^{n-1}$. Then

$$\rho_{I(K-x)}(u) = \lambda_{n-1}\big((K - x) \cap u^{\perp}\big) \leq m_K(u) = \rho_{CK}(u),$$

so $I(K - x) \subset CK$.

Suppose that $u \in S^{n-1}$ and that $\varepsilon > 0$. We have equality in the previous displayed inequality when $x \in K$ is any point on a hyperplane orthogonal to u whose intersection with K has maximal $(n-1)$-dimensional volume. Therefore we can choose $x' \in \operatorname{int} K$ sufficiently close to x to ensure that

$$\rho_{CK}(u) < \rho_{I(K-x')}(u) + \varepsilon,$$

and this shows that $CK \subset \operatorname{cl} \bigcup_{x \in \operatorname{int} K} I(K - x)$. ∎

One of the intriguing features of the cross-section body is its relation to the projection body.

Theorem 8.3.3. *If K is a convex body in $\mathbb{E}^n$, then*

$$CK \subset \Pi K.$$

Proof. Suppose the contrary. Then there is a $u \in S^{n-1}$ such that

$$\rho_{\Pi K}(u) < \rho_{CK}(u) = m_K(u).$$

It follows that $m_K(u)u \notin \Pi K$, so there exists a hyperplane separating this point from ΠK. Let $v \in S^{n-1}$ be orthogonal to this hyperplane. Then

$$\lambda_{n-1}(K|v^\perp) = h_{\Pi K}(v) < |m_K(u)u \cdot v|.$$

Denote by H a hyperplane orthogonal to u for which $\lambda_{n-1}(K \cap H) = m_K(u)$. Then $\lambda_{n-1}((K \cap H)|v^\perp) = m_K(u)|u \cdot v|$, so

$$\lambda_{n-1}(K|v^\perp) < \lambda_{n-1}((K \cap H)|v^\perp).$$

But this is impossible, since $K \cap H \subset K$. ∎

Corollary 8.3.4. *If K is a convex body in $\mathbb{E}^n$ with $o \in \operatorname{int} K$, then*

$$IK \subset \Pi K.$$

It can be shown that for $n \geq 3$ equality holds in Theorem 8.3.3 if and only if K is an ellipsoid, and from this it follows that for $n \geq 3$ we have equality in Corollary 8.3.4 if and only if K is a centered ellipsoid; see Note 8.12. In the plane, however, the cross-section body is just the projection body, as we now prove.

Theorem 8.3.5. *If K is a convex body in $\mathbb{E}^2$, then*

$$CK = \Pi K.$$

Proof. Let $u \in S^1$, and let v be a unit vector orthogonal to u. Then

$$
\begin{aligned}
\rho_{CK}(u) &= \max_{t\in\mathbb{R}} \lambda_1\big(K \cap (u^\perp + tu)\big) \\
&= \max\{\|x - y\| : x - y = sv,\ s \in \mathbb{R},\ x, y \in K\} \\
&= \max\{s \in \mathbb{R} : sv = x + (-y),\ x, y \in K\} \\
&= 2\rho_{\triangle K}(v),
\end{aligned}
$$

by Definition 3.2.2 of the central symmetral $\triangle K$. Therefore the cross-section body is the difference body $2\triangle K$, rotated by $\pi/2$ about the origin. In Theorem 4.1.4, we showed that this is just ΠK. ∎

If K is a ball, then CK is also a ball. The next theorem clarifies the behavior of the cross-section body under affine transformations, and a corollary is that if K is an ellipsoid, then CK is also an ellipsoid.

Lemma 8.3.6. *The cross-section bodies of affinely equivalent convex bodies are affinely equivalent. In fact, if $\phi \in GL_n$, then*

$$
C(\phi K) = |\det \phi| \phi^{-t}(CK).
$$

Proof. Since CK remains invariant under translations of K, the first statement follows from the second. Using Theorem 8.3.2 and Theorem 8.1.6, we obtain

$$
\begin{aligned}
C(\phi K) &= \mathrm{cl}\ \textstyle\bigcup_{x\in\mathrm{int}\,\phi K} I(\phi K - x) \\
&= \mathrm{cl}\ \textstyle\bigcup_{y\in\mathrm{int}\,K} I(\phi(K - y)), \quad (\text{where } y = \phi^{-1}x), \\
&= \mathrm{cl}\ \textstyle\bigcup_{y\in\mathrm{int}\,K} |\det \phi| \phi^{-t}\big(I(K - y)\big) \\
&= |\det \phi| \phi^{-t}\big(\mathrm{cl}\ \textstyle\bigcup_{y\in\mathrm{int}\,K} I(K - y)\big) = |\det \phi| \phi^{-t}(CK). \quad\blacksquare
\end{aligned}
$$

The next remark collects some other basic facts about cross-section bodies. References and further results can be found in Note 8.12.

Remark 8.3.7. (i) The cross-section body of a regular tetrahedron in $\mathbb{E}^3$ is a cube.
(ii) The cross-section body of a convex body in $\mathbb{E}^3$ is convex.
(iii) The cross-section body of a simplex in $\mathbb{E}^n$, $n \geq 4$, is not convex.

Open problems

Problem 8.1. (See Note 8.1.) Suppose that K is a convex body in $\mathbb{E}^n$ such that $(IK)^*$ and $\triangle K$ are homothetic. Must K be a centered ellipsoid?

Problem 8.2. (See Note 8.7.) Does Question 8.2.1 have a positive answer for centered convex bodies when $i = 2$ or 3?

Problem 8.3. (The slicing problem; see Notes 7.8, 8.9, and especially 9.8.) Is there a constant c, independent of n, such that whenever K_1 and K_2 are centered convex bodies in $\mathbb{E}^n$ satisfying

$$\lambda_{n-1}(K_1 \cap u^\perp) \le \lambda_{n-1}(K_2 \cap u^\perp),$$

for all $u \in S^{n-1}$, it follows that

$$\lambda_n(K_1) \le c\lambda_n(K_2)?$$

Problem 8.4. Which star bodies L in $\mathbb{E}^n$ are such that, for each $m \in \mathbb{N}$, $L = I^m L_m$, for some star body L_m?

Problem 8.5. (See Note 8.4.) Suppose that L is a generalized intersection body in $\mathbb{E}^n$. Are there intersection bodies L_1 and L_2 such that $L_1 = L \tilde{+} L_2$?

Problem 8.6. (See Note 8.6.) Suppose that $1 \le i \le n - 1$. Which star bodies L in $\mathbb{E}^n$ are such that $I_i^2 L$ is homothetic to L?

Problem 8.7. (See Note 8.6.) Suppose that $1 \le i \le n - 1$. Which star bodies L in $\mathbb{E}^n$ are such that $I_i L$ is homothetic to L?

Problem 8.8. (See Note 8.12.) Suppose that K is a convex body in $\mathbb{E}^n$, $n \ge 3$.

(i) Is K determined, up to a translation and reflection in the origin, by its inner section function m_K? In short, is it so determined by its cross-section body CK?

(ii) If CK is a ball, is K a ball?

Problem 8.9. (See Note 8.13.) Suppose that K is a convex body in $\mathbb{E}^n$, $n \ge 3$.

(i) Is K determined, up to a translation and reflection in the origin, by ΠK and CK?

(ii) If ΠK and CK are balls, is K a ball?

(iii) If ΠK is homothetic to either IK or CK, is K an ellipsoid?

Notes

8.1. *Intersection bodies.* One might say the history of intersection bodies began with the paper [124] of Busemann, who proved Theorem 8.1.10, which has important implications for Busemann's theory of area in Finsler spaces. (Finsler spaces are locally Minkowskian, just as Riemann spaces are locally Euclidean.) Recall (cf. Note 4.1) that for any centered convex body K in $\mathbb{E}^n$, there is a Minkowski space $(\mathbb{E}^n, K)$ with K as unit ball, and that various definitions of Minkowski surface area in $(\mathbb{E}^n, K)$ are possible. With Busemann's definition (see the works of Busemann [123], [125] and Thompson [803]), the Minkowski surface area of a face F of a polytope in $(\mathbb{E}^n, K)$ is

$$\frac{\kappa_{n-1}\lambda_{n-1}(F)}{\lambda_{n-1}(K \cap S)},$$

where F is contained in a translate of the $(n-1)$-dimensional subspace S of $\mathbb{E}^n$. Solutions to the corresponding isoperimetric problem are homothetic to $(IK)^*$. Busemann and Petty [130, Problem 5] ask whether K must be an ellipsoid, so $(\mathbb{E}^n, K)$ is Euclidean, if $n \geq 3$ and K itself is a solution to the isoperimetric problem for $(\mathbb{E}^n, K)$. (The solution for $n = 2$ is given by those K for which bd K is a Radon curve.) They also give a formulation involving maximal inscribed convex double bounded cones. See the more general Problem 8.1, surveys of Lutwak [542, Section 12] and Martini [581, Section 4], and the books of Schneider [737, p. 416] and Thompson [803].

Our proof of Theorem 8.1.10 follows the somewhat simplified version of Milman and Pajor [621, Theorem 3.9]. There is a significant generalization of Busemann's theorem, due to Barthel and Franz [43] (see also [42]), also containing Theorem 5.1.5 of Chapter 5. In order to formulate this extension, let $u_0, u_1 \in S^{n-1}$, and let S_{u_0}, S_{u_1} be the corresponding half-subspaces in the notation of Busemann's theorem, Theorem 8.1.10. The *harmonic linear combination* E_t of two sets E_0, E_1, contained in S_{u_0}, S_{u_1}, respectively, is the set of points contained in the half-subspace S_{u_t}, where u_t is the unit vector in the direction $(1-t)u_0 + tu_1$ for some $0 \leq t \leq 1$, and also lying on some line segment with one endpoint in E_0 and the other in E_1. Barthel and Franz prove the following *Busemann–Barthel–Franz inequality*:

Let E_0 and E_1 be compact subsets of $\mathbb{E}^n$ of positive λ_{n-1}-measure and $0 \leq t \leq 1$. If E_t is the corresponding harmonic linear combination, then

$$\frac{\|(1-t)u_0 + tu_1\|}{\lambda_{n-1}(E_t)} \leq \frac{(1-t)}{\lambda_{n-1}(E_0)} + \frac{t}{\lambda_{n-1}(E_1)},$$

with equality if and only if E_1 can be obtained from E_0 by a parallel (not orthogonal) projection of S_{u_0} onto S_{u_1}.

We proved this inequality (for $t = 1/2$, which suffices) in the course of establishing Busemann's theorem, Theorem 8.1.10, in the special case where E_0 and E_1 are "half-sections" of the convex body K, and without the equality condition. The strength of the Busemann–Barthel–Franz inequality lies in its freedom from convexity restrictions and its obvious analogy to the Brunn–Minkowski inequality (B.10). We note in Section B.2 that the latter can also be freed from convexity conditions.

Croft [172] found Theorems 8.1.8 and 8.1.9, solving a problem of Mazur's reported in Ulam's famous collection [807, p. 38].

Intersection bodies were first explicitly defined and named by Lutwak in the important paper [537]. It was here that the duality between intersection bodies and projection bodies was first made clear, with the powerful consequences explained in Note 8.9, and Theorems 8.1.2 and 8.1.3 and Corollary 8.1.7 were demonstrated. Lutwak proved the nontrivial Theorem 8.1.6 in [540, p. 22], and Lemma 8.1.5 is also taken from that paper.

In [537], the term intersection body is defined as in Definition 8.1.1, where the star body L contains the origin in its interior. We have usually adhered to the longer phrase "intersection body of a star body" in the text to make a distinction between this concept and the standard definition of "intersection body" in use today, namely the one given just after Theorem 8.1.20. This was introduced by Goodey, Lutwak, and Weil [312]. The fact that sufficiently smooth intersection bodies are intersection bodies of star bodies (see Remark 8.1.21(iii)) was proved by Zhang [858].

The important characterization of intersection bodies in terms of ellipsoids in Remark 8.1.21(iv) is due to Goodey and Weil [321]. (Grinberg and Zhang [337] provide a different proof of this result.) The corollary in Remark 8.1.21(vii), that a central section of an intersection body is also an intersection body, was obtained earlier by Fallert, Goodey, and Weil [221].

Lemma 8.1.12 is really a special case of the formula [390, eq. (3.10)] of Helgason. Theorem 8.1.13 was proved by Gardner [261] (see also Zhang's article

[860]) under stronger differentiability assumptions, and was proved in its present form for $n = 3$ in [263]. Note that the differentiability assumptions can be dropped completely if one adopts the newer definition of intersection body, by the closure property mentioned in Remark 8.1.21(v).

Theorem 8.1.17 is due to Gardner [263] for $n = 3$ and Zhang [862] for $n = 4$. (See Note 8.9 for more about the history of this result.) The proof given here is that of Rubin and Zhang [707], who modified an earlier version of this proof constructed by Barthe, Fradelizi, and Maurey in [40], where Lemma 8.1.14 appears for the first time. The refinement of Theorem 8.1.17 mentioned in Remark 8.1.21(ii) can also be found in [707].

Formula (8.9) in Lemma 8.1.16 was discovered by Gardner [263] (in a form related to (8.9) by integration by parts) and Zhang [862] found (8.10). Lemma 8.1.16 has the following natural generalization to n dimensions.

Let K be a centered convex body in $\mathbb{E}^n$ of class C^∞ and suppose that $\rho_K = Rg$. Let $u \in S^{n-1}$. If n is odd, then

$$g(u) = (-1)^{(n-1)/2} \frac{(n-2)!}{(2\pi)^{(n-1)}} \int_0^\infty \frac{A_K(t, u) - \sum_{i=0}^{(n-3)/2} A_K^{(2i)}(0, u) t^{2i}/(2i)!}{t^{n-1}} \, dt,$$

(8.14)

and if n is even, then

$$g(u) = (-1)^{(n-2)/2} \frac{1}{2^n \pi^{n-2}} A_K^{(n-2)}(0, u).$$

(8.15)

It follows that a centered convex body in $\mathbb{E}^n$ of class C^∞ is an intersection body if and only if the expressions in (8.14) or (8.15) (as appropriate) are nonnegative for all $u \in S^{n-1}$. The formulas (8.14) and (8.15) are taken from [40], but they first appear in a different form in [276] (see also [275]), by Gardner, Koldobsky, and Schlumprecht, where the Fourier transform is used.

The proof of Theorem 8.1.18 is that of Gardner [261]. A somewhat shorter, but less geometrically illuminating, proof can be constructed along the lines of that of Theorem 8.1.20. This was supplied by G. Zhang (private communication) and uses ideas from the paper by Rubin and Zhang [707]. These authors provide specific examples of bodies satisfying the hypotheses of Theorem 8.1.20; when $n = 5$ and 6, such examples are offered by Gardner [261].

Remark 8.1.21(i) is proved for $n \geq 4$ by Zhang [864], using techniques from his work [337] with Grinberg, and for $n = 3$ by Campi [141]. Campi also refines the results of Gardner [263] and Zhang [862] by proving that centered convex bodies in $\mathbb{E}^n$, $n \leq 4$, are intersection bodies of sets star-shaped at the origin whose radial functions satisfy conditions related to suitable noninteger Sobolev classes.

In the course of comparing Busemann surface area with Holmes–Thompson surface area, Schneider [740] (see also [742]) shows that there are centered convex bodies such that the polar body of IK is not a projection body; in particular, centered cross-polytopes and cubes have this property. It follows that there are intersection bodies in $\mathbb{E}^n$, $n \geq 3$, that are not polar projection bodies. On the other hand, Koldobsky [457] uses the Fourier transform characterization in Note 8.4 to prove that the unit ball of any n-dimensional subspace of $L^p([0, 1])$, $0 < p \leq 2$, is an intersection body. This implies the significant result of Remark 8.1.21(viii), since each polar projection body in $\mathbb{E}^n$ is the unit ball of an n-dimensional subspaces of $L^1([0, 1])$. Koldobsky [456] also proves Remark 8.1.21(ix) and obtains still more general results of this type in [458] (see also [465, Section 4.4]).

It can be shown that a generalized cylinder in $\mathbb{E}^n$ (the Cartesian product of an $(n - 1)$-dimensional convex body and a line segment) is not an intersection body when $n \geq 5$. This includes the case of the cube as well as the ordinary cylinder.

G. Zhang (private communication) has a proof that in essence requires only the methods of this chapter. In brief, one uses the definition of the spherical Radon transform of a finite Borel measure in S^{n-1} and the self-adjoint property (Theorem C.2.6) to extend Theorems C.2.8 and C.2.9 from functions to measures. With this in hand, the proof is very similar to that of Theorem 8.1.20.

In [460], Koldobsky gives a characterization of intersection bodies in terms of the concept of embedding into L^p for $p < 0$ (specifically, when $p = -1$). Koldobsky [465, Section 6.3] and Kalton and Koldobsky [415] obtain further results in this direction.

Moszyńska [638] introduces a notion of the quotient of two star bodies and notes that the quotient of intersection bodies of star bodies is again an intersection body of a star body.

The intersection body IE of a bounded Borel set E in $\mathbb{E}^n$ can be defined as in Definition 8.1.1. There is a slight abuse of notation, since IE is then generally not a body, though it is a centered Borel set star-shaped at the origin and containing it. Using this extended definition, Gardner, Vedel Jensen, and Volčič [280] obtain a version of Theorem 8.1.3 that holds for bounded Borel sets.

8.2. *The Fourier transform approach II: The section function and intersection bodies.* Suppose that L is a centered star body in $\mathbb{E}^n$ with a continuous radial function. Then ρ_L is even and positively homogeneous of degree -1 on $\mathbb{E}^n \setminus \{o\}$ (see (0.29), so ρ_L^{n-1} is even and positively homogeneous of degree $-n+1$ on $\mathbb{E}^n \setminus \{o\}$. By (C.25), we have

$$\rho_{IL}(u) = \lambda_{n-1}\left(L \cap u^{\perp}\right) = \frac{1}{n-1}\left(R\rho_L^{n-1}\right)(u) = \frac{1}{(n-1)\pi}\widehat{\rho_L^{n-1}}(u),$$

for all $u \in S^{n-1}$. See, for example, the papers of Koldobsky [454], [455]. In [457], Koldobsky uses this formula to obtain the following characterizations announced in [453].

(i) A centered star body L in $\mathbb{E}^n$ with a continuous radial function is the intersection body of a star body if and only if ρ_L is a distribution whose Fourier transform is a positive continuous function on $\mathbb{E}^n \setminus \{o\}$.

(ii) A star body L in $\mathbb{E}^n$ with a continuous radial function is an intersection body if and only if ρ_L is a positive definite distribution, that is, a distribution whose Fourier transform is a positive distribution.

Koldobsky's survey [463] and especially his book [465] contain detailed expositions of this approach.

8.3. *The map I.* It is easy to prove that the class of intersection bodies of star bodies is closed under radial linear combinations. In particular, if L_1 and L_2 are star bodies in $\mathbb{E}^n$ with continuous radial functions, then

$$I(L_1 \tilde{+}_{n-1} L_2) = IL_1 \tilde{+} IL_2,$$

in the notation introduced in Note 6.1. Ludwig [527] observes that with the operation of radial addition, the class S^n of star bodies in $\mathbb{E}^n$ with continuous radial functions is an Abelian semigroup and I is a valuation. She proves that if $\mathcal{P}_o^n$ is the class of convex polytopes in $\mathbb{E}^n$ containing the origin in their interiors and $Z : \mathcal{P}_o^n \to S^n$ is a nontrivial translation-invariant valuation that satisfies the affine invariance property of Theorem 8.1.6 (when I is replaced by Z), then there is a constant $c \geq 0$ such that $ZP = cI(P^*)$ for each $P \in \mathcal{P}_o^n$.

8.4. *Generalized intersection bodies.* It was inevitable that generalized intersection bodies would be introduced soon after intersection bodies, as the natural analogues of generalized zonoids. A *generalized intersection body* is a star body L in $\mathbb{E}^n$ with $\rho_L \in C(S^{n-1})$ such that $\rho_L = R\mu$, where μ is a *signed* finite even Borel measure in S^{n-1}. The first results were characterizations in terms of dual mixed volumes (see

Note 8.8). Every centered star body L in $\mathbb{E}^n$ with $\rho_L \in C^\infty(S^{n-1})$ is a generalized intersection body; the proof follows that of Theorem 4.1.17 for generalized zonoids, when Theorem C.2.5 is applied instead of Theorem C.2.2. (Zhang [858] shows that it is enough to assume that $\rho_L \in C^k(S^{n-1})$, where $k = [(n+1)/2] - 1$.)

8.5. *Bodies whose central sections are intersection bodies.* Consider the following question, dual to that asked in Note 4.5. Is there a centered convex body K in $\mathbb{E}^n$ that is not an intersection body such that $K \cap u^\perp$ is an intersection body in $u^\perp$ for all $u \in S^{n-1}$? Any such body must be at least 5-dimensional, by Remark 8.1.21(vi), and any centered convex body in $\mathbb{E}^5$ has this property, by Remark 8.1.21(vi) and (vii). Yaskina [851] constructs examples of such bodies in $\mathbb{E}^n$ for all $n \geq 5$.

8.6. *Intersection bodies of order i.* The concept of an intersection body has another natural extension (cf. that in Note 4.6 for the projection body). If L is a star body in $\mathbb{E}^n$ with $\rho_L \in C(S^{n-1})$, and $i \in \mathbb{R}$ is positive, the *intersection body of order i* of L is the centered star body $I_i L$ such that

$$\rho_{I_i L}(u) = \tilde{V}_{i,n-1}(L \cap u^\perp) = \frac{1}{n-1} \int_{S^{n-1} \cap u^\perp} \rho_L(v)^i \, dv,$$

for all $u \in S^{n-1}$. The term is introduced by Zhang [858]. It follows from this definition that every intersection body of order i of a star body is an intersection body of a star body, and vice versa. The generalized Funk section theorem (Theorem 7.2.6), with $k = n-1$, shows that the map I_i, when restricted to the class of centered star bodies, is injective. The corresponding versions of Theorems 8.1.2 and 8.1.3 are also easily established.

Grinberg and Zhang [337] solve the case $i = 1$ of Problems 8.6 and 8.7 by showing that the hypotheses then imply that L must be a centered ball. In connection with Problem 8.6, Lutwak [542, Section 12] suggests that centered ellipsoids may be the only star bodies L such that $I^2 L = I(IL)$ is a dilatate of L.

8.7. *k-intersection bodies and related notions.* Let $1 \leq i \leq n-1$ and let f be a continuous function on S^{n-1}. The *totally geodesic Radon transform* $R_i f$ of f is defined by

$$(R_i f)(S) = \frac{1}{i\kappa_i} \int_{S^{n-1} \cap S} f(u) \, du,$$

for each $S \in \mathcal{G}(n, i)$. Let g be a continuous function on $\mathcal{G}(n, i)$. If we define

$$(R_i^* g)(u) = \int_{u \in S} g(S) \, dS,$$

then it can be shown that

$$\int_{\mathcal{G}(n,i)} (R_i f)(S) g(S) \, dS = \int_{S^{n-1}} f(u)(R_i^* g)(u) \, du,$$

and for this reason R_i^* is called the dual transform of R_i. The duality allows the domain of R_i and R_i^* to be extended to the class of finite Borel measures in S^{n-1} and in $\mathcal{G}(n, i)$, respectively.

Let $1 \leq i \leq n-1$ and let $p \in \mathbb{R}$. A centered star body L in $\mathbb{E}^n$ with a continuous radial function is called an *(i, p)-intersection body* if there is a finite Borel measure μ in $\mathcal{G}(n, i)$ such that $\rho_L^p = R_i^* \mu$. Then $(n-1, 1)$-intersection bodies are just the usual intersection bodies defined just after Theorem 8.1.20. This generalization of intersection bodies was introduced by Rubin and Zhang in [707], where background references concerning the totally geodesic Radon transform may be found. However, the special case $p = n - i$ appeared earlier in the work of Zhang [861], who proved that if K is a centered convex body of revolution in $\mathbb{E}^n$, then K is an $(i, n-i)$-intersection body when $i = 2$ or 3. Many more facts about $(i, n-i)$-intersection bodies are discovered by Grinberg and Zhang [337]. For example, they show that every intersection body is an $(i, n-i)$-intersection body and obtain a characterization in terms of ellipsoids and invariance properties analogous to those for intersection bodies in Remark 8.1.21(iv), (v), and (vii). Milman [618] shows that if K is a centered convex body in $\mathbb{E}^n$ and $i = 2$ or 3, $n - i > 0$, then the star body whose radial function is $\rho_k^{1/(n-i)}$ is an $(i, n-i)$-intersection body.

Let L and M be star bodies in $\mathbb{E}^n$ with continuous radial functions and let $1 \le k \le n - 1$. Then L is called the k-*intersection body* of M if

$$\lambda_k(L \cap S) = \lambda_{n-k}(M \cap S^{\perp}),$$

for all $S \in \mathcal{G}(n, k)$. (Compare the notion of a (VP)-pair for projections defined in Note 4.8.) When $k = 1$, we retrieve Definition 8.1.1 of an intersection body of a star body, apart from a factor of 2. There is a more general notion of a k-intersection body that when $k = 1$ reduces to the usual intersection body defined just after Theorem 8.1.20. We shall not define this here but note the following Fourier transform characterization (compare (ii) in Note 8.2).

A centered star body L in $\mathbb{E}^n$ with a continuous radial function is a k-intersection body if and only if ρ_L^k is a positive definite distribution.

Koldobsky [459] introduced k-intersection bodies and established the previous characterization. He proves (see [465, Corollary 4.9]) that a centered convex body in $\mathbb{E}^n$ is a k-intersection body when $k = n - 3$, $n - 2$, or $n - 1$, and $k > 0$. In [460], he shows that every $(k, n - k)$-intersection body is a k-intersection body. It is not known whether the converse is true; a positive answer would solve the generalized Busemann–Petty problem (see Note 8.10). Milman [619] provides substantial evidence for a positive answer, by showing that the two classes of bodies behave similarly under the taking of kth radial sums, limits in the radial metric, central sections, and certain other operations. See Koldobsky's book [465, Chapter 4] for a detailed treatment of this topic.

8.8. *Characterizations in terms of dual mixed volumes.* Zhang [858] proves that if $1 \le i \le n - 1$, then a star body L in $\mathbb{E}^n$ with $\rho_L \in C(S^{n-1})$ is an intersection body if and only if

$$\tilde{V}(L_1, i; B, n - i - 1; L) \le \tilde{V}(L_2, i; B, n - i - 1; L),$$

for all centered star bodies L_1, L_2 with continuous radial functions such that

$$\tilde{V}_{i,n-1}(L_1 \cap u^{\perp}) \le \tilde{V}_{i,n-1}(L_2 \cap u^{\perp}),$$

for all $u \in S^{n-1}$ (or, equivalently, $I_i L_1 \subset I_i L_2$). (Lemma 8.2.7 gives one direction of this.) In fact, Zhang demonstrates that the same statement holds when L_1 is a fixed centered convex body with a positive curvature function (see [858] for the meaning of this term) and C^2 support function, and L_2 is any centered convex body.

Corresponding characterizations of generalized intersection bodies are known. For example, Goodey, Lutwak, and Weil [312] show that if $1 \le i \le n - 1$, then a star body L in $\mathbb{E}^n$ with $\rho_L \in C(S^{n-1})$ is a generalized intersection body if and only if there is a constant c, depending only on L, such that

$$|\tilde{V}(L_1, i; B, n - i - 1; L) - \tilde{V}(L_2, i; B, n - i - 1; L)|$$
$$\le c \sup_{u \in S^{n-1}} \left|\tilde{V}_{i,n-1}(L_1 \cap u^{\perp}) - \tilde{V}_{i,n-1}(L_2 \cap u^{\perp})\right|,$$

for all centered star bodies L_1, L_2 with continuous radial functions.

Other characterizations related to these can be found in [312] and [858].

The striking resemblance between characterizations of zonoids and generalized zonoids (see Note 4.8) and those of intersection bodies and generalized intersection bodies led the authors of [312] to develop a unified approach using functional analysis. A variation of the Hahn–Banach principle is used to characterize a closed convex cone of a locally convex space X by means of a dense subset A of the dual X' of X. (A similar idea was used earlier by Schneider and Weil [745, Theorem 6.1].) One can take $X = C_e(S^{n-1})$, identifying the classes of centered convex bodies and centered zonoids in $\mathbb{E}^n$ with closed convex cones in X by identifying each body with its support function. To obtain the characterization of zonoids given in Note 4.8, one can take X' to be the space $\mathcal{M}_e(S^{n-1})$ of signed finite even Borel measures in S^{n-1}, with the weak topology. Minkowski's existence theorem, Theorem A.3.2, shows that the set A of differences of surface area measures of centered convex bodies in $\mathbb{E}^n$ is dense in X' (in fact equal to X'). Other choices of A lead to different characterizations, but a stronger topology than the weak topology must sometimes be used. For characterizations of intersection bodies, one takes $X = \mathcal{M}_e(S^{n-1})$, with the weak topology, so $X' = C_e(S^{n-1})$. The class of centered star bodies with continuous radial functions can be identified with $C_e^+(S^{n-1})$, a closed convex cone in X' also embeddable as a closed convex cone in X, and the class of intersection bodies can be identified with the closed convex cone of X consisting of those members of $C_e^+(S^{n-1})$ that are images under the spherical Radon transform of finite even Borel measures in S^{n-1}. The set A of differences of functions in $C_e^+(S^{n-1})$ is dense in X' (in fact equal to X'), and from this one obtains the characterization of intersection bodies given before. The characterizations of generalized zonoids and generalized intersection bodies result from a further abstract theorem, characterizing in a similar way a vector space $C - C$ generated by a closed convex cone C of a locally convex space X.

8.9. *Larger bodies with smaller sections I: The Busemann–Petty problem.* Theorem 8.2.2 for $i = n - 1$ is implicit in unpublished work of E. Lutwak, as is Theorem 8.2.3, dealing with the easy case when $i = 1$. However, Question 8.2.1 was first answered in the negative by Busemann [129], who provides two counterexamples in $\mathbb{E}^3$, one with centered star bodies and the other with bodies convex but not centered. For the former we have chosen instead to present Theorem 8.2.4, discovered for $n = 3$ by Hadwiger [373]. The original setting of Question 8.2.1 was restricted to centered convex bodies and $i = n - 1$; this became known as the *Busemann–Petty problem.* It is the first in a list of 10 problems formulated by Busemann and Petty [130], all motivated by Minkowski geometry. The problem achieved a certain notoriety, and appears in several articles, as well as in the books of Burago and Zalgaller [112, p. 54], Croft, Falconer, and Guy [173, Problem A9, p. 22], and Schneider [737, p. 423]. Its history, which we shall now describe, is quite involved.

Hadwiger [373] shows that the answer is positive for coaxial centered convex bodies of revolution in $\mathbb{E}^3$; in fact, he only assumes a condition a little stronger than axis-convexity, and (essentially) uses the dual volume $\tilde{V}_{i,2}$ instead of λ_2. Independently, Giertz [290] obtained more or less the same result. To be precise, Giertz assumes only axis-convexity but deals exclusively with λ_2-measure of sections. The stronger Theorem 8.2.9 is due to Gardner [261]. (A version that, like Hadwiger's result, works with $\tilde{V}_{i,n-1}$ instead of λ_{n-1} can be obtained from a more general form of Theorem 8.2.8 stated in Note 8.10.)

A major breakthrough came with the paper [495] of Larman and Rogers. By randomly slicing caps of equal size, together with their antipodal caps, from a ball L_2, and then slightly expanding the resulting convex body to obtain L_1, they answered the Busemann–Petty problem in the negative for $n \geq 12$.

The next development was equally surprising. Ball [21] noted that his upper bound of $\sqrt{2}$ for the volume of an $(n-1)$-dimensional central section of the centered unit cube in $\mathbb{E}^n$ (see Note 7.8) provides a negative answer to the Busemann–Petty problem for $n \geq 10$. To see this, let L_1 be the centered unit cube in $\mathbb{E}^n$ and $L_2 = rB$, for some $r > 0$. Using (0.9) and (0.8), we see that if $r^n = \Gamma(n/2 + 1)/\pi^{n/2}$, then

$\lambda_n(L_2) = 1 = \lambda_n(L_1)$. With this value of r, we have

$$\lambda_{n-1}(L_2 \cap u^\perp) = \frac{\Gamma(\frac{n}{2} + 1)^{(n-1)/n}}{\Gamma(\frac{n-1}{2} + 1)},$$

for all $u \in S^{n-1}$. The value of this expression increases with n, and when $n = 10$, it equals $1.42039\ldots$, becoming for the first time greater than $\sqrt{2}$. A slightly smaller value of r then yields the required example in 10 dimensions.

Theorem 8.2.5, showing that 10 dimensions can be improved to 7 by replacing the cube by a suitable centered cylinder, is due to Giannopoulos [286], where the details of Remark 8.2.6(i) are also given. A completely different example is constructed by Bourgain [87], whose result is noted in Remark 8.2.6(ii). He uses spherical harmonics to prove that when $n \geq 7$, one can take L_2 to be a ball and L_1 an arbitrarily small perturbation of L_2, and also that this method will not work for $n = 3$. Perturbations of the ball and of intersection bodies are also studied by Grinberg and Rivin [336].

When $n = 5$ or 6, a centered cylinder L_1 and a suitable "rounded" centered cylinder L_2 provide counterexamples for the Busemann–Petty problem. This was discovered independently by Papadimitrakis [657] and Gardner [261].

In their original article, Busemann and Petty noted that it follows from the Busemann intersection inequality (see Corollary 9.4.5) that the answer to their question is positive if L_1 is a centered ellipsoid. Aside from work of Hadwiger and Giertz mentioned earlier, no other positive results appeared until Lutwak [537] proved Lemma 8.2.7 for $i = n - 1$ and the consequent Theorem 8.2.8. (Versions applying to bounded Borel sets are obtained by Gardner, Vedel Jensen, and Volčič [280].) By Corollary 8.1.7, Lutwak's theorem is actually a direct generalization of Busemann and Petty's observation. The first statement of Theorem 8.2.10, which is dual to Schneider's theorem of Remark 4.2.10, also appeared in [537].

The second statement of Theorem 8.2.10, and Corollary 8.2.11, can be found in [261], the latter providing a direct link between the Busemann–Petty problem and intersection bodies. This is re-proved by Zhang [858], who deduces it from his dual-mixed-volume characterization of intersection bodies (cf. Note 8.8). As Zhang observes, there is a more appealing formulation, namely: The Busemann–Petty problem has a positive answer in $\mathbb{E}^n$ if and only if each centered convex body K in $\mathbb{E}^n$ is an intersection body.

As we saw in Theorem 8.2.12, these connections yield a solution to the Busemann–Petty problem in all dimensions. The case $n \geq 5$ is considered by Gardner [261] and Zhang [860], though as we mentioned earlier, the proof given here based on Theorem 8.1.20 uses ideas of Rubin and Zhang [707]. Gardner [262], [263] discovered the connection in Lemma 8.1.16 between the radial function ρ_K of a centered convex body K and the spherical Radon transform of what is sometimes called the parallel section function A_K, leading to the positive answer when $n = 3$. Zhang's papers [857] and [860] purported to prove that a centered cube in $\mathbb{E}^4$ is not an intersection body, and hence that the answer to the Busemann–Petty problem is negative when $n = 4$. When a few years later Koldobsky [453] announced that a centered cube in $\mathbb{E}^4$ is an intersection body, Zhang found an error in his earlier result and then quickly proved in [862] that in fact the answer is positive when $n = 4$. More results arrived shortly afterwards. A unified solution is presented by Gardner, Koldobsky, and Schlumprecht [276] (see also Koldobsky's book [465, Chapter 5]) using Koldobsky's functional analysis approach. Here, both the positive answer for $n \leq 4$ and the negative answer for $n \geq 5$ are derived from the Fourier transform versions of the formulas (8.14) and (8.15) in Note 8.1. Shorter proofs along these lines that avoid the Fourier transform are constructed by Barthe, Fradelizi, and Maurey [40] and Rubin and Zhang [707]. Koldobsky [462] finds a proof of the positive answer for $n \leq 4$ using spherical harmonics.

Despite the considerable ingenuity of earlier attacks on the Busemann–Petty problem, it seems fair to say that the work [537] of Lutwak represents the beginning of its eventual solution. Consider, for example, the fact that the article [373] of Hadwiger already contains most of the ingredients: the necessary coordinate geometry for bodies of revolution, inversion of integral equations, and Hölder's inequality. What is missing is the essential use of star bodies and dual mixed volumes. The latter transforms earlier results for projections into signposts, and the dual Aleksandrov–Fenchel inequality (B.23) speeds the journey by molding Hölder's inequality into precisely the right form.

Meyer [603] shows that the answer to the Busemann–Petty problem is positive when L_1 is the cross-polytope $L_1 = \{x \in \mathbb{E}^n : |x_1| + \cdots + |x_n| \le 1\}$. (This is an easy consequence of an interesting inequality concerning sections of convex bodies by coordinate planes that was proved by Meyer [599]; see Note 9.7.) He conjectured that the answer is positive whenever L_1 is a polar projection body, and this is confirmed by Koldobsky [457] via his result in Remark 8.1.21(viii) which more generally implies that the answer is positive whenever L_1 is the unit ball of any n-dimensional subspace of $L^p([0, 1])$, $0 < p \le 2$. See also Koldobsky's book [465, Chapter 6].

Up to a constant, Theorem 8.2.13 can be found in the article [621] of Milman and Pajor, and the result itself has apparently been known for some time, but the proof given here is due to R. Schneider (private communication). Asymptotically, the result is not the best possible, since Bourgain [88] has shown that $\sqrt{n}$ can be replaced (up to a constant) by $n^{1/4}\log n$. (Dar [176] presents a relatively simple proof of this result.) It is even possible that Problem 8.3 might have a positive answer. We follow Ball [24] in naming this the *slicing problem*; it has also been called the hyperplane problem and the maximal slice problem. Several versions of the problem, all equivalent (though not obviously so), are stated in [621, Section 5] and discussed in Note 9.8. Some who work in the local theory of Banach spaces consider this the most important unsettled question in the area.

8.10. *Larger bodies with smaller sections II: Generalizations and variants of the Busemann–Petty problem.* It is worth noting that Theorem 8.2.8 has the following stronger form, dual to Theorem 4.2.6 for projections. Let $1 \le k \le n - 1$, let $i \in \mathbb{R}$, $i > 0$, and let L_1 and L_2 be star bodies in $\mathbb{E}^n$ with continuous radial functions. Suppose that $\tilde{V}_{i,k}(L_1 \cap S) \le \tilde{V}_{i,k}(L_2 \cap S)$ for all $S \in \mathcal{G}(n, k)$. If $L_1 \cap T$ is an intersection body (in T) for each $T \in \mathcal{G}(n, k + 1)$, then $\tilde{V}_{i+1,k}(L_1 \cap T) \le \tilde{V}_{i+1,k}(L_2 \cap T)$ for all $T \in \mathcal{G}(n, k + 1)$. Equality holds for all $T \in \mathcal{G}(n, k + 1)$ if and only if L_1 is a dilatate of L_2. It follows that $\tilde{V}_{n+i-k}(L_1) \le \tilde{V}_{n+i-k}(L_2)$, with the same conditions for equality. (Zhang [858] finds related results.) This can be proved by using a version of Lemma 8.2.7 in which the restriction on i is relaxed and the fact (see Remark 8.1.21(vii)) that a central section of an intersection body is again an intersection body in the corresponding subspace. A corresponding more general form of Theorem 8.2.9 can then also be obtained.

Rubin and Zhang [707] ask the following related question:

Suppose that K_1 and K_2 are centered convex bodies in $\mathbb{E}^n$ and that i and l are real numbers with $0 < i < 1$. If $\tilde{V}_{i,k}(K_1 \cap S) \le \tilde{V}_{i,k}(K_2 \cap S)$ for all $S \in \mathcal{G}(n, k)$, is it true that $\tilde{V}_l(K_1) \le \tilde{V}_l(K_2)$?

The special case $i = k = n - 1$ and $l = n$ of Question 2 is the Busemann–Petty problem, whose history was described in detail in Note 8.9. The case $i = k$ and $l = n$ has been called the *generalized Busemann–Petty problem*; it is just Question 8.2.1 restricted to centered convex bodies. This was first addressed by Zhang [861], who proved that the answer is positive if K_1 is an $(i, n - i)$-intersection body (see Note 8.7) and hence that the answer to the generalized Busemann–Petty problem is positive when $i = 2$ or 3 and K_1 is a body of revolution. Bourgain and Zhang [93] (see [707] for a correction) prove that the answer is negative when $i \ge 4$. Other partial results follow from the work of Milman [618] mentioned in Note 8.7. But Problem 8.2, concerning the case $i = 2$ or 3, remains open. By results of Koldobsky and Zhang in Note 8.7, a positive answer would follow if each k-intersection body is a $(k, n - k)$-intersection body.

Returning to the question of Rubin and Zhang just stated, Koldobsky [460] demonstrates that the answer is positive when $i = n - 1$, $k = 3$, and $l = n$. Rubin and Zhang [707] extend this positive answer to $l = i + 1$ and $k = 2$ or 3. They also prove that their question has a negative answer when $k \geq 4$ and when $k = 2$ or 3 and $l - i > n - k$. Answers are shown to be positive when $i = 2$ or 3 when K_1 is a body of revolution, but the question is open otherwise.

The remainder of this note is largely devoted to a very active program of A. Koldobsky, in which he and his students have generated an amazing number of results related to the Busemann–Petty problem.

In [459] and [461] (see also [462] and [465, Section 5.3]), Koldobsky applies formulas (8.14) and (8.15) in variations of the Busemann–Petty problem in which the hypothesis involves an inequality between derivatives of the parallel section function. Koldobsky [464] and Koldobsky, Yaskin, and Yaskina [470] consider a similar variation involving fractional powers of the Laplacian applied to the section function.

Zvavitch [870] uses Fourier transform techniques to solve a version of the Busemann–Petty problem in which λ_n is replaced by the Gaussian measure in $\mathbb{E}^n$. He generalizes this result in his remarkable paper [872] (see also Koldobsky's book [465, Chapter 6]). His *Busemann–Petty problem for general measures* asks the same question but where λ_n is replaced by an arbitrary measure μ_n in $\mathbb{E}^n$ defined in terms of a density function f that is locally integrable in each $(n - 1)$-dimensional subspace. Thus the measure $\mu_n(E)$ of a Borel subset E of $\mathbb{E}^n$ is defined by the integral of f over E with respect to λ_n, and the measure $\mu_{n-1}(E)$ of a Borel subset E of $u^{\perp}$, $u \in S^{n-1}$, is the integral of f over E with respect to λ_{n-1}. When $f(x) = 1$ for all $x \in \mathbb{E}^n$, we retrieve the usual Busemann–Petty problem, and $f(x) = \exp(-\|x\|^2/2)$ is the case of Gaussian measure. Zvavitch proves that the answer to the Busemann–Petty problem for general measures is positive when f is continuous and positive and $n \leq 4$, and negative when $f \in C^{\infty}(\mathbb{E}^n)$ and $n \geq 5$. He shows that other choices of f lead to various interesting results related to the Busemann–Petty problem. Rubin [706] treats the Busemann–Petty problem for general measures and lower-dimensional sections, using Radon transforms rather than the Fourier transform.

Versions of the Busemann–Petty problem in hyperbolic and spherical spaces are studied by Yaskin [848], [849].

8.11. *Stability results.* Groemer [355], [356, Corollary 5.6.4] obtains a dual version of the stability result for projection bodies stated in Remark 4.3.13. He proves that if $n \geq 3$ and K_1 and K_2 are centered convex bodies with $r B \subset K_1, K_2 \subset RB$, then there is a constant $c = c(n, r, R)$ such that

$$\delta(K_1, K_2) \leq c\tilde{\delta} (I K_1, I K_2)^{4/(n(n+1))},$$

where $\tilde{\delta}$ denotes the radial metric (see (0.32)). Note that this result applies to convex bodies. For a detailed investigation as to how the convexity assumption can be weakened, see the article of Campi [140].

8.12. *Cross-section bodies.* The function m_K of Definition 8.3.1 has sometimes been called *inner $(n - 1)$-quermass*, particularly by authors who prefer *outer $(n - 1)$-quermass* for the brightness function. This concept goes back at least to Bonnesen's 1926 paper [81].

Klee [447], [448] calls a value $m_K(u)$ of the inner section function an *HA-measurement*. As he points out, the term derives from a most interesting connection with Fermi surfaces of metals. The Fermi surface of a metal bounds a generally nonconvex body formed, in velocity space, by velocity states occupied at absolute zero by valence electrons of the metal. The Pauli exclusion principle allows no more than two electrons (with opposite spins) to possess the same velocity (i.e., speed and direction), so electrons can only move into unoccupied states lying outside the Fermi surface. The more electrons there are near the Fermi surface, the larger the number that can increase their energy when the metal is heated and the larger the number whose spins can be aligned with a magnetic field. In this way the Fermi surface relates to the specific heat and magnetic properties of the metal, and the concept also provides an explanation of conductivity, ductility, and so on. In fact, according to Mackintosh [555], the most meaningful definition of a metal may be "a solid with a Fermi surface." The inner section function of the body bounded by a Fermi surface may be measured by means of the de Haas–van Alphen effect, magnetism induced in the metal by a strong magnetic field at a low temperature, hence the name "HA-measurement."

Petty [660, p. 60] proves Theorem 8.3.3 and establishes the equality condition; see also Martini [576], where this result is rediscovered and Corollary 8.3.4 with its equality condition is also given. The term cross-section body appears later, however, in the works [580] and [581] of Martini and [559] of Makai and Martini. Theorem 8.3.2 is noted in [581], and Theorem 8.3.5 and Lemma 8.3.6 are proved in [559]. Many other properties of the cross-section body can be found in [559], such as the nonintuitive Remark 8.3.7(i).

In his survey paper [581], Martini raised two intriguing questions. The first, whether if K is a convex body in $\mathbb{E}^n$ such that $CK = IK$, then K must be centered, is answered positively by Makai, Martini, and Ódor [562] by means of a generalization of the spherical Radon transform. Using this method, Makai and Martini [560] show that in fact if CK is a dilatate of IK, then K is centered, and if IK and CK are both centered balls, then K is also a centered ball.

The second question Martini asked was whether CK is always convex. As was mentioned in Remark 8.3.7(iii), the answer is negative, a result of Brehm [97]. The fact (see Remark 8.3.7(ii)) that CK is convex when $n = 3$, a generalization of the 3-dimensional case of Busemann's theorem, Theorem 8.1.10, is an admirable result of Meyer [602]. In [601], Meyer proves that given any convex body K in $\mathbb{E}^n$ and any two directions u_1, u_2, there are hyperplane sections of K orthogonal to u_1 and u_2 of maximal volume that intersect.

Klee [447] posed Problem 8.8(i). This is related to an old question of Bonnesen [81, p. 80] (see also [82, pp. 133–4]), who asked if the inner section function and brightness function together suffice to determine a convex body in this sense. As Klee [447] pointed out, the answer is unknown even for a ball; cf. Problem 8.8(ii). If K is centered, then $CK = IK$, so the injectivity properties of the intersection body map I provide positive answers to these questions. Note the restriction $n \geq 3$; if $n = 2$, then $CK = \Pi K$ by Theorem 8.3.5, so a noncircular planar body of constant width (= constant brightness) provides a counterexample.

Clearly the inner section function and the cross-section body can be defined for bounded Borel sets. In this context, observe that the region in $\mathbb{E}^3$ between two centered spheres of radii r and $\sqrt{r^2 - 1}$ has the same cross-section body as the centered unit ball, so Problem 8.8 generally has a negative answer if K is not convex. Zaks [854] noted various modifications of this example.

Gardner and Giannopoulos [265], in a sequel to the paper [284] (see Note 9.5), introduce for $p > -1$ the *p-cross-section body* $C_p K$ of a convex body K in $\mathbb{E}^n$. The radial function of $C_p K$ in any direction $u \in S^{n-1}$ is the pth mean of the volumes of hyperplane sections of K orthogonal to u through points in K. As $p \to -1$, a suitable dilatate of $C_p K$ approaches the polar $D^* K$ of the difference body $DK = 2\triangle K$, and $C_\infty K = CK$. It is shown that $C_1 K$ is convex but $C_p K$ is generally not convex when $p > 1$. An inclusion of the form $a_{n,q} C_q K \subseteq a_{n,p} C_p K$, where $-1 < p < q$ and the constant $a_{n,p}$ is the best possible, is established.

Let $1 \leq k \leq n - 1$, let $S \in \mathcal{G}(n, k)$, and define

$$m_K(S) = \max_{x \in \mathbb{E}^n} \lambda_k \Big(K \cap (S + x) \Big).$$

Then $m_K(u^{\perp}) = m_K(u)$ is the inner section function. For information about $m_K(S)$ and related results, we refer the reader to the works of Makai and Martini [559], [561] and Makai, Vrećica, and Živaljević [563]. Fradelizi [246] finds a sharp inequality for the ratio of $\lambda_k(K \cap S)$ and $m_K(S)$ when K is a convex body in $\mathbb{E}^n$ with centroid at the origin. Rudelson [708] finds an upper bound for the volume of the intersection of the difference body $2\triangle K$ of a convex body K in $\mathbb{E}^n$ with $S \in \mathcal{G}(n, k)$ in terms of $m_K(S)$.

8.13. *Problems involving both projections and sections.* Problem 8.9 is raised by Klee [447]. It is of mixed type, since the given information involves both projections and sections. There are very few theorems of this sort. One is the following: If K is a convex body in $\mathbb{E}^n$ such that $\triangle K$ and $\tilde{\triangle} K$ are balls, then K is a ball. This is a consequence of Theorem 4 of [666], due to Petty and Crotty. (See the survey [154, pp. 55–6] of Chakerian and Groemer, where Fujiwara [251] is credited with the case $n = 3$.) We do not attempt to formulate all the many natural open questions such as Problem 8.9(iii); the reader will find the surveys of Lutwak [542] and Martini [581] inspirational. Note 9.10 addresses related questions.

Figure 8.9. Herbert Busemann.

8.14. *Herbert Busemann (1905–1994).* Herbert Busemann was born in Berlin. He studied at the Universities of Munich, Paris, and Rome and also spent several years in business, to please his father, one of the directors of Krupp. (According to Reid [688], he regarded the latter as time wasted.) He obtained his PhD in 1931 from the University of Göttingen and stayed on as an assistant until Hitler came to power in 1933. Busemann then left for Denmark and the University of Copenhagen; many years later, he became a member of the Royal Danish Academy. In 1936 he went to the United States, where he married in 1939 and became an American citizen in 1943. After periods spent at the Institute for Advanced Study in Princeton, Swarthmore College, the Illinois Institute of Technology, and Smith College, Busemann was appointed to a professorship at the University of Southern California in 1947. Here he stayed, as distinguished professor from 1964 until his retirement in 1970, receiving an honorary Doctor of Laws in 1971.

In 1985 Busemann was awarded the Lobachevsky Prize. (Aleksandrov (see Note 3.13) had also achieved this major distinction, but Busemann was the first

American to do so.) He received the prize especially for his invention of G-spaces (see [127, p. 37]) and for his book [127], which takes an innovative synthetic approach to classical and modern differential geometry. (No doubt Steiner (cf. Note 1.7) would have thoroughly approved!) Busemann's list of publications include five other books and more than 80 research papers.

An accomplished linguist, Busemann was able to lecture in seven languages and enjoy reading Latin and Greek classics. After retirement, he built a studio and indulged a lifelong desire to paint. An article in the July 14, 1985 *Los Angeles Times* shows Busemann with one of his abstract, geometric paintings and illustrates his independent spirit with a quote: "If I have a merit, it is that I am not influenced by what other people do."

9

Estimates from projection and section functions

In this chapter we collect results permitting the estimation of an object from projection functions, or section functions such as point X-rays. The estimates are of quantities such as volume and surface area. The topic presents an excellent opportunity to make acquaintance with centroid bodies and some fascinating and important affine isoperimetric inequalities.

Centroid bodies are introduced in the first section. A centroid body is a centered zonoid whose boundary depends on the centroids of "halves" of a compact set. The volume of a centroid body is given by a formula involving the average volumes of simplices with one vertex at the origin and the others in the associated compact set. This and Steiner symmetrization are the principal ingredients in the first affine isoperimetric inequality of Section 9.2, the Busemann random simplex inequality, Theorem 9.2.6. The Busemann–Petty centroid inequality, Corollary 9.2.7, relating volumes of a convex body and its centroid body, follows easily. This in turn leads to the Petty projection inequality, Theorem 9.2.9, an upper bound for the volume of a convex body in terms of that of the polar body of its projection body. For a centered convex body, the Blaschke–Santaló inequality, Theorem 9.2.11, provides a similar upper bound in terms of its polar body, but requires different techniques for its proof.

These affine isoperimetric inequalities have application in the remaining two sections of the chapter. In Section 9.3, bounds are given for the volume of a convex body in terms of its width or brightness function. Most of these bounds are precise, and some – those of Theorems 9.3.1 and 9.3.2 for $p = -n$ – are affine invariant. This is significant, since it means, for example, that some of the best possible estimates apply for all ellipsoids, and of course many more objects can be closely approximated by ellipsoids than by balls. When only finitely many values of the brightness function are available, it is much easier to estimate the surface area than the volume.

The main result of Section 9.4 is Theorem 9.4.4, a lower bound for the volume in terms of volumes of concurrent sections. When these are hyperplane sections, this can be recast as the Busemann intersection inequality, Corollary 9.4.5, a lower bound for the volume of a body in terms of that of its intersection body. Finding good inequalities in the opposite direction is much harder. Theorem 9.4.11 supplies a precise upper bound for the volume of a centered convex body in $\mathbb{E}^3$ in terms of the maximal area of central sections, but in general the so-called slicing problem, discussed at some length in Note 9.8, remains unsolved.

9.1. Centroid bodies

The following definition introduces the centroid bodies. We shall develop enough of their properties to serve the purposes of the next section. For convenience, we work with compact sets, though bounded measurable sets would suffice in many instances.

Definition 9.1.1. Let C be a compact set in $\mathbb{E}^n$ with $\lambda_n(C) > 0$. The *centroid body* of C is the centered compact convex set ΓC such that

$$h_{\Gamma C}(u) = \frac{1}{\lambda_n(C)} \int_C |u \cdot x| \, dx,$$

for all $u \in S^{n-1}$.

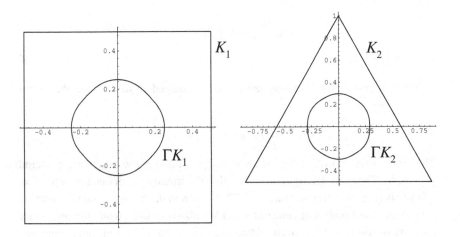

Figure 9.1. Centroid bodies.

It is easy to check that $h_{\Gamma C}$ is indeed a support function. Using (0.15), we see that when C is centered, bd ΓC is the locus of centroids of halves of C formed by slicing C by hyperplanes through the origin; hence the term "centroid body."

Generally, bd ΓC is the locus of points y_u, $u \in S^{n-1}$, obtained in the following way. Let J_u denote the closed half-space containing u and with boundary $u^\perp$, and let z_u be the centroid of $C \cap J_u$. Then y_u is the centroid of a mass $\lambda_n(C \cap J_u)/\lambda_n(C)$ at z_u and a mass $\lambda_n(C \cap J_{-u})/\lambda_n(C)$ at $-z_{-u}$.

Figure 9.1 shows the centroid bodies of the centered unit square and an equilateral triangle with centroid at o. It can be shown that the centroid body of a convex body is always of class C_+^2; see Note 9.1.

For star bodies containing the origin, the following alternative definition is available.

Lemma 9.1.2. *If L is a star body in $\mathbb{E}^n$ with $o \in L$, then*

$$h_{\Gamma L}(u) = \frac{1}{(n+1)\lambda_n(L)} \int_{S^{n-1}} \rho_L(v)^{n+1} |u \cdot v| \, dv,$$

for all $u \in S^{n-1}$.

Proof. Transforming to polar coordinates, we obtain

$$h_{\Gamma L}(u) = \frac{1}{\lambda_n(L)} \int_L |u \cdot x| \, dx$$

$$= \frac{1}{\lambda_n(L)} \int_{S^{n-1}} \int_0^{\rho_L(v)} |u \cdot tv| t^{n-1} \, dt \, dv$$

$$= \frac{1}{(n+1)\lambda_n(L)} \int_{S^{n-1}} \rho_L(v)^{n+1} |u \cdot v| \, dv,$$

for all $u \in S^{n-1}$.   ∎

The integral in the previous lemma is unchanged if we replace the factor $\rho_L(v)^{n+1}$ by

$$f(v) = \frac{1}{2} \left(\rho_L(v)^{n+1} + \rho_L(-v)^{n+1} \right).$$

Therefore every centroid body of a star body containing the origin is a centered zonoid, by Theorem 4.1.10 and the fact that the measure μ defined by $d\mu(v) = f(v)dv$ is an even Borel measure in S^{n-1}. (Theorem 9.1.5 shows that this remains true for centroid bodies of compact sets.) We also conclude, using the injectivity property of the cosine transform (Theorem C.2.1), that if L_1 and L_2 are centered star bodies in $\mathbb{E}^n$, then

$$\Gamma L_1 = \Gamma L_2 \Rightarrow L_1 = L_2.$$

Theorem 9.1.3. *The centroid bodies of linearly equivalent compact sets are linearly equivalent. In fact, if $\phi \in GL_n$, then*

$$\Gamma(\phi C) = \phi(\Gamma C).$$

Proof. Let $u \in S^{n-1}$. Then, using (0.7), (0.1), and the formula (0.27) for the change in the support function, we obtain

$$h_{\Gamma(\phi C)}(u) = \frac{1}{\lambda_n(\phi C)} \int_{\phi C} |u \cdot x| \, dx$$

$$= \frac{1}{|\det \phi| \lambda_n(C)} \int_C |u \cdot \phi y| |\det \phi| \, dy \quad \text{(where } y = \phi^{-1} x)$$

$$= \frac{1}{\lambda_n(C)} \int_C |\phi^t u \cdot y| \, dy$$

$$= h_{\Gamma C}(\phi^t u) = h_{\phi(\Gamma C)}(u).$$

This completes the proof. ∎

Corollary 9.1.4. *The centroid body of a centered n-dimensional ellipsoid is a centered n-dimensional ellipsoid. In fact,*

$$\Gamma E = \frac{2\kappa_{n-1}}{(n+1)\kappa_n} E,$$

for each such ellipsoid E in $\mathbb{E}^n$. Conversely, every centered n-dimensional ellipsoid is the centroid body of one of its dilatates.

Proof. Using Lemma 9.1.2, we see that for each $u \in S^{n-1}$, the support function of the centroid body ΓB of the unit ball satisfies

$$h_{\Gamma B}(u) = \frac{1}{(n+1)\kappa_n} \int_{S^{n-1}} |u \cdot v| \, dv = \frac{2h_{\Pi B}(u)}{(n+1)\kappa_n} = \frac{2\kappa_{n-1}}{(n+1)\kappa_n},$$

proving that

$$\Gamma B = \frac{2\kappa_{n-1}}{(n+1)\kappa_n} B.$$

(Here, we used Definition 4.1.1 of the projection body ΠB of B.)

Let E be a centered n-dimensional ellipsoid in $\mathbb{E}^n$, so that $E = \phi B$, where $\phi \in GL_n$. Then, by Theorem 9.1.3,

$$\Gamma E = \Gamma(\phi B) = \phi(\Gamma B) = \frac{2\kappa_{n-1}}{(n+1)\kappa_n} \phi B = \frac{2\kappa_{n-1}}{(n+1)\kappa_n} E.$$

This establishes the first statement, from which the second follows. ∎

We need a formula for the volume of the centroid body. To state this, some convenient notation will be introduced. Let $1 \le k \le n$. We shall denote the k-dimensional simplex in $\mathbb{E}^n$ with vertices at $o, p_1, \ldots, p_k$ by $[o, p_1, \ldots, p_k]$. Suppose that C is a compact set contained in some $S \in \mathcal{G}(n, k)$. We define, for each $m \in \mathbb{N}$,

$$g_{m,k}(C) = \int_C \cdots \int_C \lambda_k([o, p_1, \ldots, p_k])^m \, dp_1 \cdots dp_k.$$

Here, the jth integration is over all $p_j \in C$, according to our usual convention. When $\lambda_k(C) > 0$, the quantity $g_{m,k}(C)$, when divided by $\lambda_k(C)^{km}$, is the average of mth powers of volumes of simplices with one vertex at o and other vertices in C. If $k = n$, we drop the second index and write

$$g_m(C) = \int_C \cdots \int_C \lambda_n([o, p_1, \ldots, p_n])^m \, dp_1 \cdots dp_n.$$

Theorem 9.1.5. *Let C be a compact set in $\mathbb{E}^n$ with $\lambda_n(C) > 0$. Then ΓC is a centered zonoid, with volume*

$$\lambda_n(\Gamma C) = \frac{2^n}{\lambda_n(C)^n} g_1(C).$$

Proof. Let $\varepsilon > 0$ be given. We wish to approximate the integral in Definition 9.1.1 by a Riemann sum. Let $\{E_1, \ldots, E_m\}$ be a partition of C into nonempty disjoint Borel sets of diameters less than ε. As in the first paragraph of the proof of Theorem 4.1.10, for $p_j \in E_j$, $1 \le j \le m$, we have

$$\left| h_{\Gamma C}(u) - \frac{1}{\lambda_n(C)} \sum_{j=1}^m \lambda_n(E_j) |u \cdot p_j| \right| < \varepsilon,$$

for all $u \in S^{n-1}$. The polytope Z_ε such that

$$h_{Z_\varepsilon}(u) = \frac{1}{\lambda_n(C)} \sum_{j=1}^m \lambda_n(E_j) |u \cdot p_j|,$$

for $u \in S^{n-1}$, is, by (0.23) and (0.25), a centered zonotope; specifically, it is a dilatate by a factor of $1/\lambda_n(C)$ of the Minkowski sum P of the line segments $s_j = \lambda_n(E_j)[-p_j, p_j]$, $1 \le j \le m$. As $\varepsilon \to 0$, Z_ε converges to ΓC by (0.22), so ΓC is a centered zonoid.

Using Minkowski's theorem on mixed volumes, Theorem A.3.1(i), we obtain

$$\lambda_n(P) = \sum_{i_1=1}^m \cdots \sum_{i_n=1}^m V(s_{i_1}, \ldots, s_{i_n}),$$

where P is as before. By the polarization formula (A.8), and the formulas (0.12) and (0.13) for the volumes of a parallelepiped and simplex,

$$V(s_{i_1}, \ldots, s_{i_n}) = \frac{1}{n!} \lambda_n(s_{i_1} + \cdots + s_{i_n})$$

$$= 2^n \lambda_n([o, p_{i_1}, \ldots, p_{i_n}]) \lambda_n(E_{i_1}) \cdots \lambda_n(E_{i_n}).$$

Consequently,

$$\lambda_n(Z_\varepsilon) = \lambda_n\left(\frac{1}{\lambda_n(C)}P\right)$$

$$= \frac{2^n}{\lambda_n(C)^n}\sum_{i_1=1}^{m}\cdots\sum_{i_n=1}^{m}\lambda_n([o, p_{i_1}, \ldots, p_{i_n}])\lambda_n(E_{i_1})\cdots\lambda_n(E_{i_n}).$$

Since Z_ε converges to ΓC as $\varepsilon \to 0$, $\lambda_n(Z_\varepsilon)$ converges to $\lambda_n(\Gamma C)$. On the other hand, the previous equation shows that as $\varepsilon \to 0$, $\lambda_n(Z_\varepsilon)$ also converges to

$$\frac{2^n}{\lambda_n(C)^n}\int_C\cdots\int_C\lambda_n([o, p_1, \ldots, p_n])\,dp_1\cdots dp_n = \frac{2^n}{\lambda_n(C)^n}g_1(C),$$

and this completes the proof.   ∎

9.2. Some affine isoperimetric inequalities

We shall use several affine isoperimetric inequalities in the sections following this one. Our first goal here is to prove the Busemann random simplex inequality. This requires some subsidiary results of interest in their own right.

Theorem 9.2.1. *Let C be a compact subset of $\mathbb{E}^n$ with $\lambda_n(C) > 0$. There exists a unique centered n-dimensional ellipsoid, called the* Löwner *ellipsoid, of least volume containing C.*

Proof. The proof that there is a centered n-dimensional ellipsoid, containing C and of minimal volume, closely follows the first paragraph of the proof of Theorem 4.2.12, concerning the John ellipsoid. It remains to show that the Löwner ellipsoid is unique.

Suppose that E_j, $j = 1, 2$, are centered n-dimensional ellipsoids containing C and of minimal volume $a > 0$, say. Let $\phi \in SL_n$ be such that with a suitable choice of axes, $E_1' = \phi E_1$ is the ball $\{x : \sum_i x_i^2 \leq b^2\}$ and $E_2' = \phi E_2$ is the centered ellipsoid $\{x : \sum_i x_i^2/a_i^2 \leq 1\}$. Then using the formulas (0.9) and (0.11) for the volumes of the unit ball and an ellipsoid and the fact that ϕ preserves volume, we obtain

$$a = b^n\kappa_n = a_1 a_2 \cdots a_n \kappa_n.$$

Furthermore, each centered ellipsoid that contains ϕC has volume at least a. Since $\phi C \subset E_j'$, $j = 1, 2$, we also have $\sum_i x_i^2 \leq b^2$ and $\sum_i x_i^2/a_i^2 \leq 1$ for all $x \in \phi C$. Let E_3' be the centered ellipsoid

$$E_3' = \left\{x : \sum_i x_i^2(b^{-2} + a_i^{-2})/2 \leq 1\right\}.$$

Each $x \in \phi C$ satisfies the latter inequality, so $\phi C \subset E_3'$. This implies that $\lambda_n(E_3') \geq a$. Moreover

$$\lambda_n(E_3') = \kappa_n \prod_{i=1}^{n} \sqrt{2} b a_i (b^2 + a_i^2)^{-1/2}$$

$$\leq \kappa_n \prod_{i=1}^{n} (b a_i)^{1/2} = \lambda_n(E_1')^{1/2} \lambda_n(E_2')^{1/2} = a,$$

where we used the arithmetic–geometric mean inequality (B.1). Therefore $\lambda_n(E_3') = a$, and equality holds in the arithmetic–geometric mean inequality, yielding $b = a_i$ for $1 \leq i \leq n$. Consequently $E_1' = E_2'$, so $E_1 = E_2$. ∎

Lemma 9.2.2. *Let $u \in S^{n-1}$, and suppose that K is a convex body in $\mathbb{E}^n$. Then for each $m \in \mathbb{N}$,*

$$g_m(K) \geq g_m(S_u K).$$

Equality holds if and only if the midpoints of all chords of K parallel to u lie in a hyperplane containing the origin.

Proof. $S_u K$ is the Steiner symmetral of K in the direction u (cf. Definition 2.1.3). Let $p_j \in S_u K$, $1 \leq j \leq n$. For each j, there is a unique $y_j \in u^{\perp}$ and $s_j \in \mathbb{R}$ such that $p_j = y_j + s_j u$, and a unique $q_j = y_j + t_j u \in K$ transforming to p_j by the process of Steiner symmetrization in $u^{\perp}$. Let z_j be the center of the line segment $K \cap (l_u + y_j)$ containing q_j, and let $q_j' = y_j + t_j' u$ be the reflection of q_j in z_j parallel to u, so $s_j = (t_j - t_j')/2$, $1 \leq j \leq n$.

We may assume that u is parallel to the x_n-axis, so that $u^{\perp}$ is the hyperplane $x_n = 0$. Let us denote the kth component of p_j by p_{jk}, and similarly for other vectors. By (0.13),

$$\lambda_n([o, p_1, \ldots, p_n]) = \left| \det(p_{jk}) \right| / n!,$$

where $p_{jk} = y_{jk}$ for $1 \leq k \leq n-1$ and $p_{jn} = s_j$. Similarly,

$$\lambda_n([o, q_1, \ldots, q_n]) = \left| \det(q_{jk}) \right| / n!,$$

where $q_{jk} = y_{jk}$ for $1 \leq k \leq n-1$ and $q_{jn} = t_j$, and the corresponding expression holds with q_j and t_j replaced by q_j' and t_j', respectively. By the relation $s_j = (t_j - t_j')/2$ and the multilinearity of determinants,

$$\lambda_n([o, p_1, \ldots, p_n]) = \frac{1}{2n!} \left| \det(q_{jk}) - \det(q_{jk}') \right|$$

$$\leq \frac{1}{2} \left(\lambda_n([o, q_1, \ldots, q_n]) + \lambda_n([o, q_1', \ldots, q_n']) \right).$$

Jensen's inequality (B.3) for means implies that

$$\lambda_n\left([o, p_1, \ldots, p_n]\right)^m \le \frac{1}{2}\lambda_n\left([o, q_1, \ldots, q_n]\right)^m + \frac{1}{2}\lambda_n\left([o, q_1', \ldots, q_n']\right)^m.$$

Now the equations

$$g_m(S_u K) = \int_{S_u K} \cdots \int_{S_u K} \lambda_n\left([o, p_1, \ldots, p_n]\right)^m dp_1 \cdots dp_n$$

and

$$g_m(K) = \int_K \cdots \int_K \lambda_n\left([o, q_1, \ldots, q_n]\right)^m dq_1 \cdots dq_n$$

$$= \int_K \cdots \int_K \lambda_n\left([o, q_1', \ldots, q_n']\right)^m dq_1' \cdots dq_n'$$

yield the required inequality.

Suppose that equality holds. Then the continuity of the volume of a simplex as a function of its vertices forces equality in the previous inequalities for each choice of p_j, $1 \le j \le n$. This in turn implies that $\det(q_{jk}) = -\det(q_{jk}')$. But

$$\lambda_n\left([o, z_1, \ldots, z_n]\right) = \left|\det(z_{jk})\right|/n!,$$

where $z_{jk} = y_{jk}$ for $1 \le k \le n-1$ and $z_{jn} = (t_j + t_j')/2$. Therefore $\lambda_n([o, z_1, \ldots, z_n]) = 0$, meaning that the points z_j, $1 \le j \le n$, lie in a hyperplane containing the origin. ∎

The following lemma is very well known; see [52, Theorem 9.13.6] for a clear (and illustrated) proof.

Lemma 9.2.3. *Let K be a convex body in $\mathbb{E}^n$. There is a sequence of directions $u_m \in S^{n-1}$, $m \in \mathbb{N}$, such that the successive Steiner symmetrals $K_m = S_{u_m}S_{u_{m-1}} \cdots S_{u_1} K$ converge to a centered ball.*

Theorem 9.2.4. *Let K be a convex body in $\mathbb{E}^n$, and suppose that the sequence of directions $u_m \in S^{n-1}$, $m \in \mathbb{N}$, is as in the previous lemma. If for each $m \in \mathbb{N}$, the midpoints of all chords of K parallel to u_m lie in a hyperplane containing the origin, then K is a centered ellipsoid.*

Proof. Let $u \in S^{n-1}$ be such that the midpoints of all chords of K parallel to u lie in a hyperplane S containing o. We claim that there is a $\phi \in SL_n$ such that $S_u K = \phi K$. To see this, let $x \in \mathbb{E}^n$, so $x = y + su$, where $y \in u^\perp$ and $s \in \mathbb{R}$, and suppose that z is the unique point in S for which $z = y + tu$, $t \in \mathbb{R}$. Let ϕ be defined by $\phi x = x - (z - y)$. Then ϕ is linear, since y is the orthogonal projection of x on $u^\perp$, and z is a linear function of y; moreover, $\phi \in SL_n$, by the Cavalieri principle, Lemma 1.2.2, and the fact that ϕ is just a translation in each line parallel to u. From the definition of $S_u K$, it is clear that $\phi K = S_u K$.

Let $m \in \mathbb{N}$. Since K_m is obtained from K by a finite sequence of Steiner symmetrizations in directions having the property of the previous paragraph, it follows that K_m can be obtained from K by a finite sequence of volume-preserving linear transformations, and hence by one such transformation, ϕ_m, say.

Suppose that the sets K_m, $m \in \mathbb{N}$, converge to rB, $r > 0$. Then $\lambda_n(rB) = \lambda_n(K)$. Let E be the Löwner ellipsoid of K, as in Theorem 9.2.1. Then $\lambda_n(\phi_m E) = \lambda_n(E)$, and $\phi_m E$ is the Löwner ellipsoid of K_m. The ellipsoids $\phi_m E$ converge to the Löwner ellipsoid of rB, just rB itself. Therefore $\lambda_n(E) = \lambda_n(rB) = \lambda_n(K)$, and since $K \subset E$, this implies that $K = E$ is a centered ellipsoid. ∎

Corollary 9.2.5. *Let K be a convex body in $\mathbb{E}^n$. Then for each $m \in \mathbb{N}$,*

$$g_m(K) \geq g_m(rB),$$

where $\lambda_n(K) = \lambda_n(rB)$, with equality if and only if K is a centered ellipsoid.

Proof. By Lemma 9.2.3, there is a sequence of directions such that the successive Steiner symmetrals of K converge to a centered ball rB. Now $\lambda_n(K) = \lambda_n(rB)$, because each symmetral has the same volume as K, by the Cavalieri principle, Lemma 1.2.2. The inequality follows from Lemma 9.2.2 applied to each symmetrization in the sequence. If equality holds, we must have equality for each such symmetrization, and then Theorem 9.2.4 and the equality condition from Lemma 9.2.2 imply that K is a centered ellipsoid. ∎

Theorem 9.2.6 (Busemann random simplex inequality). *Let K be a convex body in $\mathbb{E}^n$. Then*

$$\frac{1}{\lambda_n(K)^n} g_1(K) \geq \left(\frac{\kappa_{n-1}}{(n+1)\kappa_n}\right)^n \lambda_n(K),$$

with equality if and only if K is a centered ellipsoid.

Proof. A dilatation of K by a factor $r > 0$ increases each side of the inequality by a factor of r^n, so we may assume that $\lambda_n(K) = \kappa_n$. Theorem 9.1.5, Corollary 9.1.4, and (0.8) then imply that

$$g_1(B) = \frac{\kappa_n^n}{2^n} \lambda_n(\Gamma B) = \frac{\kappa_n^n}{2^n} \lambda_n \left(\frac{2\kappa_{n-1}}{(n+1)\kappa_n} B\right) = \frac{\kappa_{n-1}^n \kappa_n}{(n+1)^n}.$$

Therefore, by Corollary 9.2.5 with $r = 1$,

$$\frac{1}{\lambda_n(K)^n} g_1(K) \geq \frac{1}{\kappa_n^n} g_1(B) = \frac{\kappa_{n-1}^n \kappa_n}{(n+1)^n \kappa_n^n} = \left(\frac{\kappa_{n-1}}{(n+1)\kappa_n}\right)^n \lambda_n(K),$$

with equality if and only if K is a centered ellipsoid. ∎

The previous inequality is often quoted in the form

$$\frac{1}{\lambda_n(K)^n} g_1(K) \geq \frac{2\kappa_{n+1}^{n-1}}{(n+1)!\kappa_n^{n+1}} \lambda_n(K).$$

To see that this is equivalent, observe that from (0.9) it follows that

$$\kappa_{n+1} = \frac{2\pi\kappa_{n-1}}{(n+1)} \quad \text{and} \quad \kappa_n\kappa_{n-1} = \frac{2^n\pi^{n-1}}{n!}.$$

Consequently,

$$\frac{2\kappa_{n+1}^{n-1}}{(n+1)!\kappa_n^{n+1}} = \left(\frac{\kappa_{n-1}}{(n+1)\kappa_n}\right)^n \left(\frac{2^n\pi^{n-1}}{n!\kappa_n\kappa_{n-1}}\right) = \left(\frac{\kappa_{n-1}}{(n+1)\kappa_n}\right)^n.$$

Corollary 9.2.7 (Busemann–Petty centroid inequality). *Let K be a convex body in $\mathbb{E}^n$. Then*

$$\lambda_n(\Gamma K) \geq \left(\frac{2\kappa_{n-1}}{(n+1)\kappa_n}\right)^n \lambda_n(K),$$

with equality if and only if K is a centered ellipsoid.

Proof. This follows directly from Theorems 9.2.6 and 9.1.5. ∎

We shall use the previous corollary to obtain another important inequality, the Petty projection inequality.

If K is a convex body in $\mathbb{E}^n$, we denote the polar body of the projection body of K by Π^*K and refer to this as the *polar projection body* of K.

Lemma 9.2.8. *Let K be a convex body and L a star body in $\mathbb{E}^n$ with $o \in L$. Then*

$$V(K, n-1; \Gamma L) = \frac{2}{(n+1)\lambda_n(L)} \tilde{V}_{-1}(L, \Pi^*K).$$

Proof. By (A.56) with $i = -1$, the polar relation (0.36) between support and radial function, Definition 4.1.1 of projection body, Lemma 9.1.2, and (A.32) with $i = n-1$, we have

$$\frac{2}{(n+1)\lambda_n(L)} \tilde{V}_{-1}(L, \Pi^*K)$$

$$= \frac{2}{n(n+1)\lambda_n(L)} \int_{S^{n-1}} \rho_L(u)^{n+1} \rho_{\Pi^*K}(u)^{-1} \, du$$

$$= \frac{2}{n(n+1)\lambda_n(L)} \int_{S^{n-1}} \rho_L(u)^{n+1} h_{\Pi K}(u) \, du$$

$$= \frac{1}{n(n+1)\lambda_n(L)} \int_{S^{n-1}} \int_{S^{n-1}} \rho_L(u)^{n+1} |u \cdot v| \, dS(K, v) \, du$$

$$= \frac{1}{n} \int_{S^{n-1}} \frac{1}{(n+1)\lambda_n(L)} \int_{S^{n-1}} \rho_L(u)^{n+1} |u \cdot v| \, du \, dS(K, v)$$

$$= \frac{1}{n} \int_{S^{n-1}} h_{\Gamma L}(v) dS(K, v)$$

$$= V(K, n-1; \Gamma L).　\blacksquare$$

Theorem 9.2.9 (Petty projection inequality). *Let K be a convex body in $\mathbb{E}^n$. Then*

$$\lambda_n(K)^{n-1} \lambda_n(\Pi^* K) \leq \left(\frac{\kappa_n}{\kappa_{n-1}} \right)^n,$$

with equality if and only if K is an ellipsoid.

Proof. If we put $L = \Pi^* K$ in Lemma 9.2.8, we get

$$V(K, n-1; \Gamma(\Pi^* K)) = \frac{2}{n+1}.$$

We use Corollary 9.2.7 with K replaced by $\Pi^* K$, Minkowski's first inequality (B.13), and the last equation, to obtain

$$\lambda_n(K)^{n-1} \lambda_n(\Pi^* K) \leq \left(\frac{(n+1)\kappa_n}{2\kappa_{n-1}} \right)^n \lambda_n(K)^{n-1} \lambda_n(\Gamma(\Pi^* K))$$

$$\leq \left(\frac{(n+1)\kappa_n}{2\kappa_{n-1}} \right)^n V(K, n-1; \Gamma(\Pi^* K))^n$$

$$= \left(\frac{(n+1)\kappa_n}{2\kappa_{n-1}} \right)^n \left(\frac{2}{n+1} \right)^n = \left(\frac{\kappa_n}{\kappa_{n-1}} \right)^n,$$

as required. The equality condition follows from those of Corollary 9.2.7 and Minkowski's first inequality.　$\blacksquare$

Remark 9.2.10. (i) Corollary 9.2.5 actually holds when K is an arbitrary compact set, and this allows the Busemann random simplex inequality (Theorem 9.2.6) and Busemann–Petty centroid inequality (Corollary 9.2.7) to be extended to compact sets. For star bodies containing the origin, a proof of the latter can be obtained directly from the Petty projection inequality, with the help of Lemma 9.2.8 and inequality (B.24). See Note 9.4.

(ii) The following reverse form of the Petty projection inequality is called the *Zhang projection inequality*: If K is a convex body in $\mathbb{E}^n$, then

$$\lambda_n(K)^{n-1} \lambda_n(\Pi^* K) \geq \frac{1}{n^n} \binom{2n}{n},$$

with equality if and only if K is a simplex.

Our next aim is to prove the Blaschke–Santaló inequality for centered convex bodies. Unfortunately, there does not seem to be a very easy proof of the equality condition. The one presented here appeals to Theorem 7.1.10, the false center theorem (obtained from Theorem 7.1.7, whose proof we omitted). See Note 9.4 for further comments.

Theorem 9.2.11 (Blaschke–Santaló inequality for centered convex bodies). *Let K be a centered convex body in $\mathbb{E}^n$. Then*

$$\lambda_n(K)\lambda_n(K^*) \leq \kappa_n^2,$$

with equality if and only if K is an ellipsoid.

Proof. When $n = 2$, the Petty projection inequality, Theorem 9.2.9, for a centered convex body K is precisely the inequality in the statement of the theorem. Moreover, the equality condition also follows from that of Theorem 9.2.9. We may therefore assume that $n \geq 3$.

If $r > 0$, then it follows from the definition (0.35) of the polar body that $(rK)^* = r^{-1}K^*$, and then (0.8) implies that the quantity $\lambda_n(K)\lambda_n(K^*)$ is unaffected by dilatation. Therefore we may assume that $\lambda_n(K) = \kappa_n$. Suppose that we can prove that for all $u \in S^{n-1}$,

$$\lambda_n(K^*) \leq \lambda_n(S_u^* K), \tag{9.1}$$

where $S_u^* K = (S_u K)^*$ denotes the polar body of $S_u K$. The Cavalieri principle, Lemma 1.2.2, tells us that $\lambda_n(S_u K) = \lambda_n(K)$ for each $u \in S^{n-1}$, so by Lemma 9.2.3,

$$\lambda_n(K)\lambda_n(K^*) \leq \lambda_n(B)\lambda_n(B^*) = \kappa_n^2.$$

Let us prove (9.1). We may assume that u is parallel to the x_n-axis, so that $u^{\perp}$ is the hyperplane $x_n = 0$. Now

$$S_u K = \left\{ (y, s) : y \in u^{\perp}, s = \frac{s_1 - s_2}{2}, (y, s_j) \in K, j = 1, 2 \right\},$$

where $(y, s) = y + su$. Also,

$$K^* = \{(z, t) : (z + tu) \cdot x \leq 1 \text{ for all } x \in K\}$$

$$= \{(z, t) : y \cdot z + st \leq 1 \text{ for all } (y, s) \in K\},$$

where $z \in u^{\perp}$ and $t \in \mathbb{R}$. Therefore

$$S_u^* K = \left\{ (z, t) : y \cdot z + \left(\frac{s_1 - s_2}{2} \right) t \leq 1 \text{ for all } (y, s_j) \in K, j = 1, 2 \right\}.$$

If E is a set and $t \in \mathbb{R}$, let $E_t = \{z \in u^{\perp} : (z, t) \in E\}$. Let $t \in \mathbb{R}$, and let

$$z \in \frac{1}{2}K_t^* + \frac{1}{2}K_{-t}^* = \frac{1}{2}K_t^* + \frac{1}{2}(-K^*)_t = \Delta(K_t^*),$$

where $K_t^* = (K^*)_t$ and we used the fact that K, and hence K^*, is centered to replace K_{-t}^* by $(-K^*)_t$. Then $z = (p+q)/2$, where $p \in K_t^*$ and $q \in K_{-t}^*$. Let $(y, s_j) \in K$, $j = 1, 2$. We have

$$y \cdot z + \left(\frac{s_1 - s_2}{2}\right)t = y \cdot \left(\frac{p+q}{2}\right) + \left(\frac{s_1 - s_2}{2}\right)t$$

$$= \frac{1}{2}(y \cdot p + s_1 t) + \frac{1}{2}(y \cdot q - s_2 t) \le 1.$$

Consequently,

$$\Delta(K_t^*) \subset (S_u^* K)_t,$$

for each $t \in \mathbb{R}$. By Theorem 3.2.3, we obtain

$$\lambda_{n-1}(K_t^*) \le \lambda_{n-1}(\Delta(K_t^*)) \le \lambda_{n-1}((S_u^* K)_t). \tag{9.2}$$

Integration with respect to t yields (9.1) and establishes the inequality.

Suppose that equality holds in the statement of the theorem. Then equality holds in (9.1) for all $u \in S^{n-1}$, and therefore also in (9.2) for all $u \in S^{n-1}$ and $t \in \mathbb{R}$. By the equality condition in Theorem 3.2.3, every section of K^* by a hyperplane is centrally symmetric. The false center theorem, Theorem 7.1.10, implies that K^* is an ellipsoid, so (as we noted in Section 0.8) K is also an ellipsoid. ∎

The Blaschke–Santaló inequality is actually valid for any convex body whose centroid is at the origin; see Note 9.4.

9.3. Volume estimates from projection functions

As motivation, let us consider the following situation. We have been able to find exact measurements of the volumes of all projections of a convex body K in $\mathbb{E}^n$ on hyperplanes through the origin; that is, we know the brightness function of K. What is the best way to estimate the volume of the body? Note that if K is centrally symmetric, a satisfactory reconstruction algorithm is available, even if the measurements are finite in number and noisy (see Section 4.4), but here we do not assume that K is centrally symmetric.

Such an estimate can be made, by the following method. Cauchy's surface area formula (A.49) states that

$$S(K) = \frac{1}{\kappa_{n-1}} \int_{S^{n-1}} \lambda_{n-1}(K|u^\perp) \, du,$$

where $S(K)$ is the surface area of K. Rearranging the isoperimetric inequality (B.14), we obtain

$$\lambda_n(K) \le \kappa_n \left(\frac{S(K)}{\omega_n} \right)^{n/(n-1)}.$$

A combination of these two expressions yields an upper bound for the volume of K. For reasons that will become clear later, we shall use (0.10) to write this in the form

$$\kappa_n^{1/n} \lambda_n(K)^{(n-1)/n} \le \frac{1}{n\kappa_{n-1}} \int_{S^{n-1}} \lambda_{n-1}(K|u^\perp)\,du. \tag{9.3}$$

It turns out that a considerable improvement can be made to (9.3), which is suggested by translating it into different notation. Suppose that $1 \le i \le n$. With $X = \mathcal{G}(n,i)$, and $p \in \mathbb{R}$, $p \ne 0$, the definition of the pth mean of a function $f \in C(X)$ is

$$M_p(f) = \left(\int_{\mathcal{G}(n,i)} f(S)^p\,dS \right)^{1/p}.$$

In particular, when $i = 1$ or $n-1$, this becomes

$$M_p(f) = \left(\frac{1}{n\kappa_n} \int_{S^{n-1}} f(u)^p\,du \right)^{1/p}.$$

Recall also that $M_\infty(f) = \max f$ and $M_{-\infty}(f) = \min f$. Suppose that K is a convex body in $\mathbb{E}^n$. If $S \in \mathcal{G}(n,i)$, we shall write

$$v_{i,K}(S) = \lambda_i(K|S)$$

for the ith projection function of K as defined in Chapter 3. If $i = 1$ or $n-1$, we identify $\mathcal{G}(n,i)$ with S^{n-1} and consider the width function $v_{1,K} = w_K$ and brightness function $v_{n-1,K} = v_K$ as functions of $u \in S^{n-1}$.

We can now reformulate the estimate (9.3) as follows:

$$\kappa_n^{1/n} \lambda_n(K)^{(n-1)/n} \le \frac{\kappa_n}{\kappa_{n-1}} M_1(v_K). \tag{9.4}$$

We put the inequality in this form because the quantity on the right is just $W_1(K)$, the first quermassintegral of K, by (A.26) and (A.49). Referring to Jensen's inequality (B.8) for integrals, we notice that a stronger upper bound for $\lambda_n(K)$ will result if the first mean in (9.4) can be replaced by the pth mean for some $p < 1$. With the help of the Petty projection inequality, this is accomplished by the next theorem.

Theorem 9.3.1. *Suppose that K is a convex body in $\mathbb{E}^n$ and that $p \geq -n$. Then*

$$\frac{1}{n}\left(\frac{2n}{n}\right)^{1/n} M_{-n}(v_K) \leq \kappa_n^{1/n}\lambda_n(K)^{(n-1)/n} \leq \frac{\kappa_n}{\kappa_{n-1}}M_p(v_K).$$

If $p > -n$, equality holds on the right if and only if K is a ball, whereas if $p = -n$, equality holds on the right if and only if K is an ellipsoid. Equality holds on the left if and only if K is a simplex.

Proof. Using the polar coordinate formula (A.58) for volume, the polar relation (0.36) between support and radial function, and Definition 4.1.1 of a projection body, we see that

$$\lambda_n(\Pi^*K) = \frac{1}{n}\int_{S^{n-1}} \rho_{\Pi^*K}(u)^n \, du$$

$$= \frac{1}{n}\int_{S^{n-1}} h_{\Pi K}(u)^{-n} \, du$$

$$= \frac{1}{n}\int_{S^{n-1}} v_K(u)^{-n} \, du$$

$$= \kappa_n M_{-n}(v_K)^{-n}.$$

The Petty projection inequality (Theorem 9.2.9) now gives

$$\lambda_n(K)^{(n-1)/n} \leq \frac{\kappa_n}{\kappa_{n-1}}\lambda_n(\Pi^*K)^{-1/n}$$

$$= \frac{\kappa_n^{(n-1)/n}}{\kappa_{n-1}}M_{-n}(v_K).$$

This establishes the right-hand inequality in the statement of the theorem for $p = -n$, and the corresponding equality condition follows from that of Theorem 9.2.9. Jensen's inequality (B.8) for integrals yields the right-hand inequality for all $p > -n$.

Suppose that equality holds in the statement of the theorem for some $q > -n$. By Jensen's inequality for integrals again, we see that $M_q(v_K) = M_{-n}(v_K)$, that equality holds also for $p = -n$, and that v_K is therefore constant. This means that K is both an ellipsoid and of constant brightness, and so, by Theorem 3.3.11, K is a ball.

The same method provides the left-hand inequality and its equality condition, when the Zhang projection inequality of Remark 9.2.10(ii) is used instead of the Petty projection inequality. ∎

The right-hand inequality in the previous theorem does not hold for $p < -n$. To see this, let K be any nonspherical ellipsoid. If the inequality held for some $p < -n$, we could use the previous argument for the equality condition to obtain a contradiction.

The upper bound provided by Theorem 9.3.1 when $p = -n$ not only is a strengthening of (9.4), but also has the tremendous advantage of being affine invariant; equality holds if and only if K is an ellipsoid, whereas equality in (9.4) holds, as in the isoperimetric inequality, if and only if K is a ball.

Suppose now that we have measured the width function of K instead of the brightness function, and wish to derive a similar estimate. The case $i = 1$ of the extended isoperimetric inequality (B.21) and the formula (A.50) for mean width yield the following upper bound for the volume of K:

$$\kappa_n^{(n-1)/n} \lambda_n(K)^{1/n} \leq \frac{\kappa_n}{2} M_1(w_K).$$

Here, the quantity on the right is $W_{n-1}(K)$, the $(n-1)$th quermassintegral of K, by (A.26) and (A.50). Once again, however, a significant improvement is possible, this time by means of the Blaschke–Santaló inequality.

Theorem 9.3.2. *Suppose that K is a convex body in $\mathbb{E}^n$ and that $p \geq -n$. Then*

$$\kappa_n^{(n-1)/n} \lambda_n(K)^{1/n} \leq \frac{\kappa_n}{2} M_p(w_K).$$

If $p > -n$, then equality holds if and only if K is a ball, whereas if $p = -n$, equality holds if and only if K is an ellipsoid.

Proof. We shall denote the polar body of the central symmetral of K by $\triangle^* K$. Using the formula (A.58) for volume in polar coordinates, the polar relation (0.36) between support and radial function, and Definition 3.2.2 of the central symmetral, we deduce that

$$\lambda_n(\triangle^* K) = \frac{1}{n} \int_{S^{n-1}} \rho_{\triangle^* K}(u)^n \, du$$

$$= \frac{1}{n} \int_{S^{n-1}} h_{\triangle K}(u)^{-n} \, du$$

$$= \frac{1}{n} \int_{S^{n-1}} \left(\frac{w_K(u)}{2} \right)^{-n} \, du$$

$$= 2^n \kappa_n M_{-n}(w_K)^{-n}.$$

Now using Theorem 3.2.3 and the Blaschke–Santaló inequality for centered convex bodies, Theorem 9.2.11, applied to $\triangle K$, we obtain

$$\lambda_n(K)^{1/n} \leq \lambda_n(\triangle K)^{1/n} \leq \kappa_n^{2/n} \lambda_n(\triangle^* K)^{-1/n} = \frac{\kappa_n^{1/n}}{2} M_{-n}(w_K),$$

supplying the inequality in the statement of the theorem for $p = -n$. The equality condition follows from that of the Blaschke–Santaló inequality. Jensen's inequality (B.8) for integrals implies that the inequality in the statement of the theorem holds for all $p > -n$.

Suppose that equality holds in the statement of the theorem for some $q > -n$. Then Jensen's inequality for integrals shows that $M_q(w_K) = M_{-n}(w_K)$, that equality holds also for $p = -n$, and that w_K is therefore constant. This means that K is both an ellipsoid and of constant width, and so, by Theorem 3.2.7, K is a ball.   ∎

Again, one can use the preceding argument for the equality condition to show that the inequality in the previous theorem does not hold for $p < -n$ when K is a nonspherical ellipsoid.

Theorem 9.3.2 only gives an upper bound for volume; a lower bound was not stated, since the best estimate is not known. However, the method of Theorem 9.3.2 certainly yields a lower bound for volume. To compute it, one requires two inequalities. The first is the Rogers–Shephard inequality mentioned in Note 3.5, a reverse form of the inequality of Theorem 3.2.3 between the volume of a convex body and that of its central symmetral. The second is a reverse form of the Blaschke–Santaló inequality for centrally symmetric convex bodies. Inequalities of this type are known, but the conjectured optimal inequality has not yet been established; see Problem 9.2 and Note 9.4.

At this stage it is natural to consider the general case. The following lower bound for volumes of zonoids is essentially equivalent to the case $i = k$ of Corollary 4.2.7.

Theorem 9.3.3. *Let $1 \le i \le n - 1$, and suppose that K_1 is a convex body and that K_2 is a zonoid in $\mathbb{E}^n$. Then*

$$\left(\frac{\lambda_n(K_2)}{\lambda_n(K_1)}\right)^{i/n} \ge \min_{S \in \mathcal{G}(n,i)} \frac{v_{i,K_2}(S)}{v_{i,K_1}(S)},$$

with equality if and only if K_1 and K_2 are homothetic.

Proof. If $\dim K_2 < n$, the inequality is true, since the right-hand side is zero. If not, we can choose $r > 0$ so that

$$\min_{S \in \mathcal{G}(n,i)} \frac{v_{i,rK_2}(S)}{v_{i,K_1}(S)} = 1.$$

Then $\lambda_i(K_1|S) \le \lambda_i(rK_2|S)$, for all $S \in \mathcal{G}(n,i)$, so by Corollary 4.2.7 with $i = k$, we have $\lambda_n(K_1) \le \lambda_n(rK_2)$. Therefore, using (0.8),

$$\left(\frac{\lambda_n(K_2)}{\lambda_n(K_1)}\right)^{i/n} = \frac{1}{r^i}\left(\frac{\lambda_n(rK_2)}{\lambda_n(K_1)}\right)^{i/n}$$

$$\ge \frac{1}{r^i} = \frac{1}{r^i}\min_{S \in \mathcal{G}(n,i)} \frac{v_{i,rK_2}(S)}{v_{i,K_1}(S)} = \min_{S \in \mathcal{G}(n,i)} \frac{v_{i,K_2}(S)}{v_{i,K_1}(S)}.$$

The equality condition follows from that of Corollary 4.2.7.   ∎

Corollary 9.3.4. *Let* $1 \leq i \leq n - 1$, *and suppose that* K *is a zonoid in* $\mathbb{E}^n$. *Then*

$$\kappa_n^{(n-i)/n} \lambda_n(K)^{i/n} \geq \frac{\kappa_n}{\kappa_i} M_{-\infty}(v_{i,K}),$$

with equality if and only if K *is a ball.*

Proof. This follows immediately from the previous theorem when $K_1 = B$ and $K_2 = K$. ∎

Remark 9.3.5. When $i = 1$, the preceding two results apply not only to zonoids, but to all centrally symmetric convex bodies, since Theorem 4.2.8 can be used in place of Corollary 4.2.7. Even for $i > 1$, versions hold at the cost of an additional factor depending on n. For example, let $i = n - 1$. When K is a centrally symmetric convex body in $\mathbb{E}^n$, we can argue as before, applying Theorem 4.2.13 instead of Corollary 4.2.7, to obtain

$$\kappa_n^{1/n} \lambda_n(K)^{(n-1)/n} \geq \frac{\kappa_n}{n^{(n-1)/2n} \kappa_{n-1}} M_{-\infty}(v_K).$$

In fact, Remark 4.2.14 shows that this estimate works for an arbitrary convex body K, an extra constant factor being necessary only when $n \leq 3$.

If K is a zonoid, Theorem 9.3.1 and Corollary 9.3.4 both provide a lower bound for the volume of K. Neither bound can be discarded, however. Simple computations show that for the ellipse with equation

$$\frac{x^2}{a^2} + a^2 y^2 = 1,$$

$a \leq 1$, in $\mathbb{E}^2$, the latter bound is better if and only if $a > \sqrt{6}/\pi$. In particular, for the unit disk, the estimate in the previous paragraph is better than that of Theorem 9.3.1.

Let $1 \leq i \leq n - 1$, and suppose that K is an arbitrary convex body in $\mathbb{E}^n$. From the extended isoperimetric inequality (B.21) and (A.47), we have

$$\kappa_n^{(n-i)/n} \lambda_n(K)^{i/n} \leq \frac{\kappa_n}{\kappa_i} M_1(v_{i,K}).$$

The quantity on the right is $W_{n-i}(K)$, the $(n - i)$th quermassintegral of K. By analogy with the cases $i = 1$ and $n - 1$ considered earlier, we might hope to strengthen this upper bound for the volume of K in terms of the ith projection function $v_{i,K}$ of K. This possibility motivates the following definition.

Definition 9.3.6. Let K be a convex body in $\mathbb{E}^n$. The quantities

$$\Phi_{n-i}(K) = \frac{\kappa_n}{\kappa_i} M_{-n}(v_{i,K}) = \frac{\kappa_n}{\kappa_i} \left(\int_{\mathcal{G}(n,i)} \lambda_i(K|S)^{-n} \, dS \right)^{-1/n},$$

$1 \leq i \leq n - 1$, are called the *affine quermassintegrals* of K.

By Jensen's inequality (B.8) for integrals, $\Phi_{n-i}(K) \leq W_{n-i}(K)$, and we would like to replace $W_{n-i}(K)$ by $\Phi_{n-i}(K)$ in the upper bound for the volume of K. Unfortunately, apart from the cases $i = 1$ and $n - 1$ settled in Theorems 9.3.2 and 9.3.1, and the results mentioned in Note 9.6, it is unknown whether this can be done; see Problem 9.3.

Let us now turn to the question of estimates from only finitely many values of a projection function. We shall restrict the discussion to the situation in which we know the values $v_K(u_j)$, $1 \leq j \leq m$, of the brightness function of K; see Note 9.6 for references dealing with other cases. (The reader is reminded that if K is centrally symmetric, a satisfactory reconstruction algorithm is available, even if the measurements are finite in number and noisy (see Section 4.4).) The next theorem provides upper and lower estimates for $V_{n-1}(K)$ (and therefore, by Cauchy's surface area formula (A.49), also for $S(K)$) in terms of the inradius and circumradius of a certain zonotope.

Theorem 9.3.7. *Let K be a compact convex set in $\mathbb{E}^n$, and let $u_j \in S^{n-1}$ and $a_j > 0, 1 \leq j \leq m$. Denote by Z the associated zonotope*

$$Z = \sum_{j=1}^{m} a_j[-u_j, u_j].$$

Then

$$r(Z)V_{n-1}(K) \leq \sum_{j=1}^{m} a_j v_K(u_j) \leq R(Z)V_{n-1}(K).$$

Proof. We have $r(Z)B \subset Z$, by definition of the inradius $r(Z)$ of Z. Using (A.12), the formula in Lemma 4.1.8 for the support function of Z, and the Cauchy projection formula ((A.45) with $i = n - 1$), we obtain

$$r(Z)V_{n-1}(K) = \frac{1}{2}\int_{S^{n-1}} h_{r(Z)B}(u)\,dS(K, u)$$

$$\leq \frac{1}{2}\int_{S^{n-1}} h_Z(u)\,dS(K, u)$$

$$= \frac{1}{2}\int_{S^{n-1}} \sum_{j=1}^{m} a_j|u \cdot u_j|\,dS(K, u)$$

$$= \sum_{j=1}^{m} \frac{a_j}{2}\int_{S^{n-1}} |u \cdot u_j|\,dS(K, u)$$

$$= \sum_{j=1}^{m} a_j v_K(u_j).$$

The bound involving the circumradius $R(Z)$ of Z follows in the same fashion. ∎

The lower bound (or upper bound) in the previous theorem is attained if and only if the support of $S(K, \cdot)$ is contained in the subset of S^{n-1} on which h_Z is minimal (or maximal, respectively). These subsets of S^{n-1}, and hence corresponding extremal bodies, can, in principle, be determined from Z.

A simple example will illustrate. Suppose that $n = 2$, that the vectors u_j, $1 \le j \le 3$, correspond to directions with polar angles 0, $2\pi/3$, and $4\pi/3$, and that $a_j = 1$, $1 \le j \le 3$. The zonotope Z is then a centered regular hexagon of side length 2, so $r(Z) = \sqrt{3}$ and $R(Z) = 2$. We therefore have the estimate

$$\sum_{j=1}^{3} a_j v_K(u_j) \le S(K) = 2V_1(K) \le \frac{2}{\sqrt{3}} \sum_{j=1}^{3} a_j v_K(u_j) \tag{9.5}$$

for the perimeter $S(K)$ of K, where $2/\sqrt{3} = 1.1547\ldots$. The upper bound in (9.5) corresponds to the lower bound in Theorem 9.3.7; since h_Z is minimal in the directions orthogonal to its edges, Z itself is an example where this bound is attained. This can be verified directly, since $S(Z) = 12$ and $v_Z(u_j) = 2\sqrt{3}$, $1 \le j \le 3$. The lower bound in (9.5) corresponds to the upper bound in Theorem 9.3.7. Now h_Z is maximal in the directions of the vertices of Z. Let Z' be a regular hexagon, inscribed in the unit circle and such that the outward normals to its sides are in these directions. Then Z' is an example for which this bound is attained. Again, this can be verified directly, since $S(Z') = 6$ and $v_{Z'}(u_j) = 2$, $1 \le j \le 3$.

Examples such as the previous one have been calculated for $n = 3$; see Note 9.6.

If the brightness function of K is known in the coordinate directions, any other value of v_K can be estimated by the following inequality.

Theorem 9.3.8. *Let K be a compact convex set in $\mathbb{E}^n$, and let e_j, $1 \le j \le n$, denote the unit vectors in the coordinate directions. Then*

$$v_K(u)^2 \le \sum_{j=1}^{n} v_K(e_j)^2,$$

for all $u \in S^{n-1}$.

Proof. Suppose that the projection body ΠK of K is inscribed in a box P. For any $u \in S^{n-1}$, we have $2v_K(u) = 2h_{\Pi K}(u) = w_{\Pi K}(u)$. Therefore

$$4v_K(u)^2 = w_{\Pi K}(u)^2 \le (\operatorname{diam} P)^2$$

$$= \sum_{j=1}^{n} w_P(e_j)^2$$

$$= \sum_{j=1}^{n} w_{\Pi K}(e_j)^2 = 4 \sum_{j=1}^{n} v_K(e_j)^2. \qquad \blacksquare$$

Using this in conjunction with Theorem 9.3.1, one can then obtain an upper bound for the volume of K. There is, in fact, an inequality (the Loomis–Whitney inequality) furnishing an upper bound for the volume when the brightness function is only known in the coordinate directions; see Note 9.6.

9.4. Volume estimates from section functions

Suppose that L is a star body containing the origin in $\mathbb{E}^n$ and that we know the volumes of sections of L by hyperplanes through the origin (in other words, its $(n-1)$th section function or $(n-1)$-dimensional X-ray at o). By (A.62) with $i = n - 1$, we have

$$\tilde{V}_{n-1}(L) = \frac{1}{n\kappa_{n-1}} \int_{S^{n-1}} \lambda_{n-1}(L \cap u^{\perp}) \, du,$$

from which we also see that $\tilde{V}_{n-1}(B) = \kappa_n$. Substituting into the dual isoperimetric inequality (B.28), we obtain

$$\kappa_n^{1/n} \lambda_n(L)^{(n-1)/n} \geq \frac{1}{n\kappa_{n-1}} \int_{S^{n-1}} \lambda_{n-1}(L \cap u^{\perp}) \, du, \tag{9.6}$$

providing a lower bound for the volume of L.

We shall rewrite (9.6), just as we did the dual inequality in the previous section. Suppose that C is a compact set in $\mathbb{E}^n$. For each $S \in \mathcal{G}(n, i)$, define

$$\tilde{v}_{i,C}(S) = \lambda_i(C \cap S).$$

Thus $\tilde{v}_{i,C}$ is the i-dimensional X-ray of C at o, or ith section function, as in Section 7.2. If $i = 1$ or $n - 1$, we identify $\mathcal{G}(n, i)$ with S^{n-1} and consider $\tilde{v}_{1,C}$ and $\tilde{v}_{n-1,C}$ as functions of $u \in S^{n-1}$. Furthermore, we drop one index when $i = n - 1$, and simply write $\tilde{v}_C = \tilde{v}_{n-1,C}$.

Using this notation, (9.6) becomes

$$\kappa_n^{1/n} \lambda_n(L)^{(n-1)/n} \geq \frac{\kappa_n}{\kappa_{n-1}} M_1(\tilde{v}_L). \tag{9.7}$$

Our aim is to improve this estimate, as we did the corresponding one for projections, by proving the same inequality holds when the first mean is replaced by the pth mean, for certain values of $p > 1$. It will be a consequence of Theorem 9.4.4 and Remark 9.4.6 that this can be done with $p = n$, yielding the estimate

$$\kappa_n^{1/n} \lambda_n(L)^{(n-1)/n} \geq \frac{\kappa_n}{\kappa_{n-1}} M_n(\tilde{v}_L). \tag{9.8}$$

The bound (9.8) is known as the Busemann intersection inequality, and unlike (9.7), where equality holds if and only if L is a centered ball, it is affine invariant. In contrast to the case of projections, however, we can prove an inequality applying not only to $\tilde{v}$, but also to $\tilde{v}_{i,L}$ for any i with $1 \leq i \leq n - 1$. This will require some preliminary work.

The first item is an integration formula leading to an integral expression, involving the function g_m introduced earlier, for the λ_n-measure of a compact set. As motivation, consider a star body L containing the origin in $\mathbb{E}^2$, with boundary given by $r = r(\theta)$ in polar coordinates. The area of L is given by

$$\lambda_2(L) = \frac{1}{2} \int_0^{2\pi} r(\theta)^2 \, d\theta = \frac{1}{2} \int_0^{2\pi} \int_{-r(\theta-\pi/2)}^{r(\theta+\pi/2)} |r| \, dr \, d\theta.$$

The first expression is the usual polar coordinate formula for volume, whose generalization (A.58) to $\mathbb{E}^n$ is familiar. The quantity $|r|$ in the second expression can be regarded as the 1-dimensional volume of the 1-dimensional simplex $[o, p]$, where $p = (r, \theta \pm \pi/2)$ lies in the intersection of L and a 1-dimensional subspace. The next lemma provides a generalization of this volume formula to $\mathbb{E}^n$. Each linearly independent set of i points determines an i-dimensional subspace, and the formula relates a multiple integral over $\mathbb{E}^n$ to a multiple integral over an i-dimensional subspace, followed by integration over all such subspaces. It shows that the absolute value of the Jacobian for the change of variables is a constant multiple of $\lambda_i([o, p_1, \ldots, p_i])^{n-i}$, not surprising when we remember that by (0.13) this is just a constant multiple of the absolute value of the determinant whose entries are the components of the points p_j, $1 \le j \le i$. We refer the reader to [746, Satz 6.1.3] or [616, p. 358, eq. (15)] for proof.

Lemma 9.4.1. *Let $1 \le i \le n-1$, and suppose that f is a Borel function on the product space $(\mathbb{E}^n)^i$. Then*

$$\int_{\mathbb{E}^n} \cdots \int_{\mathbb{E}^n} f(p_1, \ldots, p_i) \, dp_1 \cdots dp_i$$

$$= c \int_{\mathcal{G}(n,i)} \int_S \cdots \int_S f(p_1, \ldots, p_i) \lambda_i([o, p_1, \ldots, p_i])^{n-i} \, dp_1 \cdots dp_i \, dS,$$

where c is a constant depending only on n and i.

Corollary 9.4.2. *Let $1 \le i \le n-1$, and suppose that C is a compact set in $\mathbb{E}^n$. Then*

$$\lambda_n(C)^i = c \int_{\mathcal{G}(n,i)} g_{n-i,i}(C \cap S) \, dS,$$

where c is a constant depending only on n and i.

Proof. Set $f(p_1, \ldots, p_i) = 1_C(p_1) \cdots 1_C(p_i)$ in Lemma 9.4.1, and recall the notation introduced just before Theorem 9.1.5. ∎

Definition 9.4.3. *Let C be a compact set in $\mathbb{E}^n$. The quantities*

$$\tilde{\Phi}_{n-i}(C) = \frac{\kappa_n}{\kappa_i} M_n(\tilde{v}_{i,C}) = \frac{\kappa_n}{\kappa_i} \left(\int_{\mathcal{G}(n,i)} \lambda_i(C \cap S)^n \, dS \right)^{1/n},$$

$1 \le i \le n-1$, *are called the* dual affine quermassintegrals *of C.*

Theorem 9.4.4. *Let* $1 \le i \le n - 1$, *and suppose that* K *is a convex body in* $\mathbb{E}^n$. *Then*

$$\kappa_n^{(n-i)/n} \lambda_n(K)^{i/n} \ge \tilde{\Phi}_{n-i}(K).$$

Equality holds when $1 < i \le n - 1$ *if and only if* K *is a centered ellipsoid, and when* $i = 1$ *if and only if* K *is centered.*

Proof. We shall prove the equivalent inequality

$$\lambda_n(K)^i \ge \frac{\kappa_n^i}{\kappa_i^n} \int_{\mathcal{G}(n,i)} \lambda_i(K \cap S)^n \, dS. \tag{9.9}$$

Let $S \in \mathcal{G}(n, i)$. Applying Corollary 9.2.5 to $K \cap S$, with n replaced by i and $m = n - i$, we have

$$g_{n-i,i}(K \cap S) \ge g_{n-i,i}(r_S B \cap S),$$

for some $r_S > 0$, where

$$\lambda_i(K \cap S) = \lambda_i(r_S B \cap S).$$

By Corollary 9.4.2, we obtain

$$\lambda_n(K)^i \ge c \int_{\mathcal{G}(n,i)} g_{n-i,i}(r_S B \cap S) \, dS,$$

where c is a constant depending only on n and i. We also infer from Corollary 9.4.2 that for any $r > 0$,

$$\lambda_n(rB)^i = c \int_{\mathcal{G}(n,i)} g_{n-i,i}(rB \cap S) \, dS = c g_{n-i,i}(rB \cap S),$$

since the integrand is constant. Therefore

$$\begin{aligned} \lambda_n(K)^i &\ge \int_{\mathcal{G}(n,i)} \lambda_n(r_S B)^i \, dS \\ &= \frac{\kappa_n^i}{\kappa_i^n} \int_{\mathcal{G}(n,i)} \lambda_i(r_S B \cap S)^n \, dS \\ &= \frac{\kappa_n^i}{\kappa_i^n} \int_{\mathcal{G}(n,i)} \lambda_i(K \cap S)^n \, dS. \end{aligned}$$

Equality holds in (9.9) if and only if $g_{n-i,i}(K \cap S) = g_{n-i,i}(r_S B \cap S)$ for all $S \in \mathcal{G}(n, i)$, because all the preceding integrands are continuous functions of S. Therefore, by Corollary 9.2.5, equality holds precisely when $K \cap S$ is a centered ellipsoid in S, for all $S \in \mathcal{G}(n, i)$. If $1 < i \le n - 1$, this occurs if and only if K is a centered ellipsoid, by Theorem 7.1.5. If $i = 1$, it is straightforward to check that equality holds if and only if K is centered. ∎

Corollary 9.4.5 (Busemann intersection inequality). *Let K be a convex body in $\mathbb{E}^n$ with $o \in \operatorname{int} K$. Then*

$$\lambda_n(IK) \le \frac{\kappa_{n-1}^n}{\kappa_n^{n-2}} \lambda_n(K)^{n-1},$$

with equality if and only if K is a centered ellipsoid.

Proof. By Definition 8.1.1 of IK, and the polar coordinate formula (A.58) for volume, we have

$$\lambda_n(IK) = \frac{1}{n} \int_{S^{n-1}} \lambda_{n-1}(K \cap u^\perp)^n \, du.$$

With this expression in hand, the corollary is just a rearrangement of the case $i = n - 1$ of Theorem 9.4.4. ∎

Remark 9.4.6. By Remark 9.2.10, Corollary 9.2.5 holds for compact sets, so Theorem 9.4.4 also remains true when K is a compact set. Our statement of the Busemann intersection inequality therefore applies whenever the intersection body map I is defined, that is, for star bodies with continuous radial functions.

Corollary 9.4.7 (Furstenberg–Tzkoni formula). *Let $1 \le i \le n - 1$, and let E be a centered n-dimensional ellipsoid in $\mathbb{E}^n$. Then*

$$\lambda_n(E)^i = \frac{\kappa_n^i}{\kappa_i^n} \int_{\mathcal{G}(n,i)} \lambda_i(E \cap S)^n \, dS.$$

Proof. This is implied by the equality case of Theorem 9.4.4. ∎

Corollary 9.4.8 (dual Furstenberg–Tzkoni formula). *Let $1 \le i \le n - 1$, and let E be an n-dimensional ellipsoid in $\mathbb{E}^n$. Then*

$$\lambda_n(E)^i = \frac{\kappa_n^i}{\kappa_i^n} \left(\int_{\mathcal{G}(n,i)} \lambda_i(E|S)^{-n} \, dS \right)^{-1}.$$

Proof. Since both sides of the previous equation are invariant under translations, we can assume that E is centered. It follows from the equality case of the Blaschke–Santaló inequality (Theorem 9.2.11) that

$$\lambda_n(E^*) = \kappa_n^2 \lambda_n(E)^{-1}.$$

Applying this in $S \in \mathcal{G}(n, i)$ with E^* replaced by $E^* \cap S$, and using (0.38), we also obtain

$$\lambda_i(E^* \cap S) = \kappa_i^2 \lambda_i\big((E^* \cap S)^*\big)^{-1} = \kappa_i^2 \lambda_i(E|S)^{-1}.$$

Corollary 9.4.7 with E replaced by E^* and the previous two expressions yield the required formula.   ∎

The author is not aware of a method to obtain comparable upper bounds for volume in terms of section functions, except in the following special cases.

Theorem 9.4.9. *Let $1 \leq i \leq n - 1$. Suppose that L_1 is the intersection body of a star body and that L_2 is a star body in $\mathbb{E}^n$ with $\rho_{L_2} \in C(S^{n-1})$. Then*

$$\left(\frac{\lambda_n(L_1)}{\lambda_n(L_2)}\right)^{i/n} \leq \max_{S \in \mathcal{G}(n,i)} \frac{\tilde{v}_{i,L_1}(S)}{\tilde{v}_{i,L_2}(S)},$$

with equality if and only if L_1 is a dilatate of L_2.

Corollary 9.4.10. *Let $1 \leq i \leq n - 1$, and suppose that L is the intersection body of a star body in $\mathbb{E}^n$. Then*

$$\kappa_n^{(n-i)/n} \lambda_n(L)^{i/n} \leq \frac{\kappa_n}{\kappa_i} M_\infty(\tilde{v}_{i,L}),$$

with equality if and only if L is a centered ball.

We omit the proofs. When $i = 1$ or $i = n - 1$, they closely follow those of Theorem 9.3.3 and Corollary 9.3.4, in view of Theorems 8.2.3 and 8.2.8. (Begin by choosing $r > 0$ so that

$$\max_{S \in \mathcal{G}(n,i)} \frac{v_{i,L_1}(S)}{v_{i,rL_2}(S)} = 1$$

instead.) In fact, the latter two theorems are essentially equivalent to the cases $i = 1$ and $i = n - 1$ of Theorem 9.4.9, respectively, and for $i = 1$, it is sufficient to assume that L_1 or L is centered. For $1 < i < n$, one must apply the generalization of Theorem 8.2.8, mentioned in Note 8.10, with $i = k$. The latter also can be employed to demonstrate that Theorem 9.4.9 (and Corollary 9.4.10) remains true when L_1 (or L, respectively) is an intersection body as defined just after Theorem 8.1.20; we shall use this fact in the next theorem.

As in Remark 9.3.5 concerning projections, one can obtain a version of Corollary 9.4.10 for $i = n - 1$ applying to any centered convex body, by substituting Theorem 8.2.13 for Theorem 8.2.8, at the cost of an extra factor depending on n. In four or fewer dimensions, a better estimate is as follows.

Theorem 9.4.11. *Suppose that K is a centered convex body in $\mathbb{E}^n$, $n \leq 4$. Then*

$$\lambda_n(K)^{(n-1)/n} \leq \frac{\kappa_n^{(n-1)/n}}{\kappa_{n-1}} \max_{u \in S^{n-1}} \lambda_{n-1}(K \cap u^\perp),$$

with equality if and only if K is a ball.

Proof. Suppose that K is a centered convex body in $\mathbb{E}^n$, $n \leq 4$, for which this inequality is false. By the remarks in the last paragraph of Section 0.9, we can assume that K is of class C^∞, so K is the intersection body of a star body, by Theorem 8.1.17. This contradicts Corollary 9.4.10. The equality condition follows from the fact that each centered convex body in $\mathbb{E}^n$, $n \leq 4$ is an intersection body as defined just after Theorem 8.1.20, and the fact, mentioned earlier, that Corollary 9.4.10 remains true when L is an intersection body. ∎

Open problems

Problem 9.1 (Petty's conjectured projection inequality). (See Note 9.4.) Let K be a convex body in $\mathbb{E}^n$. Is it true that

$$\lambda_n(\Pi K) \geq \frac{\kappa_{n-1}^n}{\kappa_n^{n-2}} \lambda_n(K)^{n-1},$$

with equality if and only if K is an ellipsoid?

Problem 9.2 (Mahler's conjecture). (See Note 9.4.) Let K be a convex body in $\mathbb{E}^n$ whose centroid is at the origin. Is it true that the following reverse form of the Blaschke–Santaló inequality holds?

$$\lambda_n(K)\lambda_n(K^*) \geq (n+1)^{(n+1)}/(n!)^2,$$

with equality if and only if K is a simplex.

Problem 9.3. (See Note 9.6.) Let $1 < i < n - 1$, and suppose that K is a convex body in $\mathbb{E}^n$. Is it true that

$$\kappa_n^{(n-i)/n} \lambda_n(K)^{i/n} \leq \Phi_{n-i}(K),$$

with equality if and only if K is an ellipsoid?

Problem 9.4. (See Note 9.6.) Does Problem 9.3 have a positive answer when K is an arbitrary compact set, where Φ_{n-i} is defined as in Definition 9.3.6?

Problem 9.5. (See Note 9.6.) Let $0 \leq i \leq j \leq n - 1$, and suppose that K is a convex body in $\mathbb{E}^n$. Is it true that

$$\kappa_n^j \Phi_i(K)^{n-j} \leq \kappa_n^i \Phi_j(K)^{n-i},$$

with equality when $i < j$ if and only if K is an ellipsoid? (Here, $\Phi_0(K) = \lambda_n(K)$ and $\Phi_n(K) = \kappa_n$.)

Problem 9.6. (See Notes 9.4 and 9.8.) Let $1 \leq i \leq n - 1$, and let K be a convex body in $\mathbb{E}^n$. Find inequalities of the form

$$a_{n,i} \Phi_{n-i}(K) \leq \kappa_n^{(n-i)/n} \lambda_n(K)^{i/n} \leq \tilde{a}_{n,i} \tilde{\Phi}_{n-i}(K),$$

where the constants $a_{n,i}$ and $\tilde{a}_{n,i}$ correspond to extremal cases and are to be determined.

Notes

9.1. *Centroid bodies and polar projection bodies.* Centroid bodies were defined by Petty [661], but they appear in another guise in work of Dupin, connected with floating bodies (see Note 9.2), and of Blaschke [69]. A quite general result of Petty [661] implies that the centroid body of a convex body is always of class C_+^2. In the same paper, Petty establishes an integral representation for the mixed volume of centroid bodies, of which Theorem 9.1.5 is a special case. (More generally still, one can obtain an integral representation for the mixed volume of generalized zonoids, as in the survey [745, eq. (9.5)] of Schneider and Weil.) Centroid bodies were revived by Lutwak in [533] and [534]. Here, Lemma 9.1.2 can be found, among many other more significant results, some of which are mentioned later. Milman and Pajor also study centroid bodies in [621], where the term *zonoid of inertia* is used.

Theorem 9.1.3 is a remark of Blaschke [69]. A proof is given by Lutwak [538]. In the same paper, Lutwak defines a body $\hat{\nabla}L$, for each star body L in $\mathbb{E}^n$ with $o \in \operatorname{int} L$, by

$$\hat{\nabla}L = r\tilde{\nabla}_{n+1}L,$$

where r is a computable constant depending only on L, and proves that $\hat{\nabla}L$ is the unique centered body such that

$$\Gamma(\hat{\nabla}L) = \Gamma L.$$

He also proves that

$$\lambda_n(\hat{\nabla}L) \geq \lambda_n(L),$$

with equality if and only if L is centered. In addition, Lutwak considers an analogue of Shephard's problem (cf. Section 4.2) and of the Busemann–Petty problem (cf. Section 8.2). The results are also analogous, but polar projection bodies, first defined by Blaschke [69], play a part. For example, Theorem 9.3 of [538] is as follows:

Let L_1 and L_2 be star bodies containing the origin in $\mathbb{E}^n$ such that $\Gamma L_1 \subset \Gamma L_2$. If L_2 is a polar projection body, then $\lambda_n(L_1) \leq \lambda_n(L_2)$, with equality if and only if $L_1 = L_2$.

For extensions of the latter result in the L^p-Brunn–Minkowski theory, see the articles of Grinberg and Zhang [337] and Yaskin and Yaskina [850].

Zhang [858] finds a characterization of polar projection bodies in terms of dual mixed volumes, similar to that of intersection bodies; this was re-proved by Goodey, Lutwak, and Weil [312], using their functional-analytic approach (see Note 8.8). Lonke [519] investigates the boundary structure of zonoids whose polar bodies are also zonoids. He shows, in particular, that bodies of this type cannot have proper faces other than vertices and facets, and that there are non-smooth bodies of this type.

We observed just before Theorem 9.1.3 that centered star bodies are determined by their centroid bodies. Groemer [355], [356, Corollary 5.6.12] finds a stability version of this statement.

See Problem 4.1 for a question concerning polar projection bodies, and consult the surveys of Lutwak [542, Section 12] and Martini [581, Section 4] for comments, references, and other related problems.

9.2. *The floating body problem.* Ulam [807, p. 38] asked, "A solid S of uniform density ρ has the property that it will float in equilibrium (without turning) in water in every given orientation. Must S be a sphere?" (The density of water is assumed to be one, as it is at $4°C$ in the metric system.) Let us agree to call such a solid a *stable floating body*. Ulam notes that Auerbach [15] found 2-dimensional noncircular (both convex and nonconvex) stable floating bodies for $\rho = 1/2$, so his question refers to solids in $\mathbb{E}^3$ (or, more generally, in $\mathbb{E}^n$, $n \geq 3$).

As Falconer [217] remarks, the centroid of the part of a stable floating body lying beneath the water surface must always be at the same depth. When the body is a centrally symmetric star body L, and $\rho = 1/2$, the center of L always lies on the water surface. Taking the center of L to be o, we see that in this case a necessary and sufficient condition for L to be a stable floating body is that ΓL is a ball. By the injectivity property noted after Lemma 9.1.2, L must be a ball. (This conclusion is reached by Falconer for convex bodies, and by Gilbert [291] for star bodies of revolution.) The only other known partial result corresponds to the limiting case as $\rho \to 0$, also asked by Ulam in the form of the question, "If a body rests in equilibrium in every position on a flat horizontal surface, is it a sphere?" This was confirmed by Groemer [345] (see also the paper [628] of Montejano).

Note that some restriction on the body is necessary, since the solid between two concentric spheres is a stable floating body, and there are other examples besides. A related idea is that of a floating body, arising as follows. It can be shown that the submerged part of a stable floating body always has the same volume and that the centroid of this part always lies in a fixed sphere. For an arbitrary convex body K in $\mathbb{E}^n$, and positive constant $\delta < \lambda_n(K)$, the hyperplanes cutting off parts of K with volume δ form the envelope of a body $K_{[\delta]}$, called a *floating body*. Schütt and Werner [752] attribute the notion of a floating body to Dupin [200] and introduce the *convex floating body* K_δ of K, the set of points $x \in K$ such that there is a $u \in S^{n-1}$ for which $\lambda_n(K \cap H(x, u)) \geq \delta$, where $H(x, u)$ is the half-space containing x in its boundary and with outer normal vector $-u$. If the floating body is convex, it is the convex floating body.

For other work on floating bodies, see the papers of Meyer and Reisner [606], Schütt [750], Schütt and Werner [753], [754], Stancu [785], and Werner [840], and the references given there. The interested reader should also consult the book [173, Problem A6, p. 19] of Croft, Falconer, and Guy for related open questions.

9.3. *Affine surface area, the covariogram, and convolution and sectional bodies.* Extending an earlier result of Blaschke, Leichtweiss (see [752] for references) proved that if K is a convex body in $\mathbb{E}^n$ of class C_+^2, then

$$\lim_{\delta \to 0} \frac{\lambda_n(K) - \lambda_n\left(K_{[\delta]}\right)}{\delta^{2/(n+1)}} = c_n \int_{\mathrm{bd}\, K} \kappa(x)^{1/(n+1)}\, d\lambda_{n-1}(x),$$

where c_n is a constant that depends only on n. Here $K_{[\delta]}$ is the floating body defined in Note 9.2 and $\kappa(x)$ is the Gauss curvature of K at x. The integral on the right-hand side is an important quantity denoted by $\Omega(K)$, the *affine surface area* of K. In a remarkable paper, Ludwig and Reitzner [528] prove that every real-valued valuation on $\mathcal{K}^n$, also invariant under translations and special linear transformations and upper semicontinuous, is a linear combination of the Euler characteristic $V_0(K) = 1$, volume $V_n(K)$, and $\Omega(K)$. (Compare Hadwiger's theorem in Section A.5.) Clearly, then, affine surface area is of fundamental importance, so it is not surprising that it has many interesting properties, extensions, and connections. See the book of Schneider [737, p. 419], and papers of Schütt [750], [751] and Schütt and Werner [754] for some references.

The *covariogram* g_K of a convex body K in $\mathbb{E}^n$ is defined by

$$g_K(x) = \lambda_n\left(K \cap (K + x)\right),$$

for $x \in \mathbb{E}^n$. In 1986, G. Matheron asked whether two convex bodies having the same covariogram must be equal, up to translation and reflection in the origin. This has become known as the *covariogram problem*. Gardner, Gronchi, and Zong [272], who address an integer lattice version of the covariogram problem in the spirit of discrete tomography, note the following connection with X-rays: Two convex bodies have equal covariograms if and only if for all $u \in S^{n-1}$ their X-rays in the direction u are rearrangements of one another. It follows that the covariogram of K determines its *chord-length distribution*, that is, the distribution of the chord lengths of K based on random lines that meet K, and of course this in turn determines the chord-power integrals of K. Due to this connection, some of the references given for chord-power integrals in Note 7.12 contain results involving the covariogram; for example, Nagel [641] provides a positive answer to the covariogram problem for polygons. (A much simpler proof has been found by Bianchi [56].) Since Mallows and Clark [565] construct two noncongruent convex planar polygons with equal chord-length distributions, we see that the covariogram provides strictly more information than the chord-length distribution. Chord-length distributions can be considered for non-convex bodies and are of wide interest beyond mathematics, as Mazzolo, Roesslinger, and Gille [588] describe.

The covariogram of K can be expressed as $g_K = 1_K * 1_{-K}$, the convolution of the characteristic functions of K and $-K$. Since the Fourier transform $\widehat{1_{-K}}$ is the complex conjugate of $\widehat{1_K}$, the Fourier transform of the previous equation yields $\widehat{g_K} = |\widehat{1_K}|^2$ and the covariogram problem therefore asks for information about 1_K from knowledge of the modulus of its Fourier transform. This is a special case of the *phase retrieval problem*, where 1_K is replaced by a function with compact support. This has applications in X-ray crystallography, optics, electron microscopy, and other areas, references to which may be found in [62]. In this article, Bianchi, Segala, and Volčič obtain a positive answer to the covariogram problem for planar C_+^2 convex bodies. Bianchi [57] proves a common generalization of this result and Nagel's.

In a breakthrough, Bianchi [57] shows that if K_1 and K_2 are convex bodies in $\mathbb{E}^n$ and ϕ is a nonsingular linear transformation, then $\phi(K_1 \times K_2)$ and $\phi(K_1 \times (-K_2))$ are convex bodies in $\mathbb{E}^{2n}$ with equal covariograms. Moreover, when K_1 and K_2 are not centrally symmetric, $\phi(K_1 \times K_2)$ and $\phi(K_1 \times (-K_2))$ are not equal up to translation and reflection in the origin. It follows that for $n \geq 4$, there are pairs of convex polytopes in $\mathbb{E}^n$ that answer Matheron's question in the negative. In contrast, a positive answer for convex polytopes in $\mathbb{E}^3$ has been announced by Bianchi [59]; the proof is long and includes an unexpected application of -1-chord functions.

Quite remarkably, the covariogram problem is still open when $n = 2$; Bianchi [58] discusses this and other related open questions. In related work, Averkov and Bianchi [17] show, among other results, that most planar convex bodies are determined by the values of their covariograms in the union of a neighborhood of the boundary of its support and a finite set. Applications of the covariogram to stereology are discussed by Cabo and Baddeley [135].

If K is a convex body in $\mathbb{E}^n$, then

$$K(\delta) = \{x \in \mathbb{E}^n : g_K(x) \geq \delta\}$$

is called a *convolution body* of K. This notion is due to K. Kiener, who noted that $K(\delta)$ is convex. Building on Kiener's results, Schmuckenschläger [726] proves that if K is centrally symmetric and $\lambda_n(K) = 1$, then

$$\lim_{\delta \to 1} \frac{K(\delta)}{1 - \delta} = \Pi^* K,$$

the polar projection body of K. He also shows that if K is centrally symmetric, then

$$\lim_{\delta \to 0} \frac{\lambda_n(K) - \lambda_n(K(\delta))}{\delta^{2/(n+1)}} = c_n \Omega(K),$$

where c_n is a constant that depends only on n. Tsolomitis [804], [805], [806] defines the convolution body of a pair of convex bodies and obtains some results on these more general convolution bodies.

Fradelizi [248] introduces a new class of bodies with properties analogous to those above. If K is a convex body in $\mathbb{E}^n$, the *sectional body* $K(t)$, $t \geq 0$ is the set of all points $x \in K$ such that $\lambda_{n-1}(K \cap (x + u^\perp)) \geq t$ for every $u \in S^{n-1}$. (The notation is too similar to that for the convolution body, but it should not cause confusion here.) Fradelizi notes that $K(t)$ is always convex and proves that if K is of class C_+^2, then

$$\lim_{t \to 0} \frac{\lambda_n(K) - \lambda_n(K(t))}{t^{2/(n-1)}} = c_n \int_{\mathrm{bd}\, K} \kappa(x)^{1/(n-1)} \, d\lambda_{n-1}(x),$$

where c_n is a constant that depends only on n. He also defines a more general class of bodies and notes relations with intersection and cross-section bodies.

9.4. *Affine isoperimetric inequalities.* Affine isoperimetric inequalities abound in geometry, and have numerous applications there, as well as in other areas of mathematics, such as differential equations. New affine isoperimetric inequalities, and unexpected connections between old ones, are turning up all the time. The works of Lutwak and his coauthors are of fundamental importance in this area, and Section 9.2 is inspired by his survey article [542]. In particular, we have followed [542] in naming the fundamental inequalities. The reader should consult this paper for a lucid exposition of many other related inequalities, for more references, and for a long list of problems. We do not attempt to survey all the many generalizations of the inequalities treated in Section 9.2, a topic worthy of a book to itself. See Note 9.5, however, for a discussion of certain extensions related to the L^p-Brunn–Minkowski theory.

The Busemann random simplex inequality, Theorem 9.2.6, is a consequence of [126, p. 11, eq. (21)]. The various ingredients in the proof were also employed by Busemann. These include Theorem 9.2.1, the existence and uniqueness of the Löwner ellipsoid (which has many applications – see, for example, the articles of Berger [54] and Danzer, Laugwitz, and Lenz [175]), Lemma 9.2.2, and Theorem 9.2.4. (Bonnesen and Fenchel [83, Section 70] observe that the latter was discovered by various authors, at least in two and three dimensions.) Busemann indicates that the use of Steiner symmetrization was suggested by Blaschke's solution of *Sylvester's problem* of finding the probability that four points chosen at random in a planar convex body are the vertices of a convex quadrilateral. (See the works of Lutwak [542], Meckes [596], Milman and Pajor [621, p. 94], and Santaló [721, pp. 63–5] for discussions of this problem.)

The Busemann random simplex inequality was motivated by the Busemann intersection inequality, Corollary 9.4.5, also stated in [126, p. 2, eq. (4)]. From remarks in [126, p. 7], it appears that Busemann was aware of the integration formula of Lemma 9.4.1, though he only derives it explicitly for $i = n - 1$ and $i = n - 2$, together with corresponding versions of Corollary 9.4.2. The latter appeared in general form in the papers of Busemann and Straus [132] and Grinberg [334]; see Note 9.7 for further comments.

The Busemann–Petty centroid inequality, Corollary 9.2.7, is stated by Petty [661, p. 1544, eq. (3.18)], along with a remark that convexity is not necessary. Petty [663] used this inequality, conjectured by Blaschke [69] for centrally symmetric convex bodies in $\mathbb{E}^3$, in obtaining the Petty projection inequality, Theorem 9.2.9. Our proof of the Petty projection inequality, via Lemma 9.2.8, is taken from [542, Section 5], an approach deriving from Lutwak's papers [533], [534], and [535], where many related inequalities and generalizations are proved. The Zhang projection inequality of Remark 9.2.10(ii) is due to Zhang [856], and when $i = n - 1$ this yields the right-hand inequality of Problem 9.6. (The earlier paper of Eggleston [206] deals with the case $n = 2$.)

Bisztriczky and Böröczky, Jr. [67] conjecture a sharp complementary Busemann–Petty centroid inequality, giving an upper bound for the volume of the centroid body of a centered convex body in which equality holds precisely for parallelotopes. They note that this would imply a positive answer to the slicing problem (see Note 9.8) and use Blaschke shaking (see Note 2.4) to prove their conjecture in the planar case.

The extension of Corollary 9.2.5 to compact sets, mentioned in Remark 9.2.10(i), is due to Pfiefer [670]. More precisely, Pfiefer proves in [670] an extension of the corresponding inequality in which the simplices do not necessarily have one vertex at the origin. Pfiefer states the extension of Corollary 9.2.5 explicitly in [669, p. 70].

The Blaschke–Santaló inequality is the inequality of Theorem 9.2.11 for convex bodies with centroid at the origin. It was proved by Blaschke [70] for $n \leq 3$ and by Santaló [720] for all n. The equality condition in its general form is due to Petty [665]. A reasonably succinct self-contained proof of Theorem 9.2.11 including the equality condition was found by Saint Raymond [715]. Our proof of the inequality of Theorem 9.2.11 is due to Meyer and Pajor [605, Lemma 1], who also presented an inequality stronger than the Blaschke–Santaló inequality. For several closely related inequalities, see [542, Section 11].

Mahler's conjecture, Problem 9.2, is an outstanding question in the area. The problem is even open for centrally symmetric bodies when the constant on the right-hand side is conjectured to be $4^n/n!$. This has been proved for zonoids by Reisner (see the improved proof of Gordon, Meyer, and Reisner [328]), and for n-dimensional convex polytopes, $n \leq 8$, with at most $2n + 2$ vertices by Lopez and Reisner [522]. The best estimate currently available is due to Bourgain and Milman [92]. A very short proof of an asymptotically weaker result is provided by Kuperberg [483], who in [484] states a conjecture that if true would improve Bourgain and Milman's bound. Both the centrally symmetric and general case were proved by Mahler when $n = 2$, but even here the equality condition was not fully established until the work of Meyer [600]. A commentary and list of references is given by Lutwak [542, Section 12]. In addition, Ball [26] finds a connection between Mahler's conjecture and the theory of wavelets.

Another important conjecture is Problem 9.1, due to Petty [663]. (Compare the Busemann intersection inequality, Corollary 9.4.5.) Schneider [736] observed that extremal bodies for Petty's conjectured projection inequality must be such that $\Pi^2 K$ and K are homothetic (cf. Problem 4.4). See [542, Section 12] for further comments. Brannen [94] confirms Petty's conjectured projection inequality for (general) cylinders in $\mathbb{E}^3$. He also investigates the corresponding upper bound for $\lambda_n(\Pi K)\lambda_n(K)^{1-n}$, finding counterexamples to a conjecture of Schneider [735] for the case when K is centrally symmetric and conjecturing that for arbitrary K, the upper bound is attained if and only if K is a simplex. Brannen's conjecture is open, but Brannen [95] and Lutwak, Yang, and Zhang [548] obtain some results in this direction.

9.5. *The L^p-Brunn–Minkowski theory: centroid bodies, ellipsoids, and inequalities.* Let C be a compact set in $\mathbb{E}^n$ with $\lambda_n(C) > 0$, and suppose that $p \geq 1$. The L^p-*centroid body* $\Gamma_p C$ of C is the centered compact convex set such that

$$h_{\Gamma_p C}(u) = \left(\frac{1}{\lambda_n(C)} \int_C |u \cdot x|^p \, dx \right)^{1/p},$$

for all $u \in S^{n-1}$. Clearly $\Gamma_1 C = \Gamma C$, and the integral version of Minkowski's inequality (B.4) (see [382, Section 6.13]) can be used to show that $h_{\Gamma_p C}$ is indeed a support function. Moreover, the L^2-centroid body $\Gamma_2 C$ is always an ellipsoid, called the Fenchel ellipsoid by Petty [661] and the Legendre ellipsoid by Milman and Pajor [621], the latter name indicating early origins of this observation. A clear proof is supplied by Lutwak [539], where the notation $\Gamma_p C$ is introduced for $p = 2$.

In [546], Lutwak, Yang, and Zhang introduce an ellipsoid $\Gamma_{-2} K$ associated with a convex body K in $\mathbb{E}^n$ that is dual to the Legendre ellipsoid $\Gamma_2 K$ in its properties.

For example, if K is centered, then

$$\frac{2}{\sqrt{n}}\frac{\lambda_{n-1}(K|u^{\perp})}{\lambda_n(K)} \le h_{\Gamma_{-2}^* K}(u) \le 2\frac{\lambda_{n-1}(K|u^{\perp})}{\lambda_n(K)},$$

for all $u \in S^{n-1}$, to be compared with the known inequalities

$$\frac{\sqrt{n+2}}{2\sqrt{3}}\frac{\lambda_n(K)}{\lambda_{n-1}(K \cap u^{\perp})} \le h_{\Gamma_2 K}(u) \le \frac{n}{\sqrt{2(n+2)}}\frac{\lambda_n(K)}{\lambda_{n-1}(K \cap u^{\perp})}$$

of Milman and Pajor [621] obtained in their investigation of the slicing problem (see Note 9.8). In a remarkable sequel [549], Lutwak, Yang, and Zhang find new methods that enable them to define $\Gamma_{-2}L$ for any star body L in $\mathbb{E}^n$ and to show that $\Gamma_{-2}L \subset \Gamma_2 L$. They explain why this can be regarded as an analog of the Cramer–Rao inequality from information theory.

Work of Ludwig [525] on matrix-valued valuations indicates that the Legendre ellipsoid $\Gamma_2 K$ and the LYZ ellipsoid $\Gamma_{-2}K$ enjoy a special status among all ellipsoids defined in terms of a convex body K.

In a further study, Lutwak, Yang, and Zhang [552] introduce the L^p-John ellipsoid of a convex body K, denoted by $E_p K$, where $p > 0$. The classical John ellipsoid is $E_\infty K$ and $E_2 K = \Gamma_{-2}K$. Among many results, it is shown that $\Gamma_{-p}K \subset E_p K$ when $0 < p < 2$ and the reverse inclusion holds when $p > 2$.

Let L be a star body in $\mathbb{E}^n$ containing the origin in its interior, and denote the polar body of the L^p-centroid body $\Gamma_p L$ of L by $\Gamma_p^* L$. Lutwak and Zhang [553] prove that if $p \ge 1$, then the maximum of the volume product $\lambda_n(L)\lambda_n(\Gamma_p^* L)$ is attained if and only if L is a centered ellipsoid. This result is called the L^p-Blaschke–Santaló inequality, since if L is a centered convex body, then as $p \to \infty$, $\Gamma_p^* L \to L^*$ and Lutwak and Zhang's bound becomes the Blaschke–Santaló inequality for centered convex bodies, Theorem 9.2.11. Lutwak and Zhang also formulate an equivalent analytic inequality and mention some interesting open problems, one a generalization of the slicing problem (see Note 9.8). Lutwak, Yang, and Zhang [547] solve one of these problems by proving the L^p-Busemann–Petty centroid inequality; this gives a lower bound for $\lambda_n(\Gamma_p L)/\lambda_n(L)$, when L is a star body in $\mathbb{E}^n$ containing the origin in its interior and $1 \le p < \infty$, that is attained if and only if L is a centered ellipsoid. When $p = 1$, this becomes the Busemann–Petty centroid inequality, Corollary 9.2.7. The L^p-Busemann–Petty centroid inequality is obtained as a corollary of a remarkable L^p-Petty projection inequality that gives an upper bound for $\lambda_n(K)^{(n-p)/p}\lambda_n(P_p^* K)$, when K is a convex body in $\mathbb{E}^n$ and $1 \le p < \infty$, that is attained if and only if K is a centered ellipsoid. Here $P_p^* K$ is the polar body of the L^p-projection body $P_p K$ of K defined in Note 4.7, and the Petty projection inequality, Theorem 9.2.9, corresponds to the case $p = 1$.

Campi and Gronchi [146], [149] discover completely different proofs of the results in the previous paragraph. Their technique employs *shadow systems*, a powerful generalization of the Steiner symmetrization process first introduced by Rogers and Shephard in the course of proving the Rogers–Shephard inequality. In [147], Campi and Gronchi obtain further results, including a sharp upper bound for $\lambda_n(\Gamma_p K)/\lambda_n(K)$, when K is a centered convex body in $\mathbb{E}^2$ and $1 \le p < \infty$, that is attained if and only if K is a centered parallelogram. The case $p = 1$ was proved earlier in [67]; see Note 9.4.

The interested reader can also consult other works of Campi and Gronchi [148] and Lutwak, Yang, and Zhang [550], [551] related to those just mentioned. Yaskin and Yaskina [850] extend the definition of the polar L^p-centroid body $\Gamma_p^* L$ to include the values $-1 < p < 1$.

Gardner and Zhang [284] define for $p > -1$ the *radial pth mean body* $R_p K$ of a convex body K. The radial function of $R_p K$ is the pth mean of the values of the extended radial function $\rho_K(x, u)$ of K (see the definition in the proof of Lemma 8.1.15) with respect to points inside K. As $p \to -1$, a suitable dilatate of $R_p K$ approaches the polar projection body $\Pi^* K$ of K, and $R_\infty K = 2\Delta K$, the difference body of K. It is proved that $R_p K$ is convex for $p \geq 0$ (the convexity of $R_p K$ is an open question for $-1 < p < 0$). The authors also obtain a sharp affine inequality relating the volumes of $R_p K$ and $R_q K$, $p \neq q$, which yields as special cases both the Rogers–Shephard inequality and the Zhang projection inequality. A related notion, the p-cross-section body, is described in Note 8.12.

9.6. *Volume estimates from projection functions.* The upper bound in Theorem 9.3.1 is due to Lutwak [532]. Santaló [721, p. 189] proved Theorem 9.3.2, thereby strengthening the *Urysohn inequality* (the case $p = 1$), in turn an improvement of the *Bieberbach inequality* (the case $p = \infty$). Another proof was given by Lutwak [530]. Santaló also stated a nonoptimal lower bound.

Theorems 3.2.3 and 3.3.9 already show that the unit ball in $\mathbb{E}^n$ has maximum volume among all bodies of the same constant width or brightness. The problem of finding the minimal volume, still unsolved for $n \geq 3$ (in the plane, it is attained by a Reuleaux triangle), is discussed by Chakerian and Groemer [154, Section 7] and Croft, Falconer, and Guy [173, Problem A22].

The affine quermassintegrals $\Phi_{n-i}(K)$, $1 \leq i \leq n - 1$, were introduced by Lutwak [532]. Grinberg [334] justified the term by proving that these quantities are invariant under volume-preserving affine transformations. Problems 9.3 and 9.5 were posed by Lutwak [536]. The proof of Theorem 9.3.1 indicates that the inequality of Problem 9.3 (for $i = n - 1$) is the Petty projection inequality, while the proof of Theorem 9.3.2 shows that the case $i = 1$, for centrally symmetric bodies, is the Blaschke–Santaló inequality. The conjectured inequality of Problem 9.5 holds when affine quermassintegrals are replaced by quermassintegrals (see [737, p. 334, eq. (6.4.7)]), and would imply the inequality of Problem 9.3, since the latter is just the special case $i = 0$. Lutwak [536] contributes to these problems by showing that if K is a convex body in $\mathbb{E}^n$, and $0 \leq i \leq j \leq n - 1$, then

$$\kappa_n^j \hat{W}_i(K)^{n-j} \leq \kappa_n^i \hat{W}_j(K)^{n-i},$$

with equality when $i < j$ if and only if K is a ball. The quantities in this inequality are the *harmonic quermassintegrals* of K, defined by $\hat{W}_0(K) = \lambda_n(K)$, $\hat{W}_n(K) = \kappa_n$, and

$$\hat{W}_{n-i}(K) = \frac{\kappa_n}{\kappa_i} M_{-1}(v_{i,K}) = \frac{\kappa_n}{\kappa_i} \left(\int_{\mathcal{G}(n,i)} \lambda_i(K|S)^{-1} \, dS \right)^{-1},$$

for $1 \leq i \leq n - 1$; they were introduced by Hadwiger [370, Section 6.4.8].

In [533, Corollary 6.5], Lutwak finds the following upper bound for the volume of a convex body K in $\mathbb{E}^n$ in terms of its ith girth function: If $1 \leq i < n - 1$ and $p \geq -n$, then

$$\kappa_n^{-(n-i)/n} \lambda_n(K)^{i/n} \leq \frac{c_{n-1,i}\kappa_n}{\kappa_{n-1}} \left(\frac{1}{n\kappa_n} \int_{S^{n-1}} V_i(K|u^\perp)^p \, du \right)^{1/p},$$

with equality if and only if K is a ball. (Here, $c_{n-1,i}$ is as in (A.27).) Note that when $i = n - 1$, this is Theorem 9.3.1, except that there equality holds for ellipsoids if $p = -n$. Lutwak extends this result in [535, Corollary 5.20], obtaining a similar upper bound for $V_i(K)$ in terms of the jth girth function of K, $1 \leq j \leq i < n - 1$.

When $i = n - 1$, Theorem 9.3.3 is contained in a result of Goodey and Zhang [326].

Problem 9.4 is inspired by the paper [584] of Mattila, who uses Riesz capacities to get estimates of the form

$$\kappa_n^{(n-i)/n} \lambda_n(E)^{i/n} \leq \frac{c\kappa_n}{\kappa_i} M_{-1}(v_{i,E}),$$

$1 \leq i \leq n - 1$, for a Borel subset E of $\mathbb{E}^n$, where c is a constant depending only on n and i. When $i = n - 2$, Mattila proves that $c = 1$, so Problem 9.4 would have a positive answer for this value of i if the affine quermassintegral were replaced by the corresponding harmonic quermassintegral. For $i \neq n - 2$, however, the optimal value of c is apparently unknown.

Theorem 9.3.7 is due to Betke and McMullen [55]. The authors apply the result to recalculate examples in the earlier paper [636] of Moran, where projections in $\mathbb{E}^3$ are taken in directions orthogonal to the facets of a regular icosahedron or dodecahedron. (In the planar case, Kendall and Moran [427, p. 74] ascribe the method to H. Steinhaus.) In [55], the more general problem is considered in which $V_i(K)$ is estimated from a finite set of values of the projection function $V_i(K|S)$, $S \in \mathcal{G}(n,k)$, $1 \leq i \leq n - 1$. The problem is also solved for $i = k = 1$, but optimal results are unavailable otherwise. Betke and McMullen observe that their problem is closely connected with the approximation of the unit ball by zonotopes; see Note 4.1 for references on this topic.

When $\dim K = n - 1$, there is equality in Theorem 9.3.8 for all $u \in S^{n-1}$; this fact, called the *Binet–Cauchy formula* by Evans and Gariepy [212], has been rediscovered often. The inequality of Theorem 9.3.8 extends readily to the ith girth function, for $1 \leq i \leq n - 1$, since ΠK can be replaced by the ith projection body $\Pi_i K$ in the proof. The resulting *Pythagorean inequalities* were discovered by Firey [234].

The *Loomis–Whitney inequality* says that if E is a Borel set in $\mathbb{E}^n$, and e_j, $1 \leq j \leq n$, denote the unit vectors in the coordinate directions, then

$$\lambda_n(E)^{n-1} \leq \prod_{j=1}^{n} \lambda_{n-1}(E|e_j^\perp).$$

Observe that equality holds for boxes. The proof is given in [521] and in the book [668, Lemma 12.1.4] of Pfeffer. A more general version in which the upper bound is expressed in terms of the projections of E on lower-dimensional coordinate subspaces, stated by Burago and Zalgaller [112, Theorem 11.3.1, p. 95] and Hadwiger [370, Theorem 4.4.2, pp. 161–3], follows from the previous inequality by induction. A different generalization of the inequality is obtained by Ball [24].

Motivated by a file management problem, Schwenk and Munro [758] consider the minimum of the geometric mean of the cardinality of projections of a finite subset E of $\mathbb{E}^n$ onto coordinate subspaces of fixed dimension k. Their result in the case $k = n - 1$, for example, is that

$$\left(\prod_{j=1}^{n} |E|e_j^\perp| \right)^{1/n} \geq |E|^{(n-1)/n}.$$

However, this follows immediately from (and is equivalent to) the Loomis–Whitney inequality, as can be seen by replacing each point in E by a small cube whose facets are parallel to the coordinate hyperplanes. Their general result follows similarly from a generalization of the Loomis–Whitney inequality mentioned above. Schwenk [757] tackles a variant of the problem above in which the geometric mean is replaced by the maximum, and solves this when $n = 3$ and $k = 2$, noting that the general case appears very difficult (compare Note 9.8).

Suppose that $u_j \in S^{n-1}, 1 \leq j \leq m$, contains a basis for $\mathbb{E}^n$, and K is a convex body in $\mathbb{E}^n$. Consider the class $\mathcal{B}$ of all convex bodies K' such that $v_{K'}(u_j) = v_K(u_j)$ for $1 \leq j \leq m$. Campi, Colesanti, and Gronchi [142] use the Loomis–Whitney inequality to prove that there is a unique centered polytope in $\mathcal{B}$ of greater volume than any other member of $\mathcal{B}$. (This is in fact the polytope P in Theorem 4.4.2.) They note that $\mathcal{B}$ may contain no member of minimal volume, and if it does, there may be infinitely many members of $\mathcal{B}$ that attain the minimum; moreover, they give a criterion for the existence of members of $\mathcal{B}$ of minimal volume, and prove that these must be polytopes with no more than $m + n$ facets.

Bollobás and Thomason [80] prove that for each body E in $\mathbb{E}^n$ and each coordinate subspace S there is a box P such that $\lambda_n(P) = \lambda_n(E)$ and $\lambda_k(P|S) \leq \lambda_k(E|S)$, where dim $S = k$. (When E is convex and $k = n - 1$, the first result of Campi, Colesanti, and Gronchi [142] stated in the previous paragraph is stronger.)

Campi, Colesanti, and Gronchi [143] study the minimum volume of convex bodies of revolution of constant width or brightness. Gronchi [360] proves that for constant brightness this minimum occurs precisely for the body constructed in Theorem 3.3.15.

If a finite set of the projections themselves, rather than the projection functions, are known, one can still inquire how best to estimate the volume. Small [772] shows that the natural idea of approximating the body by intersecting cylinders over the projections is more efficient than estimates based on Cauchy's surface area formula and the isoperimetric inequality.

9.7. *Volume estimates from section functions.* The dual affine quermassintegrals $\tilde{\Phi}_{n-i}(K), 1 \leq i \leq n - 1$, are shown by Grinberg [334] to be invariant under volume-preserving linear transformations.

Theorem 9.4.4 is a consequence of a more general result established by Busemann and Straus [132, p. 70, eq. (9.4)] and independently by Grinberg [334]. (Schneider and Weil [746, Satz 6.3.3] present both this and a variant [746, Satz 6.3.5] applying to sections by all planes instead of all subspaces.) Various special cases have appeared. For example, the Busemann intersection inequality, Corollary 9.4.5, is proved by Busemann [126]; the Furstenberg–Tzkoni formula (Corollary 9.4.7) appears in [255]; and the dual Furstenberg-Tzkoni formula (Corollary 9.4.8) is due to Lutwak [536]. See [542, Section 7] for more about the Busemann intersection inequality and related matters.

When $i = n - 1$, Theorem 9.4.9 is contained in a result of Zhang [858].

Theorem 7.2.2 already shows that B has the minimal volume of all star bodies L in $\mathbb{E}^n$ with $o \in \text{int } L$ and of the same constant i-section as B. For $i = 1$, Groemer [349] finds that the (unattained) maximal volume corresponds to that of a half-ball of radius 2.

Meyer [599] establishes an analogue of the Loomis–Whitney inequality. He shows that if K is a convex body in $\mathbb{E}^n$, and $e_j, 1 \leq j \leq n$, are the unit vectors in the coordinate directions, then

$$\lambda_n(K)^{n-1} \geq \frac{\left((n-1)!\right)^n}{(n!)^{n-1}} \prod_{j=1}^{n} \lambda_{n-1}\left(K \cap e_j^{\perp}\right),$$

with equality if and only if K is a cross-polytope with vertices on the coordinate axes. Generalizations are given by Meyer and also by Scott [759]. In related work, McMullen [591] shows that if $F_j, 1 \leq j \leq m$, are mutually orthogonal planes in $\mathbb{E}^n$, with dimensions d_j summing to n, then

$$\lambda_n(K) \geq \frac{d_1! \cdots d_m!}{n!} \prod_{j=1}^{m} \lambda_{d_j}(K \cap F_j).$$

By "dualizing" the proof of Theorem 9.3.8, it is not difficult to obtain the following *dual Pythagorean inequalities*: If K is a centered convex body in $\mathbb{E}^n$, e_j, $1 \le j \le n$, denote the unit vectors in the coordinate directions, and $1 \le i \le n-1$, then

$$\tilde{v}_{i,K}(u)^{-2} \le \sum_{j=1}^{n} \tilde{v}_{i,K}(e_j)^{-2},$$

for all $u \in S^{n-1}$.

Sallee [716], extending a result of McKenney [589], finds that precisely $[(n-i+1)^2 + (n-i+1)]/2$ nonzero values of $\tilde{v}_{i,E}$ are needed to determine the volume of an n-dimensional ellipsoid E in $\mathbb{E}^n$. McKenney encountered the problem while consulting at a medical center, where medical researchers wished to determine the volume of part of a rat's lung from cross-sectional samples.

9.8. *The slicing problem.* The slicing problem, Problem 8.3, was mentioned briefly in Note 8.9, in connection with the Busemann–Petty problem. We are now in a position to examine versions of the problem in more detail. The most informative published article concerning the problem is that of Milman and Pajor [621], and in this summary we shall label the constants as they do.

Perhaps the simplest of these versions asks if there is a universal constant c_4 such that for each centered convex body K in $\mathbb{E}^n$, there is an $S \in \mathcal{G}(n, n-1)$ with

$$\lambda_n(K)^{(n-1)/n} \le c_4 \lambda_{n-1}(K \cap S). \tag{9.10}$$

("Universal" means that c_4 does not depend on K or on n.) This explains why the term "maximal slice problem" has been used. If we let L_2 be the centered unit cube in Theorem 9.4.9, we see that (9.10) holds, with $c_4 = 1$, whenever K is the intersection body of a star body. From work of Ball [23] it follows that there is a universal constant c_4 such that (9.10) holds whenever K is a projection body. Since each polar projection body is an intersection body by Remark 8.1.21(viii), (9.10) also holds when K is a polar projection body, and in this case Zhang [861] finds that one can take $c_4 < 1$. Other partial results are obtained by Junge [411], [412]. Milman [617] uses dual mixed volumes to re-prove, strengthen, and generalize the results of Ball and Junge. It seems possible that $c_4 = 1$ for all centered convex K. Bourgain, Klartag, and Milman [89], [90] prove that there is a constant $c > 0$ such that the slicing problem can be restricted to the class of convex bodies whose volume ratios are no larger than c.

Milman and Pajor begin their discussion with an apparently quite different, though equivalent, problem. A body M in $\mathbb{E}^n$ is called *isotropic* if $\lambda_n(M) = 1$ and

$$\int_M |u \cdot x|^2 \, dx = L_M^2$$

is independent of $u \in S^{n-1}$, so M has the same moments of inertia with respect to each axis through the origin. In this case L_M is called the *constant of isotropy*. The starting point in [621] is the problem of whether there is a universal constant C such that for each isotropic centered convex body K in $\mathbb{E}^n$,

$$L_K \le C. \tag{9.11}$$

Isotropic bodies are not as special as they may seem. Indeed, any body M in $\mathbb{E}^n$ is isotropic if and only if its L^2-centroid body $\Gamma_2 M$ (defined in Note 9.5) is a ball, so there is always a $\phi \in GL_n$ such that ϕM is isotropic. Proofs are also given by Petty [661] and Ball [20].

Let us now consider yet another formulation of the slicing problem. This asks if there is a universal constant c_3 such that for each centered convex body in $\mathbb{E}^n$,

$$\lambda_n(K)^{(n-1)/n} \leq (c_3 C) M_n(\tilde{v}_K) = c_3 C \left(\int_{\mathcal{G}(n,n-1)} \lambda_{n-1}(K \cap S)^n \, dS \right)^{1/n},$$
$$(9.12)$$

where the constant C is from (9.11). Note that the Busemann intersection inequality, Corollary 9.4.5, is in the opposite direction; in fact, Busemann [126, p. 11] asks for a (not necessarily universal) bound as in (9.12). In [621, Proposition 5.4], it is shown that (9.11) $\Leftrightarrow$ (9.12), that is, this formulation is equivalent to the previous one, and since $M_n(\tilde{v}_K) \leq M_\infty(\tilde{v}_K)$, we see that (9.12) $\Rightarrow$ (9.10). It is not hard to check that (9.12) is also equivalent to asking whether the constant $\tilde{a}_{n,i}$ in Problem 9.6 is universal when $i = n - 1$, as well as to the existence of a universal constant c_0 such that

$$\lambda_n(K)^{n-1} \leq c_0 \lambda_n(IK).$$

Before discussing the implication (9.10) $\Rightarrow$ (9.11), we shall explain the link to Problem 8.3, given in [621, Proposition 5.5]. The proof that a positive answer to Problem 8.3 yields (9.10) is essentially the same as that of Theorem 9.4.9 (or Theorem 9.3.3). The converse follows directly from (9.12) and the Busemann intersection inequality, Corollary 9.4.5.

The implication (9.10) $\Rightarrow$ (9.11) is a consequence of the fact that there are universal constants $c_1 > 0$ and c_2 such that for each isotropic centered convex body K in $\mathbb{E}^n$,

$$\frac{c_1}{L_K} \leq \lambda_{n-1}(K \cap S) \leq \frac{c_2}{L_K},$$
$$(9.13)$$

for all $S \in \mathcal{G}(n, n - 1)$. These estimates, first discovered by Hensley [392] (see also [621, Corollary 3.2]), can be expressed in other ways. For example, for each isotropic centered convex body K in $\mathbb{E}^n$, we have

$$\frac{\lambda_{n-1}(K \cap S_1)}{\lambda_{n-1}(K \cap S_2)} \leq \frac{c_2}{c_1} = \gamma_1,$$

for all $S_1, S_2 \in \mathcal{G}(n, n - 1)$. It is easy to prove (cf. [621, Lemma 4.1]) that there is a universal constant $c > 0$ such that $c \leq L_K$ for each isotropic centered convex body K in $\mathbb{E}^n$, so from (9.13) we obtain

$$\lambda_{n-1}(K \cap S) \leq \frac{c_2}{L_K} \leq \frac{c_2}{c} = \frac{c_2}{c} \lambda_n(K)^{(n-1)/n},$$

for each such K and *all* $S \in \mathcal{G}(n, n - 1)$. On the other hand, the question of whether there exists a universal constant c' such that for each isotropic centered convex body K in $\mathbb{E}^n$,

$$1 = \lambda_n(K)^{(n-1)/n} \leq c' \lambda_{n-1}(K \cap S),$$
$$(9.14)$$

for *all* $S \in \mathcal{G}(n, n - 1)$, is again equivalent to the slicing problem. (The reader can check that (9.10) $\Rightarrow$ (9.14), where $c' = c_4 \gamma_1$, and (9.14) $\Rightarrow$ (9.11), where $C = c' c_2$.)

Hensley [392] proved, more generally, that for each isotropic centered convex body K in $\mathbb{E}^n$, and $1 \leq i \leq n - 1$, there is a constant γ_i depending only on i such that

$$\left(\frac{\lambda_{n-i}(K \cap S_1)}{\lambda_{n-i}(K \cap S_2)} \right)^{1/i} \leq \gamma_i,$$

for all $S_1, S_2 \in \mathcal{G}(n, n-i)$. Hensley's work yields $\gamma_i = O(i!)$, which was improved to $\gamma_i = O(\sqrt{i})$ by Ball [20]. Another generalization of (9.13), proved by Milman and Pajor [621, Proposition 3.11] and implicit in [20], states that that there are universal constants $c_1 > 0$ and c_2 such that for each isotropic centered convex body K in $\mathbb{E}^n$,

$$\frac{c_1 L_H}{L_K} \le \lambda_{n-i}(K \cap S)^{1/i} \le \frac{c_2 L_H}{L_K},$$

for all $S \in \mathcal{G}(n, n-i)$, where H is an auxiliary convex body depending on i. A positive answer to the slicing problem would therefore yield universal constants γ and $\delta > 0$ (independent of i) such that

$$\delta^i \le \lambda_{n-i}(K \cap S) \le \gamma^i, \tag{9.15}$$

for each isotropic centered convex body K in $\mathbb{E}^n$ and all $S \in \mathcal{G}(n, n-i)$. Note that when $K = C^n$, the centered unit cube in $\mathbb{E}^n$, Vaaler and Ball have already verified that (9.15) holds for $\delta = 1$ and $\gamma = \sqrt{2}$ in their work on critical sections of the cube, discussed in Note 7.8.

In view of the previous remarks, a positive answer to the slicing problem implies that for each centered convex body K in $\mathbb{E}^n$, there are universal constants γ and $\delta > 0$ and a $\phi \in GL_n$ such that

$$\delta^{n-i} \lambda_n(\phi K)^{i/n} \le \lambda_i(\phi K \cap S) \le \gamma^{n-i} \lambda_n(\phi K)^{i/n}, \tag{9.16}$$

for all $S \in \mathcal{G}(n, i)$. The conjecture that $\delta = 1$ in the left-hand inequality originates in work of Vaaler [810]. Ball [22] confirms this conjecture when $i = 1$, by estimating the volume ratio $vr(K)$ of a centered convex body K in $\mathbb{E}^n$. Ball proves that

$$vr(K) \le vr(C^n) = \frac{2}{\kappa_n^{1/n}},$$

where C^n is the centered unit cube in $\mathbb{E}^n$. Compare Ball's estimate for arbitrary convex bodies given in Remark 4.2.14; Vaaler's conjecture for $i = 1$ follows by choosing ϕ so that $\phi E = B$. For more on volume ratios, see the works of Ball [23], [25], Bourgain and Milman [92], and Pisier [673, Chapter 6].

Ball [24] has proved that the left-hand inequality in (9.16) is true when sections are replaced by projections, with $\delta = 1$ and $i = n-1$; for projections, the case $i = 1$ also follows from the volume-ratio estimate in the previous paragraph.

Fradelizi [247] obtains best-possible inclusion relationships between the ellipsoid $\Gamma_2^* K$ and the intersection body IK (also the cross-section body CK) when K is a convex body in $\mathbb{E}^n$ with centroid at the origin. He uses these to show the equivalence for arbitrary convex bodies of some of the formulations of the slicing problem considered above. Klartag and Milman [443] offer yet another problem, involving Steiner symmetrization, that is equivalent (though not obviously so) to the slicing problem.

As he does for the Busemann–Petty problem (see Note 8.10), Zvavitch [871] poses a version of the slicing problem in which λ_n is replaced by the Gaussian measure in $\mathbb{E}^n$. This remains open, but Zvavitch obtains a weaker positive result.

9.9. *Central limit theorems for convex bodies.* Brehm and Voigt [99] (see also Voigt's paper [817]) ask whether there is a "central limit theorem" for convex bodies, in the following sense. Let K be an isotropic convex body in $\mathbb{E}^n$ with constant of isotropy L_K, and suppose $g_{L_K^2}$ is the Gaussian probability density function with zero mean and variance L_K^2. For each $u \in S^{n-1}$, let

$$A_K(t, u) = X_{u^\perp} K(tu) = \lambda_{n-1}(K \cap (u^\perp + tu))$$

be the $(n-1)$-dimensional X-ray of K orthogonal to u. Then Brehm and Voigt conjecture that

$$\sup_{K \in \mathcal{K}} \mathbb{E} \left(\sup_{t \in \mathbb{R}} |A_K(t,u) - g_{L_K^2}(t)| \right) \to 0$$

as $n \to \infty$, where $\mathbb{E}$ denotes the expectation with respect to normalized spherical Lebesgue measure in S^{n-1}. They formulate other related conjectures and provide supporting evidence for them. Similar ideas were pursued independently by Antilla, Ball, and Perissinaki [11]. See also further work of Brehm, Voigt, and coauthors in [100] and [101].

Koldobsky and Lifshits [466] obtain precise asymptotic estimates for the average volume of k-dimensional central sections of the unit ball in l_p^n, $0 < p \le \infty$, as $n \to \infty$. If K is a centered convex body in $\mathbb{E}^n$, let

$$H_K(t) = \frac{1}{\omega_n} \int_{S^{n-1}} A_K(t,u) \, du$$

be the average of the volumes of its sections by hyperplanes at distance $t \ge 0$ from the origin. In [466], the limit of $H_K(t)$ as $n \to \infty$ is evaluated for a centered unit cube K, and other related results are obtained. Bobkov and Koldobsky [75] give an estimate for $|H_K(t) - g_{L_K^2}(t)|$ for $0 < t \le c\sqrt{n}$ for a universal constant c, in terms of n, L_K, and a quantity depending on the variance of the square of the absolute value of a random vector uniformly distributed over K. See also the paper of Paouris [656] for related results.

9.10. *Estimates concerning both projections and sections.* Rogers and Shephard [696] showed that if K is a convex body in $\mathbb{E}^n$ with $o \in K$, and $1 \le i \le n-1$, then

$$\lambda_i(K \cap S)\lambda_{n-i}(K|S^\perp) \le \binom{n}{i} \lambda_n(K),$$

for each $S \in \mathcal{G}(n,i)$, with equality if and only if all i-dimensional sections of K parallel to S are similar and similarly situated. In the other direction, Spingarn [782] proved that if K is a convex body in $\mathbb{E}^n$ with centroid at o, and $1 \le i \le n-1$, then

$$\lambda_n(K) \le \lambda_i(K \cap S)\lambda_{n-i}(K|S^\perp),$$

for each $S \in \mathcal{G}(n,i)$, with equality if and only if all i-dimensional sections of K parallel to S are translates.

These results have several interesting consequences. When $i = 1$, we have

$$\lambda_n(K) \le \max_{x \in u^\perp} \lambda_1 \left(K \cap (l_u + x) \right) v_K(u) = 2\rho_{\triangle K}(u) h_{\Pi K}(u) \le n\lambda_n(K),$$

for all $u \in S^{n-1}$. The right-hand inequality is due to A. M. Macbeath; see Martini's paper [580] and others listed in his survey [581, Section 4], where conditions for equality in various sets of directions are studied and the relationship to Problem 4.1 is explained. Choosing u in the direction of a diameter of K, we retrieve an inequality of Firey [238]:

$$(\text{diam } K) \left(\min_{u \in S^{n-1}} v_K(u) \right) \le n\lambda_n(K). \tag{9.17}$$

(See also the work of Martini and Weissbach [582]. A generalization to ith projection functions and intrinsic volumes, with a different proof, is given by Lutwak and Oliker [545].) The case $i = n - 1$ yields

$$\lambda_n(K) \le m_K(u)w_K(u) = 2\rho_{CK}(u)h_{\triangle K}(u) \le n\lambda_n(K),$$

for all $u \in S^{n-1}$, where m_K is the inner section function and CK is the cross-section body of K (see Section 8.3). Again, see [581] for further references. If u is in the direction of the minimal width of K, we get

$$2\lambda_n(K) \leq (\text{minw}\, K)S(K),\qquad (9.18)$$

since $S(K) \geq 2v_K(u) \geq 2m_K(u)$ for all $u \in S^{n-1}$. This can also be found in [545].

9.11. *Estimates for inradius and circumradius.* Suppose that K is a convex body in $\mathbb{E}^n$. There are known inequalities of the form $R(K) \leq c\,\text{diam}\, K$, where c is a constant depending only on n, as in Jung's theorem (see Eggleston's book [205, Theorem 49, p. 111], for example). This can then be combined with (9.17) above and Theorem 9.3.1 to yield an upper bound for the circumradius of K in terms of the brightness function of K. There are also known inequalities of the form $r(K) \geq c'\text{minw}\, K$, where c' is a constant depending only on n, as in [205, Theorem 50, p. 112], for example. This can then be combined with (9.18), Theorem 9.3.1, and Cauchy's surface area formula (A.49) to arrive at a lower bound for the inradius of K in terms of the brightness function of K.

For bodies of constant width, estimates are stated by Chakerian and Groemer [154, eq. (7.3)].

Figure 9.2. Hugo Hadwiger.

9.12. *Hugo Hadwiger (1908–1981).* (Sources: [183], [364].) Hugo Hadwiger was born in Karlsruhe, Germany, but grew up in Bern, Switzerland. He studied mathematics at the University of Bern, with minors in physics and actuarial mathematics. He obtained his doctorate in 1934, and by 1937 was professor of higher analysis at the University of Bern, where he spent the rest of his life, despite many offers of positions at other universities. His work in geometry began after visiting Blaschke (cf. Note 6.4) in Hamburg in 1935. As director of the mathematical institute at his university, Hadwiger was demanding but accessible, and noted for his elegant, clear lectures and beautiful writing on the blackboard. He remained modest as his fame grew; when asked during a lecture about the origin of a result belonging to himself, he would say, "This was proved in Bern." From 1963 on Hadwiger suffered from polyneuritis without complaining, and in fact considered himself lucky to enjoy his work.

Hadwiger's 251 publications include some papers on stochastic and actuarial problems, but most contribute toward his reputation as one of the leading geometers of the twentieth century. His papers address so many different problems that it is difficult to single out any particular topic. Indeed, his predilection for open problems and problem solving led him to introduce the column *Ungelöste Probleme* (*Unsolved Problems*) in the journal *Elemente der Mathematik*. This inspired the *Research Problems* section of the *American Mathematical Monthly*, whose first article (that of Klee [446] on the equichordal problem) was dedicated to Hadwiger on his sixtieth birthday. Hadwiger wrote extensively on combinatorial geometry, and problems involving covering or lattice points, but the theme of most impact for geometric tomography is his investigation of metric properties of sets in $\mathbb{E}^n$. His book [370] was at that time a comprehensive treatise on measure and geometry, and there is still no equivalent in the English language today. It refers to 64 of Hadwiger's own papers, including such achievements as his characterization of valuations (cf. Section A.5) and his extension of the Euler–Poincaré characteristic and various integral-geometric formulas to the convex ring. The book also treats the fascinating topic of Hilbert's third problem on equidissections of polytopes, where Hadwiger extended Dehn's necessary conditions for two polytopes to be equidissectable with respect to the group of all isometries from $\mathbb{E}^3$ to $\mathbb{E}^n$, and specified a corresponding set of conditions for the group of translations.

Appendix A

Mixed volumes and dual mixed volumes

The theory of mixed volumes provides a unified treatment of various important metric quantities in geometry, such as volume, surface area, and mean width. Apart from some historical roots in the works of Steiner [788] and Brunn [109], [110], its creation is due to Minkowski [624], [626]. The theory of area measures goes a step further, and can be regarded as a localization of the theory of mixed volumes. Area measures were introduced in the late 1930s, by Aleksandrov [2] and by Fenchel and Jessen [230], independently.

Until recently, there was no adequate introduction to these important topics in English, but fortunately, this situation has changed. The primary source of information is now Schneider's book [737], a superb sequel to Bonnesen and Fenchel's treatise [83]. (The latter is still well worth consulting, though it appeared too early for area measures to be included. It is regrettable that the books of Blaschke [71] and Hadwiger [370] have not yet been translated into English.) Apart from this, Webster's text [827, Chapter 6] provides an introduction to mixed volumes, and summaries of the theory are provided by Burago and Zalgaller [112, Chapter 4] and Sangwine-Yager [718].

Most of this appendix is also a summary, tailored to our particular requirements, of the theory of mixed volumes and area measures. The first section is designed to open the door to this enchanting but labyrinthine palace. After this, the approach is taken from that of Schneider in [737], and references are generally given to this work. In particular, we refer the reader to the extensive notes in Chapters 4 and 5 of [737] for detailed historical background.

The final section is devoted to the relatively new dual mixed volumes, due to Lutwak [529]. More Bauhaus than Byzantine in structure, the theory of dual mixed volumes is no less valuable in geometric tomography, since it caters to problems about sections just as efficiently as mixed volumes deal with projections.

A.1. An example

Let K be a box in $\mathbb{E}^3$, and let $\varepsilon > 0$ be suitably small. The Minkowski sum $K + \varepsilon B$, alias the outer parallel set K_ε of K (cf. (0.5)), can be partitioned in a natural way into K itself; six congruent copies of a box P, one on each of the facets of K; twelve congruent copies of a quarter-cylinder C, one on each of the edges of K; and eight congruent copies of an eighth part O of a translate of B, one at each of the vertices of K. (See Figure A.1, where only the front of $K + \varepsilon B$ is shown, for simplicity.)

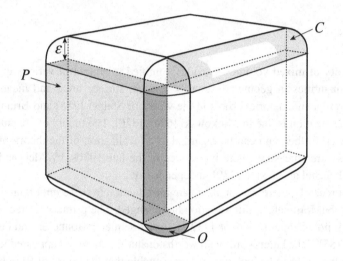

Figure A.1. The convex body $K + \varepsilon B$ (front only).

Therefore

$$\lambda_3(K + \varepsilon B) = \lambda_3(K) + 6\lambda_3(P) + 12\lambda_3(C) + 8\lambda_3(O).$$

The volumes of the boxes, quarter-cylinders, and eighth-balls vary with ε, ε^2, and ε^3, respectively. We have

$$6\lambda_3(P) = S(K)\varepsilon,$$

where $S(K)$ is the surface area of K,

$$12\lambda_3(C) = \frac{\pi}{4}T(K)\varepsilon^2,$$

where $T(K)$ denotes the total length of the edges of K, and of course

$$8\lambda_3(O) = \lambda(\varepsilon B) = \lambda_3(B)\varepsilon^3.$$

Consequently, the volume of $K + \varepsilon B$ can be expressed as the following polynomial of degree 3 in the parameter ε:

$$\lambda_3(K + \varepsilon B) = \lambda_3(K) + S(K)\varepsilon + \frac{\pi}{4}T(K)\varepsilon^2 + \lambda_3(B)\varepsilon^3.$$

Up to constant multiples, the coefficients in this polynomial are intrinsic volumes, special types of mixed volumes. Note that the first and last coefficients are just the volumes of K and B, respectively. Also, the second coefficient is the surface area of K, an important geometric quantity. The remaining coefficient is the sum of the lengths of the edges of K, each weighted by its external angle.

This polynomial in ε is an instance of Steiner's formula (A.30), obtained by repeating the process for an arbitrary compact convex set K in $\mathbb{E}^n$. The appearance of the volumes of K and B as first and last coefficients, and of the surface area $S(K)$ as second coefficient, is perfectly general (see (A.28), (A.33), and (A.35)). For a polytope in $\mathbb{E}^n$, the $(n-1)$th coefficient of the polynomial is still the sum of the lengths of the edges, each weighted by its external angle (see (A.36)), but for general compact convex sets it relates to the mean width, as in (A.50).

Let us now replace K and B by arbitrary compact convex sets K_1 and K_2, and suppose that $t_j \geq 0$, $j = 1, 2$. We might expect that the volume of the Minkowski sum $t_1 K_1 + t_2 K_2$ can, in similar fashion, be written as a polynomial of degree 3 in the parameters t_1 and t_2. By analogy with the algebraic expansion

$$(t_1 x_1 + t_2 x_2)^3 = x_1^3 t_1^3 + 3x_1^2 x_2 t_1^2 t_2 + 3x_1 x_2^2 t_1 t_2^2 + x_2^3 t_2^3,$$

we optimistically write

$$\lambda_3(t_1 K_1 + t_2 K_2) = V(K_1, K_1, K_1)t_1^3 + 3V(K_1, K_1, K_2)t_1^2 t_2$$
$$+ 3V(K_1, K_2, K_2)t_1 t_2^2 + V(K_2, K_2, K_2)t_2^3.$$

Coefficients such as $V(K_1, K_1, K_2)$ – our earlier simple example disguises the fact that the second and third coefficients will generally depend on both K_1 and K_2 – are regarded as unchanged if the order of the arguments is permuted, so

$$V(K_1, K_2, K_1) = V(K_1, K_1, K_2),$$

just as $x_1 x_2 x_1 = x_1^2 x_2$. These coefficients are called mixed volumes, and Minkowski's theorem on mixed volumes (Theorem A.3.1) validates what we have done. If we set $K_1 = K$, $K_2 = B$, $t_1 = 1$, and $t_2 = \varepsilon$, we retrieve the previous example, so $V(K, K, K) = \lambda_3(K)$, $3V(K, K, B) = S(K)$, and so on. In $\mathbb{E}^3$, the shorthand $V_2(K)$ is used for $3V(K, K, B)/2$, so the intrinsic volume $V_2(K)$ is half the surface area of K. Similarly, in $\mathbb{E}^3$ one also writes $V_1(K)$ for $3V(K, B, B)/\pi$, so if K is the unit cube, for example, then $V_1(K) = 3$.

Area measures live in the unit sphere. To illustrate, consider the convex body $K + \varepsilon B$ of Figure A.1, where for simplicity we shall now assume that K is the unit cube. If E is any Borel subset in S^2, the surface area measure $S(K + \varepsilon B, \cdot)$

associated with $K+\varepsilon B$ gives E the λ_2-measure of the set of points in the boundary of $K+\varepsilon B$ at which the outer unit normal vectors to $K+\varepsilon B$ lie in E. For example, suppose that E is the single point in S^2 corresponding to the outer unit normal vector to a facet F of $K+\varepsilon B$. Then

$$S(K+\varepsilon B, E) = \lambda_2(F) = 1,$$

and of course the surface area measure of any set containing this point must be at least this value. Naturally, $S(K+\varepsilon B, S^2)$ is just the surface area of $K+\varepsilon B$.

Because K is a polyhedron, one can see in the same way that its surface area measure $S(K, \cdot)$ is just the sum of six point masses, each of weight 1 (the area of a facet of K) and situated at the outer unit normal vector of the corresponding facet. It appears, therefore, that for any set E in S^2, $S(K+\varepsilon B, E)$ is at least $S(K, E)$, and sometimes greater. What contributes to the extra measure? There are two ingredients. Any point in a subset E of S^2 will be an outer unit normal vector to $K+\varepsilon B$ at a point in one of the rounded corners of $K+\varepsilon B$, and this contributes a factor of $\lambda_2(E)\varepsilon^2$ to $S(K+\varepsilon B, E)$. Furthermore, any point in a great circle of S^2 contained in a coordinate plane will be an outer unit normal vector to $K+\varepsilon B$ at every point in a line segment of unit length in the surface of one of the quarter-cylinders in $K+\varepsilon B$. Thus, if E is a subset of such a great circle, a further factor of $\lambda_1(E)\varepsilon$ is added to $S(K+\varepsilon B, E)$. In fact, we now see that

$$S(K+\varepsilon B, E) = S(K, E) + \mu(E)\varepsilon + \lambda_2(E)\varepsilon^2,$$

for each Borel set E in S^2, where μ is linear measure λ_1 restricted to the union of the three great circles in S^2 contained in the coordinate planes. If, by relabeling, we rewrite the previous equation in the form

$$S_2(K+\varepsilon B, E) = S_2(K, E) + 2S_1(K, E)\varepsilon + S_0(K, E)\varepsilon^2,$$

where $\mu = 2S_1(K, \cdot)$ and so on, we have a special case of the local Steiner formula (A.5). The measure $S_1(K, \cdot)$ is called the first area measure of K.

If we repeat this investigation with K and B replaced by arbitrary compact convex sets, examples of mixed area measures result. How do mixed volumes and mixed area measures relate to each other? The general relationship is given in the fundamental formula (A.12), but we can exemplify this for the special case of the unit cube K in $\mathbb{E}^3$. We saw that $V_2(K)$ is half the surface area of K, and the latter is also (in the notation just introduced) $S_2(K, S^2)$. We also computed that $V_1(K) = 3$, and this equals $S_1(K, S^2)/\pi$. These relationships between intrinsic volumes and area measures are true for any compact convex set K; see (A.34), just a very special case of (A.12).

A.2. Area measures

Let K be a convex body in $\mathbb{E}^n$. For each Borel subset E of S^{n-1}, define $g^{-1}(K, E)$
to be the set of points in bd K at which there is an outer unit normal vector in
E. (The notation suggests the connection with the Gauss map g defined in Section 0.9.) Now let

$$S(K, E) = \lambda_{n-1}\left(g^{-1}(K, E)\right). \tag{A.1}$$

Then $S(K, \cdot)$ is a finite Borel measure in S^{n-1}, called the *surface area measure* of K. (Some authors seem to imply that this is completely obvious, but it
is not. One has to show that if E_1 and E_2 are disjoint Borel subsets of S^{n-1},
then $g^{-1}(K, E_1) \cap g^{-1}(K, E_2)$ has zero λ_{n-1}-measure. To do this, one needs
Reidemeister's theorem [737, Theorem 2.2.4], which says that the set of points in
bd K with more than one outer unit normal vector has zero λ_{n-1}-measure.) Note
that $S(K, S^{n-1})$ is just the *surface area* $S(K)$ of K, defined by

$$S(K) = \lambda_{n-1}(\text{bd } K). \tag{A.2}$$

If $K = B$, then, of course, $S(K, \cdot) = \lambda_{n-1}$. The form of $S(K, \cdot)$ is also particularly simple when K is a polytope, because $S(K, \cdot)$ is then a sum of point masses
at the outer unit normal vectors to the facets of K, the weight of each being the
surface area of the corresponding facet.

The theory of area measures is a generalization of this construction. Let K be
a convex body in $\mathbb{E}^n$, and suppose that E is a Borel subset of S^{n-1}. The *brush set*
or *local parallel set* $B_\varepsilon(K, E)$ is the set of points in cl $(K_\varepsilon \setminus K)$ on rays emanating
from some $x \in g^{-1}(K, E)$ in the direction of an outer unit normal vector to K at
x (see Figure A.2) – the union of straight "hairs" of length ε growing orthogonally
out of K in the directions in E. Define

$$\mu_{\varepsilon,K}(E) = \lambda_n\left(B_\varepsilon(K, E)\right).$$

It can be shown, as in [737, Theorems 4.1.2 and 4.1.3], that $\mu_{\varepsilon,K}$ is a finite Borel
measure in S^{n-1}. Moreover, if $\{K_m\}$ is a sequence of compact convex sets converging to K, then the measures μ_{ε,K_m} converge weakly to $\mu_{\varepsilon,K}$, by [737, Theorem 4.1.1].

It can also be shown (see [737, p. 203, eq. (4.2.9)]) that for $0 \leq i \leq n - 1$,
there is a finite Borel measure $S_i(K, \cdot)$ in S^{n-1}, called the *ith area measure* of K,
such that for each Borel subset E of S^{n-1},

$$\mu_{\varepsilon,K}(E) = \frac{1}{n}\sum_{i=0}^{n-1}\binom{n}{i}S_i(K, E)\varepsilon^{n-i}. \tag{A.3}$$

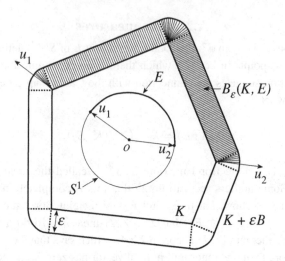

Figure A.2. A brush set.

Formula (A.3) is first obtained for a convex polytope P, for which

$$S_i(P, E) = \binom{n-1}{i}^{-1} \sum_{F \in \mathcal{F}_i(P)} \lambda_{n-i-1}\big(N(P, F) \cap E\big)\lambda_i(F), \qquad (A.4)$$

where $N(P, F)$ is the set of outer normal unit vectors to P at points in the i-dimensional face F. In other words, the ith area measure of a polytope is a finite sum of weighted copies of λ_{n-i-1}, each restricted to a region of an $(n - i - 1)$-dimensional great sphere in S^{n-1}. For example, the multiple $2S_1(K, \cdot)$ of the first area measure of the centered unit cube is linear measure λ_1 restricted to the union of the three great circles in S^2 contained in the coordinate planes, as we saw in Section A.1.

Equation (A.3) is obtained by approximating by polytopes (cf. Section 0.4) and using the weak convergence property of the measures $\mu_{\varepsilon, K}$. From (A.3), one can show that the following relation holds between the area measures:

$$S_{n-1}(K + \varepsilon B, E) = \sum_{i=0}^{n-1} \binom{n-1}{i} S_i(K, E)\varepsilon^{n-i-1}, \qquad (A.5)$$

for each Borel subset E of S^{n-1}. (See [737, p. 207]; here, $S_0(K, E)$ is just $\lambda_{n-1}(E)$.) Both (A.3) and (A.5) are known as *local Steiner formulas* (cf. (A.30)).

If $i = n - 1$, and K is a polytope, then (A.4) shows that $S_{n-1}(K, E)$ is the λ_{n-1}-measure of the set of points in bd K at which there is an outer normal unit vector in E. It follows that $S_{n-1}(K, \cdot)$ is just the surface area measure $S(K, \cdot)$ of K, and by approximation this holds for each compact convex set K, with the important consequence that

$$S_{n-1}(K, S^{n-1}) = S(K). \tag{A.6}$$

If K is of class C_+^2, then it turns out, by [737, p. 206, eq. (4.2.20)], that

$$S_i(K, E) = \int_E F_K^{(i)}(u)\, du. \tag{A.7}$$

We noted just after (0.42) that Aleksandrov proved that for an arbitrary convex body K, each $F_K^{(i)}$ can be defined for almost all $u \in S^{n-1}$ and is integrable on S^{n-1} with respect to spherical Lebesgue measure. He also showed that for an arbitrary convex body K, (A.7) holds when $S_i(K, \cdot)$ is absolutely continuous with respect to spherical Lebesgue measure. A convenient reference for this fact is [403, Theorem 3.5].

A.3. Mixed volumes and mixed area measures

The *mixed volume* $V(K_1, \ldots, K_n)$ of the compact convex subsets $K_1, \ldots, K_n$ of $\mathbb{E}^n$ is defined by

$$V(K_1, \ldots, K_n) = \frac{1}{n!} \sum_{j=1}^{n} (-1)^{n+j} \sum_{i_1 < \cdots < i_j} \lambda_n(K_{i_1} + \cdots + K_{i_j}). \tag{A.8}$$

The *mixed area measure* $S(K_1, \ldots, K_{n-1}, \cdot)$ of $K_1, \ldots, K_{n-1}$ is defined by

$$S(K_1, \ldots, K_{n-1}, \cdot) = \frac{1}{(n-1)!} \sum_{j=1}^{n-1} (-1)^{n+j-1} \sum_{i_1 < \cdots < i_j} S_{n-1}(K_{i_1} + \cdots + K_{i_j}, \cdot). \tag{A.9}$$

Equations (A.8) and (A.9) are called *polarization formulas*. They provide one method for calculating mixed volumes and mixed area measures. For example, in $\mathbb{E}^3$ the first formula reads:

$$6V(K_1, K_2, K_3) = \lambda_3(K_1 + K_2 + K_3) - \lambda_3(K_1 + K_2) - \lambda_3(K_1 + K_3)$$
$$- \lambda_3(K_2 + K_3) + \lambda_3(K_1) + \lambda_3(K_2) + \lambda_3(K_3).$$

It is important to note that by their definition, mixed volumes and mixed areas are invariant under permutations of their arguments.

The fundamental theorem for mixed volumes and mixed area measures is the following one.

Theorem A.3.1. (i) (Minkowski's theorem on mixed volumes). *Let K_j, $1 \le j \le m$, be compact convex sets in $\mathbb{E}^n$. The volume of the Minkowski linear combination*

$$K = t_1 K_1 + \cdots + t_m K_m,$$

where $t_j \geq 0$, is a homogeneous polynomial of degree n in the variables t_j, whose coefficients are the mixed volumes just defined. Specifically,

$$\lambda_n(K) = \sum_{j_1=1}^{m} \cdots \sum_{j_n=1}^{m} V(K_{j_1}, \ldots, K_{j_n}) t_{j_1} \cdots t_{j_n}. \qquad (A.10)$$

(ii) *The analogous statement holds for the mixed area measures of K:*

$$S_{n-1}(K, \cdot) = \sum_{j_1=1}^{m} \cdots \sum_{j_{n-1}=1}^{m} S(K_{j_1}, \ldots, K_{j_{n-1}}, \cdot) t_{j_1} \cdots t_{j_{n-1}}. \qquad (A.11)$$

(iii) *The following equation relates mixed volumes and mixed area measures:*

$$V(K_1, \ldots, K_{n-1}, K_n) = \frac{1}{n} \int_{S^{n-1}} h_{K_n}(u) \, dS(K_1, \ldots, K_{n-1}, u). \qquad (A.12)$$

The notation $V(K_1, i_1; \ldots; K_m, i_m)$ or $S(K_1, i_1; \ldots; K_m, i_m, \cdot)$ will be used when the set K_j appears i_j times.

Part (i) of the previous theorem is proved in [83, Section 29], but there the mixed volumes are defined as the coefficients in (A.10), and then of course the polarization formula must be proved as a theorem. We are following [737, Theorem 5.1.6]. Both proofs require approximation by polytopes.

Let us consider some instances of the theorem. If we let $t_1 = 1$ and $t_j = 0$ for $2 \leq j \leq n$, and let $K_1 = K$, we see that

$$V(K, \ldots, K) = \lambda_n(K). \qquad (A.13)$$

Now suppose that $K_1 = \cdots = K_n = P$ is a polytope in $\mathbb{E}^n$. Then Theorem A.3.1(iii) becomes, in this very special case, the well-known formula for the volume of the polytope P in $\mathbb{E}^n$ having facets F_j with outer unit normal vectors $u_j, 1 \leq j \leq m$:

$$\lambda_n(P) = \frac{1}{n} \sum_{j=1}^{m} h_P(u_j) \lambda_{n-1}(F_j).$$

(This is easy to prove directly when $o \in \operatorname{int} P$, by partitioning P into the pyramids conv $(F_j \cup \{o\})$ and using (0.14), and not difficult to prove directly in the general case, as in [737, Lemma 5.1.1].)

Though the polarization formula (A.8) has its merits, Theorem A.3.1(i) is more useful in deriving other formulas, the technique being to obtain two different expressions for the volume of a Minkowski linear combination and then to compare the coefficients in the expansion. As an illustration, consider the following sum, obtained by Theorem A.3.1(i) with $t_1 = 1$ and $t_2 = \varepsilon$:

$$\lambda_n(K_1 + \varepsilon K_2) = \sum_{i=0}^{n} \binom{n}{i} V(K_1, i; K_2, n-i)\varepsilon^{n-i}. \qquad (A.14)$$

With (A.13), this yields the formula

$$nV(K_1, n-1; K_2) = \lim_{\varepsilon \to 0+} \big(\lambda_n(K_1 + \varepsilon K_2) - \lambda_n(K_1)\big)/\varepsilon. \qquad (A.15)$$

Let us list some useful properties of mixed volumes.

(i) (*Positive multilinearity*) If $a, b \geq 0$, then

$$V(aK_1 + bK_1', K_2, \ldots, K_n) = aV(K_1, K_2, \ldots, K_n) + bV(K_1', K_2, \ldots, K_n). \qquad (A.16)$$

(ii) (*Continuity*) The mixed volume $V(K_1, \ldots, K_n)$ is a continuous function of $K_1, \ldots, K_n$.

(iii) (*Change under affine transformations*) If $\phi \in GA_n$, then

$$V(\phi K_1, \ldots, \phi K_n) = |\det \phi| V(K_1, \ldots, K_n). \qquad (A.17)$$

(iv) (*Monotonicity*) If $K_1 \subset K_1'$, then

$$V(K_1, K_2, \ldots, K_n) \leq V(K_1', K_2, \ldots, K_n). \qquad (A.18)$$

(v) (*Nonnegativity*) $V(K_1, \ldots, K_n) \geq 0$. Moreover, $V(K_1, \ldots, K_n) > 0$ if and only if there are line segments $s_j \subset K_j$, $1 \leq j \leq n$, with linearly independent directions. In particular, $V(K_1, \ldots, K_n) = 0$ if any K_j is a single point, and it follows from (A.16) that:

(vi) (*Invariance under individual translation*) If $x \in \mathbb{E}^n$, then

$$V(K_1 + x, K_2, \ldots, K_n) = V(K_1, K_2, \ldots, K_n). \qquad (A.19)$$

These properties of mixed volumes can be deduced easily from Theorem A.3.1 (see [737, pp. 276–9]). If, as in [83, Section 29], Theorem A.3.1(i) is used to define mixed volumes, then the proofs require more work; monotonicity, in particular, is not at all trivial, and its proof needs some auxiliary results leading to Theorem A.3.1(iii).

Analogous properties can be established for mixed area measures. In particular, they are finite Borel measures (i.e., nonnegative), unchanged when all the arguments are transformed by the same rotation about the origin. Another property is that the centroid of S^{n-1} with respect to any mixed area measure must be the origin, that is,

$$\int_{S^{n-1}} u \, dS(K_1, \ldots, K_{n-1}, u) = o. \qquad (A.20)$$

To see this, let $x \in \mathbb{E}^n$ and $K_n \in \mathcal{K}^n$ be arbitrary, and note that

$$x \cdot \int_{S^{n-1}} u \, dS(K_1, \ldots, K_{n-1}, u)$$

$$= \int_{S^{n-1}} x \cdot u \, dS(K_1, \ldots, K_{n-1}, u)$$

$$= V(K_1, \ldots, K_{n-1}, K_n + x) - V(K_1, \ldots, K_{n-1}, K_n) = 0.$$

where we have used (A.12), (0.24), and (A.19).

There have been various attempts to extend the theory of mixed volumes and area measures to sets more general than compact convex sets. Although these are to some extent successful, some of the properties (i) to (vi) are lost; see [112, Section 26].

We now turn to the important problem of finding necessary and sufficient conditions that a finite Borel measure in S^{n-1} be the ith area measure of some convex body K in $\mathbb{E}^n$. This is unsolved for $1 < i < n - 1$; the answer is known for $i = 1$ (cf. [737, Section 4.3]), but we do not need it in this book. The following theorem provides the full answer for $i = n - 1$.

Theorem A.3.2 (Minkowski's existence theorem). *For the finite Borel measure μ in S^{n-1} to be the surface area measure $S(K, \cdot)$ of some convex body K in $\mathbb{E}^n$, it is necessary and sufficient that μ is not concentrated on any great subsphere of S^{n-1} and that*

$$\int_{S^{n-1}} u \, d\mu(u) = o. \tag{A.21}$$

When we remember that the necessity of (A.21) has already been established in (A.20), and note that if μ were concentrated on a great sphere, then K could not be n-dimensional, we see that we could ask no more of this beautiful theorem (except, perhaps, that the proof of sufficiency be a bit simpler; see [737, Section 7.1]).

When K is a convex polytope having facets with volumes a_j and outward unit normal vectors v_j, $1 \leq j \leq m$, the condition (A.21) reduces to

$$\sum_{j=1}^{m} a_j v_j = o. \tag{A.22}$$

That (A.22) must hold also follows immediately from the divergence theorem. Indeed, let $v \in \mathbb{E}^n$, let F be the constant vector field defined by $F(x) = v$ for all $x \in \mathbb{E}^n$, and let w be the left-hand side of (A.22). Then

$$v \cdot w = \sum_{j=1}^{m} (v \cdot v_j) a_j$$

is the flux of F across the boundary of K. Since $\nabla \cdot F = 0$, the divergence theorem implies that

$$v \cdot w = \int_K \nabla \cdot F(x) \, dx = 0,$$

and since v was arbitrary, (A.22) follows. Remarkably, the fact that a convex polytope must satisfy (A.22) is actually equivalent to the divergence theorem for convex polytopes; this is proved by Klain [442].

Here is one useful consequence of Minkowski's existence theorem. If K_j is a convex body in $\mathbb{E}^n$, and $a_j \geq 0$, $j = 1, 2$, then there is a convex body K such that

$$S(K, \cdot) = a_1 S(K_1, \cdot) + a_2 S(K_2, \cdot). \tag{A.23}$$

Moreover, K is unique, up to translation, by Aleksandrov's uniqueness theorem, Theorem 3.3.1. The existence and uniqueness of the Blaschke body ∇K of a convex body (see Definition 3.3.8) hinges on this; we remark in Note 3.4 that no equivalent concept for $1 < i < n - 1$ exists.

A.4. Reconstruction from surface area measures

Minkowski's existence theorem, Theorem A.3.2, implies that given reals $a_j > 0$ and unit vectors v_j, $1 \leq j \leq m$, that span $\mathbb{E}^n$ and satisfy (A.22), there is an n-dimensional convex polytope P in $\mathbb{E}^n$ whose facets have volumes a_j and outer unit normal vectors v_j, $1 \leq j \leq m$. Moreover, P is unique, up to translation, by Aleksandrov's uniqueness theorem, Theorem 3.3.1. The following algorithm for reconstructing P was proposed by Lemordant, Tao, and Zouaki [505] and implemented (at least for $n \leq 3$) both by them and by Gardner and Milanfar [278] (see below for details).

Algorithm MinkData

Input: Natural numbers $n \geq 2$ and $m \geq n$; real numbers $a_j > 0$ and directions $v_j \in S^{n-1}$, $1 \leq j \leq m$, that span $\mathbb{E}^n$ and satisfy (A.22).

Task: Compute the k-dimensional faces, $0 \leq k \leq n - 1$, of a convex polytope P in $\mathbb{E}^n$ whose facets have volumes a_j and outer unit normal vectors v_j, $1 \leq j \leq m$.

Action: If $n = 2$:

Order the directions v_j so that the corresponding polar angles are increasing with j. Let $x_0 = o$ and for $1 \leq j \leq m$, let $x_j = x_{j-1} + a_j v_j$. In view of (A.22), $x_n = o$ and so $\{x_j : 1 \leq j \leq m\}$ form the vertices of a convex polygon P' with edges of length a_j parallel to the directions v_j, $1 \leq j \leq m$. The output polygon P is P' rotated by $\pi/2$ about o.

Otherwise, if $n \geq 3$:

1. For real numbers $h_j > 0$, $1 \leq j \leq m$, let $h = (h_1, \ldots, h_m)$ and let $P(h)$ be the convex hull of the half-spaces with outer unit normal vectors v_j and bounding hyperplanes at distances h_j from the origin, $1 \leq j \leq m$. Let

$$\bar{a} = \sum_{j=1}^{m} a_j.$$

With initial guess $h^{(0)} = (1/\bar{a}, 1/\bar{a}, \ldots, 1/\bar{a})$, if needed, solve the nonlinear optimization problem (NL):

$$\max_h \quad V(P(h))^{1/n}, \tag{A.24}$$

$$\text{subject to} \quad \sum_{j=1}^{m} a_j h_j = 1$$

$$\text{and} \quad h_j \geq 0, \quad 1 \leq j \leq m.$$

Let $\widehat{h} = (\widehat{h}_1, \ldots, \widehat{h}_m)$ be a solution to (NL), and let

$$h_0 = (nV(P(\widehat{h})))^{-1/(n-1)}\widehat{h}$$

and $P = P(h_0)$.

2. Compute the k-dimensional faces of P for $0 \leq k \leq n - 1$. ∎

Algorithm MinkData is an adaptation of the earlier algorithm of Little [514], based on Minkowski's original proof of his existence theorem (see [737, Theorem 7.1.1]). This proof shows that the vector of distances from the origin to the hyperplanes containing the facets of the convex polytope with facets of volumes a_j and outer unit normal vectors v_j, $1 \leq j \leq m$ is, up to a scaling constant, the solution of the optimization problem (NL'):

$$\min_h \quad \sum_{j=1}^{m} a_j h_j,$$

$$\text{subject to} \quad V(P(h)) = 1$$

$$\text{and} \quad h_j \geq 0, \quad 1 \leq j \leq m.$$

The transition from (NL') to the equivalent problem (NL) is achieved by programming duality; see [505, Section 2.4]. In problem (NL), the objective function in (A.24) is concave by the Brunn–Minkowski inequality (B.10). Therefore (NL) involves a concave objective function and linear constraints. As is the case for (NL'), the solution only gives the vector of distances from the origin to the hyperplanes containing the facets of the desired polytope up to a scaling factor, which however is easy to calculate and is given above in Step 1.

The computation of the volume $V(P(h))$ for the objective function in (A.24) must be carried out at each step of the optimization problem (NL). The polytope $P(h)$ is known in its $\mathcal{H}$-*representation*, that is, by the equations of the hyperplanes containing its facets. Therefore $V(P(h))$ can be calculated by Lasserre's algorithm (see [497]).

The output polytope P is also known in its $\mathcal{H}$-representation. Step 2 can be implemented by first computing the vertices of P, that is, its $\mathcal{V}$-*representation*, from its $\mathcal{H}$-representation, and then computing the convex hull of P from its $\mathcal{V}$-representation. An algorithm for this purpose is described by Barber, Dobkin, and Huhdanpaa [29].

Büeler, Enge, and Fukuda [111] provide an in-depth survey of techniques and extensive remarks concerning polytope volume computation and related matters concerning their $\mathcal{H}$- and $\mathcal{V}$-representations.

With the assistance of Chris Street, then a Western Washington University student, Gardner and Milanfar [278] implemented Algorithm MinkData in the course of their work on reconstruction from brightness functions (see Section 4.4). When $n = 2$, this is straightforward and can be done with Matlab. When $n \geq 3$, the implementation of Step 1 of Algorithm MinkData uses the "fmincon" function in Matlab's Optimization Toolbox for solving the problem (NL). At each step of (NL), the computation of the volume $V(P(h))$ is done by a free C program called Vinci obtainable from ftp://ftp.ifor.math.ethz.ch/pub/volume/Volumen.html. The conversion from $\mathcal{H}$-representation to $\mathcal{V}$-representation and computation of the convex hull in Step 2 of Algorithm MinkData are handled by the free program called qhull available from http://www.qhull.org. When $n = 3$, qhull can provide the output in the form of a Mathematica graphics object which can be displayed by the Mathematica graphics package. See Figures 4.6, 4.7, and 4.8 for some sample reconstructions that used Algorithm MinkData as part of a more elaborate algorithm.

Other algorithms for reconstruction from surface area measures are set out by Lamberg [491] and Sumbatyan and Troyan [793]. Gritzmann and Hufnagel [340] find a polynomial-time algorithm for reconstructing, in a fixed dimension, a convex polytope whose surface area measure is approximately equal to a given one.

A.5. Quermassintegrals and intrinsic volumes

Quermassintegrals and intrinsic volumes are important instances of mixed volumes. If K is a compact convex set in $\mathbb{E}^n$, the *quermassintegrals* $W_i(K)$ of K are defined for $0 \leq i \leq n$ by

$$W_i(K) = V(K, n-i; B, i). \tag{A.25}$$

We also need a weighted and relabeled version of the quermassintegrals. The *intrinsic volumes* $V_i(K)$ of K are defined for $0 \leq i \leq n$ by

$$V_i(K) = \frac{1}{c_{n,i}} W_{n-i}(K) = \frac{1}{c_{n,i}} V(K, i; B, n-i), \qquad (A.26)$$

where

$$c_{n,i} = \frac{\kappa_{n-i}}{\binom{n}{i}}. \qquad (A.27)$$

(Since $\binom{n}{0} = 1$, we have $c_{n,0} = \kappa_n$.) Intrinsic volumes, introduced by McMullen [590], have certain advantages over quermassintegrals. The relabeling is convenient, because

$$V_n(K) = W_0(K) = V(K, \ldots, K) = \lambda_n(K). \qquad (A.28)$$

Moreover, they are aptly named, since they do not depend on the dimension of the containing space; that is, if K is a compact convex set in $\mathbb{E}^k$ and $0 \leq i \leq k \leq n$, then

$$V_i(K) = \frac{1}{c_{n,i}} V(K, i; B^n, n-i) = \frac{1}{c_{k,i}} V(K, i; B^k, k-i). \qquad (A.29)$$

This is particularly useful when working with projections. Setting $i = k$ in (A.29), we see that $V_i(K) = \lambda_i(K)$ if $\dim K \leq i$.

The properties of mixed volumes listed earlier apply to quermassintegrals and intrinsic volumes. In particular, V_i is continuous in K, nonnegative, and monotone, but not, of course, linear, when $i > 1$.

Putting $K_2 = B$ in (A.14), and using (A.25) and (A.26), we can present *Steiner's formula* in the form

$$\lambda_n(K + \varepsilon B) = \sum_{i=0}^{n} \binom{n}{i} W_i(K)\varepsilon^i = \sum_{i=0}^{n} \kappa_{n-i} V_i(K)\varepsilon^{n-i}. \qquad (A.30)$$

By comparing (A.5) with (A.11), one can show that for $1 \leq i \leq n-1$,

$$S_i(K, \cdot) = S(K, i; B, n-i-1, \cdot). \qquad (A.31)$$

From (A.12) we now obtain

$$V(K_1; K_2, i; B, n-i-1) = \frac{1}{n} \int_{S^{n-1}} h_{K_1}(u) \, dS_i(K_2, u). \qquad (A.32)$$

Let us continue to investigate the meaning of quermassintegrals and intrinsic volumes. We have

$$V_0(K) = \frac{W_n(K)}{\kappa_n} = \frac{V(B, \ldots, B)}{\kappa_n} = 1. \qquad (A.33)$$

If we put $K_1 = B$ in (A.32), we see that

$$c_{n,i} V_i(K) = W_{n-i}(K) = \frac{1}{n} S_i\left(K, S^{n-1}\right),\qquad (A.34)$$

for $0 \leq i \leq n - 1$. In particular, if $i = n - 1$ and K is a convex body in $\mathbb{E}^n$, it is evident from (A.6) and (A.15) that

$$2V_{n-1}(K) = S(K) = \lim_{\varepsilon \to 0+} \left(\lambda_n(K + \varepsilon B) - \lambda_n(K)\right)/\varepsilon. \qquad (A.35)$$

Note the right-hand quantity; this is the *Minkowski surface area* of K, which we now see coincides with surface area as defined earlier. It also follows from (A.7) and (A.34) that if K is of class C_+^2, then

$$W_i(K) = \frac{1}{n} \int_{S^{n-1}} F_K^{(n-i)}(u)\, du,$$

while if P is a polytope, then by (A.4) and (A.34) we have

$$V_i(P) = \sum_{F \in \mathcal{F}_i(P)} \gamma(F, P) \lambda_i(F). \qquad (A.36)$$

Here $\gamma(F, P) = \lambda_{n-i-1}\left(N(P, F) \cap S^{n-1}\right)/\omega_{n-i}$ is the (normalized) *external angle* of P at the face F; see [737, p. 210, eq. (4.2.30)]. (The last two formulas are an indication that the different weighting of quermassintegrals and intrinsic volumes makes each more convenient in various contexts.)

Another formula for intrinsic volumes, providing the easily understood interpretation (A.50) of $V_1(K)$, is set out in (A.47).

There are many formulas for quermassintegrals and intrinsic volumes of special convex bodies; see, for example, the books of Hadwiger [370, Section 6.1.9] or Santaló [721, pp. 224–32].

A *valuation* is a function τ taking values in an Abelian semigroup and defined on a class of sets such that

$$\tau(E \cup F) + \tau(E \cap F) = \tau(E) + \tau(F),$$

whenever E, F, $E \cup F$, and $E \cap F$ all belong to the class. Every measure is a valuation on the class of measurable sets, but valuations are much more general. It is not difficult to deduce from Theorem A.3.1 that intrinsic volumes and quermassintegrals are valuations on $\mathcal{K}^n$; see the survey [594, p. 178] of McMullen and Schneider. This is relevant because *Hadwiger's theorem* says that every real-valued valuation on $\mathcal{K}^n$, also invariant under rigid motions and continuous, is a linear combination of the intrinsic volumes V_i, $0 \leq i \leq n$. In other words, intrinsic volumes form the building blocks from which is made every real-valued valuation on $\mathcal{K}^n$ reasonably tied to the geometry of $\mathbb{E}^n$. See [370, Section 6.1.10] or the short proofs by Chen [161] and Klain [438].

A.6. Projection formulas

Let K be a compact convex set in $\mathbb{E}^n$. If $u \in S^{n-1}$, then the Cavalieri principle, Lemma 1.2.2, implies that for each $\varepsilon \geq 0$,

$$\lambda_n(K + \varepsilon[o, u]) = \lambda_n(K) + \varepsilon\lambda_{n-1}(K|u^\perp).$$

See Figure A.3. On the other hand, Theorem A.3.1(i) tells us (cf. (A.14)) that

$$\lambda_n(K + \varepsilon[o, u]) = \sum_{i=0}^n \binom{n}{i} V(K, n - i; [o, u], i)\varepsilon^i.$$

Comparing coefficients, we find that

$$nV(K, n - 1; [o, u]) = \lambda_{n-1}(K|u^\perp), \tag{A.37}$$

a basic formula giving the brightness function of K in terms of mixed volumes (cf. [83, Section 30, p. 50]).

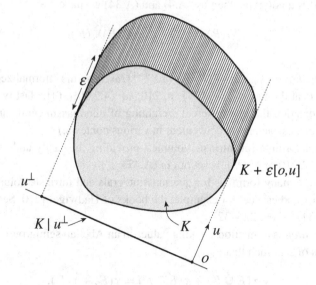

Figure A.3. The body $K + \varepsilon[o, u]$.

This formula can be considerably generalized. Let K_j, $1 \leq j \leq k$, be compact convex sets in $\mathbb{E}^n$, and let H_j, $1 \leq j \leq n - k$, be compact convex subsets of the orthogonal complement $S^\perp$ of $S \in \mathcal{G}(n, k)$. Then

$$\binom{n}{k} V(K_1, \ldots, K_k, H_1, \ldots, H_{n-k}) = V(K_1|S, \ldots, K_k|S)V(H_1, \ldots, H_{n-k}). \tag{A.38}$$

This formula is stated by Burago and Zalgaller [112, Section 19.4], who ascribe it to Fedotov [225]. Here is a sketch of the proof of (A.38). Consider first the case when $K_1 = \cdots = K_k = K$ and $H_1 = \cdots = H_{n-k} = H$, where H is an $(n-k)$-dimensional polytope. Then, if $\varepsilon > 0$, we have

$$\lambda_n(H + \varepsilon K) = \lambda_{n-k}(H)\lambda_k(K|S) + O(\varepsilon^{k+1}).$$

This can be seen by partitioning the set $H + \varepsilon K$ into parts corresponding to the faces of H; for example, draw the picture analogous to Figure A.1, where H is a square in a plane in $\mathbb{E}^3$ and $K = B$. Setting $K_1 = H$ and $K_2 = K$ in (A.14), and comparing, yields (A.38) for this case. From this and the continuity of mixed volumes, (A.38) holds for any compact convex $H \subset S^\perp$. The general case follows by the standard trick of substituting $H = t_1 H_1 + \cdots + t_{n-k} H_{n-k}$, expanding both sides, equating coefficients of $t_1 t_2 \cdots t_{n-k}$, and then repeating this procedure with $K = t_1 K_1 + \cdots + t_k K_k$.

Suppose that $1 \le i \le k$, and in (A.38) take $K_1 = \cdots = K_i = K$, $K_{i+1} = \cdots = K_k = B$, and $H_j = B \cap S^\perp$, $1 \le j \le n-k$. We obtain

$$V(K, i; B, k-i; B \cap S^\perp, n-k) = c_{n,k} V_i(K|S). \tag{A.39}$$

In particular, if $i = k$, then

$$V(K, i; B \cap S^\perp, n-i) = c_{n,i} \lambda_i(K|S) \tag{A.40}$$

expresses the ith projection function of K in terms of mixed volumes. When $i = 1$, we obtain the width function

$$w_K(u) = \lambda_1(K|l_u) = \frac{n}{\kappa_{n-1}} V(K; B \cap u^\perp, n-1), \tag{A.41}$$

and so the support function

$$h_K(u) = \frac{n}{2\kappa_{n-1}} V(K; B \cap u^\perp, n-1), \tag{A.42}$$

for centered K.

Returning to (A.38), let $u_j \in S^{n-1}$, $H_j = [o, u_j]$, $1 \le j \le n-i$, and $S = (\lim\{u_1, \ldots, u_{n-i}\})^\perp$. If $P = [o, u_1] + \cdots + [o, u_{n-i}]$, then, using (A.8), we obtain

$$\binom{n}{i} V(K_1, \ldots, K_i, [o, u_1], \ldots, [o, u_{n-i}])$$

$$= V(K_1|S, \ldots, K_i|S)\lambda_{n-i}(P)/(n-i)!. \tag{A.43}$$

An integral formula can be deduced if we set $i = n-1$. Then, using also $[-u, u] = [-u, o] + [o, u]$ and (A.16), we obtain

$$V(K_1|u^\perp, \ldots, K_{n-1}|u^\perp) = nV(K_1, \ldots, K_{n-1}, [o, u])$$

$$= \frac{n}{2} V(K_1, \ldots, K_{n-1}, [-u, u]).$$

(We met the special case where all the sets are equal in (A.37).) Therefore, by (A.12),

$$V(K_1|u^\perp, \ldots, K_{n-1}|u^\perp) = \frac{1}{2} \int_{S^{n-1}} |u \cdot v| \, dS(K_1, \ldots, K_{n-1}, v), \qquad (A.44)$$

for all $u \in S^{n-1}$. Letting $K_1 = \cdots = K_i = K$ and $K_{i+1} = \cdots = K_{n-1} = B$, we see that

$$c_{n-1,i} V_i(K|u^\perp) = \frac{1}{2} \int_{S^{n-1}} |u \cdot v| \, dS_i(K, v). \qquad (A.45)$$

We refer to (A.44) and (A.45) as the *generalized Cauchy projection formulas*.

Some of the preceding special cases of (A.43) can be found in [737, Section 5.3], where they are deduced from general integral-geometric formulas. Let K be a compact convex set in $\mathbb{E}^n$. *Kubota's integral recursion* states that for $1 \le i \le k \le n - 1$,

$$V_i(K) = \frac{\binom{n}{i} \kappa_{k-i} \kappa_n}{\binom{k}{i} \kappa_{n-i} \kappa_k} \int_{\mathcal{G}(n,k)} V_i(K|S) \, dS. \qquad (A.46)$$

The proof is given in [370, Section 6.2.5]. In particular, when $i = k$, we obtain

$$V_i(K) = \frac{\kappa_n}{\kappa_i c_{n,i}} \int_{\mathcal{G}(n,i)} \lambda_i(K|S) \, dS. \qquad (A.47)$$

(A slightly modified form is stated in [737, p. 295, eq. (5.3.27)], again deduced from general integral-geometric formulas.) This says that the ith intrinsic volume of K is, up to a factor depending only on n and i, the average of the volumes of the projections of K on i-dimensional subspaces. The same is, of course, true for the quermassintegrals. Note that the volume of a projection of K on a subspace S equals the volume of the cross-section, parallel to S, of the cylinder with base $K|S$. This is the origin of the term "quermassintegral."

When $k = n - 1$ in (A.46), we have

$$V_i(K) = \frac{n\kappa_{n-i-1}\kappa_n}{(n-i)\kappa_{n-i}\kappa_{n-1}} \int_{\mathcal{G}(n,n-1)} V_i(K|S) \, dS$$

$$= \frac{\kappa_{n-i-1}}{(n-i)\kappa_{n-i}\kappa_{n-1}} \int_{S^{n-1}} V_i(K|u^\perp) \, du. \qquad (A.48)$$

Setting $i = n - 1$ also, we obtain the special case of Kubota's integral recursion known as *Cauchy's surface area formula*:

$$S(K) = 2V_{n-1}(K) = \frac{1}{\kappa_{n-1}} \int_{S^{n-1}} \lambda_{n-1}(K|u^\perp) \, du. \qquad (A.49)$$

This says that the surface area of a convex body is, up to a factor depending only on n, the average volume of its shadows. A simple direct proof is contained in [83, Section 32], where a proof of (A.48) is also outlined.

Another special case of (A.47) worthy of note is $i = 1$, when

$$V_1(K) = \frac{n\kappa_n}{2\kappa_{n-1}} \int_{\mathcal{G}(n,1)} \lambda_1(K|S)\, dS.$$

If $u \in S^{n-1}$ and $S = l_u$, then $\lambda_1(K|S) = w_K(u)$. For this reason, the integral in the last equation is called the *mean width* of K, and we have

$$2\kappa_{n-1}V_1(K)/n\kappa_n = \text{mean width of } K. \qquad (A.50)$$

A.7. Dual mixed volumes

Dual mixed volumes were introduced by Lutwak [529]. A brief summary is given by Burago and Zalgaller [112, Section 24.3], but we require rather more, and we do not restrict the discussion to star bodies containing the origin in their interiors.

Let L_j be a star body in $\mathbb{E}^n$ with $o \in L_j$, $1 \le j \le n$. We define the *dual mixed volume* $\tilde{V}(L_1, L_2, \ldots, L_n)$ by

$$\tilde{V}(L_1, L_2, \ldots, L_n) = \frac{1}{n} \int_{S^{n-1}} \rho_{L_1}(u)\rho_{L_2}(u) \cdots \rho_{L_n}(u)\, du, \qquad (A.51)$$

where ρ_L, the radial function of L, is a bounded Borel function on S^{n-1} (see Section 0.7). Note that dual mixed volumes are unchanged if the sets are permuted. There is the following analogue of Theorem A.3.1(i).

Theorem A.7.1. *Let L_j, $1 \le j \le m$, be star bodies containing the origin in $\mathbb{E}^n$. The volume of the radial linear combination*

$$L = t_1 L_1 \tilde{+} \cdots \tilde{+} t_m L_m,$$

where $t_j \ge 0$, is a homogeneous polynomial of degree n in the variables t_j, whose coefficients are the foregoing dual mixed volumes. Specifically,

$$\lambda_n(L) = \sum_{j_1=1}^{m} \cdots \sum_{j_n=1}^{m} \tilde{V}(L_{j_1}, \ldots, L_{j_n})t_{j_1} \cdots t_{j_n}. \qquad (A.52)$$

Proof. Just use (A.51) and expand with the help of (0.31). ■

Dual mixed volumes enjoy the following properties, similar to those of mixed volumes. All of them follow easily from the definition.

(i) (*Positive multilinearity*) If $a, b \ge 0$, then

$$\tilde{V}(aL_1 \tilde{+} bL_1', L_2, \ldots, L_n) = a\tilde{V}(L_1, L_2, \ldots, L_n) + b\tilde{V}(L_1', L_2, \ldots, L_n).$$
$$(A.53)$$

(ii) (*Change under linear transformations*) If $\phi \in GL_n$, then

$$\tilde{V}(\phi L_1, \dots, \phi L_n) = |\det \phi| \tilde{V}(L_1, \dots, L_n). \tag{A.54}$$

(iii) (*Monotonicity*) If $L_1 \subset L_1'$, then

$$\tilde{V}(L_1, L_2, \dots, L_n) \leq \tilde{V}(L_1', L_2, \dots, L_n). \tag{A.55}$$

(iv) (*Nonnegativity*) $\tilde{V}(L_1, \dots, L_n) \geq 0$.

Observe, however, that dual mixed volumes are not invariant under translations.

As with mixed volumes, we use the notation $\tilde{V}(L_1, i_1; \dots; L_m, i_m)$ to denote the dual mixed volume in which the set L_j appears i_j times.

For a special case, it will be convenient to relax the restriction on the numbers i_j. Define, for all $i \in \mathbb{R}$,

$$\tilde{V}_i(L_1, L_2) = \frac{1}{n} \int_{S^{n-1}} \rho_{L_1}(u)^{n-i} \rho_{L_2}(u)^i \, du. \tag{A.56}$$

Therefore $\tilde{V}_i(L_1, L_2)$ is just $\tilde{V}(L_1, n-i; L_2, i)$, where i is now allowed to be any real number.

If L is a star body in $\mathbb{E}^n$ with $o \in L$, and $i \in \mathbb{R}$, the *dual volume* $\tilde{V}_i(L)$ and *dual quermassintegral* $\tilde{W}_{n-i}(L)$ of L are defined by

$$\tilde{V}_i(L) = \tilde{W}_{n-i}(L) = \tilde{V}(L, i; B, n-i) = \frac{1}{n} \int_{S^{n-1}} \rho_L(u)^i \, du. \tag{A.57}$$

Thus $\tilde{V}_i(L) = \tilde{V}_i(B, L)$. When $i = n$, we have

$$\tilde{V}_n(L) = \frac{1}{n} \int_{S^{n-1}} \rho_L(u)^n \, du = \lambda_n(L), \tag{A.58}$$

according to the formula for volume in polar coordinates.

Note that unlike their counterparts the intrinsic volumes, dual volumes depend on the dimension of the containing space. For this reason, and in order to work with general star sets (cf. Section 0.7), we need another definition, proposed by Gardner and Volčič [283]. If L is any star set in $\mathbb{E}^n$ contained in $S \in \mathcal{G}(n, k)$, $1 \leq k \leq n$, we define, for nonzero $i \in \mathbb{R}$, the *dual volume* $\tilde{V}_{i,k}(L)$ of L by

$$\tilde{V}_{i,k}(L) = \frac{1}{2k} \int_{S^{n-1} \cap S} \rho_{i,L}(u) \, du, \tag{A.59}$$

and drop the second index when $S = \mathbb{E}^n$. Here, $\rho_{i,L}$ is the i-chord function of L, as in Definition 6.1.1; we stress that as in that definition, i is not restricted to integer values, and it is assumed that $o \in \operatorname{relint} L$ or $o \notin L$ when $i < 0$. One can also define $\tilde{V}_{0,k}(L) = \kappa_k$ if $\dim L = k$ and $o \in L$, and $\tilde{V}_{0,k}(L) = 0$ otherwise; when L does not contain the origin in its relative boundary, these definitions ensure that $\tilde{V}_{i,k}(L)$ is continuous in i. Note that $\tilde{V}_i(L) > 0$ if and only if $\dim L = n$, and in general $\tilde{V}_{i,k}(L) > 0$ if and only if $\dim L = k$.

Suppose that L is a star body with $o \in L$. Then by (A.57),

$$\tilde{V}_i(L) = \frac{1}{n} \int_{S^{n-1}} \rho_L(u)^i \, du$$

$$= \frac{1}{2n} \int_{S^{n-1}} \left(\rho_L(u)^i + \rho_L(-u)^i \right) du$$

$$= \frac{1}{2n} \int_{S^{n-1}} \rho_{i,L}(u) \, du,$$

so (A.59) agrees with (A.57) in this special case.

Dual volumes have properties similar to those of intrinsic volumes. For example, $\tilde{V}_i$ is nonnegative, monotone, and invariant under volume-preserving linear transformations.

If i is an integer with $1 \le i \le n$, and L is any i-dimensional star set in $\mathbb{E}^n$, then by the formula for volume in polar coordinates, we have

$$\tilde{V}_{i,i}(L) = \lambda_i(L). \tag{A.60}$$

It follows straight from the definition that $2\tilde{V}_1(L)/\kappa_n$ is the average length of chords of L on lines through the origin. Unfortunately, $\tilde{V}_{n-1}$ does not appear to have the simple geometrical interpretation enjoyed by V_{n-1}. (Sangwine-Yager [719] indicates a connection with a certain generalization of outer parallel sets.)

The next theorem, the analogue of (A.46), is essentially due to Lutwak [531, Theorem 1] (see also [283]).

Theorem A.7.2 (dual Kubota integral recursion). *Let L be a star body in $\mathbb{E}^n$ with $o \in L$, let $1 \le k \le n - 1$, and suppose that $i \in \mathbb{R}$ is positive. Then*

$$\tilde{V}_i(L) = \frac{\kappa_n}{\kappa_k} \int_{\mathcal{G}(n,k)} \tilde{V}_{i,k}(L \cap S) \, dS.$$

Proof.

$$\tilde{V}_i(L) = \frac{1}{n} \int_{S^{n-1}} \rho_L(u)^i \, du$$

$$= \frac{\kappa_n}{k\kappa_k} \int_{\mathcal{G}(n,k)} \int_{S^{n-1} \cap S} \rho_L(u)^i \, du \, dS$$

$$= \frac{\kappa_n}{\kappa_k} \int_{\mathcal{G}(n,k)} \tilde{V}_{i,k}(L \cap S) \, dS.$$

(The second equality comes from the fact that λ_{n-1} is the unique Borel-regular, rotation-invariant measure in S^{n-1} such that S^{n-1} has measure $\omega_n = n\kappa_n$, as detailed in the proof of Theorem 7.2.3.) ∎

In particular, when $k = i$, we obtain

$$\tilde{V}_i(L) = \frac{\kappa_n}{\kappa_i} \int_{\mathcal{G}(n,i)} \lambda_i(L \cap S) \, dS. \tag{A.61}$$

Also, when $k = n - 1$, we get

$$\tilde{V}_i(L) = \frac{\kappa_n}{\kappa_{n-1}} \int_{\mathcal{G}(n,n-1)} \tilde{V}_{i,n-1}(L \cap S) \, dS = \frac{1}{n\kappa_{n-1}} \int_{S^{n-1}} \tilde{V}_{i,n-1}(L \cap u^\perp) \, du. \tag{A.62}$$

A partial extension of the theory of dual mixed volumes to bounded Borel sets can be found in the article [280] of Gardner, Vedel Jensen, and Volčič. If $i \in \mathbb{R}$ is positive, and C is a bounded Borel subset of $S \in \mathcal{G}(n, k)$, $1 \le k \le n$, one can define the dual volume $\tilde{V}_{i,k}(C)$ of C by

$$\tilde{V}_{i,k}(C) = \frac{i}{k} \int_C \|x\|^{i-k} \, dx, \tag{A.63}$$

dropping the second index when $S = \mathbb{E}^n$. It is proved in [280] that definition (A.63) is consistent with those given above for star sets and that Theorem A.7.2 holds for bounded Borel sets under this definition.

Klain [439], [440] takes a somewhat different approach to dual mixed volumes. He works with sets containing o and star-shaped at o, whose radial functions belong to $L^n(S^{n-1})$. Defining dual mixed volumes by A.51, he establishes the same basic properties, including Theorem A.7.2. Moreover, he shows that in this setting it is possible to obtain a natural analogue of Hadwiger's theorem in the dual theory.

Appendix B

Inequalities

This appendix collects together the principal inequalities used in the text. The first section deals with basic inequalities for means and sums, for which the classic treatise [382] of Hardy, Littlewood, and Pólya is still unparalleled. For the Aleksandrov–Fenchel inequality and its consequences, Chapter 6 of Schneider's book [737] contains a good deal of what is currently known. Further information concerning the Aleksandrov–Fenchel inequality, as well as the dual Aleksandrov–Fenchel inequality and many other geometric inequalities, can be found in Burago and Zalgaller's work [112].

In the inequalities that we consider, conditions for equality are usually listed whenever they are known. The value of such equality conditions cannot be emphasized too heavily, and they are constantly used in the text.

B.1. Inequalities involving means and sums

Suppose that $a_j \geq 0$ and that $w_j > 0$, $1 \leq j \leq m$. The *arithmetic–geometric mean inequality* (see, for example, [382, Section 2.5]) states that

$$(a_1^{w_1} \cdots a_m^{w_m})^{1/(w_1 + \cdots + w_m)} \leq \frac{w_1 a_1 + \cdots + w_m a_m}{w_1 + \cdots + w_m}, \tag{B.1}$$

with equality if and only if $a_1 = \cdots = a_m$. (This inequality is often quoted in the special form obtained by taking all the weights $w_j = 1/m$.)

There are various ways to generalize this inequality. We shall only need to consider equal weights here. Suppose that $p \in \mathbb{R}$ and that $a_j \geq 0$, $1 \leq j \leq m$. The *pth mean* $M_p = M_p(a_1, \ldots, a_m)$ is defined by

$$M_p = \begin{cases} \left(\frac{1}{m} \sum_{j=1}^{m} a_j^p \right)^{1/p} & \text{if } p \neq 0, \\ (a_1 \cdots a_m)^{1/m} & \text{if } p = 0, \end{cases} \tag{B.2}$$

except we define $M_p = 0$ if $p < 0$ and at least one of the numbers a_j is zero. Note that (B.1) implies that $M_0 \leq M_1$. More generally, if $p < q$, then *Jensen's inequality for means* (see [382, Section 2.9]) says that

$$M_p \leq M_q, \tag{B.3}$$

with equality if and only if $a_1 = \cdots = a_m$ or $q \leq 0$ and at least one of the numbers a_j is zero. It is also true that

$$\min\{a_1, \ldots, a_m\} \leq M_q \leq \max\{a_1, \ldots, a_m\},$$

with the same equality condition.

If $a = (a_1, \ldots, a_m)$, let us write $M_p(a)$ for $M_p(a_1, \ldots, a_m)$. If $p \geq 1$, then

$$M_p(a + b) \leq M_p(a) + M_p(b), \tag{B.4}$$

whereas for $p \leq 1$ the inequality is reversed, and for $p \neq 1$ equality holds if and only if there is a constant c such that $a_j = cb_j$, for $1 \leq j \leq m$, or $p \leq 0$ and $a_j = b_j = 0$ for some j; this is *Minkowski's inequality*, as in [382, Section 2.11].

The *pth sum* $S_p = S_p(a_1, \ldots, a_m)$ is defined for nonzero $p \in \mathbb{R}$ by

$$S_p = \left(\sum_{j=1}^{m} a_j^p \right)^{1/p}, \tag{B.5}$$

except we define $S_p = 0$ if $p < 0$ and at least one of the numbers a_j is zero. *Jensen's inequality for sums* states that if $0 < p < q$, then

$$\max\{a_1, \ldots, a_m\} \leq S_q \leq S_p, \tag{B.6}$$

with equalities if and only if all but one of the numbers a_j are zero; see [382, Section 2.10]. (Note that pth sums decrease with p, whereas pth means increase with p.) Moreover, if $p < q < 0$, then

$$S_q \leq S_p \leq \min\{a_1, \ldots, a_m\}, \tag{B.7}$$

with equalities if at least one of the numbers a_j is zero. (This can be seen from (B.6) by replacing each a_j by a_j^{-1}.)

Let $f \in C(X)$ be nonnegative, and suppose that μ is a Borel measure in X with $\mu(X) = 1$. If $p \neq 0$, the *pth mean* $M_p(f)$ of f is defined by

$$M_p(f) = \left(\int_X f(x)^p \, d\mu(x) \right)^{1/p},$$

where it is understood that $M_p(f) = 0$ if $p < 0$ and $f = 0$ in a set of positive μ-measure. (When $p \geq 1$, this is just the L^p norm $\|f\|_p$ of f, of course.) We also write $M_\infty(f) = \sup f$ and $M_{-\infty}(f) = \inf f$. Then if $p, q \neq 0$ and $p < q$,

$$M_{-\infty}(f) \leq M_p(f) \leq M_q(f) \leq M_\infty(f), \tag{B.8}$$

with equality anywhere if and only if f is constant (or if p or q are negative and the corresponding mean is zero); see [382, Sections 6.6 and 6.10]. We call the middle inequality *Jensen's inequality for integrals*. For nonnegative bounded Borel functions f and g on X, and positive $p, q \in \mathbb{R}$ with

$$\frac{1}{p} + \frac{1}{q} = 1,$$

Hölder's inequality states that for $p > 1$,

$$\int_X f(x)g(x)\, d\mu(x) \leq \|f\|_p \|g\|_q, \tag{B.9}$$

with equality if and only if there exist nonnegative constants a, b, not both zero, such that $af^p = bg^q$ μ-almost everywhere. For the proof, see [382, Section 6.9].

B.2. The Brunn–Minkowski inequality

The best-known inequality concerning volumes of compact convex sets is the *Brunn–Minkowski inequality*, stating that if K_0 and K_1 are compact convex sets in $\mathbb{E}^n$, and $0 \leq t \leq 1$, then

$$\lambda_n\big((1-t)K_0 + tK_1\big)^{1/n} \geq (1-t)\lambda_n(K_0)^{1/n} + t\lambda_n(K_1)^{1/n}, \tag{B.10}$$

with equality for some $0 < t < 1$ if and only if K_0 and K_1 lie in parallel hyperplanes or are homothetic.

Let us note at once that the coefficients $(1-t)$ and t can be replaced by arbitrary nonnegative t_0 and t_1. To see this, write down the inequality with $t = t_1/(t_0 + t_1)$ and use (0.8) to clear denominators.

To understand the content of the Brunn–Minkowski inequality, think of K_0 and K_1 as sitting in the parallel n-dimensional hyperplanes $x_1 = 0$ and $x_1 = 1$, respectively, in $\mathbb{E}^{n+1}$. Then it is easy to see that

$$(1-t)K_0 + tK_1 = \operatorname{conv}(K_0 \cup K_1) \cap \{x : x_1 = t\}.$$

See Figure B.1. Now for each $0 \leq t \leq 1$, let D_t be the n-dimensional ball in $\mathbb{E}^{n+1}$ contained in the hyperplane $x_1 = t$, with center on the x_1-axis and with n-dimensional volume equal to that of $(1-t)K_0 + tK_1$. The union D of these balls D_t, $0 \leq t \leq 1$, is just the Schwarz symmetral of $\operatorname{conv}(K_0 \cup K_1)$ (see Definition 2.1.3). Then the Brunn–Minkowski inequality says that D is a convex set, or, in other words, that the radius of D_t, $0 \leq t \leq 1$, is a concave function of t.

Two proofs of (B.10), each pointing to a certain type of extension, and a discussion, can be found in Schneider's book [737, Section 6.1]. Related proofs are also given by Bonnesen and Fenchel [83, Sections 41 and 48] and Pisier [673]. A completely different proof, based on approximation by boxes and the arithmetic–geometric mean inequality, is so short and charming that we shall give it here. As

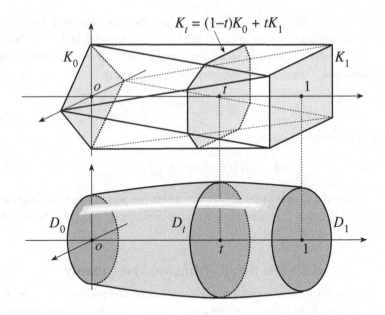

Figure B.1. Geometry of the Brunn–Minkowski inequality.

a bonus, we shall see that (B.10) still holds when K_0 and K_1 are more general sets.

Proof of (B.10). If X and Y are boxes in $\mathbb{E}^n$ with sides of length x_i and y_i, respectively, in the ith coordinate directions, then

$$\lambda_n(X) = \prod_{i=1}^{n} x_i, \quad \lambda_n(Y) = \prod_{i=1}^{n} y_i, \quad \text{and} \quad \lambda_n(X+Y) = \prod_{i=1}^{n} (x_i + y_i).$$

Now

$$\left(\prod_{i=1}^{n} \frac{x_i}{x_i + y_i}\right)^{1/n} + \left(\prod_{i=1}^{n} \frac{y_i}{x_i + y_i}\right)^{1/n} \leq \frac{1}{n}\sum_{i=1}^{n} \frac{x_i}{x_i + y_i} + \frac{1}{n}\sum_{i=1}^{n} \frac{y_i}{x_i + y_i} = 1,$$

by the arithmetic–geometric mean inequality (B.1). This gives

$$\lambda_n(X+Y)^{1/n} \geq \lambda_n(X)^{1/n} + \lambda_n(Y)^{1/n}, \tag{B.11}$$

when X and Y are boxes.

Next, we use a trick sometimes called a *Hadwiger–Ohmann cut* to obtain the inequality for finite unions X and Y of boxes, as follows. By translating X, if necessary, we can assume that a coordinate hyperplane, $\{x_n = 0\}$ say, separates two of the boxes in X. (The reader might find a picture illustrating the planar case useful at this point.) Let X_+ (or X_-) denote the union of the boxes formed by intersecting the boxes in X with $\{x_n \geq 0\}$ (or $\{x_n \leq 0\}$, respectively). Now translate Y so that

$$\frac{\lambda_n(X_\pm)}{\lambda_n(X)} = \frac{\lambda_n(Y_\pm)}{\lambda_n(Y)}, \tag{B.12}$$

where Y_+ and Y_- are defined analogously to X_+ and X_-. Note that $X_+ + Y_+ \subset \{x_n \geq 0\}$, $X_- + Y_- \subset \{x_n \leq 0\}$, and that the numbers of boxes in $X_+ \cup Y_+$ and $X_- \cup Y_-$ are both smaller than the number of boxes in $X \cup Y$. By induction on the latter number and (B.12), we have

$$\begin{aligned}
\lambda_n(X + Y) &\geq \lambda_n(X_+ + Y_+) + \lambda_n(X_- + Y_-)\\
&\geq \left(\lambda_n(X_+)^{1/n} + \lambda_n(Y_+)^{1/n}\right)^n + \left(\lambda_n(X_-)^{1/n} + \lambda_n(Y_-)^{1/n}\right)^n\\
&= \lambda_n(X_+)\left(1 + \frac{\lambda_n(Y)^{1/n}}{\lambda_n(X)^{1/n}}\right)^n + \lambda_n(X_-)\left(1 + \frac{\lambda_n(Y)^{1/n}}{\lambda_n(X)^{1/n}}\right)^n\\
&= \lambda_n(X)\left(1 + \frac{\lambda_n(Y)^{1/n}}{\lambda_n(X)^{1/n}}\right)^n = \left(\lambda_n(X)^{1/n} + \lambda_n(Y)^{1/n}\right)^n.
\end{aligned}$$

Now that (B.11) is established for finite unions of boxes, the proof of the same inequality when X and Y are compact convex sets follows by approximating these sets with finite unions of boxes. Inequality (B.10) follows on substituting $X = (1-t)K_0$ and $Y = tK_1$ and using the homogeneity (0.8) of volume. ∎

The previous proof works without any extra assumptions when K_0 and K_1 are bounded Borel sets. It does not, however, supply the equality condition for (B.10) stated above. For compact convex sets, the equality condition is established in [737, Section 6.1] and [827, Section 6.5]. We refer the reader to the comprehensive survey of Gardner [264] for a full discussion of equality conditions and references, as well as much more information about the Brunn–Minkowski inequality, its various forms, related inequalities, and applications. See also the articles of Barthe [39] and Maurey [587].

The proof of the following theorem is taken from [264, Section 5].

Theorem B.2.1. *Let K_0 and K_1 be compact convex sets in $\mathbb{E}^n$. The Brunn–Minkowski inequality implies* Minkowski's first inequality, *stating that*

$$V(K_0, n-1; K_1)^n \geq \lambda_n(K_0)^{n-1}\lambda_n(K_1), \tag{B.13}$$

with equality if and only if K_0 and K_1 lie in parallel hyperplanes or are homothetic.

Proof. Substituting $\varepsilon = t/(1-t)$ in (A.15) and using the homogeneity (0.8) of volume, we obtain

$$nV(K_0, n-1; K_1) = \lim_{t \to 0+} \frac{\lambda_n\left((1-t)K_0 + tK_1\right) - (1-t)^n \lambda_n(K_0)}{t(1-t)^{n-1}}$$

$$= \lim_{t \to 0+} \frac{\lambda_n\left((1-t)K_0 + tK_1\right) - \lambda_n(K_0)}{t}$$

$$+ \lim_{t \to 0+} \frac{(1-(1-t)^n))\lambda_n(K_0)}{t}$$

$$= \lim_{t \to 0+} \frac{\lambda_n\left((1-t)K_0 + tK_1\right) - \lambda_n(K_0)}{t} + n\lambda_n(K_0).$$

Using this new expression for $V(K_0, n - 1; K_1)$ and letting $f(t) = \lambda_n\left((1-t)K_0 + tK_1\right)^{1/n}$ for $0 \le t \le 1$, we see that

$$f'(0) = \frac{V(K_0, n-1; K_1) - \lambda_n(K_0)}{\lambda_n(K_0)^{(n-1)/n}}.$$

Therefore (B.13) is equivalent to $f'(0) \ge f(1) - f(0)$. As was noted above, the Brunn–Minkowski inequality (B.10) says that f is concave, so Minkowski's first inequality (B.13) follows.

Suppose that equality holds in (B.13). Then $f'(0) = f(1) - f(0)$. Since f is concave, we have

$$\frac{f(t) - f(0)}{t} = f(1) - f(0)$$

for $0 < t \le 1$, and this is just equality in the Brunn–Minkowski inequality (B.10). The equality condition for (B.13) follows immediately.   ∎

Klain [441] has found a way to prove both Theorem A.3.2 and (B.13) for convex polytopes simultaneously, providing a relationship between these fundamental results.

If we set $K_0 = K$ and $K_1 = B$ in (B.13) we get

$$V(K, n-1; B)^n \ge \lambda_n(K)^{n-1}\lambda_n(B),$$

leading, with the help of (A.26) and (A.35), to

$$\left(\frac{S(K)}{\omega_n}\right)^n \ge \left(\frac{\lambda_n(K)}{\kappa_n}\right)^{n-1}, \tag{B.14}$$

the *isoperimetric inequality*, with equality if and only if K is a ball.

The isoperimetric inequality has spawned a vast literature; see the survey [801] of Talenti and the references given therein. The isoperimetric inequality in the plane says that $S(K)^2 \ge 4\pi\lambda_2(K)$, with equality if and only if K is a disk. (Of course, $S(K)$ is then called the *perimeter*, rather than the surface area, of K.) Aside from the equality condition, which caused considerable difficulties in the nineteenth century (see the discussion by Gruber [364]), a completely elementary and self-contained proof can be given in a couple of pages, as in Lay's book [499,

Section 13]. Even this cannot quite be called trivial, and the fact that the isoperimetric inequality in $\mathbb{E}^n$ is but a very special instance of (B.13) is a conclusive demonstration of the power of Minkowski's theory of mixed volumes.

From the Brunn–Minkowski inequality, using only a little calculus, one can also derive *Minkowski's second inequality*:

$$V(K_1, n-1; K_2)^2 \geq \lambda_n(K_1)V(K_1, n-2; K_2, 2). \tag{B.15}$$

See, for example, [737, Theorem 6.2.1] or [83, Section 49]. Exact conditions for equality are unknown, but [737, Theorem 6.6.18] gives the known conditions when K_2 is a convex body.

Note that Minkowski's second inequality is quadratic. It is just a very special case of another quadratic inequality, to which the next section is devoted.

B.3. The Aleksandrov–Fenchel inequality

There is a quadratic inequality including not only Minkowski's second inequality (B.15), but also Minkowski's first inequality (B.13), and therefore, by Theorem B.2.1, also implying the Brunn–Minkowski inequality (B.10). This is the celebrated *Aleksandrov–Fenchel inequality*, stating that for compact convex sets $K_1, \ldots, K_n$ in $\mathbb{E}^n$,

$$V(K_1, K_2, K_3, \ldots, K_n)^2 \geq V(K_1, 2; K_3, \ldots, K_n)V(K_2, 2; K_3, \ldots, K_n).$$

$$\tag{B.16}$$

The Aleksandrov–Fenchel inequality appeared in the works of Aleksandrov [3] and Fenchel and Jessen [230], and is essentially the most powerful inequality of its type known. As well as all the inequalities in the previous section, it implies several others we list subsequently; in fact, we do not need its full force in this book. The inequality has found several important applications and connections outside convex geometry. Egorychev [208] employed it in his solution of the van der Waerden conjecture concerning the permanent of a doubly stochastic matrix; there are combinatorial applications; and recent proofs link the inequality to the Hodge index theorem. The most comprehensive general reference is [737, Chapter 6], while extensive summaries are provided in [718, Sections 6 and 7] and [112, Chapter 4], the latter including a section on the connection with algebraic geometry.

This quadratic inequality is the case $i = 2$ of the following one, valid for $1 \leq i \leq n$:

$$V(K_1, K_2, \ldots, K_n)^i \geq \prod_{j=1}^{i} V(K_j, i; K_{i+1}, \ldots, K_n). \tag{B.17}$$

Inequality (B.16) is actually just as general as (B.17). The proof, due originally to Aleksandrov [3], is reproduced by Busemann [128, pp. 49–50].

Unfortunately, a precise equality condition for the Aleksandrov–Fenchel inequality is unknown in general; see [737, Section 6.6] for a full discussion. If $K_3, \ldots, K_n$ are smooth convex bodies, however, then equality holds if and only if K_1 and K_2 are homothetic, by [737, Theorem 6.6.8].

Suppose that $1 \le j \le i$, and let $K_i = K_{i-1} = \cdots = K_{i-j+2} = K_1$ and $K_{i-j+1} = K_{i-j} = \cdots = K_2$. Then (B.17) gives

$$V(K_1, j; K_2, i - j; \mathcal{A})^i \ge V(K_1, i; \mathcal{A})^j V(K_2, i; \mathcal{A})^{i-j}, \qquad (B.18)$$

where $\mathcal{A} = \{K_{i+1}, \ldots, K_n\}$. If K_1 and K_2 have dimension at least i, and all the sets in $\mathcal{A}$ are smooth convex bodies, then equality holds if and only if K_1 and K_2 are homothetic; this follows from [737, Theorem 6.6.9], bearing in mind the equivalence of (B.16) and (B.17).

The special case $j = 1$ and $\mathcal{A} = \{B, \ldots, B\}$ of (B.18) leads to

$$V(K_1; K_2, i - 1; B, n - i)^i \ge c_{n,i}^i V_i(K_2)^{i-1} V_i(K_1). \qquad (B.19)$$

If K_1 and K_2 are of dimension at least i, then equality holds if and only if K_1 and K_2 are homothetic.

Setting $i = n$ in (B.17) gives

$$V(K_1, \ldots, K_n)^n \ge \lambda_n(K_1)\lambda_n(K_2) \cdots \lambda_n(K_n). \qquad (B.20)$$

(Note that Minkowski's first inequality (B.13) results from putting $K_2 = \cdots = K_n = K_0$ in (B.20).) Now letting $K_1 = K_2 = \cdots = K_i = K$ and $K_{i+1} = \cdots = K_n = B$ in the previous inequality, we obtain, by (A.26), the following *extended isoperimetric inequality*: If $1 \le i \le n - 1$, then

$$\left(\frac{V_i(K)}{V_i(B)} \right)^n \ge \left(\frac{\lambda_n(K)}{\kappa_n} \right)^i, \qquad (B.21)$$

with equality if and only if K is a ball.

B.4. The dual Aleksandrov–Fenchel inequality

All the inequalities in this section were derived by Lutwak [529]. In essence, they are consequences of an appropriate version of Hölder's inequality, but their power lies in the way they are cast into mirror images of corresponding inequalities from the theory of mixed volumes. It will be convenient to assume throughout this section that $n \ge 2$.

Let $f_0, f_1, \ldots, f_i$ be nonnegative bounded Borel functions on S^{n-1}, and $a_1, \ldots, a_i$ positive reals satisfying

$$\frac{1}{a_1} + \cdots + \frac{1}{a_i} = 1.$$

Then

$$\int_{S^{n-1}} f_0(u) f_1(u) \cdots f_i(u) \, du \leq \prod_{j=1}^{i} \left(\int_{S^{n-1}} f_0(u) f_j(u)^{a_j} \, du \right)^{1/a_j}, \qquad (B.22)$$

with equality if and only there exist nonnegative constants $b_1, \ldots, b_i$, not all zero, such that $b_1 f_0(u) f_1(u)^{a_1} = \cdots = b_i f_0(u) f_i(u)^{a_i}$ for λ_{n-1}-almost all $u \in S^{n-1}$. This follows easily from the standard form (B.9) of Hölder's inequality. Now let L_j be a star body in $\mathbb{E}^n$ with $o \in L_j$, $1 \leq j \leq n$, and suppose that $1 \leq i \leq n$. Then we have the following *dual Aleksandrov–Fenchel inequality*:

$$\tilde{V}(L_1, L_2, \ldots, L_n)^i \leq \prod_{j=1}^{i} \tilde{V}(L_j, i; L_{i+1}, \ldots, L_n), \qquad (B.23)$$

with equality, when $i = n$ or when $i \neq n$ and $o \in \mathrm{int}\, L_j$, $1 \leq j \leq n$, if and only if $L_1, \ldots, L_i$ are dilatates of each other. This follows immediately from (B.22) on setting $f_0 = \rho_{L_{i+1}} \cdots \rho_{L_n}$ ($f_0 = 1$ if $i = n$) and $a_j = i$ and $f_j = \rho_{L_j}$, $1 \leq j \leq i$.

Inequality (B.23) is named, of course, for its striking resemblance to the version (B.17) of the Aleksandrov–Fenchel inequality; there is a tilde over the V and the inequality is reversed. There seems to be no adequate explanation for this phenomenon at the present time.

We shall also need the following inequality between dual mixed volumes. Suppose that L_1 and L_2 are star bodies in $\mathbb{E}^n$ containing the origin in their interiors, and that i, j, and k are real numbers satisfying $i \leq j \leq k$. Then, using the notation (A.56),

$$\tilde{V}_j(L_1, L_2)^{k-i} \leq \tilde{V}_i(L_1, L_2)^{k-j} \tilde{V}_k(L_1, L_2)^{j-i}, \qquad (B.24)$$

with equality when $i = j$ or $j = k$, and otherwise, if and only if L_1 is a dilatate of L_2. This is trivial when $i = j$ or $j = k$, and for $i < j < k$ it follows immediately from (B.22), this time on setting the index i in that inequality equal to 2 and then $f_0 = \rho_{L_1}^{n-i} \rho_{L_2}^i$, $f_1 = \rho_{L_1}^{i-j} \rho_{L_2}^{j-i}$, $f_2 = 1$, $a_1 = (k-i)/(j-i)$, and $a_2 = (k-i)/(k-j)$. When i, j, and k are integers between 1 and $n-1$, this is an instance of (B.23), but the point is that this restriction on the indices has been relaxed. There is a mixed-volume version of (B.24) (see [737, p. 334, eq. (6.4.5)]), but we do not need it in this book.

From the dual Aleksandrov–Fenchel inequality, we can at once obtain dual forms of other inequalities. In particular we have

$$\tilde{V}(L_1, n-1; L_2)^n \leq \lambda_n(L_1)^{n-1} \lambda_n(L_2) \qquad (B.25)$$

(the *dual Minkowski inequality*),

$$\tilde{V}(L_1; L_2, i-1; B, n-i)^i \leq \tilde{V}_i(L_2)^{i-1} \tilde{V}_i(L_1), \qquad (B.26)$$

and

$$\tilde{V}(L_1, \ldots, L_n)^n \le \lambda_n(L_1)\lambda_n(L_2)\cdots\lambda_n(L_n), \tag{B.27}$$

with equality in (B.25) and (B.27), and in (B.26) when $o \in \operatorname{int} L_j$, $j = 1, 2$, if and only if the bodies are dilatates of each other. Also, if $1 \le i \le n-1$, we obtain an (extended) *dual isoperimetric inequality*:

$$\left(\frac{\tilde{V}_i(L)}{\tilde{V}_i(B)}\right)^n \le \left(\frac{\lambda_n(L)}{\kappa_n}\right)^i, \tag{B.28}$$

with equality if and only if L is a centered ball.

Some special cases of (B.24) are also useful. For example, when $i \in \mathbb{R}$ and $0 \le i \le n$, we have

$$\tilde{V}_i(L_1, L_2)^n \le \lambda_n(L_1)^{n-i}\lambda_n(L_2)^i, \tag{B.29}$$

and the reverse inequality also holds when $i < 0$ or $i > n$, all with equality if and only if $i = n$ or $i \ne n$ and L_1 is a dilatate of L_2.

Using (A.53) and the dual Minkowski inequality (B.25), we obtain for star bodies L, L_0, and L_1 in $\mathbb{E}^n$, containing the origin,

$$\tilde{V}(L, n-1; L_0\tilde{+}L_1) = \tilde{V}(L, n-1; L_0) + \tilde{V}(L, n-1; L_1)$$
$$\le \lambda_n(L)^{(n-1)/n}\left(\lambda_n(L_0)^{1/n} + \lambda_n(L_1)^{1/n}\right).$$

Putting $L = L_0\tilde{+}L_1$, we infer the following *dual Brunn–Minkowski inequality*:

$$\lambda_n(L_0\tilde{+}L_1)^{1/n} \le \lambda_n(L_0)^{1/n} + \lambda_n(L_1)^{1/n}, \tag{B.30}$$

with equality if and only if L_0 is a dilatate of L_1.

B.5. Other inequalities

Various refinements of the Aleksandrov–Fenchel inequality are available. We shall only find use for one such planar inequality, which does not fully generalize to higher dimensions. This is a version of *Bonnesen's inequality* (see [737, p. 324, eq. (6.2.26)], or Berger's book [53, 12.12.15] for a proof), stating that if K is a convex body in $\mathbb{E}^2$, then there are concentric circles, of radii r_0 and R_0 with $r_0 \le R_0$ enclosing bd K such that

$$\pi(R_0 - r_0)^2 \le V_1(K)^2 - \pi\lambda_2(K). \tag{B.31}$$

We turn now to a pair of inequalities of a different type. If K_1 and K_2 are convex bodies in $\mathbb{E}^n$, there is a convex body K such that

$$S(K, \cdot) = S(K_1, \cdot) + S(K_2, \cdot),$$

by (A.23). (The body K is the Blaschke sum of K_1 and K_2; see Note 3.4.) Then the *Kneser–Süss inequality* states that

$$\lambda_n(K)^{(n-1)/n} \geq \lambda_n(K_1)^{(n-1)/n} + \lambda_n(K_2)^{(n-1)/n}, \tag{B.32}$$

with equality if and only if K_1 and K_2 are homothetic. The proof, straightforward given (A.12) and (B.13), can be found in [737, Theorem 7.1.3].

The previous inequality admits the following dual form due to Lutwak [537, Proposition 6.3].

Theorem B.5.1 (dual Kneser–Süss inequality). *Let L_1 and L_2 be star bodies in $\mathbb{E}^n$ containing the origin in their interiors, and let L be the star body defined by*

$$\rho_L^{n-1} = \rho_{L_1}^{n-1} + \rho_{L_2}^{n-1}.$$

Then

$$\lambda_n(L)^{(n-1)/n} \leq \lambda_n(L_1)^{(n-1)/n} + \lambda_n(L_2)^{(n-1)/n}, \tag{B.33}$$

with equality if and only if L_1 is a dilatate of L_2.

Proof. Let M be any star body in $\mathbb{E}^n$ with $o \in M$. Using (A.51), the definition of L, and (B.25), we have

$$\tilde{V}(L, n-1; M) = \tilde{V}(L_1, n-1; M) + \tilde{V}(L_2, n-1; M)$$
$$\leq \left(\lambda_n(L_1)^{(n-1)/n} + \lambda_n(L_2)^{(n-1)/n}\right)\lambda_n(M)^{1/n},$$

with equality if and only if L_1 and L_2 are dilatates of M. The result follows immediately once we put $M = L$. ∎

The body L in the last theorem is the $(n-1)$th radial sum of L_1 and L_2; see Note 6.1.

Appendix C

Integral transforms

The purpose of this appendix is to present some basic material concerning integral transforms employed in the text. The two of most interest here, the cosine and spherical Radon transforms, have a section to themselves.

C.1. X-ray transforms

Throughout this section, f will denote a bounded measurable (meaning λ_n-measurable) function on $\mathbb{E}^n$ vanishing outside a bounded set.

Although we shall not study them directly, let us begin by introducing two well-known integral transforms. The *n-dimensional Fourier transform* $\hat{f}$ of f is defined for $y \in \mathbb{E}^n$ by

$$\hat{f}(y) = \int_{\mathbb{E}^n} e^{-ix\cdot y} f(x)\, dx. \tag{C.1}$$

If $t \in \mathbb{R}$ and $u \in S^{n-1}$, let

$$\tilde{f}(t, u) = \int_{u^\perp + tu} f(x)\, dx. \tag{C.2}$$

This is the *Radon transform* of f, named for the pioneering work [685] of Radon. For each $u \in S^{n-1}$, Fubini's theorem guarantees that $\tilde{f}(t, u)$ exists for almost all t. Of course, the set $H = u^\perp + tu$ is a hyperplane in $\mathbb{E}^n$, and so one can rewrite the Radon transform in the form

$$\tilde{f}(H) = \int_H f(x)\, dx, \tag{C.3}$$

where it becomes a function on the space of hyperplanes in $\mathbb{E}^n$. (Note that (t, u) and $(-t, -u)$ represent the same H.)

Let $u \in S^{n-1}$. For each $x \in u^{\perp}$, define

$$X_u f(x) = \int_{-\infty}^{\infty} f(x + tu) \, dt, \tag{C.4}$$

where dt denotes integration with respect to λ_1. The function $X_u f$ is called the *X-ray transform*, or the (parallel) *X-ray, of f in the direction u.*

Now let $p \in \mathbb{E}^n$. For each $u \in S^{n-1}$, define

$$D_p f(u) = \int_0^{\infty} f(p + tu) \, dt. \tag{C.5}$$

This function, called the *divergent beam transform*, or the *directed X-ray, of f at the point p,* was introduced by Hamaker et al. [379]. We also define, for each $u \in S^{n-1}$,

$$X_p f(u) = \int_{-\infty}^{\infty} f(p + tu) \, dt. \tag{C.6}$$

The function $X_p f$ is called the *line transform*, or the *X-ray, of f at the point p.* See, for example, the papers of Finch [233] or Smith and Keinert [773]. The frequently quoted paper of Leahy, Smith, and Solmon [500] was presented at a conference whose proceedings never appeared; see the article [777] of Solmon.

One can also consider the following generalization of the X-ray transform. Let $1 \leq k \leq n - 1$ and $S \in \mathcal{G}(n, k)$. For each $y \in S^{\perp}$, define

$$X_S f(y) = \int_S f(x + y) \, dx. \tag{C.7}$$

The function $X_S f$ is called the *k-plane transform*, or the *X-ray, of f parallel to the subspace S.* This goes back at least to work of Helgason [387]. A good comprehensive reference is Keinert's survey [419]; see also the works of Smith, Solmon, and Wagner [774] and Solmon [775], [776].

These transforms are all related. Fixing $u \in S^{n-1}$, we can regard $\tilde{f}(t, u)$ as a function of t. Then, if $u^{\perp} = S \in \mathcal{G}(n, n-1)$, we have $\tilde{f}(t, u) = X_S f(y)$, where $y = tu$. In this sense, therefore, the k-plane transform is a common generalization of the X-ray transform and the Radon transform, which correspond to $k = 1$ and $k = n - 1$, respectively. In particular, the X-ray transform and Radon transform are the same when $n = 2$. There is an easily established but important relationship between the Fourier transform of f and that of its k-plane transform. If $S \in \mathcal{G}(n, k)$, then

$$(X_S f)\hat{}(y) = \hat{f}(y), \tag{C.8}$$

for all $y \in S^{\perp}$. See, for example, [419, Theorem 5.1], where this is called the *central slice theorem*, or [775, Lemma 1.4]. (Note that the Fourier transform is defined differently by these authors.)

Smith and Keinert [773] point out the simple relations expressed by

$$X_p f(u) = D_p f(u) + D_p f(-u) = X_u f(p|u^\perp),$$

for each $u \in S^{n-1}$.

We also employ the following more general version of the line transform. Let $p \in \mathbb{E}^n$, and suppose that $i \in \mathbb{R}$. We define

$$X_{i,p} f(u) = \int_{-\infty}^{\infty} f(p + tu)|t|^{i-1} dt, \qquad (C.9)$$

for each $u \in S^{n-1}$ such that the integral exists. We call $X_{i,p} f$ the *line transform of order i of f at p*. (This does not seem to be named in the literature; see [419, eq. (5.10)] and [645, p. 31], for example. It can be inverted by the method of Quinto [682].)

It is worth mentioning that Keinert [419] also introduces the divergent k-plane transform, where integration in lines through the point p in the line transform is replaced by integration in k-dimensional planes through p. For characteristic functions, this reduces to the k-dimensional X-ray at the point p (see Chapter 7).

The basic uniqueness theorem for the X-ray transform is the following one.

Theorem C.1.1. *Let f be a bounded measurable function on $\mathbb{E}^n$, vanishing outside a bounded set. Suppose that $D \subset S^{n-1}$ is an infinite set. If $X_u f = 0$ for all $u \in D$, then $f = 0$, λ_n-almost everywhere.*

A proof is given by Solmon [775, Theorem 1.7], for example. (It is also stated for C^∞ functions in Natterer's book [645, Theorem 3.5].) The proof is not difficult, given some basic facts about the Fourier transform. By (C.8) with $k = 1$, the hypotheses imply that $\hat{f}(y) = 0$ whenever $y \in u^\perp$ and $u \in D$. The Fourier transform $\hat{f}$ is an analytic function on $\mathbb{E}^n$, and an analytic function vanishing on an infinite set of hyperplanes must be identically zero. The result follows, since $\hat{f} = 0$ implies that f is zero λ_n-almost everywhere.

Theorem C.1.1 has a long and complicated history, which we shall not attempt to unravel. For differentiable functions, $n = 2$, and $D = S^{n-1}$, Radon's paper [685] contains a proof.

A consequence of the paragraph following Theorem 2.2.5 is that when $1 < k \le n - 1$, it is possible that $X_S f = 0$ for infinitely many $S \in \mathcal{G}(n, k)$, but f is not zero λ_n-almost everywhere. However, there are uniqueness theorems for the k-plane transform. The most general seems to be Theorem 5.2 of [419]. This says that if $X_{S_m} f = 0$ for $m \in \mathbb{N}$, where $S_m \in \mathcal{G}(n, k)$ and the subspaces $S_m^\perp$, $m \in \mathbb{N}$, are not contained in a proper algebraic variety, then f is zero λ_n-almost everywhere. When $k = 1$, each $S_m^\perp$ is a hyperplane, and the subspaces $S_m^\perp, m \in \mathbb{N}$, cannot be contained in a proper algebraic variety, since no nonzero polynomial on $\mathbb{E}^n$ can vanish on an infinite number of hyperplanes; this yields Theorem C.1.1 as a special case.

For the divergent beam transform, the following uniqueness theorem is available. Notice that the hypotheses imply that it applies equally to the line transform.

Theorem C.1.2. *Let f be a bounded measurable function on $\mathbb{E}^n$ whose domain is contained in the unit ball B. Suppose that $P \subset \mathbb{E}^n \setminus B$ is an infinite set. If $D_p f = 0$ for all $p \in P$, then $f = 0$, λ_n-almost everywhere.*

This was first proved in [379, Theorem 5.1]; see also [419, Theorem 5.5]. For C^∞ functions, one can also consult [645, Theorem 3.6]. The proof takes a little more work than that of Theorem C.1.1. Since P is infinite, there is an accumulation point u_0 of the set of directions of vectors in P. One shows that

$$\int_{S^{n-1}} D_p f(v)(u \cdot v)^{1-n} \, dv = \int_{-\infty}^{\infty} \tilde{f}(t, u)(t - u \cdot p)^{1-n} \, dt,$$

for all $u \in S^{n-1}$. The hypotheses of the theorem are used to avoid singularities and conclude that the right-hand integral over the range $-1 \leq t \leq 1$ vanishes whenever $p \in P$ and u is sufficiently close to u_0. Expanding the integrand, one can show that $\tilde{f} = 0$. Then (C.8) with $k = n - 1$ implies that $\hat{f} = 0$, and it follows that f is zero λ_n-almost everywhere.

To achieve full symmetry between the last two theorems, it would be satisfying to have a positive answer to Problem C.1.

A further uniqueness theorem, in which it is assumed that $D_p f = 0$ for all p in a suitable C^1 curve, was proved by Leahy, Smith, and Solmon [500] using the line transform of order i; see Natterer's book [645, Theorem 3.3].

C.2. The cosine and spherical Radon transforms

Suppose that f is a Borel function on S^{n-1}. The *cosine transform* Cf of f is defined by

$$Cf(u) = \int_{S^{n-1}} |u \cdot v| f(v) \, dv, \tag{C.10}$$

for all $u \in S^{n-1}$.

An important property of the cosine transform is that it is injective on $C_e(S^{n-1})$. More generally, we have the following theorem.

Theorem C.2.1. *If μ is a signed finite even Borel measure in S^{n-1} such that*

$$\int_{S^{n-1}} |u \cdot v| \, d\mu(v) = 0,$$

for each $u \in S^{n-1}$, then $\mu = 0$.

Proof. We shall show that the linear span H of the functions $|u \cdot v|$, $v \in S^{n-1}$, is dense in $C_e(S^{n-1})$ with the sup norm (i.e., the L^∞ norm); the theorem will follow immediately.

Applying Cauchy's projection formula ((A.45) with $i = n - 1$) to the unit ball, we have the formula (also easily proved directly)

$$\kappa_{n-1} = \frac{1}{2} \int_{S^{n-1}} |u \cdot v| \, dv,$$

for each $u \in S^{n-1}$. Therefore, if $\phi \in GL_n$,

$$\kappa_{n-1} = \frac{1}{2\|\phi u\|} \int_{S^{n-1}} |\phi u \cdot v| \, dv,$$

for each $u \in S^{n-1}$. It follows, using (0.1), that

$$\|\phi u\| = \frac{1}{2\kappa_{n-1}} \int_{S^{n-1}} |\phi u \cdot v| \, dv = \frac{1}{2\kappa_{n-1}} \int_{S^{n-1}} |u \cdot \phi^t v| \, dv,$$

for each $u \in S^{n-1}$.

Let us denote the last integral by $f(u)$. It can be uniformly approximated by finite sums of the form

$$f_m(u) = \frac{1}{2\kappa_{n-1}} \sum_{i=1}^{m} |u \cdot \phi^t v_i| = \frac{1}{2\kappa_{n-1}} \sum_{i=1}^{m} |c_i| |u \cdot w_i|,$$

where $w_i \in S^{n-1}$ and $c_i \in \mathbb{R}$ has been chosen to satisfy $\phi^t v_i = c_i w_i$, $1 \le i \le m$. Since each $f_m \in H$, we see that $f \in \operatorname{cl} H$.

Now if $u = (u_1, \ldots, u_n)$, then

$$f(u) = \|\phi u\| = \left(\sum_{i,j} a_{ij} u_i u_j \right)^{1/2},$$

where $A = (a_{ij})$ is a symmetric positive-definite $n \times n$ matrix. Moreover, if A is any such matrix, there is a ϕ such that the previous equation holds, and then the expression on the right belongs to $\operatorname{cl} H$. Differentiating with respect to a_{ij}, we obtain

$$\frac{1}{2} \left(\sum_{i,j} a_{ij} u_i u_j \right)^{-1/2} u_i u_j.$$

This derivative, by its definition as a limit, is a uniform limit of functions in $\operatorname{cl} H$, and so itself belongs to $\operatorname{cl} H$. If we now take A to be the identity matrix, we see that $u_i u_j \in \operatorname{cl} H$, for $1 \le i, j \le n$. Repeated differentiation shows that every monomial of even degree, and therefore every even polynomial, in the components u_i of u belongs to $\operatorname{cl} H$. The Stone–Weierstrass theorem (see [14, Section A7] or [710, p. 162]) implies that an arbitrary even continuous function on S^{n-1} belongs to $\operatorname{cl} H$, completing the proof. ∎

Aleksandrov [3] first proved Theorem C.2.1, but the clever proof given here, with the advantage of avoiding spherical harmonics, was found by Choquet [164,

p. 53], [165, p. 171]. The result has been rediscovered many times; see the references in [737, p. 192]. (Short introductions to spherical harmonics in geometry are given in [353] and [737, Appendix], but Groemer's book [356, Section 3.4] provides a detailed and exhaustive account of this topic, including a proof of the previous theorem and a stability version of it.)

A different proof of the previous theorem is given as part of Theorem 3.5.3 of [737], using spherical harmonics. The rest of Theorem 3.5.3 of [737], applying spherical harmonics again, proves that if $k \geq n + 2$ is even and $f \in C_e^k(S^{n-1})$, then there is a $g \in C_e^k(S^{n-1})$ such that $f = Cg$. This implies the following theorem.

Theorem C.2.2. *The cosine transform is a continuous bijection of $C_e^{\infty}(S^{n-1})$ to itself.*

A final property of the cosine transform is an immediate consequence of Fubini's theorem.

Theorem C.2.3. *The cosine transform is self-adjoint, in the sense that if f and g are bounded Borel functions on S^{n-1}, then*

$$\int_{S^{n-1}} f(u) C g(u) \, du = \int_{S^{n-1}} Cf(u) g(u) \, du.$$

Suppose that f is a Borel function on S^{n-1}. The *spherical Radon transform* Rf of f is defined by

$$Rf(u) = \int_{S^{n-1} \cap u^{\perp}} f(v) \, dv, \qquad (C.11)$$

for all $u \in S^{n-1}$.

The spherical Radon transform shares some of the important properties of the cosine transform. There is, in fact, a direct connection between the two transforms. If we let

$$\square = (\Delta_S + n - 1)/2,$$

then it can be shown that

$$\square C = R. \qquad (C.12)$$

Goodey and Weil [318, Proposition 2.1] note that this follows from a result of Berg [51], and attribute the case $n = 3$ to Blaschke.

The next theorem deals with the injectivity properties of R. See Groemer's book [356, Section 3.4] for a proof and a stability version.

Theorem C.2.4. *If f is a bounded even Borel function on S^{n-1} such that for λ_{n-1}-almost all $u \in S^{n-1}$,*

$$\int_{S^{n-1} \cap u^{\perp}} f(v)\, dv = 0,$$

then $f = 0$, λ_{n-1}-almost everywhere.

With the stronger assumption that $Rf(u) = 0$ for all $u \in S^{n-1}$, Theorem C.2.4 for continuous functions f and $n = 3$ was first proved by Minkowski [625] (see also [83, Section 67]), who used it to show that a convex body in $\mathbb{E}^3$ of constant girth also has constant width (cf. Theorem 3.3.13). For continuous functions and general n, the injectivity follows from results of Helgason [387]; alternative proofs are supplied by Petty [661] and Schneider [728], [729]. The injectivity actually extends to (equivalence classes of) bounded even λ_{n-1}-measurable functions, and further to even distributions. This is obtained from Theorem C.2.4 by standard methods by Goodey and Weil [318, pp. 676–7]. It is the latter extension that also allows the weaker assumption that $Rf(u) = 0$ for λ_{n-1}-almost all $u \in S^{n-1}$. (Alternatively, this follows easily from Theorem C.2.4 with the stronger assumption and Theorem C.2.6.)

Goodey and Weil [317, p. 677] observe that the next two theorems are consequences of some very general results of Helgason [389, pp. 144, 161]. Alternatively, Groemer [356, Propositions 3.4.12 and 3.6.4 and Theorem 3.4.14] supplies a direct proof of the first; for the second, proofs avoiding spherical harmonics are provided by Groemer [356, Lemma 1.3.3] and by Schneider and Weil [746, p. 150].

Theorem C.2.5. *The spherical Radon transform is a continuous bijection of $C_e^{\infty}(S^{n-1})$ to itself.*

Theorem C.2.6. *The spherical Radon transform is self-adjoint, in the sense that if f and g are bounded Borel functions on S^{n-1}, then*

$$\int_{S^{n-1}} f(u) Rg(u)\, du = \int_{S^{n-1}} Rf(u) g(u)\, du.$$

We shall also need some other properties of R.

Lemma C.2.7. *The spherical Radon transform commutes with rotations.*

Proof. Let $\phi \in SO_n$, let f be a Borel function on S^{n-1}, and let $u \in S^{n-1}$. Then

$$\phi(Rf)(u) = Rf(\phi^{-1}u) = \int_{S^{n-1} \cap (\phi^{-1}u)^{\perp}} f(v)\, dv.$$

Substitute $w = \phi v$. By (0.1) and the orthogonality of ϕ, we have $\phi^{-1}u \cdot v = 0$ if and only if $u \cdot \phi v = u \cdot w = 0$. Using the SO_n-invariance of λ_{n-2}, the last integral becomes

$$\int_{S^{n-1}\cap u^{\perp}} f(\phi^{-1}w)\,dw = R(\phi f)(u),$$

as required.   ∎

Suppose that f is a bounded Borel function on S^{n-1}. Let φ denote the vertical angle of spherical polar coordinates. For $0 \le \varphi \le \pi$, let $\overline{f}(\varphi)$ be the average of f over the sphere of latitude with angle φ from the north pole in S^{n-1}. Let us identify $S^{n-1} \cap \{x : x_n = 0\}$ with S^{n-2}. Then, identifying each $u \in S^{n-1}$ with the pair (v, φ), where $0 \le \varphi \le \pi$ and $v \in S^{n-2}$ corresponds to the projection of u on $x_n = 0$, we have

$$\overline{f}(\varphi) = \frac{1}{\omega_{n-1}} \int_{S^{n-2}} f(v, \varphi)\,dv.$$

We can regard $\overline{f}$ as a function defined on S^{n-1} by writing $\overline{f}(v, \varphi) = \overline{f}(\varphi)$ for all $v \in S^{n-2}$. The following observation was made by Funk [252, VI, p. 285] for $n = 3$.

Theorem C.2.8. *If* $f = Rg$, *then* $\overline{f} = R\overline{g}$.

Proof. We have

$$\overline{f}(\varphi) = \lim_{m\to\infty} \frac{1}{m} \sum_{i=1}^{m} f(v_i, \varphi),$$

for suitable choices of v_i, $1 \le i \le m$, and a similar expression for $\overline{g}(\varphi)$. Let v_0 be any fixed member of S^{n-2}, and denote by ϕ_i a rotation in S^{n-2} taking v_0 to v_i. Then

$$\overline{g}(v_0, \varphi) = \overline{g}(\varphi) = \lim_{m\to\infty} \frac{1}{m} \sum_{i=1}^{m} g(v_i, \varphi) = \lim_{m\to\infty} \frac{1}{m} \sum_{i=1}^{m} (\phi_i^{-1}g)(v_0, \varphi).$$

Also, using Lemma C.2.7,

$$f(v_i, \varphi) = (\phi_i^{-1}f)(v_0, \varphi)$$
$$= (\phi_i^{-1}(Rg))(v_0, \varphi) = (R(\phi_i^{-1}g))(v_0, \varphi).$$

Consequently,

$$\overline{f}(\varphi) = \lim_{m\to\infty} \frac{1}{m} \sum_{i=1}^{m} (R(\phi_i^{-1}g))(v_0, \varphi)$$

$$= R\left(\lim_{m\to\infty} \frac{1}{m} \sum_{i=1}^{m} (\phi_i^{-1}g)\right)(v_0, \varphi)$$

$$= R\overline{g}(\varphi).   ∎$$

There are some useful formulas involving rotationally symmetric functions. A *rotationally symmetric function* f on S^{n-1} is one that can be defined via a function $f(\varphi)$ of the vertical angle φ in spherical polar coordinates by setting $f(v, \varphi) = f(\varphi)$, for all $v \in S^{n-2}$. The results are due to Helgason [390, eqs. (3.10), (3.11)].

Theorem C.2.9. *Suppose that f and g are rotationally symmetric bounded even Borel functions on S^{n-1}, $n \geq 3$, such that $f = Rg$. Then*

$$f(u) = f(\varphi) = \frac{2\omega_{n-2}}{\sin\varphi} \int_{\frac{\pi}{2}-\varphi}^{\frac{\pi}{2}} g(\psi) \left(1 - \frac{\cos^2\psi}{\sin^2\varphi}\right)^{(n-4)/2} \sin\psi \, d\psi, \quad \text{(C.13)}$$

for each $u \in S^{n-1}$ whose angle from the x_n-axis is φ, $0 < \varphi \leq \pi/2$.

Proof. Let p be a point in the $(n-2)$-sphere $S^{n-1} \cap u^\perp$ such that the line l through p and o has minimal angle η with the x_n-axis. Then $\varphi = \pi/2 - \eta$. Let q be an arbitrary point in $S^{n-1} \cap u^\perp$, and suppose that ψ and α are its angles at o with the x_n-axis and l, respectively. (See Figure 8.4 for a picture of these angles.) The set of points in $S^{n-1} \cap u^\perp$ having the same angle α (at o with l) as q is an $(n-3)$-sphere of radius $\sin\alpha$, and g is constant and equal to $g(\psi)$ on this sphere. Therefore, using the fact that g is even, we obtain

$$f(u) = 2\omega_{n-2} \int_0^{\frac{\pi}{2}} g(\psi) \sin^{n-3}\alpha \, d\alpha.$$

The angle between l and the x_n-axis is η, so

$$\cos\psi = \cos\alpha \cos\eta,$$

where for our purposes η is fixed. Using this equation to substitute for α in the integral, we find that for $0 \leq \eta < \pi/2$,

$$f(u) = f(\pi/2 - \eta) = 2\omega_{n-2} \int_\eta^{\frac{\pi}{2}} g(\psi) \left(1 - \frac{\cos^2\psi}{\cos^2\eta}\right)^{(n-4)/2} \left(\frac{\sin\psi}{\cos\eta} d\psi\right).$$

The substitution $\varphi = \pi/2 - \eta$ completes the proof. ∎

The techniques of the next theorem go back at least to Abel, when $n = 3$, and similar formulas have been employed many times in geometry, tomography, and other subjects; see, for example, the works of Helgason [390], John [408, p. 82], and Natterer [645, p. 23].

Theorem C.2.10. *Suppose that $n \in \mathbb{N}$, $n \geq 3$, $G \in C([0, 1])$, and*

$$F(x) = \frac{2\omega_{n-2}}{x^{n-3}} \int_0^x G(t)(x^2 - t^2)^{(n-4)/2} dt, \quad \text{(C.14)}$$

for $0 < x \le 1$. Then

$$G(t) = \frac{1}{(n-3)!\omega_{n-1}} t \left(\frac{1}{t}\frac{d}{dt}\right)^{n-2} \int_0^t F(x)x^{n-2}(t^2 - x^2)^{(n-4)/2} dx, \quad (C.15)$$

for $0 < t \le 1$.

Proof. We shall only outline the proof, leaving the calculus details to the reader. Multiply both sides of (C.14) by $x^{n-2}(s^2 - x^2)^{(n-4)/2}$, integrate with respect to x from 0 to s, and change the order of integration on the right-hand side, to obtain

$$\int_0^s F(x)x^{n-2}(s^2 - x^2)^{(n-4)/2} dx$$

$$= \omega_{n-2} \int_0^s G(t) \int_t^s 2x\left((x^2 - t^2)(s^2 - x^2)\right)^{(n-4)/2} dx\, dt.$$

Now

$$2\int_t^s x\left((x^2 - t^2)(s^2 - x^2)\right)^{(n-4)/2} dx$$

$$= 2^{3-n}(s^2 - t^2)^{n-3} \int_{-1}^1 (1 - w^2)^{(n-4)/2} dw$$

$$= \frac{\omega_{n-1}}{2^{n-3}\omega_{n-2}}(s^2 - t^2)^{n-3}.$$

Here, the first equation is achieved by the substitution

$$w = \frac{t^2 + s^2 - 2x^2}{t^2 - s^2},$$

and the second equation results from putting $w = \sin\theta$ and using the recursion formula for the integral of $\cos^k \theta$, $k \in \mathbb{N}$, (0.9), and (0.10). Next, note that when $n \ge 4$,

$$\left(\frac{1}{s}\frac{d}{ds}\right)\int_0^s G(t)(s^2 - t^2)^{n-3} dt = 2(n-3)\int_0^s G(t)(s^2 - t^2)^{n-4} dt.$$

Equation (C.15) then results from applying the differential operator in brackets $(n-3)$ times, with one further differentiation with respect to s. ∎

Corollary C.2.11. *Suppose that $f = Rg$, where $f, g \in C_e(S^{n-1})$, $n \ge 3$, are rotationally symmetric functions. Then*

$$f(\sin^{-1} x) = \frac{2\omega_{n-2}}{x^{n-3}} \int_0^x g(\cos^{-1} t)(x^2 - t^2)^{(n-4)/2} dt, \quad (C.16)$$

for $0 < x \le 1$. This equation may be inverted, to yield

$$g(\cos^{-1} t) = \frac{1}{(n-3)!\omega_{n-1}} t \left(\frac{1}{t}\frac{d}{dt}\right)^{n-2} \int_0^t f(\sin^{-1} x)x^{n-2}(t^2 - x^2)^{(n-4)/2} dx,$$

$$(C.17)$$

for $0 < t \le 1$.

Proof. Put $x = \sin \varphi$ and $t = \cos \psi$ in Theorem C.2.9 to obtain (C.16), and then let $F(x) = f(\sin^{-1} x)$ and $G(t) = g(\cos^{-1} t)$ in Theorem C.2.10 to obtain (C.17). ∎

When $n = 3$, we need to summarize some extra information. In this case, (C.14) becomes

$$F(x) = 4 \int_0^x \frac{G(t)}{\sqrt{x^2 - t^2}} \, dt, \qquad (C.18)$$

for $0 < x \leq 1$. Equation (C.18) is an *Abel integral equation*. The existence and integrability of a solution G, unique in $L^1(0, 1)$, was proved by L. Tonelli, under the assumption that F is absolutely continuous; see the book of Gorenflo and Vessella [329, Chapter 1], especially Theorem 1.2.1. One form of the solution is the one given by Theorem C.2.10 (cf. [329, (1.B.5i), p. 24]), namely,

$$G(t) = \frac{1}{2\pi} \frac{d}{dt} \int_0^t \frac{x F(x)}{\sqrt{t^2 - x^2}} \, dx, \qquad (C.19)$$

for $0 < t \leq 1$. An alternative form is (cf. [329, (1.B.5ii), p. 24])

$$G(t) = \frac{1}{2\pi} \left(F(0) + t \int_0^t \frac{F'(x)}{\sqrt{t^2 - x^2}} \, dx \right), \qquad (C.20)$$

for $0 < t \leq 1$. If $F(0) = 2\pi G(0)$, then $G(0) = F(0)/2\pi$, and (C.20) is valid for $0 \leq t \leq 1$. The existence of a unique *continuous* G of either form follows from the assumption that the function F satisfies $F \in C^1([0, 1])$. Indeed, by substituting (C.20) into (C.18), we see that

$$4 \int_0^x \frac{G(t)}{\sqrt{x^2 - t^2}} \, dt = \frac{2}{\pi} \int_0^x \frac{F(0) \, dt}{\sqrt{x^2 - t^2}} + \frac{2}{\pi} \int_0^x \frac{t}{\sqrt{x^2 - t^2}} \int_0^t \frac{F'(y)}{\sqrt{t^2 - y^2}} \, dy \, dt$$

$$= F(0) + \frac{2}{\pi} \int_0^x F'(y) \int_y^x \frac{t}{\sqrt{(x^2 - t^2)(t^2 - y^2)}} \, dt \, dy$$

$$= F(0) + \int_0^x F'(y) \, dy = F(x),$$

so (C.20) is a solution of (C.18). (It is also not difficult to obtain (C.20) from (C.19), by integrating by parts and differentiating under the integral sign.) According to [329, Theorem 5.1.5], with $m = 1$, $K(x, t) \equiv 1$, $\beta = 0$, $\mu = 2$, and $\alpha = 1/2$, there is a continuous solution G of (C.18) if $F \in C^1([0, 1])$; since the solution is unique, we conclude that the G given by (C.19) for $0 < t \leq 1$ and $G(0) = F(0)/2\pi$, or by (C.20), is continuous for $0 \leq t \leq 1$. (A direct proof of the continuity of G is fairly straightforward using (C.20).)

Setting $t = 1$ in (C.20), and substituting $x = \sin \varphi$, we obtain

$$2\pi G(1) = F(0) + \int_0^{\frac{\pi}{2}} F'(\sin \varphi) \, d\varphi. \qquad (C.21)$$

Suppose that f is a Borel function on S^{n-1} and $p > -1$ is real. The *p-cosine transform* $C_p f$ of f is defined by

$$C_p f(u) = \int_{S^{n-1}} |u \cdot v|^p f(v) \, dv, \tag{C.22}$$

for all $u \in S^{n-1}$. Thus the usual cosine transform Cf of f is just $C_1 f$. The p-cosine transform provides an interesting bridge between the cosine transform and spherical Radon transform R, since

$$R = \lim_{p \to -1+} \left(\frac{p+1}{2} \right) C_p. \tag{C.23}$$

See the articles of Grinberg and Zhang [337, Corollary 8.3] and Koldobsky [454], for example.

Koldobsky [454] and Rubin [704], [705] provide much information about the p-cosine and related transforms and their properties, as well as some historical remarks. In particular, it is known that C_p is injective on even functions if and only if p is not an even integer. Under special but important circumstances, these transforms have a clear connection to the Fourier transform, as follows.

Theorem C.2.12. *Let $n \geq 2$ and let $p < -n + 1$ be such that $-n - p$ is not an even integer. Let f be an even Borel function, positively homogeneous of degree p on $\mathbb{E}^n \setminus \{o\}$, such that its restriction to S^{n-1} is in $L^1(S^{n-1})$. Then*

$$\hat{f}(u) = \frac{\pi}{\alpha_{-n-p}} C_{-n-p} f(u),$$

for $u \in S^{n-1}$, where

$$\alpha_p = 2^{p+1} \frac{\sqrt{\pi}\,\Gamma((p+1)/2)}{\Gamma(-p/2)}.$$

The statement of the previous theorem follows Koldobsky [454], who rediscovered similar earlier results of V. I. Semyanistyi (see the remarks of Koldobsky [465, Section 3.5] and Rubin [705]).

If we take $p = -n - 1$ in Theorem C.2.12, we obtain the relation

$$\hat{f}(u) = -\frac{\pi}{2} Cf(u), \tag{C.24}$$

for $u \in S^{n-1}$, where f is an even Borel function, positively homogeneous of degree $-n - 1$ on $\mathbb{E}^n \setminus \{o\}$, such that its restriction to S^{n-1} is in $L^1(S^{n-1})$. A separate treatment of the case $p = -n + 1$ (see, for example, [454, Lemma 2.5] or [465, Lemma 3.7]) gives

$$\hat{f}(u) = \pi R f(u), \tag{C.25}$$

for $u \in S^{n-1}$, where f is an even Borel function, positively homogeneous of degree $-n + 1$ on $\mathbb{E}^n \setminus \{o\}$, such that its restriction to S^{n-1} is in $L^1(S^{n-1})$.

Open problem

Problem C.1. (See Theorem C.1.2.) Let f be a bounded measurable function on $\mathbb{E}^n$, vanishing outside a bounded set. Suppose that P is any infinite set in $\mathbb{E}^n$. If $D_p f = 0$ for all $p \in P$, is f zero λ_n-almost everywhere?

REFERENCES

[1] P. W. Aitchison, C. M. Petty, and C. A. Rogers. A convex body with a false centre is an ellipsoid. *Mathematika* **18** (1971), 50–9.

[2] A. D. Aleksandrov. On the theory of mixed volumes of convex bodies. I. Extensions of certain concepts in the theory of convex bodies [in Russian]. *Mat. Sb.* **2** (1937), 947–72.

[3] ———. On the theory of mixed volumes of convex bodies. II. New inequalities between mixed volumes and their applications [in Russian]. *Mat. Sb.* **2** (1937), 1205–38.

[4] ———. On the theory of mixed volumes of convex bodies. III. Extensions of two theorems of Minkowski on convex polyhedra to arbitrary convex bodies [in Russian]. *Mat. Sb.* **3** (1938), 27–46.

[5] ———. On the surface area function of a convex body [in Russian]. *Mat. Sb.* **6** (1939), 167–74.

[6] J. Alonso and P. Martin. Some characterizations of ellipsoids by sections. *Discrete Comput. Geom.* **31** (2004), 643–54.

[7] A. Alpers and P. Gritzmann. On stability, error correction, and noise compensation in discrete tomography. *SIAM J. Discrete Math.* **20** (2006), 227–39.

[8] A. Alpers, P. Gritzmann, and L. Thorens. Stability and instability in discrete tomography. In *Dagstuhl Seminar: Digital and Image Geometry 2000*, Lecture Notes in Computer Science 2243. Springer, Berlin, 2001, pp. 175–86.

[9] D. Amir. *Isometric Characterizations of Inner Product Spaces*. Birkhäuser, Basel, 1986.

[10] Yu. E. Anikonov. Uniqueness theorem for convex surfaces. *Math. Notes Acad. Sci. USSR* **6** (1969), 528–9.

[11] M. Antilla, K. Ball, and I. Perissinaki. The central limit problem for convex bodies. *Trans. Amer. Math. Soc.* **335** (2003), 4723–35.

[12] E. M. Arkin, M. Held, J. S. Mitchell, and S. S. Skiena. Recognizing polygonal parts from width measurements. *Comput. Geom.* **9** (1998), 237–46.

[13] J. L. Arocha, L. Montejano, and E. Morales. A quick proof of Höbinger-Burton-Mani's theorem. *Geom. Dedicata* **63** (1996), 331–5.

[14] R. B. Ash. *Measure, Integration and Functional Analysis*. Academic Press, New York, 1972.

[15] H. Auerbach. Sur un problème de M. Ulam concernant l'équilibre des corps flottants. *Studia Math.* **7** (1938), 121–42.

[16] G. Aumann. On a topological characterization of compact convex point sets. *Ann. of Math.* (2) **37** (1936), 443–7.

[17] G. Averkov and G. Bianchi. Retrieving convex bodies from restricted covariogram functions. Preprint.

[18]   M. Baake and C. Huck. Discrete tomography of cyclotomic model sets. Preprint.

[19]   K. M. Ball. Cube slicing in $\mathbf{R}^n$. *Proc. Amer. Math. Soc.* **97** (1986), 465–73.

[20]   _____ . Logarithmically concave functions and sections of convex sets in $\mathbf{R}^n$. *Studia Math.* **88** (1988), 69–84.

[21]   _____ . Some remarks on the geometry of convex sets. In *Geometric Aspects of Functional Analysis*, ed. by J. Lindenstrauss and V. D. Milman, Lecture Notes in Mathematics 1317. Springer, Heidelberg, 1988, pp. 224–31.

[22]   _____ . Volumes of sections of cubes and related matters. In *Geometric Aspects of Functional Analysis*, ed. by J. Lindenstrauss and V. D. Milman, Lecture Notes in Mathematics 1376. Springer, Heidelberg, 1989, pp. 251–60.

[23]   _____ . Normed spaces with a weak-Gordon–Lewis property. In *Functional Analysis*, Austin, Texas 1987/1989, Lecture Notes in Mathematics 1470. Springer, Heidelberg, 1991, pp. 36–47.

[24]   _____ . Shadows of convex bodies. *Trans. Amer. Math. Soc.* **327** (1991), 891–901.

[25]   _____ . Volume ratios and a reverse isoperimetric inequality. *J. London Math. Soc.* (2) **44** (1991), 351–9.

[26]   _____ . Mahler's conjecture and wavelets. *Discrete Comput. Geom.* **13** (1995), 271–7.

[27]   _____ . An elementary introduction to modern convex geometry. In *Flavors of Geometry*, ed. by S. Levy, MSRI Publications 31. Cambridge University Press, New York, 1997, pp. 1–58.

[28]   C. Bandle. *Isoperimetric Inequalities and Applications*. Pitman, London, 1980.

[29]   C. B. Barber, D. P. Dobkin, and H. Huhdanpaa. The quickhull algorithm for convex hulls. *ACM Trans. Math. Software* **22**(1996), 469–83.

[30]   E. Barcucci, A. Del Lungo, M. Nivat, and R. Pinzani. Reconstructing convex polyominoes from their horizontal and vertical projections. *Theoret. Comput. Sci.* **155** (1996), 321–47.

[31]   _____ . Reconstructing convex polyominoes from their horizontal and vertical projections. II. In *Discrete Geometry for Computer Imagery (Lyons, 1996)*, Lecture Notes in Computer Science 1176. Springer, Berlin, 1996, pp. 295–306.

[32]   _____ . X-rays characterizing some classes of digital pictures. *Linear Algebra Appl.* **339** (2001), 3–21.

[33]   E. Barcucci, S. Brunetti, A. Del Lungo, and M. Nivat. Reconstruction of lattice sets from their horizontal, vertical and diagonal X-rays. *Discrete Math.* **241** (2001), 65–78.

[34]   J. A. Barker and D. G. Larman. Determination of convex bodies by certain sets of sectional volumes. *Discrete Math.* **241** (2001), 79–96.

[35]   S. Barov, J. Cobb, and J. J. Dijkstra. On closed sets with convex projections. *J. London Math. Soc.* (2) **65** (2002), 155–66.

[36]   S. Barov and J. J. Dijkstra. More on compacta with convex projections. *Real Anal. Exchange* **26** (2000/01), 277–84.

[37]   F. Barthe. Mesures unimodales et sections des boules B_p^n. *C. R. Acad. Sci. Paris Sér. I Math.* **321** (1995), 865–8.

[38]   _____ . An extremal property of the mean width of the simplex. *Math. Ann.* **310** (1998), 685–93.

[39]   _____ . Autour de l'inégalité de Brunn–Minkowski. *Ann. Fac. Sci. Toulouse Math.* (6) **12** (2003), 127–78.

[40]   F. Barthe, M. Fradelizi, and B. Maurey. A short solution to the Busemann-Petty problem. *Positivity* **3** (1999), 95–100.

[41]   F. Barthe and A. Naor. Hyperplane projections of the unit ball of l_p^n. *Discrete Comput. Geom.* **27** (2002), 215–26.

[42]   W. Barthel. Zum Busemannschen und Brunn–Minkowskischen Satz. *Math. Z.* **70** (1959), 407–29.

[43] W. Barthel and G. Franz. Eine Verallgemeinerung des Busemannschen Satzes vom Brunn–Minkowskischen Typ. *Math. Ann.* **144** (1961), 183–98.

[44] S. Basu and Y. Bresler. Uniqueness of tomography with unknown view angles. *IEEE Trans. Image Process.* **9** (2000), 1094–106.

[45] ———. Feasibility of tomography with unknown view angles. *IEEE Trans. Image Process.* **9** (2000), 1107–22.

[46] K. J. Batenburg. A network flow algorithm for reconstructing binary images from discrete X-rays. Preprint.

[47] C. Bauer. Intermediate surface area measures and projection functions of convex bodies. *Arch. Math. (Basel)* **64** (1995), 69–74.

[48] ———. Determination of convex bodies by projection functions. *Geom. Dedicata* **72** (1998), 309–24.

[49] F. Beauvais. Reconstructing a set or measure with finite support from its images. PhD dissertation, Dept. of Mathematics, University of Rochester, Rochester, NY, 1987.

[50] L. Belcastro, W. C. Karl, and A. S. Willsky. Tomographic reconstruction of polygons from knot locations and chord length measurements. *Graph. Models Image Process.* **58** (1996), 233–45.

[51] C. Berg. Corps convexes et potentiels sphériques. *Mat.-Fys. Medd. Danske Vid. Selsk.* **37** (1969), 1–64.

[52] M. Berger. *Geometry I*. Springer, Berlin, 1987.

[53] ———. *Geometry II*. Springer, Berlin, 1987.

[54] ———. Convexity. *Amer. Math. Monthly* **97** (1990), 650–78.

[55] U. Betke and P. McMullen. Estimating the sizes of convex bodies from projections. *J. London Math. Soc.* (2) **27** (1983), 525–38.

[56] G. Bianchi. Determining convex polygons from their covariograms. *Adv. in Appl. Probab.* **34** (2002), 261–6.

[57] ———. Matheron's conjecture for the covariogram problem. *J. London Math. Soc.* (2) **71** (2005), 203–20.

[58] ———. Some open problems regarding the determination of a set from its covariogram. *Matematiche (Catania)* **60** (2005), 200–10.

[59] ———. The covariogram of three-dimensional convex polytopes. Preprint.

[60] G. Bianchi and P. M. Gruber. Characterizations of ellipsoids. *Arch. Math. (Basel)* **49** (1987), 344–50.

[61] G. Bianchi and M. Longinetti. Reconstructing plane sets from projections. *Discrete Comput. Geom.* **5** (1990), 223–42.

[62] G. Bianchi, F. Segala, and A. Volčič. The solution of the covariogram problem for plane C_+^2 bodies. *J. Differential Geom.* **60** (2002), 177–98.

[63] G. Bianchi and A. Volčič. Hammer's X-ray problem is well-posed. *Ann. Mat. Pura Appl.* (4) **155** (1989), 205–11.

[64] H. Bieri and W. Nef. A sweep-plane algorithm for computing the volume of polyhedra represented in Boolean form. *Linear Algebra Appl.* **52/53** (1983), 69–97.

[65] L. J. Billera and B. Sturmfels. Fiber polytopes. *Ann. of Math.* (2) **135** (1992), 527–49.

[66] R. H. Bing. Spheres in E^3. *Amer. Math. Monthly* **71** (1964), 353–64.

[67] T. Bisztriczky and K. Böröczky, Jr. About the centroid body and the ellipsoid of inertia. *Mathematika* **48** (2001), 1–13.

[68] W. Black, J. Kimble, D. Koop, and D. C. Solmon. Functions that are the directed X-ray of a planar convex body. *Rend. Istit. Mat. Univ. Trieste* **35** (2003), 81–115.

[69] W. Blaschke. Über affine Geometrie VII: Neue Extremeigenschaften von Ellipse und Ellipsoid. *Ber. Verh. Sächs. Akad. Wiss. Leipzig Math.-Phys. Kl.* **69** (1917), 306–18.

[70] ———. Über affine Geometrie IX: Verschiedene Bemerkungen und Aufgaben. *Ber. Verh. Sächs. Akad. Wiss. Leipzig Math.-Phys. Kl.* **69** (1917), 412–20.

[71] ———. *Kreis und Kugel*. Chelsea, New York, 1949.

[72] ———. Zur Affingeometrie der Eilinien und Eiflächen. *Math. Nachr.* **15** (1956), 258–64.

[73] W. Blaschke and G. Hessenberg. Lehrsätze über konvexe Körper. *Jber. Deutsche Math.-Verein.* **26** (1917), 215–20.

[74] W. Blaschke, H. Rothe, and W. Weitzenböck. Aufgabe 552. *Arch. Math. Phys.* **27** (1917), 82.

[75] S. G. Bobkov and A. Koldobsky. On the central limit property of convex bodies. In *Geometric Aspects of Functional Analysis*, ed. by V. D. Milman and G. Schechtman, Lecture Notes in Mathematics 1807. Springer, Berlin, 2003, pp. 44–52.

[76] A. Bocconi. Riconstruzione Tomografica di Anelli Convessi. Tesi de Laurea, Corso di Laurea in Matematica, University of Florence, Florence, Italy, 1994.

[77] J. Boissonnat. Shape reconstruction from planar cross-sections. *Comput. Vision Graphics Image Process.* **44** (1988), 1–29.

[78] G. Bol. Über Eikörper mit Vieleckschatten. *Math. Z.* **48** (1942), 227–46.

[79] E. D. Bolker. A class of convex bodies. *Trans. Amer. Math. Soc.* **145** (1969), 323–45.

[80] B. Bollobás and A. Thomason. Projections of bodies and hereditary properties of hypergraphs. *Bull. London Math. Soc.* **27** (1995), 417–24.

[81] T. Bonnesen. Om Minkowskis uligheder for konvekse legemer. *Mat. Tidsskr. B* (1926), 74–80.

[82] ———. *Les Problèmes des Isopérimètres et des Isépiphanes*. Gauthier-Villars, Paris, 1929.

[83] T. Bonnesen and W. Fenchel. *Theory of Convex Bodies*. BCS Associates, Moscow, ID, 1987. German original: Springer, Berlin, 1934.

[84] K. Böröczky, Jr. and M. Henk. Random projections of regular polytopes. *Arch. Math. (Basel)* **73** (1999), 465–73.

[85] K. Borsuk. *Multidimensional Analytic Geometry*. Polish Scientific Publishers, Warsaw, 1969.

[86] A. Bottino and A. Laurentini. Introducing a new problem: Shape-from-silhouette when the relative position of the viewpoints is unknown. *IEEE Trans. Pattern Anal. and Machine Intell.* **25** (2003), 1484–93.

[87] J. Bourgain. On the Busemann–Petty problem for perturbations of the ball. *Geom. Funct. Anal.* **1** (1991), 1–13.

[88] ———. On the distribution of polynomials on high dimensional convex sets. In *Geometric Aspects of Functional Analysis*, ed. by J. Lindenstrauss and V. D. Milman, Lecture Notes in Mathematics 1469. Springer, Heidelberg, 1991, pp. 127–37.

[89] J. Bourgain, B. Klartag, and V. Milman. A reduction of the slicing problem to finite volume ratio bodies. *C. R. Acad. Sci. Paris Sér. I Math.* **336** (2003), 331–4.

[90] ———. Symmetrization and isotropic constants of convex bodies. In *Geometric Aspects of Functional Analysis*, ed. by V. D. Milman and G. Schechtman, Lecture Notes in Mathematics 1850. Springer, Berlin, 2004, pp. 101–15.

[91] J. Bourgain and J. Lindenstrauss. Projection bodies. In *Geometric Aspects of Functional Analysis*, ed. by J. Lindenstrauss and V. D. Milman, Lecture Notes in Mathematics 1317. Springer, Heidelberg, 1988, pp. 250–70.

[92] J. Bourgain and V. D. Milman. New volume ratio properties for convex symmetric bodies in $\mathbb{R}^n$. *Invent. Math.* **88** (1987), 319–40.

[93] J. Bourgain and G. Zhang. On a generalization of the Busemann-Petty problem. In *Convex Geometric Analysis*, ed. by K. M. Ball and V. Milman. Cambridge University Press, New York, 1999, pp. 65–76.

[94] N. S. Brannen. Volumes of projection bodies. *Mathematika* **43** (1996), 255–64.

[95] ———. Three-dimensional projection bodies. *Adv. Geom.* **5** (2005), 1–13.

[96] U. Brehm. Nonuniqueness results for X-ray problems with point sources. *Discrete Comput. Geom.* **8** (1992), 153–69.

[97] ——— . Convex bodies with nonconvex cross-section bodies. *Mathematika* **46** (1999), 127–9.

[98] ——— . X-ray problems with point sources for grid polygons. Preprint.

[99] U. Brehm and J. Voigt. Asymptotics of cross sections for convex sets. *Beiträge Algebra Geom.* **41** (2000), 437–54.

[100] U. Brehm, H. Vogt, and J. Voigt. Permanence of moment estimates for p-products of convex bodies. *Studia Math.* **150** (2002), 243–60.

[101] U. Brehm, P. Hinow, H. Vogt, and J. Voigt. Moment inequalities and central limit properties of isotropic convex bodies. *Math. Z.* **240** (2002), 37–51.

[102] A. L. Brown. Best n-dimensional approximation to set functions. *Proc. London Math. Soc.* (3) **14** (1964), 577–94.

[103] S. Brunetti and A. Daurat. An algorithm reconstructing lattice convex sets. *Theoret. Comput. Sci.* **304** (2003), 35–57.

[104] ——— . Stability in discrete tomography: Some positive results. *Discrete Appl. Math* **147** (2005), 207–26.

[105] ——— . Determination of Q-convex bodies by X-rays. *Electronic Notes in Discrete Mathematics* **20** (2005), 67–81.

[106] S. Brunetti, A. Daurat, and A. Del Lungo. Approximate X-rays reconstruction of special lattice sets. *Pure Math. Appl.* **11** (2000), 409–25.

[107] S. Brunetti, A. Del Lungo, and Y. Gerard. On the computational complexity of reconstructing three-dimensional lattice sets from their two-dimensional X-rays. *Linear Algebra Appl.* **339** (2001), 59–73.

[108] S. Brunetti, A. Del Lungo, F. Del Ristoro, A. Kuba, and M. Nivat. Reconstruction of 4- and 8-connected convex discrete sets from row and column projections. *Linear Algebra Appl.* **339** (2001), 37–57.

[109] H. Brunn. Über Ovale und Eiflächen. Dissertation, University of Munich, Munich, 1887.

[110] ——— . Über Curven ohne Wendepunkte. Habilitationsschrift, University of Munich, Munich, 1889.

[111] B. Büeler, A. Enge, and K. Fukuda. Exact volume computation for polytopes: A practical study. In *Polytopes—Combinatorics and Computation (Oberwolfach, 1997)*. Birkhäuser, Basel, 2000, pp. 131–54.

[112] Yu. D. Burago and V. A. Zalgaller. *Geometric Inequalities*. Springer, New York, 1988. Russian original: 1980.

[113] J. J. Burckhardt. Steiner, Jakob. In *Biographical Dictionary of Mathematicians*. Scribner, New York, 1991, pp. 2318–27.

[114] T. Burger and P. Gritzmann. Finding optimal shadows of polytopes. *Discrete Comput. Geom.* **24** (2000), 219–39.

[115] T. Burger, P. Gritzmann, and V. Klee. Polytope projection and projection polytopes. *Amer. Math. Monthly* **103** (1996), 742–55.

[116] G. R. Burton. On the sum of a zonotope and an ellipsoid. *Comment. Math. Helv.* **51** (1976), 369–87.

[117] ——— . Sections of convex bodies. *J. London Math. Soc.* (2) **12** (1976), 331–6.

[118] ——— . Some characterisations of the ellipsoid. *Israel J. Math.* **28** (1977), 339–49.

[119] ——— . Congruent sections of a convex body. *Pacific J. Math.* **81** (1979), 303–16.

[120] ——— . Skeleta and sections of convex bodies. *Mathematika* **27** (1980), 97–103.

[121] G. R. Burton and D. G. Larman. On a problem of J. Höbinger. *Geom. Dedicata* **5** (1976), 31–42.

[122] G. R. Burton and P. Mani. A characterisation of the ellipsoid in terms of concurrent sections. *Comment. Math. Helv.* **53** (1978), 485–507.

[123] H. Busemann. The isoperimetric problem for Minkowski area. *Amer. J. Math.* **71** (1949), 743–62.

442   References

[124] _____ . A theorem on convex bodies of the Brunn–Minkowski type. *Proc. Natl. Acad. Sci.*
       USA **35** (1949), 27–31.

[125] _____ . The foundations of Minkowskian geometry. *Comment. Math. Helv.* **24** (1950), 156–
       86.

[126] _____ . Volume in terms of concurrent cross-sections. *Pacific. J. Math.* **3** (1953), 1–12.

[127] _____ . *The Geometry of Geodesics.* Academic Press, New York, 1955.

[128] _____ . *Convex Surfaces.* Interscience, New York, 1958.

[129] _____ . Volumes and areas of cross-sections. *Amer. Math. Monthly* **67** (1960), 248–50, 671.

[130] H. Busemann and C. M. Petty. Problems on convex bodies. *Math. Scand.* **4** (1956), 88–94.

[131] H. Busemann and G. C. Shephard. Convexity on nonconvex sets. In *Proc. Colloquium Convex-
       ity, Copenhagen, 1965,* ed. by W. Fenchel. Københavns Univ. Mat. Inst., Copenhagen, 1967,
       pp. 20–33.

[132] H. Busemann and E. G. Straus. Area and normality. *Pacific J. Math.* **10** (1960), 35–72.

[133] D. Butcher, A. Medin, and D. C. Solmon. Planar convex bodies with a common directed X-ray.
       Rend. Istit. Univ. Trieste, in press.

[134] A. J. Cabo. Set Functionals in Stochastic Geometry. PhD dissertation, Technische Universiteit
       Delft, Delft, Netherlands, 1994.

[135] A. Cabo and A. Baddeley. Estimation of mean particle volume using the set covariance func-
       tion. *Adv. in Appl. Probab.* **35** (2003), 27–46.

[136] S. Campi. On the reconstruction of a function on a sphere by its integrals over great circles.
       Boll. Un. Mat. Ital. C(5) **18** (1981), 195–215.

[137] _____ . On the reconstruction of a star-shaped body from its "half-volumes." *J. Austral. Math.*
       Soc. Ser. A **37** (1984), 243–57.

[138] _____ . Reconstructing a convex surface from certain measurements of its projections. *Boll.*
       Un. Mat. Ital. B (6) **5** (1986), 945–59.

[139] _____ . Recovering a centred convex body from the areas of its shadows: a stability estimate.
       Ann. Mat. Pura Appl. (4) **151** (1988), 289–302.

[140] _____ . Stability estimates for star bodies in terms of their intersection bodies. *Mathematika*
       45 (1998), 287–303.

[141] _____ . Convex intersection bodies in three and four dimensions. *Mathematika* **46** (1999),
       15–27.

[142] S. Campi. A. Colesanti, and P. Gronchi. Convex bodies with extremal volumes having pre-
       scribed brightness in finitely many directions. *Geom. Dedicata* **57** (1995), 121–33.

[143] _____ . Minimum problems for volumes of convex bodies. In *Partial Differential Equations
       and Applications,* ed. by P. Marcellini, G. T. Talenti, and E. Vesentini. Marcel Dekker, New
       York, 1996, pp. 43–55.

[144] _____ . Blaschke-decomposable convex bodies. *Israel J. Math* **105** (1998), 185–95.

[145] _____ . Shaking compact sets. *Beiträge Algebra Geom.* **42** (2001), 123–36.

[146] S. Campi and P. Gronchi. The L^p-Busemann–Petty centroid inequality. *Adv. Math.* **167** (2002),
       128–41.

[147] _____ . On the reverse L^p-Busemann–Petty centroid inequality. *Mathematika* **49** (2002), 1–
       11.

[148] _____ . Extremal convex sets for Sylvester–Busemann type functionals. *Appl. Anal.* **85**
       (2006), 129–41.

[149] _____ . On volume product inequalities for convex sets. *Proc. Amer. Math. Soc.,* in press.

[150] S. Campi, D. Haas, and W. Weil. Approximation of zonoids by zonotopes in fixed directions.
       Discrete Comput. Geom. **11** (1994), 419–31.

[151] G. D. Chakerian. The affine image of a convex body of constant breadth. *Israel J. Math.* **3**
       (1965), 19–22.

[152] _____. Sets of constant relative width and constant relative brightness. *Trans. Amer. Math. Soc.* **129** (1967), 26–37.

[153] _____. Is a body spherical if all its projections have the same I.Q.? *Amer. Math. Monthly* **77** (1970), 989–92.

[154] G. D. Chakerian and H. Groemer. Convex bodies of constant width. In *Convexity and Its Applications*, ed. by P. M. Gruber and J. M. Wills. Birkhäuser, Basel, 1983, pp. 49–96.

[155] G. D. Chakerian and D. Logothetti. Cube slices, pictorial triangles, and probability. *Math. Mag.* **64** (1991), 219–41.

[156] G. D. Chakerian and E. Lutwak. On the Petty–Schneider theorem. *Contemp. Math.* **140** (1992), 31–7.

[157] _____. Bodies with similar projections. *Trans. Amer. Math. Soc.* **349** (1997), 1811–20.

[158] S.-K. Chang. The reconstruction of binary patterns from their projections. *Commun. ACM* **14** (1971), 21–4.

[159] S.-K. Chang and C. K. Chow. The reconstruction of three-dimensional objects from two orthogonal projections and its application to cardiac cineangiography. *IEEE Trans. Comput.* **22** (1973), 18–28.

[160] S.-K. Chang and Y. R. Wang. Three dimensional object reconstruction from orthogonal projections. In *Pattern Recognition*. Pergamon Press, Oxford, 1975, pp. 167–76.

[161] B. Chen. A simplified elementary proof of Hadwiger's volume theorem. *Geom. Dedicata* **105** (2004), 107–20.

[162] B. L. Chilton and H. S. M. Coxeter. Polar zonohedra. *Amer. Math. Monthly* **70** (1963), 946–51.

[163] W. G. Chinn and N. E. Steenrod. *First Concepts of Topology*. Random House, New York, 1966.

[164] G. Choquet. *Lectures on Analysis*. Vol. 3. Benjamin, Reading, MA, 1969.

[165] _____. Mesures coniques, affines et cylindriques. In *Proc. Symposia Math., INDAM*, vol. 2. Academic Press, London, 1969, pp. 145–82.

[166] M. Chrobak and C. Dürr. Reconstructing polyatomic structures from discrete X-rays: NP-completeness proof for three atoms. *Inform. Process. Lett.* **69** (1999), 283–9.

[167] R. V. Churchill. *Complex Variables and Applications*. 2d ed. McGraw-Hill, New York, 1960.

[168] D. L. Cohn. *Measure Theory*. Birkhäuser, Basel, 1980.

[169] R. Cole and C. K. Yap. Shape from probing. *J. Algorithms* **8** (1987), 19–38.

[170] M. C. Costa, D. de Werra, C. Piouleau, and D. Schindl. A solvable case of image reconstruction in discrete tomography. *Discrete Appl. Math.* **148** (2005), 240–5.

[171] H. S. M. Coxeter. *Regular Polytopes*. MacMillan, New York, 1963.

[172] H. T. Croft. Two problems on convex bodies. *Proc. Cambridge Philos. Soc.* **58** (1962), 1–7.

[173] H. T. Croft, K. J. Falconer, and R. K. Guy. *Unsolved Problems in Geometry*. Springer, New York, 1991.

[174] L. Dalla and D. G. Larman. Convex bodies with almost all k-dimensional sections polytopes. *Math. Proc. Cambridge Philos. Soc.* **88** (1980), 395–401.

[175] L. Danzer, D. Laugwitz, and H. Lenz. Über das Löwnersche Ellipsoid und sein Analogon unter den einem Eikörper einbeschriebenen Ellipsoiden. *Arch. Math. (Basel)* **8** (1957), 214–19.

[176] S. Dar. Remarks on Bourgain's problem on slicing of convex bodies. In *Geometric Aspects of Functional Analysis*, ed. by J. Lindenstrauss and V. D. Milman, *Operator Theory: Adv. and Appl.* **77** (1995), 61–6.

[177] M. G. Darboux. Sur un problème de géométrie élémentaire. *Bull. Sci. Math.* **2** (1878), 298–304.

[178] M. Dartmann. Rekonstruktion konvexer, homogener Gebiete aus wenigen Punktquellen. Diplomarbeit, University of Münster, Münster, 1991.

[179] Y. Das and W. M. Boerner. On radar target shape estimation using algorithms for reconstruction from projections. *IEEE Trans. Antennas and Propagation* **26** (1978), 274–9.

[180] A. Daurat. Determination of Q-convex sets by X-rays. *Theoret. Comput. Sci.* **332** (2005), 19–45.

[181] R. J. Daverman. Slicing theorems for n-spheres in Euclidean $(n+1)$-space. *Trans. Amer. Math. Soc.* **166** (1972), 479–89.

[182] C. de Boor, K. Höllig, and S. Riemenschneider. *Box Splines*. Springer, New York, 1993.

[183] H. E. Debrunner, P. Mani, C. Meier, J. Rätz, and F. Streit. Professor Dr. Hugo Hadwiger, 1908–1981. *Elem. Math.* **37** (1982), 65–78.

[184] A. Del Lungo, M. Nivat, and R. Pinzani. The number of convex polyominoes reconstructible from their orthogonal projections. *Discrete Math.* **157** (1996), 65–78.

[185] J. Dieudonné. Minkowski, Hermann. In *Biographical Dictionary of Mathematicians*. Scribner, New York, 1991, pp. 1720–23.

[186] J. J. Dijkstra, T. L. Goodsell, and D. W. Wright. On compacta with convex projections. *Topology Appl.* **94** (1999), 67–74.

[187] A. Dinghas. Über das Verhalten der Entfernung zweier Punktmengen bei gleichzeitiger Symmetrisierung derselben. *Arch. Math. (Basel)* **8** (1957), 46–51.

[188] G. A. Dirac. Ovals with equichordal points. *J. London Math. Soc.* **27** (1952), 429–37.

[189] D. P. Dobkin, H. Edelsbrunner, and C. K. Yap. Probing convex polytopes. In *Proc. 18th ACM Symp. Theory of Computing*. Association for Computing Machinery, New York, 1986, pp. 424–32.

[190] ——. Probing convex polytopes. In *Autonomous Robot Vehicles*, ed. by I. J. Cox and G. T. Wilfong. Springer, New York, 1990, pp. 328–41.

[191] J. Dugundji. *Topology*. Allyn and Bacon, Boston, 1966.

[192] P. Dulio. Geometric tomography in a graph. *Rend. Circ. Mat. Palermo* (2) Suppl. No. 77, in press.

[193] P. Dulio, R. J. Gardner, and C. Peri. Discrete point X-rays. *SIAM J. Discrete Math.* **20** (2006), 171–88.

[194] P. Dulio and C. Peri. Invariant valuations on spherical star sets. *Rend. Circ. Mat. Palermo* (2) Suppl. No. 65, part II (2000), 81–92.

[195] ——. On Hammer's X-ray problem in spaces of constant curvature. *Rend. Circ. Mat. Palermo* (2) Suppl. No. 70, part I (2002), 229–36.

[196] ——. Uniqueness theorems for convex bodies in non-Euclidean spaces. *Mathematika* **49** (2002), 13–31.

[197] ——. Point X-rays of convex bodies in planes of constant curvature. *Rocky Mountain J. Math.*, in press.

[198] ——. On the geometric structure of lattice U-polygons. Preprint.

[199] N. Dunford and J. Schwartz. *Linear Operators*. Vol. 1. Interscience, New York, 1958.

[200] C. Dupin. *Application de Géometrie et de Méchanique à la Marine, aux Ponts et Chaussées*. Bachelier, Paris, 1822.

[201] A. Dvoretzky. Some results on convex bodies and Banach spaces. In *Proc. Internat. Symp. on Linear Spaces (Jerusalem)*. 1961, pp. 123–60.

[202] ——. Some near-sphericity results. In *Proc. Symp. Pure Math., Vol. VII: Convexity (Providence, RI)*. Amer. Math. Soc., Providence, RI, 1963, pp. 203–10.

[203] H. Edelsbrunner and S. S. Skiena. Probing convex polygons with X-rays. *SIAM. J. Comput.* **17** (1988), 870–82.

[204] N. V. Efrimov, V. A. Zalgaller, and A. V. Pogorelov. Aleksander Danilovich Aleksandrov (on his fiftieth birthday). *Russian Math. Surveys* **17** (1962), 127–41.

[205] H. G. Eggleston. *Convexity*. Cambridge University Press, Cambridge, 1958.

[206] ——. Note on a conjecture of L. A. Santalo. *Mathematika* **8** (1961), 63–5.

[207] H. G. Eggleston, B. Grünbaum, and V. Klee. Some semicontinuity theorems for convex polytopes and cell-complexes. *Comment. Math. Helv.* **39** (1964), 165–88.

[208] G. P. Egorychev. The solution of van der Waerden's problem for permanents. *Adv. Math.* **42** (1981), 299–305.

[209] E. Ehrhart. Un ovale a deux point isocordes? *Enseign. Math.* (2) **13** (1967), 119–24.

[210] D. Eppstein. Geometry Junkyard website at http://www.ics.uci.edu/~eppstein.

[211] C. L. Epstein. *Mathematics of Medical Imaging.* Prentice Hall, Upper Saddle River, NJ, 2003.

[212] L. C. Evans and R. F. Gariepy. *Measure Theory and Fine Properties of Functions.* CRC Press, Boca Raton, FL, 1992.

[213] K. J. Falconer. A result on the Steiner symmetral of a compact set. *J. London Math. Soc.* (2) **14** (1976), 385–6.

[214] ———. Consistency conditions for a finite set of projections of a function. *Math. Proc. Cambridge Philos. Soc.* **85** (1979), 61–8.

[215] ———. Continuity properties of *k*-plane integrals and Besicovitch sets. *Math. Proc. Cambridge Philos. Soc.* **87** (1980), 221–6.

[216] ———. Function space topologies defined by sectional integrals and applications to an extremal problem. *Math. Proc. Cambridge Philos. Soc.* **87** (1980), 81–96.

[217] ———. Applications of a result on spherical integration to the theory of convex sets. *Amer. Math. Monthly* **90** (1983), 690–3.

[218] ———. Hammer's X-ray problem and the stable manifold theorem. *J. London Math. Soc.* (2) **28** (1983), 149–60.

[219] ———. On the equireciprocal point problem. *Geom. Dedicata* **14** (1983), 113–26.

[220] ———. X-ray problems for point sources. *Proc. London Math. Soc.* (3) **46** (1983), 241–62.

[221] H. Fallert, P. R. Goodey, and W. Weil. Spherical projections and centrally symmetric sets. *Adv. Math.* **129** (1997), 301–22.

[222] I. Fáry. A characterization of convex bodies. *Amer. Math. Monthly* **69** (1962), 25–31.

[223] H. Fast. Inversion of the Crofton transform for sets in the plane. *Real Anal. Exchange* **19** (1993/4), 1–22.

[224] E. S. Fedorov. *Elements in the Theory of Figures* [in Russian]. Imp. Acad. Sci., St. Petersburg, 1885. Annotated edition: Akad. Nauk SSSR, 1953.

[225] V. P. Fedotov. The sum of *p*th surface functions [in Russian]. *Ukrain. Geom. Sb.* **21** (1978), 125–31.

[226] ———. A counterexample to Firey's hypothesis. *Math. Notes Acad. Sci. USSR* **26** (1979), 626 –9(1980).

[227] ———. Polar representation of a convex compactum [in Russian]. *Ukrain. Geom. Sb.* **25** (1982), 137–8.

[228] G. Fejes Tóth and A. Kemnitz. Characterization of centrally symmetric convex domains in planes of constant curvature. In *Colloquia Mathematica Soc. János Bolyai, vol. 48. Intuitive Geometry (Siófok 1985)*, ed. by K. Böröczky and G. Fejes Tóth. North-Holland, Amsterdam, 1985, pp. 179–89.

[229] L. Fejes Tóth. *Regular Figures.* Macmillan, New York, 1964.

[230] W. Fenchel and B. Jessen. Mengenfunktionen und konvexe Körper. *Mat.-Fys. Medd. Danske Vid. Selsk.* **16** (1938), 1–31.

[231] P. Filliman. Exterior algebra and projections of polytopes. *Discrete Comput. Geom.* **5** (1990), 305–22.

[232] ———. The volume of duals and sections of polytopes. *Mathematika* **39** (1992), 67–80.

[233] D. V. Finch. Cone beam reconstruction with sources on a curve. *SIAM J. Appl. Math.* **45** (1985), 665–73.

[234] W. J. Firey. Pythagorean inequalities for convex bodies. *Math. Scand.* **8** (1960), 168–70.

[235] ———. Polar means of convex bodies and a dual to the Brunn–Minkowski theorem. *Canad. J. Math.* **13** (1961), 444–53.

[236] ———. *p*-means of convex bodies. *Math. Scand.* **10** (1962), 17–24.

[237] ———. The brightness of convex bodies. Technical Report 19, Oregon State University, Corvallis, OR, 1965.

[238] ———. Lower bounds for volumes of convex bodies. *Arch. Math. (Basel)* **16** (1965), 69–74.

[239] ———. Blaschke sums of convex bodies and mixed bodies. In *Proc. Colloquium Convexity, Copenhagen, 1965*, ed. by W. Fenchel. Københavns Univ. Mat. Inst., Copenhagen, 1967, pp. 94–101.

[240] ———. Convex bodies of constant outer *p*-measure. *Mathematika* **17** (1970), 21–7.

[241] ———. Intermediate Christoffel–Minkowski problems for figures of revolution. *Israel J. Math.* **8** (1970), 384–90.

[242] P. C. Fishburn, J. C. Lagarias, J. A. Reeds, and L. A. Shepp. Sets uniquely determined by projections on axes. I. Continuous case. *SIAM J. Appl. Math.* **50** (1990), 288–306.

[243] ———. Sets uniquely determined by projections on axes. II. Discrete case. *Discrete Math.* **91** (1991), 149–59.

[244] P. C. Fishburn, P. Schwander, L. A. Shepp, and J. Vanderbei, The discrete Radon transform and its approximate inversion via linear programming. *Discrete Appl. Math.* **75** (1997), 39–62.

[245] N. I. Fisher, P. Hall, B. Turlach, and G. S. Watson. On the estimation of a convex set from noisy data on its support function. *J. Amer. Statist. Assoc.* **92** (1997), 84–91.

[246] M. Fradelizi. Sections of convex bodies through their centroid. *Arch. Math. (Basel)* **69** (1997), 515–22.

[247] ———. Hyperplane sections of convex bodies in isotropic position. *Beiträge Algebra Geom.* **40** (1999), 163–83.

[248] ———. Sectional bodies associated with a convex body. *Proc. Amer. Math. Soc.* **128** (2000), 2735–44.

[249] M. Frantz. On Sierpiński's nonmeasurable set. *Fund. Math.* **139** (1991), 17–22.

[250] M. Fujiwara. Unendlichviele Systeme der linearen Gleichungen mit unendlichvielen Variabeln und eine Eigenschaft der Kugel. *Sci. Rep. Tohoku Univ.* **3** (1914), 199–216.

[251] ———. Über die Mittelkurve zweier geschlossenen konvexen Kurven in Bezug auf einen Punkt. *Tohoku Math. J.* **10** (1916), 99–103.

[252] P. Funk. Über Flächen mit lauter geschlossenen geodätischen Linien. *Math. Ann.* **74** (1913), 278–300.

[253] ———. Über eine geometrische Anwendung der Abelschen Integralgleichung. *Math. Ann.* **77** (1916), 129–35.

[254] ———. Nachruf auf Prof. Johann Radon. *Monatsh. Math.* **62** (1958), 189–99.

[255] H. Furstenberg and I. Tzkoni. Spherical harmonics and integral geometry. *Israel J. Math.* **10** (1971), 327–38.

[256] R. J. Gardner. Symmetrals and X-rays of planar convex bodies. *Arch. Math. (Basel)* **41** (1983), 183–9.

[257] ———. Chord functions of convex bodies. *J. London Math. Soc.* (2) **36** (1987), 314–26.

[258] ———. Measure theory and some problems in geometry. *Atti. Sem. Mat. Fis. Univ. Modena* **39** (1991), 51–72.

[259] ———. Sets determined by finitely many X-rays. *Geom. Dedicata* **43** (1992), 1–16.

[260] ———. X-rays of polygons. *Discrete Comput. Geom.* **7** (1992), 281–93.

[261] ———. Intersection bodies and the Busemann–Petty problem. *Trans. Amer. Math. Soc.* **342** (1994), 435–45.

[262] ———. On the Busemann–Petty problem concerning central sections of centrally symmetric convex bodies. *Bull. Amer. Math. Soc.* **30** (1994), 222–6.

[263] ———. A positive answer to the Busemann–Petty problem in three dimensions. *Ann. of Math.* (2) **140** (1994), 435–47.

[264] _____. The Brunn–Minkowski inequality. *Bull. Amer. Math. Soc.* **39** (2002), 355–405.

[265] R. J. Gardner and A. A. Giannopoulos. *p*-Cross-section bodies. *Indiana Univ. Math. J.* **48** (1999), 593–613.

[266] R. J. Gardner and P. Gritzmann. Successive determination and verification of polytopes by their X-rays. *J. London Math. Soc.* (2) **50** (1994), 375–91.

[267] _____. Discrete tomography: Determination of finite sets by X-rays. *Trans. Amer. Math. Soc.* **349** (1997), 2271–95.

[268] R. J. Gardner, P. Gritzmann and P. Prangenberg. On the reconstruction of binary images from their discrete Radon transforms. In *Vision Geometry V*, ed. by R. A. Melter, A. Y. Wu, and L. Latecki. Society of Photo-Optical Instrumentation Engineers Proceedings 2826, 1996, pp. 121–32.

[269] _____. On the computational complexity of reconstructing lattice sets from their X-rays. *Discrete Math.* **202** (1999), 45–71.

[270] _____. On the complexity of determining polyatomic structures by X-rays. *Theoret. Comput. Sci.* **233** (2000), 91–106.

[271] R. J. Gardner and P. Gronchi. A Brunn–Minkowski inequality for the integer lattice. *Trans. Amer. Math. Soc.* **353** (2001), 3995–4024.

[272] R. J. Gardner, P. Gronchi, and C. Zong. Sums, projections, and sections of lattice sets, and the discrete covariogram. *Discrete Comput. Geom.* **34** (2005), 391–409.

[273] R. J. Gardner and M. Kiderlen. An algorithm for reconstructing convex bodies from four X-rays. Preprint.

[274] R. J. Gardner, M. Kiderlen, and P. Milanfar. Convergence of algorithms for reconstructing convex bodies and directional measures. *Ann. Statist.*, in press.

[275] R. J. Gardner, A. Koldobsky, and T. Schlumprecht. An analytic solution to the Busemann-Petty problem. *C. R. Acad. Sci. Paris Sér. I Math.* **328** (1999), 29–34.

[276] _____. An analytical solution to the Busemann-Petty problem on sections of convex bodies. *Ann. of Math.* (2) **149** (1999), 691–703.

[277] R. J. Gardner and P. McMullen. On Hammer's X-ray problem. *J. London Math. Soc.* (2) **21** (1980), 171–5.

[278] R. J. Gardner and P. Milanfar. Reconstruction of convex bodies from brightness functions. *Discrete Comput. Geom.* **29** (2003), 279–303.

[279] R. J. Gardner, A. Soranzo, and A. Volčič. On the determination of star and convex bodies by section functions. *Discrete Comput. Geom.* **21** (1999), 69–85.

[280] R. J. Gardner, E. B. Vedel Jensen, and A. Volčič. Geometric tomography and local stereology. *Adv. in Appl. Math.* **30** (2003), 397–423.

[281] R. J. Gardner and A. Volčič. Determination of convex bodies by their brightness functions. *Mathematika* **40** (1993), 161–8.

[282] _____. Convex bodies with similar projections. *Proc. Amer. Math. Soc.* **121** (1994), 563–8.

[283] _____. Tomography of convex and star bodies. *Adv. Math.* **108** (1994), 367–99.

[284] R. J. Gardner and G. Zhang. Affine inequalities and radial mean bodies. *Amer. J. Math.* **120** (1998), 493–504.

[285] H. Gericke. Wilhelm Süss, der Gründer des Mathematischen Forschungsinstitutes Oberwolfach. *Jber. Deutsche Math.-Verein.* **69** (1968), 161–83.

[286] A. A. Giannopoulos. A note on a problem of H. Busemann and C. M. Petty concerning sections of symmetric convex bodies. *Mathematika* **37** (1990), 239–44.

[287] A. A. Giannopoulos and V. D. Milman. Asymptotic convex geometry: A short overview. In *Different Faces of Geometry, Int. Math. Ser. (N. Y.)*. Kluwer/Plenum, New York, 2004, pp. 87–162.

[288] O. Giering. Bestimmung von Eibereichen und Eikörpern durch Steiner-Symmetrisierungen. *Bayer. Akad. Wiss. Math.-Natur. Kl.S.-B.* (1962), 225–53.

[289] _____. Drei- und Viereckspaare, für deren Drei- und Vierecke jeweils zwei Steiner-Symmetrisierungen übereinstimmen. *Elem. Math.* **40** (1985), 1–10.

[290] M. Giertz. A note on a problem of Busemann. *Math. Scand.* **25** (1969), 145–8.

[291] E. N. Gilbert. How things float. *Amer. Math. Monthly* **98** (1991), 201–16.

[292] S. G. Gindikin. Some notes on the Radon transform and integral geometry. *Monatsh. Math.* **113** (1992), 23–32.

[293] E. Gluskin. On the multivariable version of Ball's slicing cube theorem. In *Geometric Aspects of Functional Analysis*, ed. by V. D. Milman and G. Schechtman, Lecture Notes in Mathematics 1850. Springer, Berlin, 2004, pp. 117–21.

[294] V. P. Golubyatnikov. On reconstructing the shape of a body from its projections. *Soviet Math. Dokl.* **25** (1982), 62–3.

[295] _____. On the tomography of polyhedra [in Russian]. In *Questions of Well-posedness in Inverse Problems in Mathematical Physics*. Computer Center Novosibirsk, Novosibirsk, 1982, pp. 75–6.

[296] _____. Unique determination of visible bodies from their projections. *Siberian Math. J.* **29** (1988), 761–4.

[297] _____. On unique recoverability of convex and visible compacta from their projections. *Math. USSR-Sb.* **73** (1991), 1–10.

[298] _____. Questions of stability in recovering certain compact sets from their projections. *Soviet Math. Dokl.* **45** (1992), 12–14.

[299] _____. Stability problems in certain inverse problems of reconstruction of convex compacta from their projections. *Siberian Math. J.* (1992), 409–14.

[300] _____. On unique reconstructibility of convex and visible compact sets from their projections. II. *Siberian Math. J.* **36** (1995), 265–9.

[301] _____. On the unique determination of compact convex sets from their projections. The complex case. *Siberian Math. J.* **40** (1999), 678–81.

[302] V. P. Golubyatnikov, I. Karaca, E. Ozyilmaz, and B. Tantay. On determinating the shapes of hypersurfaces from the shapes of their apparent contours and symplectic geometry measurements. *Siberian Adv. Math.* **10** (2000), 9–15.

[303] A. B. Goncharov. Methods of integral geometry and recovering a function with compact support from its projections in unknown directions. *Acta Appl. Math.* **11** (1988), 213–22.

[304] P. R. Goodey. Centrally symmetric convex sets and mixed volumes. *Mathematika* **24** (1977), 193–8.

[305] _____. Instability of projection bodies. *Geom. Dedicata* **20** (1986), 295–305.

[306] _____. Applications of representation theory to convex bodies. *Rend. Circ. Mat. Palermo* (2) Suppl. No. 50 (1997), 179–87.

[307] _____. Minkowski sums of projections of convex bodies. *Mathematika* **45** (1998), 253–68.

[308] _____. Radon transforms of projection functions. *Math. Proc. Camb. Phil. Soc.* **123** (1998), 161–70.

[309] P. R. Goodey and H. Groemer. Stability results for first order projection bodies. *Proc. Amer. Math. Soc.* **109** (1990), 1103–14.

[310] P. R. Goodey and W. Jiang. Minkowski sums of three dimensional projections of convex bodies. *Rend. Circ. Mat. Palermo* Suppl. No. 65 (2000), 105–119.

[311] P. R. Goodey, M. Kiderlen, and W. Weil. Section and projection means of convex bodies. *Monatsh. Math.* **126** (1998), 37–54.

[312] P. R. Goodey, E. Lutwak, and W. Weil. Functional analytic characterization of classes of convex bodies. *Math. Z.* **222** (1996), 363–81.

[313] P. R. Goodey and R. Schneider. On the intermediate area functions of convex bodies. *Math. Z.* **173** (1980), 185–94.

[314] P. R. Goodey, R. Schneider, and W. Weil. Projection functions on higher rank Grassmannians. In *Geometric Aspects of Functional Analysis*, ed. by J. Lindenstrauss and V. D. Milman, *Oper. Theory Adv. Appl.* **77** (1995), 75–90.

[315] ———. On the determination of convex bodies by projection functions. *Bull. London Math. Soc.* **29** (1997), 82–8.

[316] ———. Projection functions of convex bodies. In *Intuitive Geometry (Budapest 1995)*. Bolyai Mathematical Studies 6, 1997, pp. 23–53.

[317] P. R. Goodey and W. Weil. Centrally symmetric convex bodies and Radon transforms on higher order Grassmanians. *Mathematika* **38** (1991), 117–33.

[318] ———. Centrally symmetric convex bodies and the spherical Radon transform. *J. Differential Geom.* **35** (1992), 675–88.

[319] ———. The determination of convex bodies from the mean of random sections. *Math. Proc. Cambridge Philos. Soc.* **112** (1992), 419–30.

[320] ———. Zonoids and generalizations. In *Handbook of Convex Geometry*, ed. by P. M. Gruber and J. M. Wills. North-Holland, Amsterdam, 1993, pp. 1297–326.

[321] ———. Intersection bodies and ellipsoids. *Mathematika* **42** (1995), 295–304.

[322] ———. Translative and kinematic integral formulae for support functions II. *Geom. Dedicata* **99** (2003), 103–25.

[323] ———. Average section functions for star-shaped sets. *Adv. in Appl. Math.* **36** (2006), 70–84.

[324] ———. Directed projection functions of convex bodies. *Monatsh. Math.*, in press.

[325] ———. Determination of convex bodies by directed projection functions. Preprint.

[326] P. R. Goodey and G. Zhang. Characterizations and inequalities for zonoids. *J. London Math. Soc.* (2) **53** (1996), 184–96.

[327] ———. Inequalities between projection functions of convex bodies. *Amer. J. Math.* **120** (1998), 345–67.

[328] Y. Gordon, M. Meyer, and S. Reisner. Zonoids with minimal volume-product – a new proof. *Proc. Amer. Math. Soc.* **104** (1988), 273–6.

[329] R. Gorenflo and S. Vessella. *Abel Integral Equations*. Lecture Notes in Mathematics 1461. Springer, Berlin, 1991.

[330] A. Gray. *Modern Differential Geometry of Curves and Surfaces*. CRC Press, Boca Raton, FL, 1993.

[331] J. W. Green. Sets subtending a constant angle on a circle. *Duke Math. J.* **17** (1950), 263–7.

[332] J. Gregor and F. R. Rannou. Three-dimensional support function estimation and application for projection magnetic resonance imaging. *Internat. J. Imaging Syst. Tech.* **12** (2002), 43–50.

[333] J. P. Greschak. Reconstructing convex sets. PhD dissertation, Department of Electrical Engineering and Computer Science, Massachusetts Institute of Technology, Cambridge, MA, 1985.

[334] E. L. Grinberg. Isoperimetric inequalities and identities for k-dimensional cross-sections of convex bodies. *Math. Ann.* **291** (1991), 75–86.

[335] E. L. Grinberg and E. T. Quinto. Analytic continuation of convex bodies and Funk's characterization of the sphere. *Pacific J. Math.* **201** (2001), 309–22.

[336] E. L. Grinberg and I. Rivin. Infinitesimal aspects of the Busemann–Petty problem. *Bull. London Math. Soc.* **22** (1990), 478–84.

[337] E. L. Grinberg and G. Zhang. Convolutions, transforms, and convex bodies. *Proc. London Math. Soc.* (3) **78** (1999), 77–115.

[338] P. Gritzmann and S. De Vries. On the algorithmic inversion of the discrete Radon Transform. *Theoret. Comput. Sci.* **281** (2002), 455–69.

[339] P. Gritzmann, S. De Vries, and M. Wiegelmann. Approximating binary images from discrete X-rays. *SIAM J. Optim.* **11** (2000), 522–46.

[340] P. Gritzmann and A. Hufnagel. On the algorithmic complexity of Minkowski's reconstruction problem. *J. London Math. Soc.* (2) **59** (1999), 1081–1100.

[341] P. Gritzmann and V. Klee. On the complexity of some basic problems in computational convexity: II. Volume and mixed volumes. In *Polytopes: Abstract, Convex, and Computational*, ed. by T. Bistriczky, P. McMullen, R. Schneider, and A. Ivic Weiss. Kluwer, Boston, 1994, pp. 373–466.

[342] P. Gritzmann, V. Klee, and J. Westwater. Polytope containment and determination by linear probes. *Proc. London Math. Soc.* (3) **70** (1995), 691–720.

[343] P. Gritzmann, D. Prangenberg, S. De Vries, and M. Wiegelmann. Success and failure of certain reconstruction and uniqueness algorithms in discrete tomography. *Internat. J. Imaging Syst. Tech.* **9** (1998), 101–9.

[344] H. Groemer. Abschätzungen für die Anzahl der konvexen Körper, die einen konvexen Körper berühren. *Monatsh. Math.* **65** (1961), 74–81.

[345] ———. Eine kennzeichnende Eigenschaft der Kugel. *Enseign. Math.* (2) **7** (1961), 275–6.

[346] ———. Ein Satz über konvexe Körper und deren Projektionen. *Portugal. Math.* **21** (1962), 41–3.

[347] ———. On plane sections and projections of convex sets. *Canad. J. Math.* **21** (1969), 1331–7.

[348] ———. Stability theorems for projections of convex sets. *Israel J. Math.* **60** (1987), 177–90.

[349] ———. On the volume of equichordal sets. *Monatsh. Math.* **106** (1988), 1–8.

[350] ———. Stability theorems for convex domains of constant width. *Canad. Math. Bull.* **31** (1988), 328–37.

[351] ———. Stability properties of geometric inequalities. *Amer. Math. Monthly* **97** (1990), 382–94.

[352] ———. Stability theorems for projections and central symmetrization. *Arch. Math. (Basel)* **56** (1991), 394–9.

[353] ———. Fourier series and spherical harmonics in convexity. In *Handbook of Convexity*, ed. by P. M. Gruber and J. M. Wills. North-Holland, Amsterdam, 1993, pp. 1259–95.

[354] ———. Stability of geometric inequalities. In *Handbook of Convexity*, ed. by P. M. Gruber and J. M. Wills. North-Holland, Amsterdam, 1993, pp. 125–50.

[355] ———. Stability results for convex bodies and related spherical integral transforms. *Adv. Math.* **109** (1994), 45–74.

[356] ———. *Geometric Applications of Fourier Series and Spherical Harmonics*. Cambridge University Press, New York, 1996.

[357] ———. On the determination of convex bodies by translates of their projections. *Geom. Dedicata* **66** (1997), 265–79.

[358] ———. On a spherical integral transformation and sections of star bodies. *Monatsh. Math.* **126** (1998), 117–24.

[359] M. L. Gromov. A geometrical conjecture of Banach. *Math. USSR-Izv.* **1** (1967), 1055–64.

[360] P. Gronchi. Bodies of constant brightness. *Arch. Math. (Basel)* **70** (1998), 489–98.

[361] P. Gronchi and M. Longinetti. Affinely regular polygons as extremals of area functionals. Preprint.

[362] P. M. Gruber. Radon's contribution to convexity. In *Johann Radon, Collected Works*, ed. by P. M. Gruber, E. Hlawka, W. Nöbauer, and L. Schmetterer. Austrian Academy of Sciences, Vienna, 1987, pp. 336–42.

[363] ———. Convex billiards. *Geom. Dedicata* **33** (1990), 205–26.

[364] ———. History of convexity. In *Handbook of Convex Geometry*, ed. by P. M. Gruber and J. M. Wills. North-Holland, Amsterdam, 1993, pp. 1–15.

[365] B. Grünbaum. On a conjecture of H. Hadwiger. *Pacific J. Math.* **11** (1961), 215–19.

[366] _____ . Measures of symmetry for convex sets. In *Proc. Symp. Pure Math., Vol. VII: Convexity (Providence, RI)*. Amer. Math. Soc., Providence, RI, 1963, pp. 233–70.

[367] _____ . *Convex Polytopes*. 2d ed. Springer, New York, 2003.

[368] S. Gutman, J. H. B. Kemperman, J. A. Reeds, and L. A. Shepp. Existence of probability measures with given marginals. *Ann. Probab.* **19** (1991), 1781–97.

[369] F. Haab. Convex bodies of constant brightness and a new characterisation of spheres. *J. Differential Geom.* **52** (1999), 117–44.

[370] H. Hadwiger. *Vorlesungen über Inhalt, Oberfläche und Isoperimetrie*. Springer, Berlin, 1957.

[371] _____ . Vollständige stetige Umwendung ebener Eibereiche im Raum. In *Studies in Mathematical Analysis and Related Topics*. Stanford University Press, Stanford, CA, 1962, pp. 128–31.

[372] _____ . Seitenrisse konvexer Körper und Homothetie. *Elem. Math.* **18** (1963), 97–8.

[373] _____ . Radialpotenzintegrale zentralsymmetrischer Rotationskörper und Ungleichheitsaussagen Busemannscher Art. *Math. Scand.* **23** (1968), 193–200.

[374] _____ . Gitterperiodische Punktmengen und Isoperimetrie. *Monatsh. Math.* **76** (1972), 410–18.

[375] L. Hajdu. Unique reconstruction of bounded sets in discrete tomography. *Electronic Notes in Discrete Mathematics* **20** (2005), 15–25.

[376] L. Hajdu and R. Tijdeman. Algebraic aspects of discrete tomography. *J. Reine Angew. Math.* **534** (2001), 119–28.

[377] _____ . An algorithm for discrete tomography. *Linear Algebra Appl.* **339** (2001), 147–69.

[378] P. Hall and B. Turlach. On the estimation of a convex set with corners. *IEEE Trans. Pattern Anal. and Machine Intell.* **21** (1999), 225–34.

[379] C. Hamaker, K. T. Smith, D. C. Solmon, and S. L. Wagner. The divergent beam x-ray transform. *Rocky Mountain J. Math.* **10** (1980), 253–83.

[380] P. C. Hammer. Problem 2. In *Proc. Symp. Pure Math., vol. VII: Convexity (Providence, RI)*. Amer. Math. Soc., Providence, RI, 1963, pp. 498–9.

[381] J. Hansen and M. Reitzner. Electromagnetic wave propagation and inequalities for moments of chord lengths. *Adv. in Appl. Probab.* **36** (2004), 987–95.

[382] G. H. Hardy, J. E. Littlewood, and G. Pólya. *Inequalities*. Cambridge University Press, Cambridge, 1959.

[383] P. Hartman and A. Wintner. On pieces of convex surfaces. *Amer. J. Math.* **75** (1953), 477–87.

[384] T. L. Heath. *The Thirteen Books of Euclid's Elements*. Vol. 2. Dover, New York, 1956.

[385] G. Hector and U. Hirsch. *Introduction to the Geometry of Foliations*. Part A. Vieweg, Brunswick, 1981.

[386] V. E. Heil and H. Martini. Special convex bodies. In *Handbook of Convex Geometry*, ed. by P. M. Gruber and J. M. Wills. North-Holland, Amsterdam, 1993, pp. 347–85.

[387] S. Helgason. Differential operators on homogeneous spaces. *Acta Math.* **102** (1959), 239–99.

[388] _____ . *The Radon Transform*. Birkhäuser, Boston, 1980.

[389] _____ . *Groups and Geometric Analysis*. Academic Press, Orlando, 1984.

[390] _____ . The totally-geodesic Radon transform on constant curvature spaces. *Contemp. Math.* **113** (1990), 141–9.

[391] D. Hensley. Slicing the cube in $\mathbf{R}^n$ and probability (bounds for the measure of a central cube slice in $\mathbf{R}^n$ by probability methods). *Proc. Amer. Math. Soc.* **73** (1979), 95–100.

[392] _____ . Slicing convex bodies – bounds for slice area in terms of the body's covariance. *Proc. Amer. Math. Soc.* **79** (1980), 619–25.

[393] A. Heppes. On the determination of probability distributions of more dimensions by their projections. *Acta Math. Acad. Sci. Hungar.* **7** (1956), 403–10.

[394]   G. T. Herman and A. Kuba (eds.). *Discrete Tomography: Foundations, Algorithms, and Appli-cations.* Birkhäuser, Boston, 1999.

[395]   ———. *Advances in Discrete Tomography and Its Applications.* Birkhäuser, Boston, 2006.

[396]   E. Hlawka. Remembering Johann Radon. In *Johann Radon, Collected Works*, ed. by P. M. Gruber, E. Hlawka, W. Nöbauer, and L. Schmetterer. Austrian Academy of Sciences, Vienna, 1987, pp. 16–26.

[397]   R. D. Holmes and A. C. Thompson. N-dimensional area and content in Minkowski spaces. *Pacific J. Math.* **85** (1979), 77–110.

[398]   B. K. P. Horn. *Robot Vision.* McGraw-Hill, New York, 1986.

[399]   H. Hornich. Paul Funk. *Almanach Österr. Akad. Wiss.* (1969), 272–7.

[400]   R. Howard. Convex bodies of constant width and constant brightness. *Adv. Math.*, in press.

[401]   R. Howard and D. Hug. Smooth convex bodies with proportional projection functions. *Israel J. Math.*, in press.

[402]   ———. Nakajima's problem: Convex bodies of constant width and constant brightness. Preprint.

[403]   D. Hug. Absolute continuity for curvature measures of convex sets. I. *Math. Nachr.* **195** (1998), 139–58.

[404]   D. Hug and R. Schneider. Stability results involving surface area measures of convex bodies. *Rend. Circ. Mat. Palermo* (2) Suppl. No. 70, part II (2002), 21–51.

[405]   R. Huotari. Continuity of metric projection, Pólya algorithm, strict best approximation, and tubularity of convex sets. *J. Math. Anal. Appl.* **182** (1994), 836–56.

[406]   E. B. Jensen and H. J. G. Gundersen. Fundamental stereological formulae based on isotropi-cally orientated probes through fixed points with applications to particle analysis. *J. Microsc.* **153** (1988), 249–67.

[407]   F. John. Extremum problems with inequalities as subsidiary conditions. In *Studies and Essays Presented to R. Courant on his 60th Birthday.* Interscience, New York, 1948, pp. 187–204.

[408]   ———. *Plane Waves and Spherical Means.* Springer, New York, 1981.

[409]   K. Johnson and A. C. Thompson. On the isoperimetric mapping in Minkowksi spaces. In *Colloquia Mathematica Soc. János Bolyai, vol. 48. Intuitive Geometry (Siófok 1985)*, ed. by K. Böröczky and G. Fejes Tóth. North-Holland, Amsterdam, 1985, pp. 273–87.

[410]   F. Jones. *Lebesgue Integration on Euclidean Space.* Jones and Bartlett, Boston, 1993.

[411]   M. Junge. Hyperplane conjecture for spaces of L_p. *Forum Math.* **6** (1994), 617–35.

[412]   ———. Proportional subspaces of spaces with unconditional basis have good volume proper-ties. In *Geometric Aspects of Functional Analysis*, ed. by J. Lindenstrauss and V. D. Milman. *Oper. Theory Adv. Appl.* **77** (1995), 121–9.

[413]   M. Kaasalainen and L. Lamberg. Inverse problems of generalized projection operators. *Inverse Problems* **22** (2006), 749–69.

[414]   A. C. Kak and M. Slaney. *Principles of Computerized Tomographic Imaging.* SIAM, Philadel-phia, 2001.

[415]   N. J. Kalton and A. Koldobsky. Intersection bodies and L_p-spaces. *Adv. Math.* **196** (2005), 257–75.

[416]   W. C. Karl. Reconstructing objects from projections. PhD dissertation, Department of Elec-trical Engineering and Computer Science, Massachusetts Institute of Technology, Cambridge, MA, 1991.

[417]   W. C. Karl, S. R. Kulkarni, G. C. Verghese, and A. S. Willsky. Local tests for consistency of support hyperplane data. *J. Math. Imaging Vision* **6** (1995), 249–67.

[418]   W. C. Karl and G. C. Verghese. Curvatures of surfaces and their shadows. *Linear Algebra Appl.* **130** (1990), 231–55.

[419] F. Keinert. Inversion of k-plane transforms and applications in computer tomography. *SIAM Rev.* **31** (1989), 273–98.

[420] H. G. Kellerer. Masstheoretische Marginalprobleme. *Math. Ann.* **153** (1964), 168–98.

[421] ———. Schnittmass-Funktionen in mehrfachen Produkträumen. *Math. Ann.* **155** (1964), 369–91.

[422] ———. Uniqueness in bounded moment problems. *Trans. Amer. Math. Soc.* **336** (1993), 727–57.

[423] J. B. Kelly. Power points. *Amer. Math. Monthly* **53** (1946), 395–6.

[424] P. J. Kelly. Curves with a kind of constant width. *Amer. Math. Monthly* **64** (1957), 333–6.

[425] J. H. B. Kemperman. Sets of uniqueness and systems of inequalities having a unique solution. *Pacific J. Math.* **148** (1991), 275–301.

[426] ———. On the lack of uniqueness when reconstructing a set from finitely many projections. Unpublished manuscript.

[427] M. G. Kendall and P. A. P. Moran. *Geometrical Probability*. Hafner, New York, 1963.

[428] M. Kiderlen. Non-parametric estimation of the directional distribution of stationary line and fibre processes. *Adv. in Appl. Probab.* **33** (2001), 6–24.

[429] ———. Determination of the mean normal measure from flat sections I. *Adv. in Appl. Probab.* **34** (2002), 505–19.

[430] ———. Determination of a convex body from Minkowski sums of its projections. *J. London Math. Soc.* (2) **70** (2004), 529–44.

[431] ———. Blaschke and Minkowski endomorphisms of convex bodies. *Trans. Amer. Math. Soc.*, in press.

[432] M. Kiderlen and A. Pfrang. Algorithms to estimate the rose of directions of a spatial fiber system. *J. Microsc.* **219** (2005), 50–60.

[433] J. Kincses. On the determination of a convex set from its angle function. *Discrete Comput. Geom.* **30** (2003), 287–97.

[434] ———. An example of a stable, even order quadrangle which is determined by its angle function. In *Discrete Geometry, Monogr. Textbooks Pure Appl. Math.* 253. Marcel Dekker, New York, 2003, pp. 367–72.

[435] J. Kincses and Á. Kurusa. Can you recognize the shape of a figure from its shadows? *Beiträge Algebra Geom.* **36** (1995), 25–35.

[436] C. O. Kiselman. How smooth is the shadow of a convex body? *J. London Math. Soc.* (2) **33** (1986), 101–9.

[437] ———. Smoothness of vector sums of plane convex sets. *Math. Scand.* **60** (1987), 239–52.

[438] D. A. Klain. A short proof of Hadwiger's characterization theorem. *Mathematika* **42** (1996), 329–39.

[439] ———. Star valuations and dual mixed volumes. *Adv. Math.* **121** (1996), 80–101.

[440] ———. Invariant valuations on star-shaped sets. *Adv. Math.* **125** (1997), 95–113.

[441] ———. The Minkowski problem for polytopes. *Adv. Math.* **185** (2004), 270–88.

[442] ———. On the Minkowski condition and divergence. Preprint.

[443] B. Klartag and V. D. Milman. Rapid Steiner symmetrization of most of a convex body and the slicing problem. *Combin. Probab. Comput.* **14** (2005), 829–43.

[444] V. L. Klee. Some characterizations of convex polyhedra. *Acta Math.* **102** (1959), 79–107.

[445] ———. On a conjecture of Lindenstrauss. *Israel J. Math.* **1** (1963), 1–4.

[446] ———. Can a plane convex body have two equichordal points? *Amer. Math. Monthly* **76** (1969), 54–5.

[447] ———. Is a body spherical if its HA-measurements are constant? *Amer. Math. Monthly* **76** (1969), 539–42.

[448] ———. Shapes of the future. *Amer. Scientist* **59** (1971), 84–91.

[449] V. L. Klee and S. Wagon. *Old and New Unsolved Problems in Plane Geometry and Number Theory.* Math. Assoc. of Amer., 1991.

[450] M. Kline. *Mathematical Thought from Ancient to Modern Times.* Oxford University Press, New York, 1972.

[451] S. W. Knox. The Number of Facets of a Projection of a Convex Polytope. PhD dissertation, University of Illinois, Urbana, IL, 1996.

[452] S. Kobayashi and K. Nomizu. *Foundations of Differential Geometry.* Vol. 2. Interscience, New York, 1969.

[453] A. Koldobsky. Intersection bodies and the Busemann–Petty problem. *C. R. Acad. Sci. Paris Sér. I Math.* **325** (1997), 1181–6.

[454] _____ . Inverse formula for the Blaschke–Levy representation. *Houston J. Math.* **23** (1997), 95–108.

[455] _____ . An application of the Fourier transform to sections of star bodies. *Israel J. Math.* **106** (1998), 157–64.

[456] _____ . Intersection bodies in $\mathbb{R}^4$. *Adv. Math.* **136** (1998), 1–14.

[457] _____ . Intersection bodies, positive definite distributions and the Busemann-Petty problem. *Amer. J. Math.* **120** (1998), 827–40.

[458] _____ . Second derivative test for intersection bodies. *Adv. Math.* **136** (1998), 15–25.

[459] _____ . A generalization of the Busemann–Petty problem on sections of convex bodies. *Israel J. Math.* **110** (1999), 75–91.

[460] _____ . A functional analytic approach to intersection bodies. *Geom. Funct. Anal.* **10** (2000), 1507–26.

[461] _____ . On the derivatives of X-ray functions. *Arch. Math. (Basel)* **79** (2002), 216–22.

[462] _____ . The Busemann–Petty problem via spherical harmonics. *Adv. Math.* **177** (2003), 105–114.

[463] _____ . Sections of star bodies and the Fourier transform. *Contemp. Math.* **320** (2003), 225–47.

[464] _____ . Comparison of volumes by means of the areas of central sections. *Adv. in Appl. Math.* **33** (2005), 728–32.

[465] _____ . *Fourier Analysis in Convex Geometry.* Amer. Math. Soc., Providence, RI, 2005.

[466] A. Koldobsky and M. Lifshits. Average volume of sections of star bodies. In *Geometric Aspects of Functional Analysis*, ed. by V. D. Milman and G. Schechtman, Lecture Notes in Mathematics 1745. Springer, Berlin, 2000, pp. 119–46.

[467] A. Koldobsky, D. Ryabogin, and A. Zvavitch. Projections of convex bodies and the Fourier transform. *Israel J. Math.* **139** (2004), 361–80.

[468] _____ . Fourier analytic methods in the study of projections and sections of convex bodies. *Fourier analysis and convexity, Appl. Numer. Harmon. Anal.*, Birkhäuser, Boston, MA, 2004, pp. 119–30.

[469] A. Koldobsky and C. Shane. The determination of convex bodies from derivatives of section functions. Preprint.

[470] A. Koldobsky, V. Yaskin, and M. Yaskina. Modified Busemann-Petty problem on sections of convex bodies. *Israel J. Math.*, in press.

[471] D. Kölzow, A. Kuba, and A. Volčič. An algorithm for reconstructing convex bodies from their projections. *Discrete Comput. Geom.* **4** (1989), 205–37.

[472] A. Kosinski. Note on starshaped sets. *Proc. Amer. Math. Soc.* **13** (1962), 931–3.

[473] _____ . A theorem on families of acyclic sets and its applications. *Pacific J. Math.* **12** (1962), 317–25.

[474] H. Kramer and A. B. Németh. Triangles inscribed in smooth closed arcs. *Rev. Anal. Numér. Théorie Approximation* **1** (1972), 63–71.

[475] W. Krautwald. Kennzeichnungen der affinen Bilder von Körpern konstanter Breite. *J. Geom.* **15** (1980), 140–8.

[476] A. Kuba. The reconstruction of two-directionally connected binary patterns from their two orthogonal projections. *Comput. Vision Graphics Image Process.* **27** (1984), 249–65.

[477] ———. Reconstruction of measurable plane sets from their two projections taken in arbitrary directions. *Inverse Problems* **7** (1991), 101–7.

[478] A. Kuba and A. Volčič. Characterisation of measurable plane sets which are reconstructible from their two projections. *Inverse Problems* **4** (1988), 513–27.

[479] ———. The structure of the class of non-uniquely reconstructible sets. *Acta Sci. Math.* **58** (1993), 363–88.

[480] T. Kubota. Einfache Beweise eines Satzes über die konvexe, geschlossene Fläche. *Sci. Rep. Tohoku Univ.* **3** (1914), 235–55.

[481] ———. Einige Probleme über konvex-geschlossene Kurven und Flächen. *Tohoku Math. J.* **17** (1920), 351–62.

[482] ———. Obituary note: Matsusaburo Fujiwara (1881–1946). *Tohoku Math. J.* (2) **1** (1949), 1–2.

[483] G. Kuperberg. A low-technology estimate in convex geometry. *Internat. Math. Res. Notices* **1992** (1992), 181–3.

[484] ———. The bottleneck conjecture. *Geom. Topol.* **3** (1999), 119–35.

[485] Á. Kurusa. Generalized X-ray pictures. *Publ. Math. Debrecen* **48** (1996), 193–8.

[486] ———. The shadow picture problem for nonintersecting curves. *Geom. Dedicata* **59** (1996), 103–12.

[487] ———. You can recognize the shape of a figure from its shadows! *Geom. Dedicata* **59** (1996), 113–25.

[488] A. V. Kuz'minykh. The isoprojection property of the sphere. *Soviet Math. Dokl.* **14** (1973), 891–5.

[489] ———. Recovery of a convex body from the set of its projections. *Siberian Math. J.* **25** (1984), 284–8.

[490] D. Lam and D. C. Solmon. Reconstructing convex polygons in the plane from one directed X-ray. *Discrete Comput. Geom.* **26** (2001), 105–46.

[491] L. Lamberg. On the Minkowski problem and the lightcurve operator. *Ann. Acad. Sci. Fenn. Ser. A I Math. Dissertationes* **87** (1993), 107 pp.

[492] P. Lancaster and M. Tismenetsky. *The Theory of Matrices.* Academic Press, London, 1985.

[493] D. G. Larman. A note on the false centre problem. *Mathematika* **21** (1974), 216–27.

[494] D. G. Larman and P. Mani. Almost ellipsoidal sections and projections of convex bodies. *Math. Proc. Cambridge Philos. Soc.* **77** (1975), 529–46.

[495] D. G. Larman and C. A. Rogers. The existence of a centrally symmetric convex body with central sections that are unexpectedly small. *Mathematika* **22** (1975), 164–75.

[496] D. G. Larman and N. K. Tamvakis. A characterization of centrally symmetric convex bodies in E^n. *Geom. Dedicata* **10** (1981), 161–76.

[497] J. B. Lasserre. An analytical expression and an algorithm for the volume of a convex polyhedron in $\mathbb{R}^n$. *J. Optim. Theory Appl.* **39** (1983), 363–77.

[498] J. Lawrence. Polytope volume computation. *Math. Comput.* **57** (1991), 259–71.

[499] S. R. Lay. *Convex Sets and Their Applications.* Krieger, Malabar, FL, 1992.

[500] J. V. Leahy, K. T. Smith, and D. C. Solmon. Uniqueness, nonuniqueness and inversion in the X-ray and Radon problems. Paper presented at the International Symposium on Ill-posed Problems (Newark, Delaware), 1979.

[501] S.-N. Lee and M.-H. Shih. A volume problem for an n-dimensional ellipsoid intersecting with a hyperplane. *Linear Algebra Appl.* **132** (1990), 93–102.

[502] K. Leichtweiss. Über die affine Exzentrizität konvexer Körper. *Arch. Math. (Basel)* **10** (1959), 187–99.

[503] ———. Blaschkes Arbeiten zur Geometrie der konvexen Körper. In *Wilhelm Blaschke, Gesammelte Werke*, ed. by W. Burau et al. Thales, Essen, 1985, pp. 21–36.

[504] A. S. Lele, S. R. Kulkarni, and A. S. Willsky. Convex polygon estimation from support line measurements and applications to target reconstruction from laser radar data. *J. Opt. Soc. Amer. A* **9** (1992), 1693–714.

[505] J. Lemordant, P. D. Tao, and H. Zouaki. Modélisation et optimisation numérique pour la reconstruction d'un polyèdre à partir de son image gaussienne généralisée. *RAIRO Modél. Math. Anal. Numér.* **27** (1993), 349–74.

[506] S.-Y. R. Li. Reconstruction of polygons from projections. *Inform. Process. Lett.* **28** (1988), 235–40.

[507] I. M. Lifshitz and A. V. Pogorelov. On the determination of Fermi surfaces and electron velocities in metals by the oscillation of magnetic susceptibility [in Russian]. *Dokl. Akad. Nauk. SSSR* **96** (1954), 1143–5.

[508] M. Lindenbaum and A. Bruckstein. Reconstruction of polygonal sets by constrained and unconstrained double probing. *Ann. Math. Artif. Intell.* **4** (1991), 345–62.

[509] ———. Blind Approximation of Planar Convex Sets. *IEEE Trans. Robotics and Automation* **10** (1994), 517–29.

[510] J. Lindenstrauss, Almost spherical sections, their existence and applications. *Jber. Deutsche Math.-Verein., Jubiläumstagung 1990.* Teubner, Stuttgart, 1992, pp. 39–61.

[511] J. Lindenstrauss and V. D. Milman. The local theory of normed spaces and its applications to convexity. In *Handbook of Convex Geometry*, ed. by P. M. Gruber and J. M. Wills. North-Holland, Amsterdam, 1993, pp. 1149–220.

[512] N. F. Lindquist. Representations of central convex bodies. Technical Report 38, Oregon State University, Corvallis, OR, 1968.

[513] ———. Approximation of convex bodies by sums of line segments. *Portugal. Math.* **34** (1975), 233–40.

[514] J. J. Little. An iterative method for reconstructing convex polyhedra from extended Gaussian images. In Proc. AAAI, National Conf. Artificial Intelligence (Washington, D.C., 1983), pp. 247–50.

[515] M. Longinetti. Some questions of stability in the reconstruction of plane convex bodies from projections. *Inverse Problems* **1** (1985), 87–97.

[516] ———. Una proprietà di massimo dei poligoni affinemente regolari. *Rend. Circ. Mat. Palermo* (2) **34** (1985), 448–59.

[517] ———. An isoperimetric inequality for convex polygons and convex sets with the same symmetrals. *Geom. Dedicata* **20** (1986), 27–41.

[518] Y. Lonke. On the degree of generating distributions of centrally symmetric convex bodies. *Arch. Math. (Basel)* **69** (1997), 343–9.

[519] ———. On zonoids whose polars are zonoids. *Israel J. Math.* **102** (1997), 1–12.

[520] ———. On random sections of the cube. *Discrete Comput. Geom.* **23** (2000), 157–69.

[521] L. H. Loomis and H. Whitney. An inequality related to the isoperimetric inequality. *Bull. Amer. Math. Soc.* **55** (1949), 961–2.

[522] M. A. Lopez and S. Reisner. A special case of Mahler's conjecture. *Discrete Comput. Geom.* **20** (1998), 163–77.

[523] G. G. Lorentz. A problem of plane measure. *Amer. J. Math.* **71** (1949), 417–26.

[524] M. Ludwig. Projection bodies and valuations. *Adv. Math.* **172** (2002), 158–68.

[525] ———. Ellipsoids and matrix valued valuations. *Duke Math. J.* **119** (2003), 159–88.

[526] ———. Minkowski valuations. *Trans. Amer. Math. Soc.* **357** (2005), 4191–213.

[527] ———. Intersection bodies and valuations. *Amer. J. Math.*, in press.

[528] M. Ludwig and M. Reitzner. A characterization of affine surface area. *Adv. Math.* **147** (1999), 138–72.

[529] E. Lutwak. Dual mixed volumes. *Pacific J. Math.* **58** (1975), 531–8.

[530] ———. A general Bieberbach inequality. *Math. Proc. Cambridge Philos. Soc.* **78** (1975), 493–5.

[531] ———. Mean dual and harmonic cross-sectional measures. *Ann. Mat. Pura Appl.* (4) **119** (1979), 139–48.

[532] ———. A general isepiphanic inequality. *Proc. Amer. Math. Soc.* **90** (1984), 415–21.

[533] ———. Mixed projection inequalities. *Trans. Amer. Math. Soc.* **287** (1985), 91–106.

[534] ———. On some affine isoperimetric inequalities. *J. Differential Geom.* **23** (1986), 1–13.

[535] ———. Volume of mixed bodies. *Trans. Amer. Math. Soc.* **294** (1986), 487–500.

[536] ———. Inequalities for Hadwiger's harmonic Quermassintegrals. *Math. Ann.* **280** (1988), 165–75.

[537] ———. Intersection bodies and dual mixed volumes. *Adv. Math.* **71** (1988), 232–61.

[538] ———. Centroid bodies and dual mixed volumes. *Proc. London Math. Soc.* (3) **60** (1990), 365–91.

[539] ———. On a conjectured projection inequality of Petty. *Contemp. Math.* **113** (1990), 171–82.

[540] ———. On some ellipsoid formulas of Busemann, Furstenberg and Tzkoni, Guggenheimer, and Petty. *J. Math. Anal. Appl.* **159** (1991), 18–26.

[541] ———. Inequalities for mixed projection bodies. *Trans. Amer. Math. Soc.* **339** (1993), 901–16.

[542] ———. Selected affine isoperimetric inequalities. In *Handbook of Convex Geometry*, ed. by P. M. Gruber and J. M. Wills. North-Holland, Amsterdam, 1993, pp. 151–76.

[543] ———. The Brunn–Minkowski–Firey theory I: Mixed volumes and the Minkowksi problem. *J. Differential Geom.* **38** (1993), 131–50.

[544] ———. The Brunn–Minkowski–Firey theory II: Affine and geominimal surface areas. *Adv. Math.* **118** (1996), 244–94.

[545] E. Lutwak and V. Oliker. On the regularity of the solution of a generalization of the Minkowski problem. *J. Differential Geom.* **41** (1995), 227–46.

[546] E. Lutwak, D. Yang, and G. Zhang. A new ellipsoid associated with convex bodies. *Duke Math. J.* **104** (2000), 375–90.

[547] ———. L_p affine isoperimetric inequalities. *J. Differential Geom.* **56** (2000), 111–32.

[548] ———. A new affine invariant for polytopes and Schneider's projection problem. *Trans. Amer. Math. Soc.* **353** (2001), 1767–79.

[549] ———. The Cramer-Rao inequality for star bodies. *Duke Math. J.* **112** (2002), 59–81.

[550] ———. Moment-entropy inequalities. *Ann. Probab.* **32** (2004), 757–74.

[551] ———. Volume inequalities for subspaces of L_p. *J. Differential Geom.* **68** (2004) 159–84.

[552] ———. L_p John ellipsoids. *Proc. London Math. Soc.* (3) **90** (2005), 497–520.

[553] E. Lutwak and G. Zhang. Blaschke-Santaló inequalities. *J. Differential Geom.* **47** (1997), 1–16.

[554] L. A. Lyusternik. *Convex Figures and Polyhedra*. Dover, New York, 1963. Russian original: 1956.

[555] A. R. Mackintosh. The Fermi surface of metals. *Sci. Amer.* **207** (1963), 110–20.

[556] MacTutor History of Mathematics archive website at http://www-groups.dcs.st-and.ac.uk/~history.

[557] W. R. Madych and S. A. Nelson. Polynomial based algorithms for computed tomography. *SIAM J. Appl. Math.* **43** (1983), 157–85.

[558] G. Mägerl and A. Volčič. On the well-posedness of the Hammer X-ray problem. *Ann. Mat. Pura Appl.* (4) **144** (1986), 173–82.

458   References

458   References

[559] E. Makai and H. Martini. The cross-section body, plane sections of convex bodies and approximation of convex bodies, I. *Geom. Dedicata* **63** (1996), 267–96. Part II. *Geom. Dedicata* **70** (1998), 283–303.

[560] ———. On bodies associated with a given body. *Canad. Math. Bull.* **39** (1996), 448–59.

[561] ———. On maximal k-sections and related common transversals of convex bodies. *Canad. Math. Bull.* **47** (2004), 246–56.

[562] E. Makai, H. Martini, and T. Ódor. Maximal sections and centrally symmetric bodies. *Mathematika* **47** (2000), 19–30.

[563] E. Makai, S. Vrećica, and R. Živaljević. Plane sections of convex bodies of maximal volume. *Discrete Comput. Geom.* **25** (2001), 33–49.

[564] V. V. Makeev. On the approximation of plane sections of a convex body. *J. Math. Sci. (New York)* **100** (2000), 2297–302.

[565] C. L. Mallows and J. M. C. Clark. Linear-intercept distributions do not characterize plane sets. *J. Appl. Probab.* **7** (1970), 240–4.

[566] E. Mammen, J. S. Marron, B. A. Turlach, and M. P. Wand. A general projection framework for constrained smoothing. *Statist. Sci.* **16** (2001), 232–48.

[567] P. Mani. Fields of planar bodies tangent to spheres. *Monatsh. Math.* **74** (1970), 145–9.

[568] P. Mani-Levitska. Characterizations of convex sets. In *Handbook of Convexity*, ed. by P. M. Gruber and J. M. Wills. North-Holland, Amsterdam, 1993, pp. 19–41.

[569] C. Manni. Sulla ricostruzione tomografica di un corpo convesso. *Calcolo* (II) **23** (1986), 139–60.

[570] J. E. Marsden. *Elementary Classical Analysis*. Freeman, San Francisco, 1974.

[571] Y. Martinez-Maure. Hedgehogs of constant width and equichordal points. *Ann. Polon. Math.* **67** (1997), 285–8.

[572] ———. Hedgehogs and zonoids. *Adv. Math.* **158** (2001), 1–17.

[573] H. Martini. Zur Bestimmung konvexer Polytope durch die Inhalte ihrer Projektionen, *Beiträge Algebra Geom.* **18** (1984), 75–85.

[574] ———. Some results and problems around zonotopes. In *Colloquia Mathematica Soc. János Bolyai, vol. 48. Intuitive Geometry (Siófok 1985)*, ed. by K. Böröczky and G. Fejes Tóth. North-Holland, Amsterdam, 1985, pp. 383–417.

[575] ———. Determining classes of convex bodies by restricted sets of Steiner symmetrizations. *Geom. Dedicata* **30** (1989), 247–54.

[576] ———. On inner quermass of convex bodies. *Arch. Math. (Basel)* **52** (1989), 402–6.

[577] ———. A contribution to the light field theory. *Beiträge Algebra Geom.* **30** (1990), 193–201.

[578] ———. A new view on some characterizations of simplices. *Arch. Math. (Basel)* **55** (1990), 389–93.

[579] ———. Convex bodies whose projection bodies and difference sets are polars. *Discrete Comput. Geom.* **6** (1991), 83–91.

[580] ———. Extremal equalities for cross-sectional measures of convex bodies. In *Proc. 3d. Geometry Congress*. Aristoteles University, Thessaloniki, 1992, pp. 285–96.

[581] ———. Cross-sectional measures. In *Colloquia Mathematica Soc. János Bolyai, vol. 63. Intuitive Geometry (Szeged 1991)*, ed. by K. Böröczky and G. Fejes Tóth. North-Holland, Amsterdam, 1994, pp. 269–310.

[582] H. Martini and B. Weissbach. On Quermasses of simplices. *Studia Sci. Math. Hungar.* **27** (1992), 213–21.

[583] S. Matsuura. A problem of solid geometry. *J. Math. Osaka City Univ.* **A 12** (1961), 89–95.

[584] P. Mattila. Orthogonal projections, Riesz capacities and Minkowski content. *Indiana Univ. Math. J.* **39** (1990), 185–98.

[585] R. D. Mauldin. *The Scottish Book*. Birkhäuser, Boston, 1981.

[586] ———. On sets which meet each line in exactly two points. *Bull. London Math. Soc.* **30** (1998), 397–403.

[587] B. Maurey. Inégalité de Brunn–Minkowski–Lusternik, et autres inégalités géométriques et fonctionnelles. Séminaire Bourbaki. Vol. 2003/2004. *Asterisque* No. 299 (2005), Exp. No. 928, vii, 95–113.

[588] A. Mazzolo, B. Roesslinger, and W. Gille. Properties of chord length distributions of nonconvex bodies. *J. Math. Phys.* **44** (2003), 6195–208.

[589] J. E. McKenney. Finding the volume of an ellipsoid using cross-sectional slices. *Math. Mag.* **64** (1991), 32–4.

[590] P. McMullen. Non-linear angle-sum relations for polyhedral cones and polytopes. *Math. Proc. Cambridge Philos. Soc.* **78** (1975), 247–61.

[591] ———. The volume of certain convex sets. *Math. Proc. Cambridge Philos. Soc.* **91** (1982), 91–7.

[592] ———. Volumes of projections of unit cubes. *Bull. London Math. Soc.* **16** (1984), 278–80.

[593] ———. Volumes of complementary projections of convex polytopes. *Monatsh. Math.* **104** (1987), 265–72.

[594] P. McMullen and R. Schneider. Valuations on convex bodies. In *Convexity and Its Applications*, ed. by P. M. Gruber and J. M. Wills. Birkhäuser, Basel, 1983, pp. 170–247.

[595] P. McMullen and G. C. Shephard. *Convex Polytopes and the Upper Bound Conjecture.* Cambridge University Press, Cambridge, 1971.

[596] M. W. Meckes. Sylvester's problem for symmetric convex bodies and related problems. *Monatsh. Math.* **145** (2005), 307–19.

[597] H. Meijer and S. S. Skiena. Reconstructing polygons from X-rays. *Geom. Dedicata* **61** (1996), 191–204.

[598] G. H. Meisters and S. M. Ulam. On visual hulls of sets. *Proc. Natl. Acad. Sci. USA* **57** (1967), 1172–4.

[599] M. Meyer. A volume inequality concerning sections of convex sets. *Bull. London Math. Soc.* **20** (1988), 151–5.

[600] ———. Convex bodies with minimal volume product in $\mathbb{R}^2$. *Monatsh. Math.* **112** (1991), 297–301.

[601] ———. Two maximal volume hyperplane sections of a convex body generally intersect. *Period. Math. Hungar.* **36** (1998), 191–7.

[602] ———. Maximal hyperplane sections of convex bodies. *Mathematika* **46** (1999), 131–6.

[603] ———. On a problem of Busemann and Petty. Unpublished manuscript.

[604] M. Meyer and A. Pajor. Sections of the unit ball of l_p^n. *J. Functional Analysis* **80** (1988), 109–23.

[605] ———. On the Blaschke–Santaló inequality. *Arch. Math. (Basel)* **55** (1990), 82–93.

[606] M. Meyer and S. Reisner. A geometric property of the boundary of symmetric convex bodies and convexity of flotation surfaces. *Geom. Dedicata* **37** (1991), 327–37.

[607] G. Michelacci. A negative answer to the equichordal problem for not too small eccentricities. *Boll. Un. Mat. Ital. A* (7) **2** (1988), 203–12.

[608] ———. An iterative algorithm for reconstructing inscribed triangles. *Calcolo* **26** (1989), 107–19.

[609] ———. Inscribed polygons and fixed points of homeomorphisms on the circle. *Geom. Dedicata* **40** (1991), 103–10.

[610] ———. Reconstructing boundary points of convex sets from X-ray pictures. *Geom. Dedicata* **66** (1997), 357–68.

[611] G. Michelacci and A. Volčič. A better bound for the eccentricities not admitting the equichordal body. *Arch. Math. (Basel)* **55** (1990), 599–609.

[612] P. Milanfar. Geometric estimation and reconstruction from tomographic data. PhD disserta-
       tion, Department of Electrical Engineering and Computer Science, Massachusetts Institute of
       Technology, Cambridge, MA, 1993.

[613] P. Milanfar, W. C. Karl, and A. S. Willsky. Reconstructing binary polygonal objects from pro-
       jections: A statistical view. *Graph. Models Image Process.* **56** (1994), 371–91.

[614] _____. A moment-based variational approach to tomographic reconstruction. *IEEE Trans.
       Image Process.* **5** (1996), 459–70.

[615] P. Milanfar, G. C. Verghese, W. C. Karl, and A. S. Willsky. Reconstructing polygons from
       moments with connections to array processing. *IEEE Trans. Signal Process.* **43** (1995), 432–
       43.

[616] R. E. Miles. Isotropic random simplices. *Adv. in Appl. Probab.* **3** (1971), 353–82.

[617] E. Milman. Dual mixed volumes and the slicing problem. *Adv. Math.*, in press.

[618] _____. A comment on the low-dimensional Busemann–Petty problem. Preprint.

[619] _____. Generalized intersection bodies. Preprint.

[620] V. D. Milman. New proof of the theorem of Dvoretzky on sections of convex bodies. *Funct.
       Anal. Appl.* **5** (1971), 28–37.

[621] V. D. Milman and A. Pajor. Isotropic position and inertia ellipsoids and zonoids of the unit
       ball of a normed *n*-dimensional space. In *Geometric Aspects of Functional Analysis*, ed. by
       J. Lindenstrauss and V. D. Milman, Lecture Notes in Mathematics 1376. Springer, Heidelberg,
       1989, pp. 64–104.

[622] V. D. Milman and G. Schechtman. *Asymptotic Theory of Finite Dimensional Normed Spaces.*
       Lecture Notes in Mathematics 1200. Springer, Berlin, 1986.

[623] H. Minkowski. Allgemeine Lehrsätze über die konvexen Polyeder. *Nachr. Ges. Wiss. Göttingen*
       (1897), 198–219. Reprinted in *Gesammelte Abhandlungen.* Vol. 2. Teubner, Leipzig, 1911,
       pp. 103–21.

[624] _____. Volumen und Oberfläche. *Math. Ann.* **57** (1903), 447–95. Reprinted in *Gesammelte
       Abhandlungen.* Vol. 2. Teubner, Leipzig, 1911, pp. 230–76.

[625] _____. On bodies of constant width [in Russian]. *Mat. Sb.* **25** (1904), 505–8. German trans-
       lation: In *Gesammelte Abhandlungen.* Vol. 2. Teubner, Leipzig, 1911, pp. 277–9.

[626] _____. Theorie der konvexen Körper, inbesondere Begründung ihres Oberflächenbegriffs. In
       Gesammelte Abhandlungen. Vol. 2. Teubner, Leipzig, 1911, pp. 131–229.

[627] R. E. Molzon, B. Shiffman, and N. Siborny. Average growth estimates for hyperplane sections
       of entire analytic sets. *Math. Ann.* **257** (1981), 43–59.

[628] L. Montejano. On a problem of Ulam concerning a characterization of the sphere. *Stud. Appl.
       Math.* **53** (1974), 243–8.

[629] _____. About a problem of Ulam concerning flat sections of manifolds. *Comment. Math.
       Helv.* **65** (1990), 462–73.

[630] _____. A characterization of the Euclidean ball in terms of concurrent sections of constant
       width. *Geom. Dedicata* **37** (1991), 307–16.

[631] _____. Convex bodies with homothetic sections. *Bull. London Math. Soc.* **23** (1991), 381–6.

[632] _____. Recognizing sets by means of some of their sections. *Manuscripta Math.* **76** (1992),
       227–39.

[633] _____. Orthogonal projections of convex bodies and central symmetry. *Bol. Soc. Mat. Mexi-
       cana* (2) **38** (1993), 1–7.

[634] L. Montejano and E. Shchepin. Topological tomography in convexity. *Bull. London Math. Soc.*
       34 (2002), 353–8.

[635] E. Morales and L. Montejano. Variations of classic characterizations of ellipsoids and a short
       proof of the False Centre Theorem. Preprint.

[636] P. A. P. Moran. Measuring the surface area of a convex body. *Ann. of Math.* (2) **45** (1944),
       793–9.

[637] F. Morgan. *Geometric Measure Theory*. Academic Press, San Diego, 1988.

[638] M. Moszyńska. Quotient star bodies, intersection bodies, and star duality. *J. Math. Anal. Appl.* **232** (1999), 45–60.

[639] M. E. Munroe. *Measure and Integration*. Addison-Wesley, Reading, MA, 1971.

[640] H. Murrell. Computer-aided tomography. *Mathematica J.* **6** (1996), 60–5.

[641] W. Nagel. Orientation-dependent chord length distributions characterize convex polygons. *J. Appl. Probab.* **30** (1993), 730–6.

[642] S. Nakajima. Eine charakteristische Eigenschaft der Kugel. *Jber. Deutsche Math.-Verein.* **35** (1926), 298–300.

[643] ———. Eiflächenpaare gleicher Breiten und gleicher Umfänge. *Japan J. Math.* **7** (1930), 225–6.

[644] ———. Eine Kennzeichnung homothetischer Eiflächen. *Tohoku Math. J.* **35** (1932), 285–6.

[645] F. Natterer. *The Mathematics of Computerized Tomography*. SIAM, Philadelphia, 2001.

[646] ———. Book review. *Bull. Amer. Math. Soc.* **42** (2005), 89–91.

[647] F. Natterer and E. L. Ritman. Past and future directions in X-ray computed tomography (CT). *Internat. J. Imaging Syst. Tech.* **12** (2002), 175–87.

[648] F. Natterer and F. Wübbeling. *Mathematical Methods in Image Reconstruction*. SIAM, Philadelphia, 2001.

[649] F. L. Nazarov and A. N. Podkorytov. Ball, Haagerup, and distribution functions. In *Complex Analysis, Operators, and Related Topics, Oper. Theory Adv. Appl.* **113** (2000), 247–67.

[650] J. C. C. Nitsche. Isoptic characterization of a circle (proof of a conjecture of M. S. Klamkin). *Amer. Math. Monthly* **97** (1990), 45–7.

[651] V. I. Oliker. Hypersurfaces in $\mathbb{R}^{n+1}$ with prescribed Gaussian curvature and related equations of Monge–Ampère type. *Comm. Partial Differential Equations* **9** (1984), 807–38.

[652] S. Onn and E. Vallejo. Permutohedra and minimal matrices. *Linear Algebra Appl.* **412** (2006), 471–89.

[653] M. I. Ostrovskii. Minimal-volume projections of cubes and totally unimodular matrices. *Linear Algebra Appl.* **364** (2003), 91–103.

[654] V. Palamodov. *Reconstructive integral geometry*. Birkhäuser, Berlin, 2004.

[655] O. Palmon. The only convex body with extremal distance from the ball is the simplex. *Israel J. Math.* **80** (1992), 337–49.

[656] G. Paouris. Concentration of mass and central limit properties of isotropic convex bodies. *Proc. Amer. Math. Soc.* **133** (2005), 565–75.

[657] M. Papadimitrakis. On the Busemann–Petty problem about convex, centrally symmetric bodies in $\mathbf{R}^n$. *Mathematika* **39** (1992), 258–66.

[658] S. Pax. Appropriate cross-sectionally simple four-cells are flat. *Pacific J. Math.* **108** (1983), 379–84.

[659] S. Pennell and J. Deignan. Computing the projected area of a cone. *SIAM Rev.* **31** (1989), 299–302.

[660] C. M. Petty. On Minkowski Geometries. PhD dissertation, University of Southern California, Los Angeles, 1952.

[661] ———. Centroid surfaces. *Pacific J. Math.* **11** (1961), 1535–47.

[662] ———. Projection bodies. In *Proc. Colloquium Convexity, Copenhagen, 1965*, ed. by W. Fenchel. Københavns Univ. Mat. Inst., Copenhagen, 1967, pp. 234–41.

[663] ———. Isoperimetric problems. In *Proceedings, Conf. Convexity and Combinatorial Geometry. Univ. Oklahoma, 1971.* University of Oklahoma, Norman, OK, 1972, pp. 26–41.

[664] ———. Ellipsoids. In *Convexity and Its Applications*, ed. by P. M. Gruber and J. M. Wills. Birkhäuser, Basel, 1983, pp. 264–76.

[665] ———. Affine isoperimetric problems. *Ann. New York Acad. Sci.* **440** (1985), 113–27.

[666] C. M. Petty and J. M. Crotty. Characterizations of spherical neighbourhoods. *Canad. J. Math.* **2** (1970), 431–5.

[667] C. M. Petty and J. R. McKinney. Convex bodies with circumscribing boxes of constant volume. *Portugal. Math.* **44** (1987), 447–55.

[668] W. F. Pfeffer. *The Riemann Approach to Integration: Local Geometric Theory.* Cambridge University Press, New York, 1993.

[669] R. E. Pfiefer. The Extrema of Geometric Mean Values. PhD dissertation, Department of Mathematics, University of California, Davis, CA, 1982.

[670] ———. Maximum and minimum sets for some geometric mean values. *J. Theoret. Probab.* **3** (1990), 169–79.

[671] B. B. Phadke. Sublinearity and convexity on the Grassmann cone G_2^4. *Math. Scand.* **30** (1972), 249–52.

[672] J. Philip. Plane sections of simplices. *Math. Program.* **3** (1972), 312–25.

[673] G. Pisier. *The Volume of Convex Bodies and Banach Space Geometry.* Cambridge University Press, Cambridge, 1989.

[674] W. F. Pohl. The probability of linking of random closed curves. In *Proc. Geometry Symp., Utrecht, 1980*, ed. by E. Looijenga, D. Siersma, and F. Takens, Lecture Notes in Mathematics 894. Springer, Berlin, 1981, pp. 113–26.

[675] F. Pointet. Separation of hypersurfaces. *J. Geometry* **59** (1997), 114–24.

[676] G. Pólya and G. Szegö. *Isoperimetric Inequalities in Mathematical Physics.* Princeton University Press, Princeton, NJ, 1951.

[677] A. Poonawala, P. Milanfar, and R. J. Gardner. Shape estimation from support and diameter functions. *J. Math. Imaging Vision* **24** (2006), 229–44.

[678] J. L. Prince. Geometric model-based estimation from projections. PhD dissertation, Department of Electrical Engineering and Computer Science, Massachusetts Institute of Technology, Cambridge, MA, 1988.

[679] J. L. Prince and A. S. Willsky. Estimating convex sets from noisy support line measurements. *IEEE Trans. Pattern Anal. and Machine Intell.* **12** (1990), 377–89.

[680] ———. Convex set reconstruction using prior shape information. *Graph. Models Image Process.* **53** (1991), 413–27.

[681] ———. Hierarchical reconstruction using geometry and sinogram restoration. *IEEE Trans. Image Process.* **2** (1993), 401–16.

[682] E. T. Quinto. The invertibility of rotation invariant Radon transforms. *J. Math. Anal. Appl.* **91** (1983), 510–22.

[683] A. G. Ramm and A. I. Katsevich. *The Radon Transform and Local Tomography.* CRC Press, Boca Raton, FL, 1996.

[684] A. S. Rao and K. Y. Goldberg. Shape from diameter: Recognizing polygonal parts with a parallel-jaw gripper. *Internat. J. Robotics Research* **13** (1994), 16-37.

[685] J. Radon. Über die Bestimmung von Funktionen durch ihre Integralwerte längs gewisser Mannigfaltigkeiten. *Ber. Verh. Sächs. Akad. Wiss. Leipzig Math.-Phys. Kl.* **69** (1917), 262–7.

[686] H. Reichardt. Wilhelm Blaschke. In *Wilhelm Blaschke, Gesammelte Werke*, ed. by W. Burau et al. Thales, Essen, 1985, pp. 11–19.

[687] C. Reid. *Hilbert.* Springer, Berlin, 1970.

[688] ———. *Courant in Göttingen and New York.* Springer, New York, 1976.

[689] A. Rényi. On projections of probability distributions. *Acta Math. Acad. Sci. Hungar.* **3** (1952), 131–42.

[690] T. J. Richardson. Planar rectifiable curves are determined by their projections. *Discrete Comput. Geom.* **16** (1996), 21–31.

[691] ———. Total curvature and intersection tomography. *Adv. Math.* **130** (1997), 1–33.

[692] ———. Approximation of planar convex sets from hyperplane probes. *Discrete Comput. Geom.* **18** (1997), 151–77.

[693] C. A. Rogers. Sections and projections of convex bodies. *Portugal. Math.* **24** (1965), 99–103.

[694] ———. *Hausdorff Measures*. Cambridge University Press, Cambridge, 1970.

[695] ———. An equichordal problem. *Geom. Dedicata* **10** (1981), 73–8.

[696] C. A. Rogers and G. C. Shephard. Convex bodies associated with a given convex body. *J. London Math. Soc.* (2) **33** (1958), 270–81.

[697] J. C. W. Rogers. Remarks on the equichordal problem. Unpublished manuscript.

[698] D. J. Rossi. Reconstruction from projections based on detection and estimation of objects. PhD dissertation, Department of Electrical Engineering and Computer Science, Massachusetts Institute of Technology, Cambridge, MA, 1982.

[699] D. J. Rossi and A. S. Willsky. Reconstruction from projections based on detection and estimation of objects – Parts I and II: Performance analysis and robustness analysis. *IEEE Trans. Signal Process.* **32** (1984), 886–906.

[700] H. L. Royden. *Real Analysis*. 3d ed. Macmillan, New York, 1988.

[701] B. Rubin. Fractional calculus and wavelet transforms in integral geometry. *Fract. Calc. Appl. Anal.* **1** (1998), 193–219.

[702] ———. Inversion and characterization of the hemispherical transform. *J. Anal. Math.* **77** (1999), 105–28.

[703] ———. Generalized Minkowski–Funk transforms and small denominators on the sphere. *Fract. Calc. Appl. Anal.* **3** (2000), 177–203.

[704] ———. Inversion formulas for the spherical Radon transform and the generalized cosine transform. *Adv. in Appl. Math.* **29** (2002), 471–97.

[705] ———. Notes on Radon transforms in integral geometry. *Fract. Calc. Appl. Anal.* **6** (2003), 25–72.

[706] ———. The generalized Busemann–Petty problem with weights. Preprint.

[707] B. Rubin and G. Zhang. Generalizations of the Busemann–Petty problem for sections of convex bodies. *J. Funct. Anal.* **213** (2004), 473–501.

[708] M. Rudelson. Sections of the difference body. *Discrete Comput. Geom.* **23** (2000), 137–46.

[709] ———. Extremal distances between sections of convex bodies. *Geom. Funct. Anal.* **14** (2004), 1063–88.

[710] W. Rudin. *Principles of Mathematical Analysis*. 3d ed. McGraw-Hill, New York, 1976.

[711] D. Ryabogin and A. Zvavitch. The Fourier transform and Firey projections of convex bodies. *Indiana Univ. Math. J.* **53** (2004), 667–82.

[712] ———. Reconstruction of convex bodies of revolution from the areas of their shadows. *Arch. Math. (Basel)* **83** (2004), 450–60.

[713] M. Rychlik. A complete solution to the equichordal point problem of Fujiwara, Blaschke, Rothe, and Weizenböck. *Invent. Math.* **129** (1997), 141–212.

[714] H. J. Ryser. *Combinatorial Mathematics*. Mathematical Association of America and Quinn & Boden, Rahway, NJ, 1963.

[715] J. Saint Raymond. Sur le volume des corps convexes symétriques. In *Séminaire Choquet – Initiation à l'Analyse 80/81 Exp. No. 11 (Paris)*. Université P. and M. Curie, Paris, 1981, pp. 1–25.

[716] G. T. Sallee. Determining ellipsoids by intersections with flats. *Geom. Dedicata* **34** (1990), 139–44.

[717] P. M. Salzberg. Binary tomography in lattices. In *Proc. 26th. Southeastern Int. Conf. on Combinatorics, Graph Theory, and Computing, Boca Raton, FL, 1995. Congr. Numer.* **111** (1995), 185–92.

[718] J. R. Sangwine-Yager. Mixed volumes. In *Handbook of Convex Geometry*, ed. by P. M. Gruber and J. M. Wills. North-Holland, Amsterdam, 1993, pp. 43–71.

[719] ———. A generalization of outer parallel sets of a convex set. *Proc. Amer. Math. Soc.* **123** (1995), 1559–64.

[720] L. A. Santaló. Un invariante afin para los cuerpos convexos del espacio de *n* dimensiones. *Portugal. Math.* **8** (1949), 155–61.

[721] ———. *Integral Geometry and Geometric Probability*. Addison-Wesley, Reading, MA, 1976.

[722] S. Sasaki. Obituary note: Tadahiko Kubota (1885–1952). *Tohoku Math. J.* (2) **4** (1952), 318–19.

[723] R. Schäfke and H. Volkmer. Asymptotic analysis of the equichordal problem. *J. Reine Angew. Math.* **425** (1992), 9–60.

[724] P. Schapira. Tomography of constructible functions. In *Proc. 11th. Int. Symp. Applied Algebra, Algebraic Algorithms and Error-Correcting Codes, Paris, 1995*, ed. by G. Cohen, M. Giusti, and T. Mora, Lecture Notes in Computer Science 948. Springer, Heidelberg, 1995, pp. 427–35.

[725] L. Schmetterer. The Radon transform. In *Johann Radon, Collected Works*, ed. by P. M. Gruber, E. Hlawka, W. Nöbauer, and L. Schmetterer. Austrian Academy of Sciences, Vienna, 1987, pp. 7–9.

[726] M. Schmuckenschläger. The distribution function of the convolution square of a convex symmetric body in $\mathbb{R}^n$. *Israel J. Math* **78** (1992), 309–34.

[727] R. Schneider. Zu einem Problem von Shephard über die Projektionen konvexer Körper. *Math. Z.* **101** (1967), 71–82.

[728] ———. Functions on a sphere with vanishing integrals over certain subspheres. *J. Math. Anal. Appl.* **26** (1969), 381–4.

[729] ———. Functional equations connected with rotations and their geometric applications. *Enseign. Math.* (2) **16** (1970), 297–305.

[730] ———. On the projections of a convex polytope. *Pacific J. Math.* **32** (1970), 799–803.

[731] ———. Über eine Integralgleichung in der Theorie der konvexen Körper. *Math. Nachr.* **44** (1970), 55–75.

[732] ———. Rekonstruktion eines konvexen Körpers aus seinen Projektionen. *Math. Nachr.* **79** (1977), 325–9.

[733] ———. Boundary structure and curvature of convex bodies. In *Contributions to Geometry. Proc. Geometry Symp. Siegen, 1978*, ed. by J. Tölke and J. M. Wills. Birkhäuser, Basel, 1979, pp. 13–59.

[734] ———. Convex bodies with congruent sections. *Bull. London Math. Soc.* **12** (1980), 52–4.

[735] ———. Random hyperplanes meeting a convex body. *Z. Wahrsch. Verw. Gebiete* **61** (1982), 379–87.

[736] ———. Geometric inequalities for Poisson processes of convex bodies and cylinders. *Results Math.* **11** (1987), 165–85.

[737] ———. *Convex Bodies: The Brunn–Minkowski Theory*. Cambridge University Press, Cambridge, 1993.

[738] ———. Polytopes and Brunn–Minkowski theory. In *Polytopes: Abstract, Convex, and Computational*, ed. by T. Bisztriczky et al., NATO ASI Series, vol. 40. Kluwer, Dordrecht, 1994, pp. 273–99.

[739] ———. Volumes of projections of polytope pairs. *Rend. Circ. Mat. Palermo* (2) Suppl. No. 41 (1996), 217–25.

[740] ———. On areas and integral geometry in Minkowski spaces. *Beiträge Algebra Geom.* **38** (1997), 73–86.

[741] ———. On the determination of convex bodies by projection and girth functions. *Results Math.* **33** (1998), 155–60.

[742] _____. On the Busemann area in Minkowski spaces. *Beiträge Algebra Geom.* **42** (2001), 263–73.

[743] _____. Stable determination of convex bodies from projections. *Monatsh. Math.*, in press.

[744] R. Schneider and W. Weil. Über die Bestimmung eines konvexen Körpers durch die Inhalte seiner Projektionen. *Math. Z.* **116** (1970), 338–48.

[745] _____. Zonoids and related topics. In *Convexity and Its Applications*, ed. by P. M. Gruber and J. M. Wills. Birkhäuser, Basel, 1983, pp. 296–317.

[746] _____. *Integralgeometrie*. Teubner, Stuttgart, 1992.

[747] R. Schneider and J. A. Wieacker. Integral geometry. In *Handbook of Convex Geometry*, ed. by P. M. Gruber and J. M. Wills. North-Holland, Amsterdam, 1993, pp. 1349–90.

[748] U. Schnell. Volumes of projections of parallelotopes. *Bull. London Math. Soc.* **26** (1994), 181–5.

[749] A. Schrijver. *Theory of Linear and Nonlinear Programming*. Wiley, New York, 1986.

[750] C. Schütt. On the affine surface area. *Proc. Amer. Math. Soc.* **118** (1993), 1213–18.

[751] _____. Floating body, illumination body, and polytopal approximation. In *Convex Geometric Analysis*, ed. by K. M. Ball and V. Milman. Cambridge University Press, New York, 1999, pp. 203–29.

[752] C. Schütt and E. Werner. The convex floating body. *Math. Scand.* **66** (1990), 275–90.

[753] _____. Homothetic floating bodies. *Geom. Dedicata* **49** (1994), 335–48.

[754] _____. Surface bodies and p-affine surface area. *Adv. Math.* **187** (2004), 98–145.

[755] P. Schwander, C. Kisielowski, M. Seibt, F. H. Baumann, Y. Kim, and A. Ourmazd. Mapping projected potential, interfacial roughness, and composition in general cyrstalline solids by quantitative transmission electron microscopy. *Phys. Rev. Lett.* **71** (1993), 4150–3.

[756] H. A. Schwarz. Beweis des Satzes, dass die Kugel kleinere Oberfläche besitzt, als jeder andere Körper gleichen Volumens. *Nachr. Ges. Wiss. Göttingen* (1884), 1–13. Reprinted in *Gesammelte Abhandlungen*. Vol. 2. Springer, Berlin, 1890, pp. 327–40.

[757] A. J. Schwenk. How to minimize the largest shadow of a finite set. In *Graphs and Applications, Boulder, CO, 1982*, Wiley Interscience, New York, 1985, pp. 279–94.

[758] A. J. Schwenk and J. I. Munro. How small can the mean shadow of a set be? *Amer. Math. Monthly* **90** (1983), 325–9.

[759] P. R. Scott. Volume inequalities using sections of convex sets. *Bull. Austral. Math. Soc.* **43** (1991), 393–7.

[760] C. J. Scriba. Blaschke, Wilhelm Johann Eugen. In *Biographical Dictionary of Mathematicians*. Scribner, New York, 1991, pp. 271–3.

[761] V. I. Semyanistyi. Some integral transformations and integral geometry in elliptic space [in Russian]. *Tr. Semin. Vector. Tenzor. Anal.* **12** (1963), 297–411.

[762] G. C. Shephard. Shadow systems of convex bodies. *Israel J. Math.* **2** (1964), 229–36.

[763] G. C. Shephard and R. J. Webster. Metrics for sets of convex bodies. *Mathematika* **12** (1965), 73–88.

[764] L. A. Shepp. Computerized tomography and nuclear magnetic resonance. *J. Comput. Assisted Tomography* **4** (1980), 94–107.

[765] Siberian Math. J. editorial board. On the sixtieth birthday of Aleksandr Danilovich Aleksandrov. *Siberian Math. J.* **14** (1973), 243–9.

[766] W. Sierpiński. Sur un problème concernant les ensembles mesurables superficiellement. *Fund. Math.* **1** (1920), 112–15.

[767] E. Silverman. A Minkowski area having no convex extension. *Michigan Math. J.* **18** (1971), 231–3.

[768] S. S. Skiena. Geometric Probing. PhD dissertation, Dept. of Computer Science, University of Illinois, Urbana, IL, 1988.

[769] ———. Problems in geometric probing. *Algorithmica* **4** (1989), 599–605.

[770] ———. Interactive reconstruction via geometric probing. *Proc. IEEE* **80** (1992), 1364–83.

[771] ———. Geometric reconstruction problems. In *Handbook of Discrete and Computational Geometry*, ed. by J. E. Goodman and J. O'Rourke. CRC Press, Boca Raton, FL, 1997, pp. 481–90.

[772] C. G. Small. Reconstructing convex bodies from random projected images. *Canad. J. Statist.* **19** (1991), 341–7.

[773] K. T. Smith and F. Keinert. Mathematical foundations of computed tomography. *Appl. Optics* **24** (1985), 3950–7.

[774] K. T. Smith, D. C. Solmon, and S. L. Wagner. Practical and mathematical aspects of the problem of reconstructing objects from radiographs. *Bull. Amer. Math. Soc.* **83** (1977), 1227–70.

[775] D. C. Solmon. The X-ray transform. *J. Math. Anal. Appl.* **56** (1976), 61–83.

[776] ———. A note on k-plane integral transforms. *J. Math. Anal. Appl.* **71** (1979), 351–8.

[777] ———. Nonuniqueness and the null space of the divergent beam x-ray transform. *Contemp. Math.* **113** (1990), 243–9.

[778] G. Sonnevend. An optimal sequential algorithm for the uniform approximation of convex functions on $[0, 1]^2$. *Appl. Math. Optim.* **10** (1983), 127–42.

[779] A. Soranzo. Determination of convex bodies from $\pm\infty$-chord functions. *Rend. Istit. Mat. Univ. Trieste* **30** (1998), 129–39.

[780] A. Soranzo and A. Volčič. When do sections of different dimensions determine a convex body? *Mathematika* **50** (2003), 35–55.

[781] C. Soussen and A. Mohammad-Djafari. Polygonal and polyhedral contour reconstruction in computed tomography. *IEEE Trans. Image Process.* **13** (2004), 1507–23.

[782] J. E. Spingarn. An inequality for sections and projections of a convex set. *Proc. Amer. Math. Soc.* **118** (1993), 1219–24.

[783] M. Spivak. *A Comprehensive Introduction to Differential Geometry*. Vol. 3. Publish or Perish, Boston, 1975.

[784] K. Spriestersbach. Determination of a convex body from the average of projections and stability results. *Math. Proc. Camb. Phil. Soc.* **123** (1998), 561–70.

[785] A. Stancu. The floating body problem. *Bull. London Math. Soc.*, in press.

[786] S. Stein. Continuous choice functions and convexity. *Fund. Math.* **45** (1958), 182–5.

[787] J. Steiner. Einfache Beweis der isoperimetrischen Hauptsätze. *J. Reine Angew. Math.* **18** (1838), 289–96. Reprinted in *Gesammelte Abhandlungen*. Vol. 2. Reimer, Berlin, 1882, pp. 77–91.

[788] ———. Über parallele Flächen. *Monatsber. Preuss. Akad. Wiss., Berlin* (1840), 114–18. Reprinted in *Gesammelte Abhandlungen*. Vol. 2, Reimer, Berlin, 1882, pp. 173–6.

[789] D. Stoyan, W. S. Kendall, and J. Mecke. *Stochastic Geometry and Its Applications*. Wiley, New York, 1987.

[790] E. G. Straus. Two comments on Dvoretzky's sphericity theorem. *Israel J. Math.* **1** (1963), 221–3.

[791] R. Strichartz. *A Guide to Distribution Theory and Fourier Transforms*. CRC Press, Boca Raton, FL, 1994.

[792] K. Strubecker. Wilhelm Blaschkes mathematisches Werk. *Jber. Deutsche Math.-Verein.* **88** (1986), 146–57.

[793] M. A. Sumbatyan and E. A. Troyan. Reconstruction of the shape of a convex defect from a scattered wave field in the ray approximation. *J. Appl. Maths. Mechs.* **56** (1992), 464–8.

[794] W. Süss. Eibereiche mit ausgezeichneten Punkten; Sehnen-, Inhalts- und Umfangspunkte. *Tohoku Math. J.* **25** (1925), 86–98.

[795] ———. Eine charakteristische Eigenschaft der Kugel. *Jber. Deutsche Math.-Verein.* **34** (1926), 245–7.

[796] _____. Zusammensetzung von Eikörpern und homothetische Eiflächen. *Tohoku Math. J.* **35** (1932), 47–50.

[797] _____. Kennzeichnende Eigenschaften der Kugel als Folgerung eines Brouwerschen Fixpunktsatzes. *Comment. Math. Helv.* **20** (1947), 61–4.

[798] _____. Eine elementare kennzeichnende Eigenschaft des Ellipsoids. *Math.-Phys. Semesterber.* **3** (1953), 57–8.

[799] J. Székely. The orthogonal projections and homothety of convex bodies [in Hungarian]. *Math. Lapok* **16** (1965), 52–6.

[800] G. Talenti. On isoperimetric theorems of mathematical physics. In *Handbook of Convexity*, ed. by P. M. Gruber and J. M. Wills. North-Holland, Amsterdam, 1993, pp. 1131–47.

[801] _____. The standard isoperimetric theorem. In *Handbook of Convexity*, ed. by P. M. Gruber and J. M. Wills. North-Holland, Amsterdam, 1993, pp. 73–123.

[802] J.-P. Thirion. Segmentation of tomographic data without image reconstruction. *IEEE Trans. Medical Imaging* **11** (1992), 102–10.

[803] A. C. Thompson. *Minkowski Geometry*. Cambridge University Press, Cambridge, 1996.

[804] A. Tsolomitis. On the convolution body of two convex bodies. *C. R. Acad. Sci. Paris Sér. I Math.* **322** (1996), 63–7.

[805] _____. Convolution bodies and their limiting behavior. *Duke Math. J.* **87** (1997), 181–203.

[806] _____. A note on the M^*-limiting convolution body. In *Convex Geometric Analysis*, ed. by K. M. Ball and V. Milman. Cambridge University Press, New York, 1999, pp. 231–6.

[807] S. Ulam. *A Collection of Mathematical Problems*. Interscience, New York, 1960.

[808] E. E. Underwood. *Quantitative Stereology*. Addison-Wesley, Reading, MA, 1970.

[809] P. Ungar. Freak theorem about functions on a sphere. *J. London Math. Soc.* (2) **29** (1954), 100–3.

[810] J. D. Vaaler. A geometric inequality with applications to linear forms. *Pacific J. Math.* **83** (1979), 543–53.

[811] F. A. Valentine. *Convex Sets*. McGraw-Hill, New York, 1964.

[812] E. Vallejo. Reductions of additive sets, sets of uniqueness and pyramids. *Discrete Math.* **173** (1997), 257–67.

[813] _____. Sets of uniqueness and minimal matrices. *J. Algebra* **208** (1998), 444–51.

[814] _____. A characterization of additive sets. *Discrete Math.* **259** (2002), 201–10.

[815] E. B. Vedel Jensen. *Local Stereology*. World Scientific, Singapore, 1997.

[816] P. Vincensini. *Corps convexes. Séries linéaires. Domaines vectoriels.* Mem. des Sciences Math., vol. 44, Gauthiers-Villars, Paris, 1938.

[817] J. Voigt. A concentration of mass phenomenon for isotropic convex bodies in high dimensions. *Israel J. Math.* **115** (2000), 235–51.

[818] A. Volčič. Well-posedness of the Gardner–McMullen reconstruction problem. In *Proc. Conf. Measure Theory, Oberwolfach, 1983*, Lecture Notes in Mathematics 1089. Springer, Berlin, 1984, pp. 199–210.

[819] _____. Ghost convex bodies. *Boll. Un. Mat. Ital. A* (6) **4** (1985), 287–92.

[820] _____. A new proof of the Giering theorem. *Rend. Circ. Mat. Palermo* (2) **8** (1985), 281–95.

[821] _____. A three-point solution to Hammer's X-ray problem. *J. London Math. Soc.* (2) **34** (1986), 349–59.

[822] _____. Tomography of convex bodies and inscribed polygons. *Ricerche Mat. Suppl.* **36** (1987), 185–92.

[823] A. Volčič and T. Zamfirescu. Ghosts are scarce. *J. London Math. Soc.* (2) **40** (1989), 171–8.

[824] P. Waksman. Plane polygons and a conjecture of Blaschke's. *Adv. in Appl. Probab.* **17** (1985), 774–93.

[825]   D. W. Walkup. A simplex with a large cross section. *Amer. Math. Monthly* **75** (1968), 34–6.

[826]   S. Webb. Central slices of the regular simplex. *Geom. Dedicata* **61** (1996), 19–28.

[827]   R. Webster. *Convexity*. Oxford University Press, New York, 1994.

[828]   B. Wegner. Über eine kennzeichnende Eigenshaft der Kugel von W. Süss und deren Verallgemeinerung. *Acta Math. Acad. Sci. Hungar.* **24** (1973), 135–8.

[829]   E. R. Weibel. *Stereological Methods*. Vol. 2. Academic Press, London, 1980.

[830]   W. Weil. Über die Projektionenkörper konvexer Polytope. *Arch. Math. (Basel)* **22** (1971), 664–72.

[831]   ———. Centrally symmetric convex bodies and distributions. *Israel J. Math.* **24** (1976), 352–67.

[832]   ———. Kontinuierliche Linearkombination von Strecken. *Math. Z.* **148** (1976), 71–84.

[833]   ———. Blaschkes Problem der lokalen Charakterisierung von Zonoiden. *Arch. Math. (Basel)* **29** (1977), 655–9.

[834]   ———. Centrally symmetric convex bodies and distributions. II. *Israel J. Math.* **32** (1979), 173–82.

[835]   ———. Zonoide und verwandte Klassen konvexer Körper. *Monatsh. Math.* **94** (1982), 73–84.

[836]   ———. Stereology: A survey for geometers. In *Convexity and Its Applications*, ed. by P. M. Gruber and J. M. Wills. Birkhäuser, Basel, 1983, pp. 360–421.

[837]   ———. Translative and kinematic integral formulae for support functions. *Geom. Dedicata* **57** (1995), 91–103.

[838]   ———. On the mean shape of particle processes. *Adv. in Appl. Probab.* **29** (1997), 890–908.

[839]   A. Weir. *Lebesgue Integration and Measure*. Cambridge University Press, Cambridge, 1973.

[840]   E. Werner. Illumination bodies and affine surface area. *Studia Math.* **110** (1994), 257–69.

[841]   J. Wills. Symmetrisierung an Unterräumen. *Arch. Math. (Basel)* **20** (1969), 169–72.

[842]   E. Wirsing. Zur Analytizität von Doppelspeichenkurven. *Arch. Math. (Basel)* **9** (1958), 300–7.

[843]   H. S. Witsenhausen. A support characterization of zonotopes. *Mathematika* **25** (1978), 13–16.

[844]   G. J. Woeginger. The reconstruction of polyominoes from their orthogonal projections. *Inform. Process. Lett.* **77** (2001), 225–9.

[845]   I. M. Yaglom and V. G. Boltyanskiĭ. *Convex Figures*. Holt, Rinehart, and Winston, New York, 1961. Russian original: 1951.

[846]   K. Yanagihara. On a characteristic property of the circle and the sphere. *Tohoku Math. J.* **10** (1916), 142–3.

[847]   ———. Second note on a characteristic property of the circle and the sphere. *Tohoku Math. J.* **11** (1917), 55–7.

[848]   V. Yaskin. The Busemann–Petty problem in hyperbolic and spherical spaces. *Adv. Math.*, in press.

[849]   ———. A solution to the lower-dimensional Busemann–Petty problem in the hyperbolic space. Preprint.

[850]   V. Yaskin and M. Yaskina. Centroids and comparisons of volumes. *Indiana Univ. Math. J.*, in press.

[851]   M. Yaskina. Non-intersection bodies all of whose central sections are intersection bodies. *Proc. Amer. Math. Soc.*, in press.

[852]   D. Yost. Irreducible convex sets. *Mathematika* **38** (1991), 134–55.

[853]   L. Young. *Mathematicians and Their Times*. North-Holland, Amsterdam, 1981.

[854]   J. Zaks. Nonspherical bodies with constant HA-measurements exist. *Amer. Math. Monthly* **78** (1971), 513–16.

[855]   L. Zalcman. Offbeat integral geometry. *Amer. Math. Monthly* **87** (1980), 161–75.

[856] G. Zhang. Restricted chord projection and affine inequalities. *Geom. Dedicata* **39** (1991), 213–22.

[857] _____. Intersection bodies and the four-dimensional Busemann–Petty problem. *Duke Math. J.* **71** (1993), 233–40.

[858] _____. Centered bodies and dual mixed volumes. *Trans. Amer. Math. Soc.* **345** (1994), 777–801.

[859] _____. Geometric inequalities and inclusion measures of convex bodies. *Mathematika* **41** (1994), 95–116.

[860] _____. Intersection bodies and the Busemann–Petty inequalities in $\mathbb{R}^4$. *Ann. of Math.* (2) **140** (1994), 331–46.

[861] _____. Sections of convex bodies. *Amer. J. Math.* **118** (1996), 319–40.

[862] _____. A positive answer to the Busemann–Petty problem in four dimensions. *Ann. of Math.* (2) **149** (1999), 535–43.

[863] _____. Dual kinematic formulas. *Trans. Amer. Math. Soc.* **351** (1999), 985–95.

[864] _____. Intersection bodies and polytopes. *Mathematika* **46** (1999), 29–34.

[865] J. Zhu, S. W. Lee, Y. Ye, S. Zhao, and G. Wang. X-ray transform and Radon transform for ellipsoids and tetrahedra. *J. X-ray Sci. Tech.* **12** (2004), 215–29.

[866] J. Zhu, S. Zhao, Y. Ye, and G. Wang. Computed tomography simulation with superquadrics. *Medical Physics* **32** (2005), 3136–43.

[867] S. Zopf and A. Kuba. Reconstruction of measurable sets from two generalized projections. *Electronic Notes in Discrete Mathematics* **20** (2005), 47–66.

[868] L. Zuccheri. Approximation of convex bodies from chord functions. *Geom. Dedicata* **27** (1988), 335–48.

[869] _____. Characterization of the circle by equipower properties. *Arch. Math. (Basel)* **58** (1992), 199–208.

[870] A. Zvavitch. Gaussian measure of sections of convex bodies. *Adv. Math.* **188** (2004), 124–36.

[871] _____. An isomorphic version of the Busemann–Petty problem for Gaussian measure. In *Geometric Aspects of Functional Analysis*, ed. by V. D. Milman and G. Schechtman, Lecture Notes in Mathematics 1850. Springer, Berlin, 2004, pp. 277–83.

[872] _____. The Busemann–Petty problem for arbitrary measures. *Math. Ann.* **331** (2005), 867–87.

NOTATION

Symbol	Definition	Page where introduced
$\mathbb{E}^n$	n-dimensional Euclidean space	1
o	origin in $\mathbb{E}^n$	1
$\|\cdot\|$	Euclidean norm	1
$x \cdot y$	scalar product of x and y	1
$[x, y]$	line segment joining x and y	1
$\mathbb{N}$	natural numbers	1
$\mathbb{R}$	real numbers	1
$\mathbb{C}$	complex numbers	1
B or B^n	unit ball in $\mathbb{E}^n$	2
S^{n-1}	unit n-sphere $\mathbb{E}^n$	2
u	a unit vector	2
$u^\perp$	hyperplane through o orthogonal to u	2
l_u	line through o parallel to u	2
$S^\perp$	orthogonal complement of S	2
$\mathcal{G}(n, k)$	k-dimensional subspaces of $\mathbb{E}^n$	2
$\operatorname{lin} E$	linear hull of E	2
$\operatorname{aff} E$	affine hull of E	2
$\dim E$	dimension of E	2
$\mathbb{P}^n$	n-dimensional projective space	2
$|E|$	cardinality of E	2
$\operatorname{co} E$	complement of E	2
$\operatorname{cl} E$	closure of E	2
$\operatorname{int} E$	interior of E	2
$\operatorname{bd} E$	boundary of E	2
$\operatorname{relint} E$	relative interior of E	2
$E \bigtriangleup F$	symmetric difference of E and F	3

AUTHOR INDEX

Manni, C., 56
Martin, P., 298
Martinez-Maure, Y., 129, 182, 265
Martini, H., 88, 129, 132, 138, 181–2, 266, 292, 298, 338, 347–8, 376, 388
Matsuura, S., 229
Mattila, P., 383
Mauldin, R. D., 299
Maurey, B., 339, 344, 417
Mazzolo, A., 378
McKenney, J. E., 385
McKinney, J. R., 127, 290
McMullen, P., 6, 55, 185, 383–4, 404–5
Mecke, J., 9
Meckes, M. W., 379
Medin, A., 226
Meijer, H., 94
Meisters, G. H., 127
Meyer, M., 295, 345, 347, 377, 380, 384
Michelacci, G., 56, 228, 266
Milanfar, P., 55–6, 58, 93, 136, 190–1, 401, 403
Milman, E., 342, 346, 385
Milman, V. D., 188, 296, 338, 345, 376, 379–81, 385, 387
Minkowski, H., 129–30, 181, 391, 430
Mohammad-Djafari, A., 93
Molzon, R. E., 298
Montejano, L., 127–9, 229, 289–90, 297–8, 377
Morales, E., 229, 289
Moran, P. A. P., 383
Morgan, F., 10, 12, 300
Moszyńska, M., 340
Munro, J. I., 383
Munroe, M. E., 10, 15
Murrell, H., 53–4

Nagel, W., 300, 378
Nakajima, S., 126, 129, 132
Naor, A., 138
Natterer, F., 53, 426–7, 432
Nazarov, F. L., 295
Nef, W., 88
Németh, A. B., 56
Nitsche, J. C. C., 229

Ódor, T., 347
Oliker, V., 186, 388
Onn, S., 91
Ostrovskii, M. I., 138

Pajor, A., 295, 338, 345, 376, 379–81, 385, 387
Palamodov, V., 299
Palmon, O., 188
Paouris, G., 388

Papadimitrakis, M., 344
Pax, S., 297
Pennell, S., 188
Peri, C., 90, 227–8, 265, 291
Perissinaki, I., 388
Petty, C. M., 127, 130–1, 181–2, 187–8, 266, 289–91, 298, 338, 343–4, 347–8, 376, 379–80, 385, 430
Pfeffer, W. F., 383
Pfiefer, R. E., 380
Phadke, B. B., 182
Philip, J., 298
Pisier, G., 296, 387, 415
Podkorytov, A. N., 295
Pogorelov, A. V., 291
Pohl, W. F., 300
Pointet, F., 137
Pólya, G., 91, 413
Poonawala, A., 136
Prangenberg, D., 89, 91
Prince, J. L., 58, 135–6

Quinto, E. T., 291, 426

Radon, J., 53, 96, 424, 426
Rannou, F. R., 136
Rao, A. S., 135
Reid, C., 193, 348
Reisner, S., 377, 380
Reitzner, M., 300, 377
Rényi, A., 88
Richardson, T. J., 135, 300
Ritman, E. L., 53
Rivin, I., 344
Roesslinger, B., 378
Rogers, C. A., 12, 127, 131, 226, 289, 343, 388
Rogers, J. C. W., 266
Rossi, D. J., 57–8
Rothe, H., 266
Royden, H. L., 10
Rubin, B., 292, 339, 341, 344–6, 435
Rudelson, M., 296, 348
Ryabogin, D., 138, 183, 186, 189, 191
Rychlik, M., 267
Ryser, H. J., 88–9

Saint Raymond, J., 380
Sallee, G. T., 385
Salzberg, P. M., 89
Sangwine-Yager, J. R., 391, 411
Santaló, L. A., 139, 299, 379–80, 382, 405
Schäfke, R., 266
Schapira, P., 297
Schechtman, G., 296

SUBJECT INDEX

Note: Principal references are denoted by bold type. References to figures are indicated by "fig."

Printed in the United States
By Bookmasters